# 10th International Conference on 3D Radiation Dosimetry (IC3DDose)



Journal of Physics: Conference Series Volume 1305

Kunshan, China
16 - 19 September 2018

ISBN: 978-1-5108-9970-4
ISSN: 1742-6588

**Printed from e-media with permission by:**

Curran Associates, Inc.
57 Morehouse Lane
Red Hook, NY 12571



Institute of Physics
Dirac House, Temple Back
Bristol BS1 6BE UK

Phone: 44 1 17 929 7481
Fax: 44 1 17 920 0979

techtracking@iop.org

**Additional copies of this publication are available from:**

Curran Associates, Inc.
57 Morehouse Lane
Red Hook, NY 12571 USA
Phone: 845-758-0400
Fax: 845-758-2633
Email: curran@proceedings.com
Web: www.proceedings.com

# TABLE OF CONTENTS

## CLINICAL CHALLENGES AND OPPORTUNITIES



## DOSIMETRY AND MRI LINACS

## OPTICAL AND CHERENKOV DOSIMETRY



## CLINICAL CHALLENGES AND OPPORTUNITIES



## SMALL FIELDS, RADIOSURGERY AND PRE-CLINICAL IRRADIATORS

**CHEMICAL DOSIMETRY**



**HEAVY PARTICLES AND NEW DOSIMETRY SYSTEMS**

## NEW DOSIMETRY SYSTEMS



## BRACHYTHERAPY AND MOTION

IOP Conf. Series: Journal of Physics: Conf. Series **1305** (2019) 011001 doi:10.1088/1742-6596/1305/1/011001

## Preface

We have yet to meet a clinical therapy physicist who does not readily acknowledge the great desirability for more comprehensive and efficient dosimetry tools, including 3D dosimetry systems, for the verification of modern radiation treatments. The innovations and progress toward this goal are remarkably captured in the proceedings of nine prior conferences, the last five of which are freely available in the Journal of Physics: Conference Series.

This 10th IC3DDose meeting was the first to be held in China and was exciting to see the meeting expand and new researchers and ideas enter the field. The scientific program was carefully crafted to try to meet the objectives listed below and we were fortunate that many leading speakers shared their experience and perspectives to achieve them.

## Conference Objectives

- To provide a forum to discuss the latest research and developments in 3D and advanced radiation dosimetry.
- To elevate the quality of radiation therapy treatments and quality assurance (QA) through improved clinical dosimetry.
- To explore the dosimetric challenges posed by modern radiation treatment techniques.
- To energize and diversify dosimetry research and clinical practice by encouraging interaction and synergy between advanced, 3D, and semi-3D dosimetry techniques.

A very diverse group of clinical and research physicists from many parts of the world and it was truly a pleasure to welcome them all.

Finally we would like to acknowledge the many people who made the meeting possible. Special thanks go to the Local Organizing Committee (listed below) who put in many hours on the difficult challenge of arranging a meeting in a different continent and language. The Scientific Committee reviewed all the conference abstracts and participated in many planning conference calls, and meeting activities. We are also very grateful to our sponsors, both academic and industrial, who's support was vital to enabling this meeting.

## Co-Chairs

Mark Oldham PhD, FAAPM
Professor, Radiation Oncology,
Duke University Medical Center,
Durham, NC, USA

Fang-Fang Yin, PhD, FAAPM
Chief of Physics, Radiation Oncology,
Durham, NC, USA
Director of Medical Physics
Duke Kunshan University, Kunshan, China

## Duke Kunshan University Local Organizing Committee

Ying-Chiang (David) Huang
James Bowsher
Claire Kaimei Luo

Zhidong Xue
Hongrui Wang

**International Scientific Committee**

Sven Back (Sweden)
Clive Baldock (Australia)
Sam Beddar (USA)
Yves De Deene (Australia)
Simon Doran (UK)
Geoffrey Ibbott (USA)
Andrew Jirasek (Canada)
Kevin Jordan (Canada)
Ben Mijnheer (Netherlands)
Mark Oldham (USA)
John Schreiner (Canada)
Cheng-Shie Wuu (USA)

IOP Conf. Series: Journal of Physics: Conf. Series **1305** (2019) 011002 doi:10.1088/1742-6596/1305/1/011002

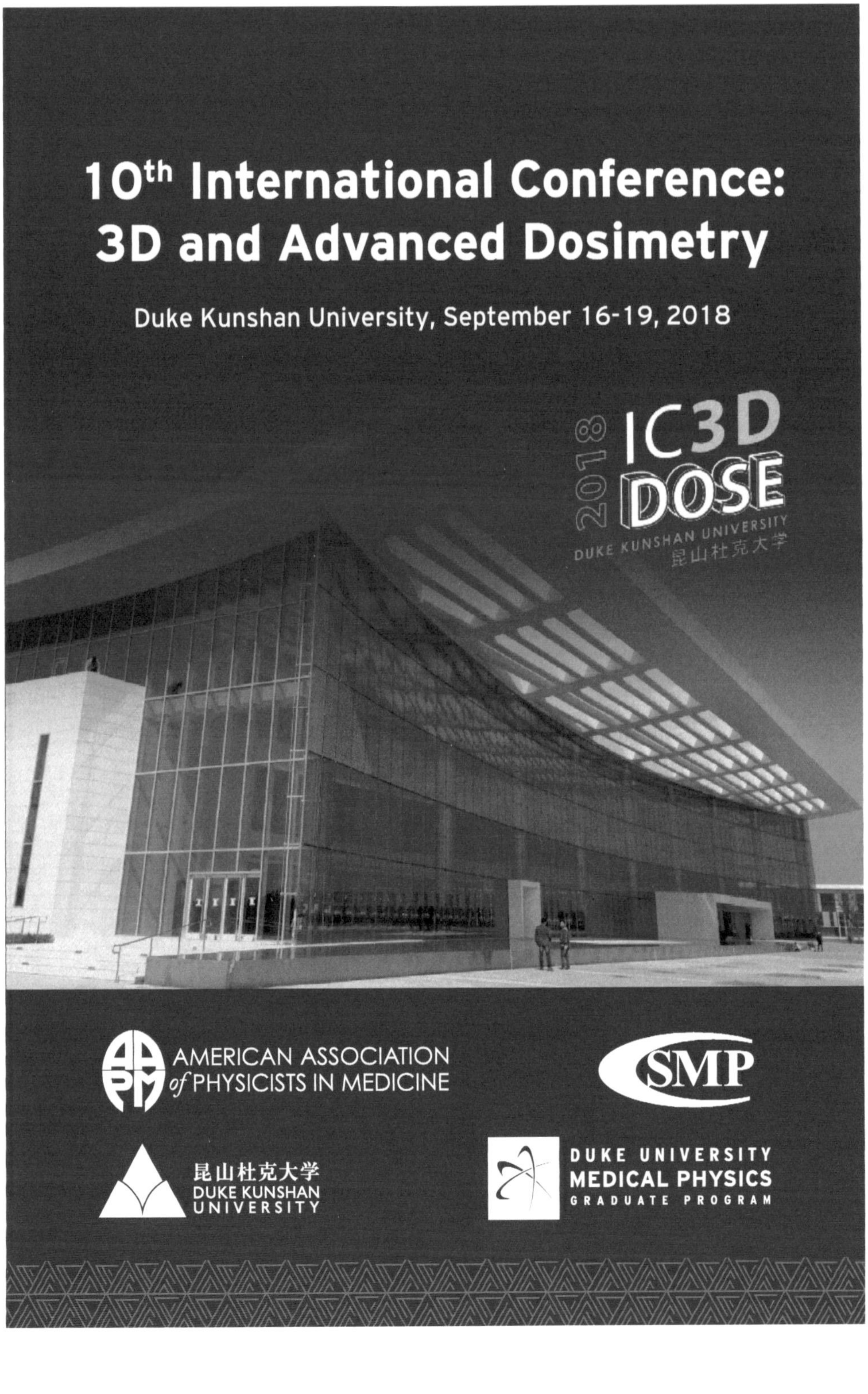

IOP Conf. Series: Journal of Physics: Conf. Series **1305** (2019) 011002 doi:10.1088/1742-6596/1305/1/011002

*10th IC3DDose Meeting Program*

9/1/2018

**Welcome Message:**

We have yet to meet a clinical therapy physicist who does not readily acknowledge the great desirability for more comprehensive and efficient dosimetry tools, including 3D dosimetry systems, for the verification of modern radiation treatments. The innovations and progress toward this goal are remarkably captured in the proceedings of nine prior conferences, the last five of which are freely available in the Journal of Physics: Conference Series. The current 10th IC3DDose meeting is the first to be held in China, and it is an exciting prospect to see the meeting expand and new researchers and ideas enter the field. The scientific program has been carefully crafted to try to meet the objectives listed below, and we are fortunate that many leading speakers will be sharing their experience and perspectives to achieve them.

**Conference Objectives:**

1. ***To provide a forum*** *to discuss the latest research and developments in 3D and advanced radiation dosimetry.*
2. ***To elevate the quality*** *of radiation therapy treatments and quality assurance (QA) through improved clinical dosimetry.*
3. ***To explore*** *the dosimetric challenges posed by modern radiation treatment techniques*
4. ***To energize and diversify*** *dosimetry research and clinical practice by encouraging interaction and synergy between advanced, 3D, and semi-3D dosimetry techniques*

On a personal note we would like to extend a very warm welcome to all conference attendees ! We have a very diverse group of clinical and research physicists from many parts of the world, and it is truly a pleasure to welcome you all.

Finally we would like to acknowledge the many people who have made the meeting possible. Special thanks go to the Local Organizing Committee (listed below) who put in many hours on the difficult challenge of arranging a meeting in a different continent and language. The Scientific Committee reviewed all the conference abstracts and participated in many planning conference calls, and meeting activities. We are also very grateful to our sponsors, both academic and industrial, who's support was vital to enabling this meeting.

Mark Oldham PhD, FAAPM
Professor, Radiation Oncology,
Duke University Medical Center,
Durham, NC, USA
Mark.Oldham@duke.edu

Fang-Fang Yin, PhD, FAAPM
Chief of Physics, Radiation Oncology,
Durham, NC, USA
Director of Medical Physics
Duke Kunshan University
FangFang.Yin@duke.edu

IOP Conf. Series: Journal of Physics: Conf. Series **1305** (2019) 011002 doi:10.1088/1742-6596/1305/1/011002

*10th IC3DDose Meeting Program*

## Program at a glance

| Time | *Sunday Sept 16th* | *Monday Sept 17th* | *Tuesday Sept 18th* | *Wednesday Sept 19th* |
|---|---|---|---|---|
| 08:00-08:30 | | *ITR 1: 3D Fundamentals* | *ITR 2: MFG 3D Dosimeters* | *ITR 3: Dose Readout* |
| 08:30 - 08:55 | IR: Opportunities | IR: Pre-clinical | ICR: Carbon ions | ICR: EPIDS |
| 08:55 - 09:20 | IR: Med Phys China | IR: SRS & SBRT | ICR: Protons | ICR: IGRT 3D & QA |
| 09:20 - 10:30 | Proffered Talks: Clinical Opportunities | Proffered Talks: SRS and SBRT | Proffered Talks: Protons / new dosimetry systems | Proffered Talks: EPID and general clinical applications |
| 10:30 - 10:50 | Break | | | |
| 10:50 - 12:00 | IR: Dosimetry & MR<br>Proffered Talks: MRI | IR: Chem Dosimetry<br>Proffered Talks | Proffered Talks: New Dosimetry Sys | Proffered Talks:<br>Concluding Comments |
| 12:00 - 13:00 | **Lunch and Posters** | | | |
| 13:00 - 13:25 | ICR: Dosimetry & MRI Linacs | Proffered Talks: Chemical Dosimetry | ICR: Motion and 4D Dosimetry | **Bon Voyage !** |
| 13:25 - 13:50 | IR: MD Anderson Experience | | Proffered Talks: Motion and Brachy | |
| 13:50 - 15:00 | Proffered Talks: MRI Linacs | Social Event Tour | | |
| 15:00 - 15:20 | Break | | Break | |
| 15:20 - 15:45 | IR: Cherenkov and optical Dosimetry | | IR: Developing Countries | |
| 15:45 - 16:45 | Proffered Talks: optical and cherenkov | | IR: Innovation in Education<br>Proffered: education | |
| 16:45 + | | | Workshops I and II | |
| Late Period | | | | |


## Co-Chairs:

Mark Oldham PhD[1], Fang-Fang Yin PhD[1,2],
[1]Duke University Medical Center, NC, USA
[2]Duke Kunshan University, Kunshan, China

## Local Organizing Committee:

Ying-Chiang (David) Huang PhD, James Bowsher PhD,
Claire Kaimei Luo, Zhidong Xue and Hongrui Wang.
Duke Kunshan University, Kunshan, China

IOP Conf. Series: Journal of Physics: Conf. Series **1305** (2019) 011002 doi:10.1088/1742-6596/1305/1/011002

*10$^{th}$ IC3DDose Meeting Program*

## International Scientific Committee:

Sven Back (Sweden), Clive Baldock (AUS), Sam Beddar (USA), Yves De Deene (AUS), Simon Doran (UK), Geoffrey Ibbott (USA), Andrew Jirasek (CA), Kevin Jordan (Canada), Ben Mijnheer (NL), Mark Oldham (USA), John Schreiner (CA), Cheng-Shie Wuu (USA)

## Invited Speaker Listing:

Sven Back, PhD, Associate Professor, Medical Radiation Physics, Lund University, Sweden Clive Baldock, PhD, Professor, Pro-Vice Chancellor, Dean of Graduate Research, University of Tasmania, Australia

Andy Beavis, PhD Head of Radiation Physics, Castle Hill Hospital, Cottingham, UK Sam Beddar, PhD, Professor and Chief of Research, Department of Radiation Physics, UT MD Anderson Cancer Center, TX USA

Petr Bruza, PhD, Thayer School of Engineering, Dartmouth College, NH, USA

Sha Chang PhD, Professor, UNC School of Medicine, Chapel Hill, NC, USA

Jianron Dai, PhD, Vice Chairman, Department of Radiation Oncology, Cancer Institute (Hospital), Beijing, CHINA

Yves DeDeene, PhD, Professor of BME, Faculty of Science and Engineering, Macquarie University, Sydney, Australia

Simon Doran, PhD, Senior Staff Scientist, CRUK Cancer Imaging Center, Institute of Cancer Research, Sutton, UK

Geoffrey Ibbott, PhD, Professor and Deputy Head, Radiation Oncology, UT MD Anderson Cancer Center, TX, USA

Urszula Jelen PhD, Medical Physics Post-doctoral Scientist, Medical Physics Department, Ingham Institute for Applied Medical Research, Liverpool, NSW, Australia

Andrew Jirasek, PhD, Associate Professor, Director, CAMPEP Graduate Program in Medical Physics, Mathematics, Statistics, Physics, and Computer Science, I.K. Barber School of Arts and Sciences, The University of British Columbia, Okanagan, BC Canada

Kevin Jordan, PhD, Physicist, Physics and Engineering, London Regional Cancer Program, London , ON, CANADA

Dan Low, Professor and Vice Chair of Physics, Radiation Oncology, University of California and Los Angeles, USA

*10th IC3DDose Meeting Program*

Jiade Lu, Professor, Shanghai Proton and Heavy Ion Center (SPHIC), Fudan University Shanghai Cancer Center (FUSCC), CHINA

Ben Mijnheer PhD, FIOMP, DEPT of Radiation Oncology, Netherlands Cancer Institute, Amsterdam , NETHERLANDS

John Schreiner, PhD, FAAPM, FCCPM, FCOMP, Chief Medical Physics, Cancer Center of Southeastern Ontario, Kingston , ON CANADA

Cheng-Shie Wuu, PhD, Professor, Director of Medical Physics, Department of Radiation Oncology, Columbia University, NEW YORK , NY

IOP Conf. Series: Journal of Physics: Conf. Series **1305** (2019) 011002 doi:10.1088/1742-6596/1305/1/011002

# Program at a Glance

| Time | Sunday Sept 16th | Monday Sept 17th | Tuesday Sept 18th | Wednesday Sept 19th |
|---|---|---|---|---|
| 08:00-08:30 | | *ITR 1: 3D Fundamentals* | *ITR 2: MFG 3D Dosimeters* | *ITR 3: Dose Readout* |
| 08:30 - 08:55 | IR: Opportunities | IR: Pre-clinical | ICR: Carbon ions | ICR: EPIDS |
| 08:55 - 09:20 | IR: Med Phys China | IR: SRS & SBRT | ICR: Protons | ICR: IGRT 3D & QA |
| 09:20 - 10:30 | Proffered Talks: Clinical Opportunities | Proffered Talks: SRS and SBRT | Proffered Talks: Protons / new dosimetry systems | Proffered Talks: EPID and general clinical applications |
| 10:30 - 10:50 | Break | | | |
| 10:50 - 12:00 | IR: Dosimetry & MR<br>Proffered Talks: MRI | IR: Chem Dosimetry<br>Proffered Talks | Proffered Talks: New Dosimetry Sys | Proffered Talks:<br>Concluding Comments |
| 12:00 - 13:00 | Lunch and Posters | | | |
| 13:00 - 13:25 | ICR: Dosimetry & MRI Linacs | Proffered Talks: Chemical Dosimetry | ICR: Motion and 4D Dosimetry | Bon Voyage ! |
| 13:25 - 13:50 | IR: MD Anderson Experience | | Proffered Talks: Motion and Brachy | |
| 13:50 - 15:00 | Proffered Talks: MRI Linacs | Social Event Tour | | |
| 15:00 - 15:20 | Break | | Break | |
| 15:20 - 15:45 | IR: Cherenkov and optical Dosimetry | | IR: Developing Countries | |
| 15:45 - 16:45 | Proffered Talks: optical and cherenkov | | IR: Innovation in Education<br>Proffered: education | |
| 16:45 + | | | Workshops I and II | |
| Late Period | | | | |

IR = Invited Review (ICR = Clinical, ITR = Technical)

| | | |
|---|---|---|
| ITR | Baldock/Schreiner | Review of Fundamentals of 3D and Advanced Dosimetry |
| ITR | Jordan / DeDeene | Making your own 3D Dosimeters |
| ITR | Doran / Jirasek | 3D Dosimetry Read-Out Techniques |
| ICR: China Med Phys | Jianrong Dai | Medical Physics and Dosimetry in China |
| IR: Opportunities | Daniel Low | Dosimetry challenges and opportunities in modern RT |
| ICR: SRT/SBRT | Mark Oldham | SRT/SBRT: QA dosimetry and 3D |
| ICR: Pre-Clinical | Cheng-Shie Wuu | Pre-clinical and small field dosimetry |
| ICR: Protons | Sam Beddar | 3D dosimetry for proton therapy |
| ICR: Carbon | Jiade Lu | 3D dosimetry for heavy ion therapy therapy |
| ICR: 4D & Gating | Sven Back | 4D and Motion: Dosimetry and motion management |
| ICR: IGRT and QA | John Schreiner | IGRT 3D & QA: End to end QA IGRT |
| WrkShp1 | Kevin Jordan | Data Challenge: optical-CT, x-ray-CT, MRI |
| WrkShp 2 | John Schreiner | Data Analysis Tools: 3D Slicer and CERR |
| ICR: Dosimetry & B fields | Simon Doran | Radiotherapy in the presence of Magnetic Fields |
| IR: Dosimetry 4 MRI Linacs | Urszula Jelen | Dosimetry needs for MRI Linacs |
| ICR: MR Linac Dosimetry | Geoff Ibbott | MRI Linacs: the MD Anderson Experience |
| IR: Chemical Dosimetry | Kevin Jordan | Chemical 3D Dosimetry |
| IR:Cherenkov Dosimetry | Petr Bruza | Optical and Cherenkov Dosimetry |
| ICR: EPID dosimetry I | Ben Mijnheer | EPIDS and QA of advanced treatments |
| ICR:Developing Countries | Sha Chang | Unmet needs and opportunities of 3D radiation dosimetry in the low and middle income countries |
| IR Innovation in Educaton | Andy Beavis | Innovation in Education of Advanced Dosimetry: Computer simulation in physics training |

IOP Conf. Series: Journal of Physics: Conf. Series **1305** (2019) 011002 doi:10.1088/1742-6596/1305/1/011002

*10th IC3DDose Meeting Program*

## 10th IC3DDose Conference Program, Sept. 2018

**Sunday 16 September 2018**

| | |
|---|---|
| **8:15-8:30** | **Welcome: Mark Oldham*, Fang-Fang Yin*, Ying Chiang Huang#**<br>***Radiation Oncology, Duke University, NC, USA**<br>**#Duke Kunshan University, Kunshan, China** |
| **8:30-10:30** | **Session: Clinical Challenges and Opportunities**<br>**Moderators - John Schreiner and Mark Oldham** |
| 8:30-8:55 | **Invited Review - Dosimetry challenges and opportunities in modern RT**<br>*Dan Low PhD,*<br>Department of Radiation Oncology, UCLA, California, USA |
| 8:55-9:20 | **Invited Review - Medical Physics and Clinical Dosimetry in China**<br>*Jianrong Dai PhD,*<br>Department of Radiation Oncology, Cancer Institute, Beijing, China |
| 9:20-9:35 | Evaluation of a Clinical Dose Accumulation Algorithm Using Deformable Gel Dosimetry *Charles K Matrosic*[1], Shannon Holmes, Bryan Bednarz, and Wesley Culberson<br>[1]Department of Medical Physics, School of Medicine and Public Health, University of<br>Wisconsin - Madison, WI |
| 9:35-9:50 | Investigation of lung tumour peripheral doses using normoxic polymer gel and film dosimetry techniques<br>A Venning[1], M Mundayadan Chandroth, B Chick, B Waller and C Morgan [1]Mid-North Coast Cancer Institute, Port Macquarie Base Hospital, Port Macquarie, NSW, Australia |
| 9:50-10:05 | Feasibility of radiosurgery dosimetry using NIPAM 3D dosimeters and x-ray CT<br>*Justus Adamson*[1], Jaclyn Carroll, Michael Trager, Paul Yoon, Jacob Kodra, Fang-Fang Yin, Evan Maynard, Michelle Hilts, Mark Oldham, Andrew Jirasik<br>[1]Department of Radiation Oncology, Duke University Medical Center, Durham, North Carolina, USA |
| 10:05-10:20 | Surface Dose Accuracy in VMAT Head & Neck Radiation Treatment Using Bolus<br>*KM Alexander*[1], J Gooding, LJ Schreiner and T Olding<br>[1]Department of Physics, Queen's University, Kingston, Ontario, Canada |
| 10:20-10:50 | ***Tea/Coffee Break*** |

| | |
|---|---|
| **10:50-12:00** | **Session: Dosimetry and MRI Linacs**<br>**Moderators - Yves De Deene and Geoffrey Ibbott** |
| 10:50-11:15 | **Invited Review - Radiotherapy in the presence of magnetic fields**<br>*Simon Doran PhD,*<br>Institute of Cancer Research, Sutton, Surrey, UK |
| 11:15-11:30 | Polymer gel-based measurements of the isocenter accuracy in an MR-LINAC<br>*S Dorsch*[1], P Mann, A Elter, A Runz, S Klüter and C P Karger<br>[1]Department of Medical Physics in Radiation Oncology, German Cancer Research Center (DKFZ), Heidelberg, Germany |
| 11:30-11:45 | Dose rate and fractionation dependence of methacrylic acid based polymer gels using optical and MRI techniques<br>*Hannah J. Lee*[1], Yvonne Roed, Geoffrey S. Ibbott<br>[1]Department of Radiation Physics, UT MD Anderson Cancer Center, Houston, TX |
| 11:45-12:00 | Characterization of small PRESAGE® samples for measurements near the dosimeter edges<br>*Filipa Costa*[1], Simon Doran, John Adamovics, Simeon Nill, Ian M Hanson and Uwe Oelfke<br>[1]Joint Department of Physics, The Institute of Cancer Research and The Royal<br>Marsden NHS Foundation Trust, London, UK |
| **12:00-1:00pm** | ***Lunch Break*** |
| **1:00-3:00pm** | **Session: Dosimetry and MRI Linacs II**<br>**Moderators - Dan Low and Andy Beavis** |
| 1:00-1:25 | **Invited Review - Dosimetry requirements for MRI Linacs**<br>*Urszula Jelen PhD,*<br>Medical Physics Department, Ingham Institute for Applied Medical Research, Liverpool, NSW, Australia |
| 1:25-1:50 | **Invited Review - Dosimetry requirements for MRI Linacs**<br>*Geoff Ibbott PhD,*<br>MD Anderson Cancer Center, Houston, TX |
| 1:50-2:05 | Evaluation of a lung-equivalent gel dosimeter for MR image-guided radiation therapy<br>*BA McDonald*[1], HJ Lee, and GS Ibbott<br>[1]Department of Radiation Physics, UT MD Anderson Cancer Center, Houston, TX |

IOP Conf. Series: Journal of Physics: Conf. Series **1305** (2019) 011002 doi:10.1088/1742-6596/1305/1/011002



2:05-2:20 MRI-based iPAGAT polymer gel dosimetry using fast recovery spin echo sequences
K Fujino[1], K Ono, S Hayashi, K Hioki, M Miyazawa, Y Akagi and Y Hirokawa
[1]High-precision Radiotherapy Center, Hiroshima Heiwa Clinic, Hiroshima, Japan

2:20-2:35 Polymer gel dosimetry in the presence of a strong magnetic fields
*Y Roed*[1], L Pinsky, and G Ibbott
[1]Department of Physics, University of Houston, Houston, TX, USA

2:35-3:00 ***BREAKOUT ACTIVITY !***

3:00-3:20 ***Tea/Coffee Break***

**3:20-5:00pm Session: Optical and Cherenkov Dosimetry (60mins: 4-5)**
**Moderators - Kevin Jordan and Simon Doran**

3:20-3:45 **Invited Review - Dosimetry from Cherenkov Imaging**
*Petr Bruza PhD,*
Thayer School of Engineering, Dartmouth College, NH, USA

3:45-4:00 Cherenkov imaging of total skin electron irradiation (TSEI)
*Timothy C. Zhu*[1],
[1]Dept. of Radiation Oncology, University of Pennsylvania, Philadelphia, PA, USA

4:00-4:15 Measurement of build-up region dose with optical cone-beam computed tomography
Scanner
*Sarah Garisto*[1], Kevin Jordan
[1]London Regional Cancer Program, London Health Sciences Centre, Canada

4:15-4:30 Feasibility study of a dry optical CT scanner using aspherical lenses
*Yves De Deene,*
School of Engineering, Macquarie University, North Ryde, Sydney, Australia

4:30-4:45 Preliminary characterization of the Duke Integrated-Lens Optical-CT scanner (DIOS)
*Cielle Collins*[1], Suk Whan Yoon, Jacob Kodra, John Adamovics, & Mark Oldham
[1]Duke University, Durham, North Carolina, USA

| | |
|---|---|
| 4:45-5:00 | Cause of cupping artifacts from radiochromic micelle gel dosimeters used in optical CT scanner measurement<br>*Takaoki Takanashi*[1], Kazuya Hayashi, Mikio Nemoto, Hiraku Kawamura, Shin-ichiro Hayashi and Hiroaki Gotoh<br>[1]3D Gel Dosimeter Research Laboratory, Cluster for Science, Technology and Innovation Hub, RIKEN, 2-1,Hirosawa, Wako, Saitam, Japan |

**Monday 17 September 2018**

| | |
|---|---|
| 8:00-8:30 | **Invited Technical Review (Refresher): Fundamentals of 3D Dosimetry**<br>*J Schreiner PhD,*<br>Cancer Center of South Eastern Ontario, Kingston, ON, Canada |
| **8:30-10:30** | **Session: Small Fields, Radiosurgery and Pre-Clinical Irradiators**<br>**Moderators - Sven Back and Sam Beddar** |
| 8:30-8:55 | **Invited Review – Pre-clinical and Small Field Dosimetry**<br>*Cheng Shie Wuu PhD*<br>Department of Radiation Oncology, Columbia University, New York, USA |
| 8:30-9:20 | **Invited Review - SRS and SBRT and Small Fields**<br>*Mark Oldham PhD,*<br>Duke University Medical Center, Durham, USA |
| 9:20-9:35 | Verification of stereotactic cranial radiotherapy treatments with MR-based gel dosimeters: practical aspects<br>*Filipa Costa*[1], Evanthia Kousi, Anne Gasnier, Emma Wells, Caroline Lamb,<br>Maria A Schmidt and Rollo Moore<br>[1]Joint Department of Physics, The Institute of Cancer Research and The Royal Marsden NHS Foundation Trust, London, UK |
| 9:35-9:50 | Assessing CBCT-based patient positioning accuracy on the Gamma Knife IconTM via Presage® 3D absolute dosimetry<br>*Andy Y. Xu*[1], Yi-Fang Wang, John Admovics, and Cheng-Shie Wuu<br>[1]Department of Radiation Oncology, Columbia University, New York, USA |
| 9:50-10:05 | Initial Commissioning Measurements of Respiratory Gated Liver VMAT Stereotactic Ablative Body Radiotherapy<br>*KM Alexander*[1], A Kerr and T Olding<br>[1]Department of Physics, Queen's University, Kingston, Ontario, Canada |

IOP Conf. Series: Journal of Physics: Conf. Series **1305** (2019) 011002 doi:10.1088/1742-6596/1305/1/011002



10:05-10:20 **BREAKOUT ACTIVITY !**
The utility of 3D Dosimetry for small fields, SRS/SBRT and Pre-clinical

10:20-10:50 **Tea/Coffee Break**

**10:50-12:00** **Session: Chemical Dosimeters I**
**Moderators - Cheng Shie Wuu and Yves De Deene**

10:50-11:15 **Invited Review - Chemical Dosimetry**
Kevin Jordan PhD and John Adamovics PhD,
Rider University, NJ, USA

11:15-11:30 Optimization for stability of the deformable FlexyDos3D radiation dosimeter and curing effects
*M J Wheatley*[1], A S Balatinac, J T Booth and Y De Deene
[1]School of Engineering, Macquarie University, North Ryde, Sydney, Australia

11:30-11:45 Influence of the components on the dose response of a radiochromic gel dosimeter based on a polyvinyl alcohol - iodide complex
Shin-ichiro Hayashi[1]), Kaoru Ono, Keisuke Fujino, Sachie Fujimoto
[1]Department of Clinical Radiology, Faculty of Health Sciences, Hiroshima International University, Higashi-Hiroshima, Hiroshima, Japan

11:45-12:00 Preliminary investigation of a reusable radiochromic sheet for radiation dosimetry
Cielle Collins[1], Jacob Kodra, Suk Whan Yoon, Robert Coakley, John Adamovics, Mark Oldham
[1]Duke University, Durham, North Carolina, USA

**12:00-1:00pm** **LUNCH BREAK**

**1:00-2:30pm** **Session: Chemical Dosimetry II**
**Moderators - Mark Oldham and Kevin Jordan**

1:00-1:15 Benzothiazole-containing tetrazolium salts as radiochromic indicators in gel dosimetry
*Kalin I Penev*[1] and Kibret Mequanint
[1]Department of Chemical and Biochemical Engineering, the University of Western Ontario, London, ON, Canada

1:15-1:30 Development of a reusable PVA-GTA-I gel dosimeter for 3D radiation dose assessments
*J Taño*[1], S Hayashi, S Hirota, CA Gonzales, H Yasuda
[1]Department of Radiation Biophysics, Research Institute for Radiation Biology and Medicine, Hiroshima University, 1 Kasumi 2-3, Minami-ku, Hiroshima, Japan

1:30-1:45 Three-dimensional radiochromic and polymer gel dosimeters with Pluronic F-127 matrix - a review of current research M Kozicki[1], M Jaszczak, K Kwiatos, P Maras, S Kadlubowski, R Wach and M Dudek
[1] Department of Man-Made Fibres, Lodz University of Technology, Lodz, Poland

1:45-2:00 Initial performance evaluation of a 3D gel dosimeter based on modified tetrazolium compounds ClearView2
*Rubin Hazarika*[1], Kalin I Penev, Kibret Mequanint and Kevin Jordan
[1]London Regional Cancer Program, London Health Sciences Centre, London, Canada

**Afternoon and evening SOCIAL EXCURSION TO ZHOUZHUANG WATERTOWN**

**Tuesday 18 September 2018**

**8:00-8:30** **Invited Technical Review (Refresher): Making 3D Dosimeters**
*Kevin Jordan and Yves De Deene*
London Regional Cancer Program, ON, Canada and Macquarie University, Sydney, Australia

**8:30-10:30** **Session: Heavy Particles and New Dosimetry Systems**
**Moderators - Geoff Ibbott and Sha Chang**

**8:30-8:55** **Invited Review: Advanced Dosimetry for Heavy Ion Therapy**
*Jiade Lu PhD*,
Professor, Shanghai Proton and Heavy Ion Center (SPHIC), Fudan University Shanghai Cancer Center (FUSCC), CHINA

8:55-9:20 **Invited Review: Proton Therapy and the potential for 3D Dosimetry**
*Sam Beddar PhD,*
Professor and Chief of Research, Department of Radiation Physics, UT MD Anderson Cancer Center, TX USA

9:20-9:35 Comparison of low dose proton and photon irradiation induced polymerization processes in advanced nMAG gels using Raman spectroscopy
*N Šeperienė*[1] and D Adlienė
[1]Physics Department, Kaunas University of Technology, Studentu St.50, Kaunas,
LT51368 Lithuania

9:35-9:50 Clear micelle gel dosimeter with nanoclay
*Kazuya Hayashi*[1], Mikio Nemoto, Takaoki Takanashi, Yoosuk Kang, Haruki Togo, Jun'ichi Kotoku, Takenori Kobayashi, Shin-ichiro Hayashi and Hiroaki Gotoh
[1]Department of Chemistry and Life Science, Graduate School of Engineering Science, Yokohama National University, Yokohama, Kanagawa, Japan

*10th IC3DDose Meeting Program*

9:50-10:05 An Investigation of Dosimetric Accuracy of A Novel PRESAGE Radiochromic Sheet and Its Clinical Applications
*Yi-Fang Wang*[1], Kevin Liu, John Adamovics, Cheng-Shie Wuu
[1]Department of Radiation Oncology, Columbia University Medical Center, New York, NY

10:05-10:20 Basic characteristics of an AQUAJOINT®-based VIPET polymer gel dosimeter
*Mikio Nemoto*[1], Ayumi Oe, Tomokazu Kotani, Daniel Antonio Sahade, and Toshimasa Hamada
[1]3D Gel Dosimeter Research Laboratory, Cluster for Science, Technology and Innovation Hub, RIKEN, 2-1, Hirosawa, Wako, Saitama, Japan

10:20-10:50 **Tea/Coffee Break**

**10:50-12:00 Session: New Dosimetry Systems**
**Moderators - Simon Doran and Sam Beddar**

10:50-11:05 Novel Dual-Wavelength Optical-CT Imaging Method for Gel Dosimeter Readout
Yi Du [1], Xiangang Wang, Xincheng Xiang, Yves, De Deene
[1]Key Laboratory of Carcinogenesis and Translational Research (Ministry of Education/Beijing), Department of Radiotherapy, Peking University Cancer Hospital & Institute, Beijing, China

11:05-11:20 Evaporation and diffusion of chloroform with the deformable FlexyDos3D radiation dosimeter
*M J Wheatley*[1], J T Booth and Y De Deene
[1]School of Engineering, Macquarie University, North Ryde, Sydney, Australia

11:20-11:35 Radiation induced degradation of rhodamine 6G and 7-Diethylamino-4-methylcoumarin in nano-clay gel for use in dosimeter
*T. Maeyama*[1], T. Takanashi
[1]Department of Chemistry, School of Science, Kitasato University, Kanagawa, Japan

11:35-11:50 Gel dosimetry measurement of dose enhancement bismuth-based nanoparticles in radiation therapy
*Azimeh Rajaee*[1], Lingyun Zhao[2], Shi Wang[1], Yaqiang Liu[1]
[1]Institute of Medical Physics and Engineering, Department of Engineering Physics, Tsinghua University, Beijing, China

IOP Conf. Series: Journal of Physics: Conf. Series **1305** (2019) 011002 doi:10.1088/1742-6596/1305/1/011002



| | |
|---|---|
| 11:50-12:10 | **BREAKOUT ACTIVITY !** |
| **12:10-1:00pm** | ***LUNCH BREAK and Varian Presentation*** |
| **1:00-3:00pm** | **Session: Brachytherapy and Motion**<br>**Moderators - John Schreiner and Ben Mijnheer** |
| 1:00-1:25 | **Invited Review: Dosimetry and Motion Management**<br>*Sven Back PhD,*<br>Associate Professor, Medical Radiation Physics, Lund University, Sweden |
| 1:25-1:30 | Validation of an Ultrasound-Guided Prostate HDR Brachytherapy Dose Delivery<br>*T Olding*[1], KM Alexander, C Joshi and LJ Schreiner<br>[1]Department of Physics, Queen's University, Kingston, Ontario, Canada |
| 1:30-1:45 | High dose rate brachytherapy three-dimensional gel dosimetry using optical computed tomography readout<br>*DA DeVries*[1], C Joshi and LJ Schreiner<br>[1]Department of Physics, Queen's University, Kingston, Ontario, Canada |
| 1:45-2:00 | Spatial dose distribution analysis of Co-60 HDR brachytherapy of cervical cancer using an AQUAJOINT®-based VIPET polymer gel dosimeter, Ayumi Oe[1], Mikio Nemoto, Masanori Miyazawa, Daniel Antonio Sahade and Toshimasa Hamada,<br>[1]Department of Radiology, Jichi Medical University Hospital, Yakushiji, Shimotsuke, Tochigi, Japan |
| 2:00-2:15 | Dose reconstruction including dynamic six-degree of freedom motion during prostate radiotherapy<br>*C G Muurholm*[1], T Ravkilde, S Skouboe, T Eade, D T Nguyen, J Booth, P J Keall and P R Poulsen<br>[1]Department of Physics and Astronomy, Aarhus University, Denmark |
| 2:15-2:30 | Development of an experimental 3-D tool based on radiochromic films to determine normal tissue doses in external radiotherapy<br>*J Colnot*[1], G Garnier, S Zefkili, J-L Dumas, R Gschwind, C Huet<br>[1]Institut de Radioprotection et de Sûreté Nucléaire (IRSN), Service de Recherche en Dosimétrie, Laboratoire de Dosimétrie des Rayonnements Ionisants, Fontenay-aux-Roses, France |

IOP Conf. Series: Journal of Physics: Conf. Series **1305** (2019) 011002 doi:10.1088/1742-6596/1305/1/011002



| | |
|---|---|
| 2:30-2:45 | Deformable gel dosimeter containing an X-ray visible dose sensitive target region<br>*C J Watson*[1], A U Yeo, J R Supple, M Geso, T Kron and R D Franich<br>[1]School of Science, RMIT University, Melbourne, Australia |
| 2:45-3:05 | **Tea/Coffee Break** |
| **3:05-4:00pm** | **Session: Dosimetry and Education**<br>**Moderators - Mark Oldham and Urszula Jelen** |
| 3:05-3:20 | **Invited Review:** Unmet needs and opportunities of 3D radiation therapy dosimetry in the low- and middle-income countries.<br>*Sha Chang PhD,*<br>Professor, UNC School of Medicine, Chapel Hill, NC, USA |
| 3:20-3:35 | **Invited Review:** Innovation in Dosimetry Education: Computer Simulation in Physics training.<br>Andy Beavis PhD,<br>Head of Radiation Physics, Castle Hill Hospital, Cottingham, UK |
| 3:35-3:45 | Teaching the principles of X-ray CT and SPECT using optical CT, glowsticks and a scaled anthropomorphic phantom.<br>*Yves De Deene,*<br>School of Engineering, Macquarie University, North Ryde, Sydney, Australia [WIP] |
| 3:45-4:00 | How important is the dose rate sensitivity of 2D and 3D radiation dosimeters?<br>*Yves De Deene,*<br>School of Engineering, Macquarie University, North Ryde, Sydney, Australia |
| 4:00-4:15 | Exact MLC Control and Dosimetric Effects in Dynamic MLC,<br>Shidong Li,<br>Department of Radiation Oncology, Fox Chase Cancer Center at Temple University Hospital, Philadelphia, PA, USA |
| **4:15-5:15 pm** | **WORKSHOPS ! Kevin Jordan and John Schreiner** |

**Wednesday 19 September 2018**

| | |
|---|---|
| **8:00-8:30** | **Invited Technical Review (Refresher): Dosimetry Read-Out Techniques**<br>**Simon Doran and Andy Jirasek** |
| **8:30-10:30** | **Session: Dosimetry with EPIDS and end-to-end QA (70 mins: 4-5 talks)**<br>**Moderators - Andy Beavis and Ben Mijnheer** |

*10th IC3DDose Meeting Program*

| | |
|---|---|
| 8:30-8:55 | **Invited Review:** EPIDs and QA of advanced treatments<br>*Ben Mijnheer PhD, FIOMP,*<br>DEPT of Radiation Oncology, Netherlands Cancer Institute, Amsterdam , NETHERLANDS |
| 8:55-9:20 | **Invited Review:** IGRT 3D & QA: End to end QA IGRT<br>John Schreiner, PhD,<br>Chief Medical Physics, Cancer Center of Southeastern Ontario, Kingston, ON CANADA |
| 9:20-9:35 | EPID-based beam matching for linear accelerators using pixel sensitivity map<br>Baozhou. Sun[1], Sreekrishna M. Goddu, Sasa Mutic, Bin Cai<br>[1]Department of Radiation Oncology, Washington University, St. Louis, MO, USA |
| 9:35-9:50 | Quantitative evaluation of transmission EPID daily imaging on a Halcyon Linac<br>P Jin[1], Y H Xie, M Huang, T C. Zhu<br>[1]Department of Radiation Oncology, University of Pennsylvania, Philadelphia, PA USA |
| 9:50-10:05 | A sliding-window approach for improved VMAT dose calculation accuracy<br>J G Li[1], J-Y Park, N J Potter, B Lu, G Yan, C-R Liu and H N Alahmad<br>[1]Department of Radiation Oncology, University of Florida College of Medicine, Gainesville, FL USA |
| 10:05-10:20 | ***Tea/Coffee Break*** |
| **10:20-12:10** | **Session: Multi-scale and Novel 3D Dosimetry**<br>**Moderators - Cheng Shie Wuu and Simon Doran** |
| 10:20-10:35 | Multi-scale dosimetry with multi-scale Chinese reference phantoms<br>*Rui Qiu*[1], Zhen Wu, Chunyan Li, Li Ren, Wenjing Wang, Ruiyao Ma, An kang Hu, Hongyu Zhu, Junli Li.<br>[1]Department of Engineering Physics, Tsinghua University, Beijing, China |
| 10:35-10:55 | A chemical evolution of NVP-containing VIPAR-family 3D polymer gel dosimeters - a brief overview<br>M Kozicki[1], M Jaszczak, P Maras and M Dudek<br>[1]Department of Man-Made Fibres, Lodz University of Technology, Lodz, Poland |

IOP Conf. Series: Journal of Physics: Conf. Series **1305** (2019) 011002 doi:10.1088/1742-6596/1305/1/011002

*10$^{th}$ IC3DDose Meeting Program*

| | |
|---|---|
| 10:55-11:10 | Dose verification of dynamic MLC-tracked radiotherapy using small PRESAGE® 3D dosimeters and a motion phantom<br>*Filipa Costa*[1], Martin J Menten, Simon Doran, John Adamovics, Ian M Hanson, Simeon Nill and Uwe Oelfke<br>[1]Joint Department of Physics, The Institute of Cancer Research and The Royal Marsden NHS Foundation Trust, London, UK |
| 11:10-11:35 | A benchtop UV irradiator for 3D dosimetry laboratories with dosimetric considerations in a spinning NMR test tube<br>*Yves De Deene*<br>School of Engineering, Macquarie University, North Ryde, Sydney, Australia |
| 11:45 | **Concluding Remarks,** John Schreiner and Mark Oldham |

IOP Conf. Series: Journal of Physics: Conf. Series **1305** (2019) 011002 doi:10.1088/1742-6596/1305/1/011002

*10$^{th}$ IC3DDose Meeting Program*

**SPONSORS: -**

**(Please see sponsor material available in the root directory of the USB pre-conference proceedings.)**

| | |
|---|---|
| **Varian -** | **Platinum** |
| **Elekta -** | **Gold** |
| **SunNuclear -** | **Gold** |
| **HGPT-** | **Gold** |
| **PTW -** | **Gold** |

***Special thanks also to the following donations for conference raffle and student prizes....***

Dr Ben Mijnheer for donating 10 electronic licenses for the book "Clinical 3D Dosimetry in Modern Radiation Therapy", 2017, Edited by B Mijnheer PhD, CRC Press, Taylor and Francis Group, ISBN: 978-1-4822-5221-7

Polygevero.com and and Marek Kozicki PhD, for donating three 2-year licenses for best student paper award and three 1-year licenses for the raffle.

IOP Conf. Series: Journal of Physics: Conf. Series **1305** (2019) 011002 doi:10.1088/1742-6596/1305/1/011002

# Peer review statement

All papers published in this volume of *Journal of Physics: Conference Series* have been peer reviewed through processes administered by the proceedings Editors. Reviews were conducted by expert referees to the professional and scientific standards expected of a proceedings journal published by IOP Publishing.

# Dosimetry challenges and opportunities in modern radiation therapy

**D A Low**
Department of Radiation Oncology, UCLA, Los Angeles, CA 90025 USA

E-mail: dlow@mednet.ucla.edu

**Abstract**. Modern radiation therapy (RT) includes conformal therapy, intensity modulated radiation therapy, proton therapy, image-guided radiation therapy, adaptive radiation therapy (ART), and recently magnetic-resonance guided radiation therapy and, upcoming, $4\pi$ radiation therapy. These techniques show the continual increase in complexity of radiation therapy techniques which, coupled with a stagnant medical physics workforce, means that the amount and complexity of work per physicist has increased and is likely to increase in the foreseeable future. Three distinct challenges can be identified that need to be addressed. First, is the recent commercial development of automated multi-metastases stereotactic radiosurgery (SRS) techniques. These techniques plan and irradiate a number (up to approximately 20) brain lesions in one treatment session, typically employing one isocenter. The spatial accuracy specifications for SRS imply that attention to the angular accuracy is more critical for these treatments than conventional SRS or other treatment methods. In parallel, our and other groups are developing $4\pi$ techniques, which is a proposed method for optimizing both beam angles and intensity fluence to provide x-ray based dose distributions with unparalleled compactness and conformality. One cost to deliver these dose distributions is the added requirement to not only rotate the couch but also shift the couch to increase the number of available beam angles. These two techniques will require efficient and quantitative dose distribution measurements of relatively large volumes for, at least, end to end testing of multi-metastases and $4\pi$ treatments. Finally, magnetic resonance guided radiation therapy has led to a resurgence in the development of on-table ART, which requires that the medical physicist compare two calculated treatment plans and efficiently and effectively determine if differences between those treatment plans are clinically significant. Modifying and improving existing tools will be critical to the safe and effective on-table ART.

## 1. Introduction

What are the characteristics of modern radiation therapy (RT)? If we define modern RT to include those techniques in current clinical practice as well as those that are on the immediate horizon, it encompasses 3D conformal therapy, intensity modulated radiation therapy (IMRT), brachytherapy, proton therapy, and magnetic-resonance guided radiation therapy (MRgRT). There is even a company that is developing a PET-guided radiation therapy machine. On the surface, dosimetry challenges for these techniques appear relatively straightforward. We have been conducting dosimetry measurements for all but the PET-guided system for at least a few years, and in the case of conformal therapy, for decades, so are there any significant new challenges to manage? I would postulate that the answer to this is yes, and it comes from a confluence of factors that include increasing dose delivery complexity, decreasing manpower resources, increasing automation, and decreasing time to manage and evaluate results.

Increased complexity can be seen by comparing conformal therapy treatment plans to fixed-field IMRT to VMAT type deliveries. Sophisticated motion coordination is required to deliver an integrated dose distribution with an absolute dose to within 5%, and the subsequent dose delivery complexity increases the chances that errors will exist within the system, errors that need to be detected before they cause a clinically relevant dose delivery error. Complexity is also increasing with respect to how we manage the patients on the table. Modern linear accelerators are essentially robots, capable of simultaneously moving multiple systems, including the patient couch, gantry, and multileaf collimator, in a choreographed dance that is intended to further increase either dose conformation and/or treatment speed. The addition of routine couch motion makes measurement-based dosimetry more complicated, yet more important than before.

Decreasing manpower resources is a function of two features, a stable workforce, and an increasing workload. In our clinic, we have found that physics-based responsibilities continue to increase, while the number of physicists in our staff does not increase. Given the expected growth in treatment options, this deficit is not likely to stabilize. More complex treatment planning and delivery options, coupled with an increase in both on-table and off-table adaptive therapy, will increase the workload for medical physicists.

Part of the complexity increase is due to the ability to automate complex workflows. All modern linear accelerators are essentially robots. They operate using a centralized computer system that manages multiple subsystems in a way that allows these subsystems to choreograph their function. This has been coupled with newly developed optimization tools and advances in computer hardware, to enable computations that a few years ago would have been at best impractical.

Much of the rationale in advancing the hardware and software in RT is to enable the process of adaptive radiation therapy (ART), which can be defined as adjusting the treatment plan due to changes in patient morphology or functional status. ART is typically subdivided into online and offline (or on-table and off-table), describing whether the new treatment plan will be applied to the fraction in which the new information was acquired, or at some point in the future, typically the next fraction [1]. In either case, the time allotted to evaluate dose distributions is greatly reduced relative to the typical workflows.

The increasing complexity of RT leads to increased demands to conduct ever more sophisticated dosimetry analyses. They are required to be quantitative comparisons that yield the data needed for safety evaluation, while being conducted within the workflow's time constraints. Figure 1 shows a schematic of the challenge we are dealing with. The amount of time allowed to measure or calculate and evaluate doses has decreased steadily. Most measurements in the 2D era were conducted as part of the systematic management of the machines. The amount of time to review the measurements could be measured in days to weeks. 3D conformal radiation therapy brought with it the use of multileaf collimators [2] and their attendant dosimetric complexity. IMRT presented the most rapid increase in dosimetric complexity, initiating mandatory per-patient measurements that had a turn-around time of one or a few days [3]. Early on, the number of data points that were being measured was dozens (exclusive of films), but modern systems acquire on the order of 1000 data points. On-table adaptive radiation therapy does not allow for pre-treatment measurement-based quality assurance, so it compares multiple calculations in an attempt to determine if the instructions for the linear accelerator will produce the desired dose distribution [4]. These calculations can have more than 1,000,000 points to compare, at the same time that this evaluation has to be conducted in less than a few minutes to be inserted into the clinical workflow. This evolution from few to many points and the reduction of available time will stress existing resources and will require new support tools for the physicist.

In summary, there are two principal challenges to address in dosimetry: 1) Increased planning and delivery complexity to manage and 2) Increased evaluation challenges without increased resources.

## 2. Complexity

As previously stated, the complexity of RT has increased dramatically from the late 1980s to the present. Dosimetry tools have been developed to aid us in managing this increased complexity, but there are new challenges that have been recently introduced or are about to be introduced that may make the existing systems insufficient to assure that dose distribution delivery has sufficient accuracy.

On the first front, there has been a recent development of commercial multi-metastases irradiation that treats many brain metastases in a single or few treatments. The treatment planning aspects of this are not substantially different than the planning of single or few metastases, but due to limitations in the MLC design, in order to isolate the radiation dose distribution in each metastasis, the planning system sometimes has to split the treatment into multiple passes. Still, as long as the planned dose distributions are clinically acceptable, are there any fundamental differences in the QA of these treatments than there are for other stereotactic treatments? Figures 2 and 3 shows examples of multi-metastasis treatment plans from vendors that are commercializing this technique.

The QA specifications for traditional stereotactic radiosurgery of brain metastases can be found in the AAPM report 54 [5]. They stated that the accuracy of a linear accelerator (as of 1995) set up for stereotactic radiosurgery was 1.0 mm. More recently, the AAPM published the report from Task Group 142 [6], within which it also specified that for stereotactic radiosurgery, the isocenter accuracy be 1.0 mm. Couch angle was specified to be accurate to within 1°. When treatment planning is conducted to irradiate a single lesion, or a few closely spaced lesions, the ultimate spatial accuracy of the treatment is dominated by the isocenter tolerance. However, when a large number of metastases are being treated, these may be distributed throughout the brain. Assuming that the volume being treated extends approximately 15 cm (the typical size of the human brain), if the isocenter is placed in the middle of the brain, some lesions may be as far as 7.5 cm from isocenter. A 1° couch angle error will move the dose distribution farthest from isocenter 1.3 mm from its intended location, much greater than the accuracy specified at isocenter. Similar errors would be caused by errors in the gantry angle, although given that most stereotactic treatments include couch rotation, the impact on the location accuracy of the resulting dose distribution is more complex and would include degrading the dose falloff as well as localization accuracy. Similarly, an error in the collimator angle would yield a combination of localization inaccuracy and dose falloff degradation.

Given that the current specifications for gantry, collimator, and couch accuracy can lead to unacceptable spatial errors in the delivered dose distributions, the specifications for these quantities needs to be updated for the single isocenter irradiation of multiple metastases. In addition, the delivery of these complex dose distributions requires an additional level of scrutiny than more traditional radiation therapy methods. 3D dosimeters are uniquely suited for such measurements. They can simultaneously determine spatial localization accuracy and dose distribution conformality (i.e. the dose falloff) for all of the irradiated lesions. There are no other techniques that would efficiently and definitively determine the delivered dose distribution end-to-end accuracy.

Similarly, there will soon be a class of radiation dose distributions that will take advantage of the robotic nature of modern commercial linear accelerators. Tran *et al* [7] determined recently that if a dose optimizer was allowed to simultaneously optimize beam directions and fluences, the resulting dose distributions would be not only conformal, but much more compact. They termed this new technique $4\pi$, referring to the number of steradians in a sphere and suggesting that the beams can come from all directions. $4\pi$ therapy employs a complicated dance between the couch and gantry motions, including couch shifts to expand the number of beam directions that can be employed without collision between the gantry and the couch or patient. The added dynamic degree of freedom of couch shifting, as well as the range and number of non-coplanar beams, will greatly stress dosimetry validation. As with multi-metastasis therapy, a technique that can validate the delivered dose distribution accuracy, including the spatial accuracy and the dose distribution conformality, will be critical to safely introducing $4\pi$ therapy.

Finally, the recent commercial introduction of MR-guided radiation therapy introduces limits in the use of traditional dosimetry systems (such as scanning ionization chambers) and in the processes used to align and localize the detector systems (since there is a relatively deep bore and the alignment lasers are displaced from isocenter). The impact of the magnetic field on ionization chambers and other dosimetry systems is being determined and is likely a strong focus of this meeting.

## 3. More work, less time

The commercial introduction of MRgRT brought along with it a new emphasis on the clinical potential of on-table adaptive radiation therapy (ART). ART requires that up-to-date images be acquired, resegmented, and the planned treatment be recalculated using the new images, or at least the new image segmentation. If the treatment plan is deemed unacceptable, it is reoptimized and the new plan used for that fraction. Given that these treatments are universally IMRT, a patient-specific QA would be required, but a measurement-based approach is unfeasible. Therefore, clinics have employed a calculation-based method to assure that the radiation dose distribution to be delivered will closely match the planned dose [4]. The operational parameters being sent to the linac are transmitted to an independent dose calculation engine that recomputes the dose distribution and compares it to the planned dose. The medical physicist then determines whether the doses are clinically similar and approves the treatment to proceed. As mentioned above, dose distributions can have millions of points to compare, and the physicist has seconds to conduct the comparison, so improved dose comparison techniques will be required.

The comparison is being made between two dose distributions that are computed on the patient model. This provides opportunities not present in most dose distribution comparisons, namely the ability to compare doses within organ systems [8]. The dose comparison can be made with the clinical tolerances in mind. For example, if a critical structure dose is well below tolerance, the dose comparison specification could be relaxed, since even under the relaxed criteria, an overdose would not be occurring. On the other hand, much of the body is not segmented, so care would be needed to assure that the dose distributions are evaluated in all tissues, not just those near tolerance. There are no generally accepted methods for conducting these types of evaluations, but extensions of current systems are often employed, such as dose distribution overlays, dose differences, and $\gamma$ evaluations. Since these are conducted in 3-dimensions, the limitations in human vision and computer displays, means that the physicist would need to scroll through the multiple slices to evaluate the entire dose distribution.

Tools, such as the $\gamma$ tool, were developed to efficiently sift through dose distribution comparisons and highlight regions of discrepancy. Unfortunately, most dose distribution comparisons were unable to pass all measurement points for even reasonable passing criteria, so compromise techniques were developed, such as to allow a predetermined fraction of points to fail the dose comparison test [9]. While this approach was efficient, it turned out that the sensitivity to clinically relevant dose errors was quite poor, in part due to the lack of dose and spatial specificity of such criteria. Given that the medical physicist will have very little time to compare two high spatial resolution dose distributions, optimizing the dose comparison workflow will be critical to detecting subtle but clinically relevant errors. Scrolling and closely inspecting the dose distribution is an impractical method for determining whether the dose distribution differences are clinically relevant, especially given that most calculation algorithms are unable to meet the typical $\gamma$ comparison criteria in all conditions.

The analysis of dose distributions for on-table adaptive therapy is further complicated by the fact that for MRgRT, the dose distributions will be calculated using Monte Carlo, which by its nature has statistical noise. As previously published by Low and Dempsey [10], $\gamma$ has sensitivity to noise that depends on the assignment of the reference and evaluated distributions, and if the evaluated distribution has significant noise, then $\gamma$ is underestimated. $\gamma$, or other dose comparison criteria will need to be adjusted to manage the statistical noise.

## 4. Conclusions

There are three domains in which 3D dosimetry will need development, the implementation of SRS for non-isocentric lesions, $4\pi$ radiation therapy, and on-table adaptive radiation therapy. The first two will require development of relatively large 3D dosimeters for, at least, end to end system tests. The last will require a re-tuning of dose distribution comparison techniques to enable the effective comparison of hundreds of thousands to millions of dose points in a very short time. These challenges will provide opportunities for further research and development in the area of 3D dosimetry.

IOP Conf. Series: Journal of Physics: Conf. Series **1305** (2019) 012001 doi:10.1088/1742-6596/1305/1/012001

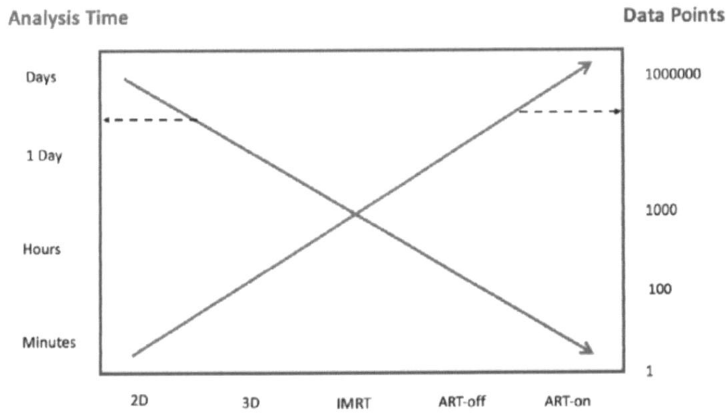


**Figure 1.** Schematic showing the increased amount of data to analyse and the decreased amount of time allocated to analyse those data. This illustrates the increasing workload that physicists have to measure and analyse dosimetry data.

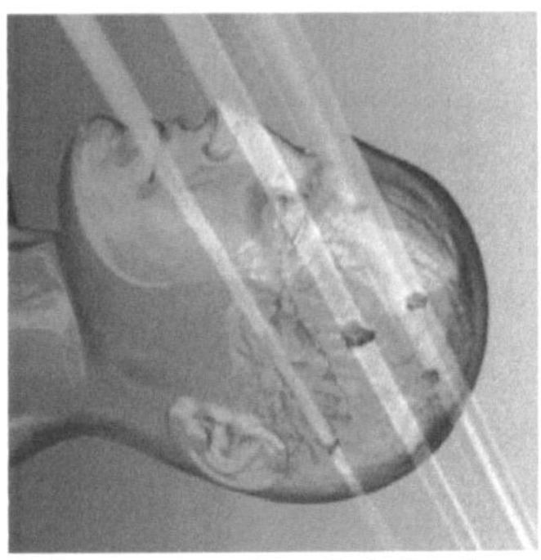

**Figure 2.** Image from the Varian website representing their implementation of multi-metastasis irradiation therapy (https://www.varian.com/sites/default/files/resource_attachments/HyperArcBriefOverview_RAD10424A_SecuredMarch2017.pdf). Note that lesions throughout the brain are being simultaneously irradiated. Meeting current stereotactic radiosurgery accuracy specifications will require improvements in 3D dosimetry.

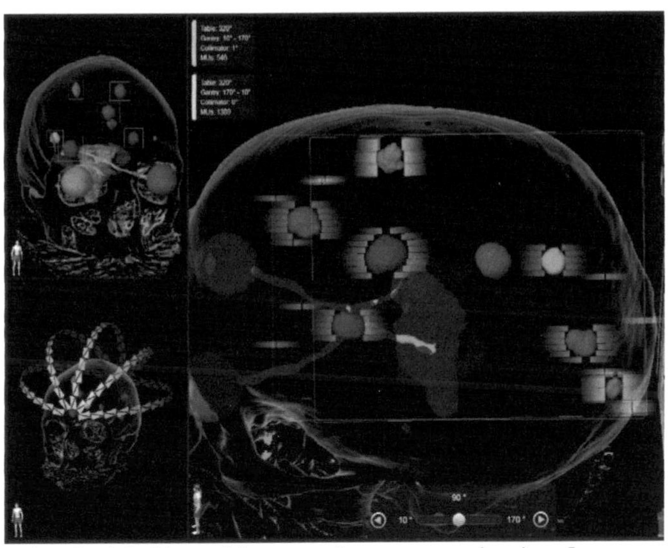

**Figure 3.** Image showing BrainLab's multi-metastasis treatment planning. Image courtesy of Dr. Nzhde Agazaryan, Ph.D.

IOP Conf. Series: Journal of Physics: Conf. Series **1305** (2019) 012001 doi:10.1088/1742-6596/1305/1/012001

## 5. References

[1] Yan D *et al* 1997 *Phys. Med. Biol.* **42** 123-32
[2] Vial P *et al* 2008 *Med. Phys.* **35** 4362-74
[3] Ezzell G A *et al* 2003 *Med. Phys.* **30** 2089-115
[4] Acharya S *et al* 2016 *Int. J. Radiat. Oncol. Biol. Phys.* **94** 394-403
[5] Schell M *et al* AAPM Report No. 54. Medical Physics Publishing: Madison, WI, 1995.
[6] Klein E E *et al* 2009 *Med. Phys.* **36** 4197-212
[7] Tran A *et al* 2017 *Radiat. Oncol.* **12** 10
[8] Zhen H *et al* 2011 *Med. Phys.* **38** 5477-89
[9] Ezzell G A et al 2009 *Med. Phys.* **36** 5359-73
[10] Low D A and Dempsey J F 2003 *Med. Phys.* **30** 2455-64

IOP Conf. Series: Journal of Physics: Conf. Series **1305** (2019) 012002 doi:10.1088/1742-6596/1305/1/012002

# Evaluation of a clinical dose accumulation algorithm using deformable gel dosimetry

**Charles K Matrosic[1], Shannon Holmes[2], Bryan Bednarz[1] and Wesley Culberson[1]**
[1]Department of Medical Physics, School of Medicine and Public Health, University of Wisconsin - Madison, Madison, WI, USA
[2]Standard Imaging Inc, Middleton, WI, USA

E-mail: cmatrosic@wisc.edu

**Abstract**. Deformable 3D dosimetry represents a robust method of verifying the accuracy of clinical deformable dose accumulation algorithms used to monitor interfraction anatomical changes during radiotherapy treatments. For this study, a deformable abdominal phantom was developed incorporating a deformable nPAG gel dosimeter for the dose verification of Adaptivo™, a commercial software program with a deformable dose accumulation algorithm. A comparison was made for three single fraction irradiations of gel dosimeters, each with a different deformation state. Additionally, a comparison was made for the cumulative dose over a three-fraction treatment of a single gel dosimeter with individual fraction deformations matching those of the single fraction measurements. The single fraction irradiations resulted in target contour dose volume histograms (DVH) created by Adaptivo™ that were in close agreement with those determined by gel dosimeter measurements for doses similar to and higher than the planned target dose, with two of the three cases matching to within 5%. Discrepancies are attributed to a deformed contour compression during analysis in the cases where the phantom was deformed. The three-fraction treatment resulted in very close agreement between the DVHs determined through the gel dosimeter measurements and Adaptivo™ calculations across the full range of doses, with an average absolute discrepancy of 2.0% and a maximum absolute discrepancy of 6.3%.

## 1. Introduction

Deformable dose accumulation algorithms have been developed to manage patient anatomy changes and other forms of interfractional patient motion during the course of radiotherapy treatments. These algorithms make use of patient imaging data during treatment to deform the delivered patient dose distribution back to the original planning images, providing clinicians with information of how closely the delivered patient treatment matches the planned treatment and allowing for subsequent treatment fractions to be adapted accordingly. To implement these systems clinically, AAPM Task Group 132 recommends the use of physical phantoms and measurements to verify the accuracy of calculations [1]. Although these measurements can be done on a point-by-point basis, a deformable 3D dosimeter allows for the most robust testing of systems.

Deformable 3D dosimetry has been investigated by several groups to address this need for robust deformable dose accumulation algorithm evaluation. Studies have shown that when a gel dosimeter is incorporated into a deformable mold or a radiochromic 3D dosimeter itself is deformable, it can be used as a physical deformable phantom to test these algorithms [2-5]. These dosimeters can be placed in different deformation states during irradiation and imaging, allowing for a measurement of the dose

IOP Conf. Series: Journal of Physics: Conf. Series **1305** (2019) 012002 doi:10.1088/1742-6596/1305/1/012002

related back to the undeformed state. Although some studies have been performed with these dosimeters placed in large, rigid motion phantoms, none have been performed within fully deformable anthropomorphic phantoms. The goal of this study was to use a deformable, anthropomorphic abdominal phantom which incorporates deformable polymer gel dosimetry for the purpose of testing a deformable dose accumulation program.

## 2. Methods

### *2.1. Deformable Abdominal Phantom*

A deformable abdominal phantom (Figure 1) was developed to evaluate Adaptivo™, which is a commercially available deformable dose accumulation program. The phantom was fabricated using polyvinyl chloride plastisol (PVCP) (M-F Manufacturing, Fort Worth, TX), a liquid plastic phantom material which can be mixed with a hardener or softener during fabrication to change the density and radiological properties of the material [6]. The phantom featured a higher density bulk phantom material. Lower density sections of the phantom were developed to represent abdominal organs at risk, including the colon, kidneys, and spleen. Additionally, a spine surrogate was created from a high-density casting material. Finally, a cavity was created in the phantom for the placement of a deformable gel dosimeter representing the treatment target. Phantom motion and deformation was driven by the s-axis 1D stage of the Washington University 4D Phantom [7].

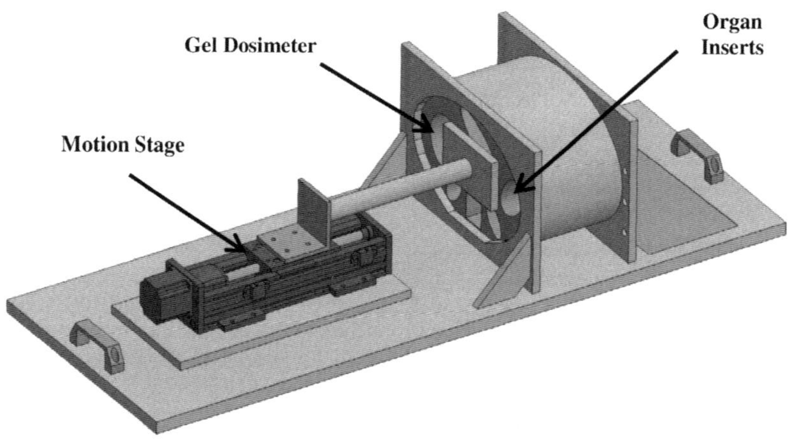


**Figure 1.** Model of the deformable phantom apparatus. The phantom features a deformable abdominal phantom made of high density PVCP bulk material and low density PVCP organ sections. It also features a deformable gel dosimeter cavity and a programmable 1D motion stage.

### *2.2. Gel Dosimeter Fabrication*

Five deformable gel dosimeters were created in-house at the University of Wisconsin Medical Radiation Research Center (UWMRRC) using a 6%T normoxic polyacrylamide gel (nPAG) formula. The nPAG formula was fabricated using the mixing methods described by Yeo et al. [5]. Deformable molds were made using PVCP to house the gel dosimeters (Figure 2). The exterior of the deformable mold conformed to the cylindrical cavity in the deformable abdominal phantom, while the inner cavity of the mold was shaped using an asymmetric 3D printed insert (Figure 2). Molds were capped and sealed using additional PVCP and injected with nPAG during dosimeter fabrication. The final dosimeter mold design allowed for a more accurate representation of an asymmetric tumor volume made up of nPAG. Fiducial markers were drilled into the top and the bottom of the outer shell of each dosimeter mold after fabrication to assist with image co-registration.

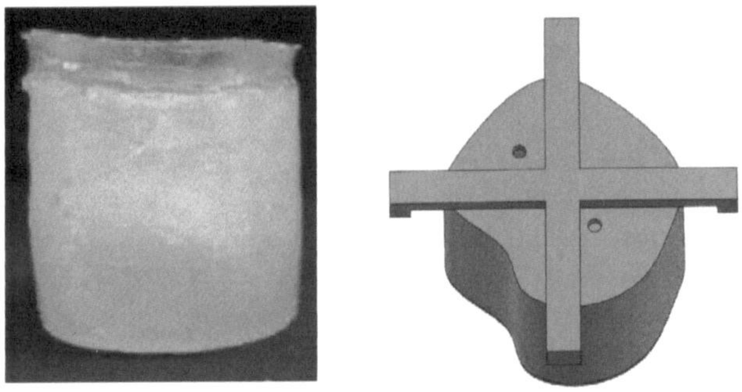

**Figure 2**. (Left) Deformable gel dosimeter created with a PVCP mold containing nPAG polymer gel. (Right) A 3D printed insert used to create the asymmetric inner cavity of the gel dosimeter.

### *2.3. Deformable Dose Accumulation Program*

The deformable dose accumulation program evaluated in this study was Adaptivo™ (Standard Imaging, Middleton, WI), a commercial software product developed to monitor dose to patient structures on a fraction-by-fraction basis using deformable image registration (DIR) to deform the structure doses back to the original planning CT. Dose calculations are performed using a reference dose perturbation algorithm [8]. To perform these calculations the software uses cone-beam CT (CBCT) images gathered before each fraction to calculate the differences between the treatment setup and the planning CT, which is then used to estimate the dose distribution during that fraction. The software outputs dose-volume histograms (DVH) of the delivered treatment for each organ or target contour deformed from the CBCT to the planning CT. These DVHs can be viewed for each fraction along with a cumulative DVH of all fractions up to a specified fraction.

### *2.4. Phantom Irradiation*

The same liver VMAT treatment plan was delivered to four of the gel dosimeters while a separate dosimeter used for calibration was irradiated with a simple four-field plan. The liver VMAT treatment plan was created using Eclipse™ (Varian Medical Systems, Palo Alto, CA) with 3 Gy per fraction dose to the target volume, which was a 1.5 cm subtraction of the nPAG contour. This subtraction of the gel contour was used to avoid oxygen and other PVCP plasticizer related contaminants in the target region to ensure accurate gel measurements. Organ dose limits were based on established clinical protocols scaled to the target dose per fraction. Of the four gels irradiated with the VMAT treatment plan, three received only a single fraction, while the remaining dosimeter received all three fractions of the treatment.

All dosimeters were irradiated using a Clinac 21EX medical linear accelerator at the UWMRRC using the phantom setup shown in Figure 3. The three gels irradiated with single fractions were placed under different deformation states using the motion stage: no deformation, 1 cm of deformation, and 2 cm of deformation. Deformation was estimated by the amount of deflection by the side of the phantom opposite of the motion stage plunger. The gel dosimeter that received three fractions underwent one fraction without deformation, one fraction with 1 cm of deformation, and one fraction with 2 cm of deformation to show the effect of dose integration over all three fractions. A CBCT was performed before the delivery of each treatment fraction to allow for the deformable dose accumulation algorithm to estimate the delivered dose for each irradiation.

**Figure 3.** Irradiation setup on the Clinac 21EX at the UWMRRC.

### 2.5. *Data Analysis*

All dosimeter dose data were gathered using a multiple spin-echo (MSE) $R_2$ mapping sequence on a 3T MRI scanner (SIGNA PET/MR, GE Healthcare, Waukesha, WI) using an 8-channel receive-only head coil. The sequence scanning parameters are shown in Table 1. The echoes from each scan were fit to exponential decay curves on a voxel-by-voxel basis to create $R_2$ maps. $R_2$ maps and target contours were co-registered to respective planning CTs using fiducial markers with Amira™ (Thermo Fisher Scientific, Waltham, MA).

**Table 1.** Multiple spin-echo pulse sequence parameters.

| | |
|---|---|
| Repetition Time (TR) | 5000 ms |
| Echo Spacing (ESP) | 40 ms |
| Echoes (NE) | 16 |
| Averages (NEX) | 1 |
| Acquisition Matrix | 128 x 128 |
| Field of View (FOV) | 128 x 128 mm$^2$ |
| Spatial Resolution | 1 mm$^2$ |
| Number of Slices (NS) | 28 - Coronal |
| Slice Thickness | 3 mm |

Gel dose-response calibrations were created separately for the lower dose single fraction dosimeters and the high dose three fraction dosimeter. The single fraction dosimeters were calibrated using the $R_2$ map and planned dose distribution of the dosimeter that did not undergo a deformation to allow for accurate calibration over the small dose range. Due to the much larger dose range, the three-fraction dosimeter was calibrated using the separate gel dosimeter that underwent a four-field irradiation to allow for a larger range of dose values to be calibrated accurately. Each of the dosimeter dose maps were used to calculate a dose volume histogram (DVH) of the target region of the dosimeter.

The CBCT data gathered before each fraction delivery was imported into Adaptivo™ and used to calculate the cumulative doses for each of the irradiated dosimeters. The cumulative DVH for each dosimeter treatment course was exported to be compared to the DVH calculated by the corresponding dosimeter.

## 3. Results and Discussion

The resulting DVHs from the single fraction experiments are shown in Figure 4. In each case both DVHs matched very well at the prescribed dose of 3 Gy per fraction. For the undeformed dosimeter and the dosimeter that was deformed 2 cm, the two DVHs matched within 5% for all doses above 2.5 Gy. In the low dose region, the gel dosimeters showed lower DVH values than Adaptivo™ did, especially for the two deformed cases. This was attributed primarily to the deformation of the target contour by the DIR algorithm slightly compressing the contour to be a smaller volume than contoured during treatment planning. The main cause of this compression was that the DIR algorithm was presented an extremely difficult case to deform – a contour based on a subtraction of another contour and not physical image intensity boundaries. This compression of the target contour caused the deformable dose accumulation algorithm to lose some lower dose components, raising its estimated low dose DVH values.

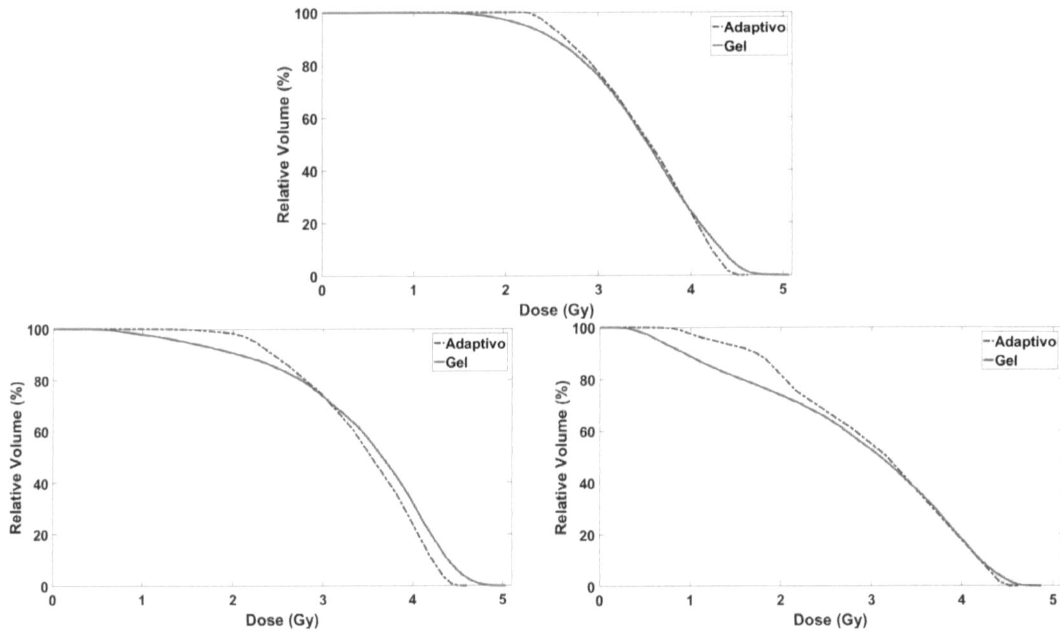


**Figure 4.** The target contour DVHs determined by both the nPAG gel dosimeter and the Adaptivo™ deformable dose accumulation algorithm over a single fraction with no deformation (Top), 1 cm of deformation (Bottom Left), and 2 cm of deformation (Bottom Right).

The resulting DVHs from the multiple fraction experiment along with point-by-point DVH differences are shown in Figure 5. Over the entire dose range, the DVHs agreed within 6.3%. The average absolute deviation between the two DVHs was 2%. In this experiment, the DIR algorithm showed very little target contour compression, causing the low dose DVH values to match well.

It is important to note that uncertainties arise in the DVHs calculated in both experiments from gel dosimetry and from deformable dose accumulation. It can be conservatively estimated that the nPAG gel dosimeter has a voxel-to-voxel combined uncertainty (k=1) of approximately 5.2%. This combined uncertainty is based on the 3.7% average Type A uncertainty of the nPAG dose measurements used for calibration, the maximum 3.6% of Type B MRI polymer gel dosimetry related uncertainties [9], and the 0.9% of linac related Type B uncertainties [10]. This combined uncertainty can be propagated to assume a 5.2% maximum uncertainty on each measured nPAG DVH value. Additionally, based on a previous study, the reference dose perturbation method used to calculate dose in the deformable dose accumulation algorithm can be estimated to have an uncertainty of approximately 3% [8]. There is also significant additional uncertainty in the deformation of the target contour by the DIR algorithm due to the challenging case presented, as previously discussed. Considering the combination of these uncertainties, the observed agreement between the nPAG gel measurements and the deformable dose accumulation algorithm calculations was considered reasonable for this study.

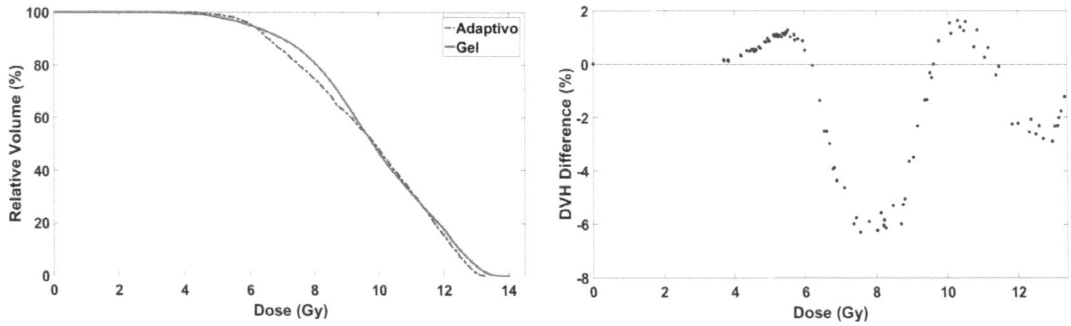


**Figure 5.** (Left) The cumulative target contour DVHs determined by both the nPAG gel dosimeter and the Adaptivo™ deformable dose accumulation algorithm over three fractions which underwent static deformation states of no deformation, 1 cm of deformation, and 2 cm of deformation. (Right) A point-by-point DVH difference of the two DVHs calculated by subtracting the gel DVH value from the Adaptivo™ DVH value.

## 4. Conclusions

In this study, a deformable abdominal phantom was paired with a deformable gel dosimeter to compare phantom measurements to the output of Adaptivo™. The first comparison made was over single fraction irradiations in which each gel was irradiated in a different deformation state. The target DVHs created by the two methods matched closely at doses at or above the 3 Gy target dose, but deviated some at low dose values due to the compression of the target contour during DIR. The second comparison was made for the cumulative dose following the delivery of three treatment fractions with different deformation states. The resultant DVHs for the two methods matched very closely, with a maximum deviation of 6.3% and an average deviation of 2.0%. In conclusion, this study provided a methodology to test and compare deformable dose accumulation algorithm calculations to physical measurements and found good agreement between the Adaptivo™ deformable dose accumulation algorithm and measurements made with a gel phantom.

## 5. Acknowledgements

This work was partially funded by NIH grant R01CA190298 and Standard Imaging Inc.

## 6. References

[1] Brock K *et al* 2017 *Med. Phys.* **44** e43-76
[2] De Deene Y *et al* 2015 *Phys. Med. Biol.* **60** 1543-63
[3] Juang T *et al* 2013 *Int. J. Radiat. Oncol. Biol. Phys.* **87** 414-21
[4] Niu C J *et al* 2012 *Med. Phys.* **39** 765-76
[5] Yeo UJ *et al* 2012 *Med. Phys.* **39** 2203-13
[6] Liao Y *et al* 2017 *Med. Phys.* **44** 2369-78
[7] Malinowski K *et al* 2007 *Proc. of SPIE* **6510** 65100E1-9
[8] Lu W *et al* 2014 *J. Phys.: Conf. Ser.* **489** 012016
[9] De Deene Y and Jirasek A 2015 *J. Phys.: Conf. Ser.* **573** 012008
[10] McEwen M 2011 COMP/AAPM Joint Meeting, Vancouver, BC, Canada

# Investigation of lung tumour peripheral doses using normoxic polymer gel and film dosimetry techniques

**A Venning[1], M Mundayadan Chandroth[1], B Chick[1], B Waller[1] and C Morgan[2]**
[1]Mid-North Coast Cancer Institute, Port Macquarie Base Hospital, Port Macquarie, NSW, Australia
[2]Mid-North Coast Diagnostic Imaging, Port Macquarie Base Hospital, Port Macquarie, NSW, Australia

Email: tony.venning@ncahs.health.nsw.gov.au

**Abstract**. This paper describes the investigation of lung tumour peripheral doses for 6MV, 6MV FFF, 10MV FFF and 15MV conformal arc therapy (CAT) beams calculated in the Monaco® TPS and delivered using a flattening filter free capable Elekta® Agility™ linac. Two independent high resolution dosimetry techniques were used for investigations. Measurements were performed using the normoxic PAG polymer gel dosimeter and GAFchromic™ EBT3 film and compared against the calculated dose plane from the Monaco treatment planning system. Both measurement methodologies indicate that Monaco is overestimating the lung tumour peripheral dose for the beams investigated in this study. It is shown that when using 10MV FFF in lung for arc deliveries, there is a dosimetric compromise compared with 6MV and 6MV FFF.

## 1. Introduction

Radiotherapy has progressed rapidly over the last few years with technologies such as flattening filter free (FFF), stereotactic ablative body radiotherapy (SABR), dynamic conformal arc therapy (DCAT) and volumetric modulated arc therapy (VMAT) with 4D image guidance (4D-IGRT). While these treatments provide a significant improvement in the quality of patient care and outcomes [1-3], the consequences of misadministration can be severe. Mainstream quality assurance tools are often utilised for measuring dosimetric and geometric accuracy to a homogenous medium or phantom. Such 2D methods provide partial-reassurance to the clinical physicist that the patient will receive the intended dose [4]. However, estimating delivered dose within patient target volumes under circumstances such as lung SABR is challenging. Confidence must be placed in the planning system's commissioning and capability to calculate and predict dose accurately within tight dosimetric and geometric tolerances in non-homogenous medium. The advantages of treating high dose regimes with FFF is clear, with significant delivery time savings which leads to less chance of intra-fraction motion. In our clinic, there has been a desire to use 10MV FFF for lung SABR due to its 2200 MU/min nominal dose rate, compared to 1400MU/min for 6MV FFF and 470MU/min for 6MV. However, there is concern with using higher energies to treat lung tumours due to the extended build-up region in the tumour periphery [5]. This phenomenon is investigated closely for the range of photon beams available in our clinic. Regions of electronic disequilibrium are often too challenging for many dosimeters to provide accurate

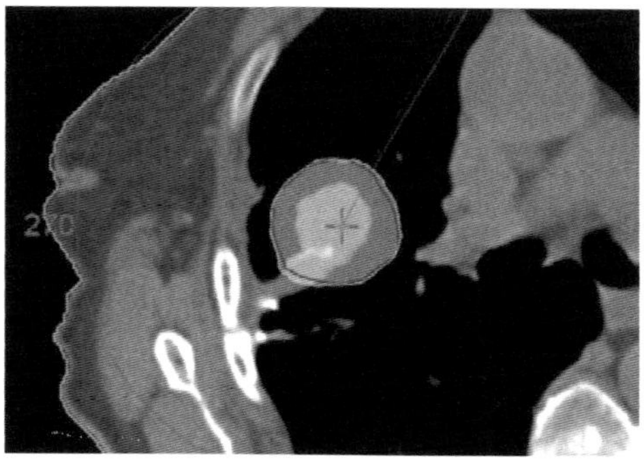

**Figure 1.** Example of PTV in lung = 100% Rx using 10MV FFF.

measurements. Monaco by default calculates dose to medium and applies human equivalent tissue density mapping. Modern day dosimetric phantoms, such as plastic water and end-to-end phantoms, mimic electron-densities of human equivalent tissue while their composition varies significantly from that of human tissue composition. This can become problematic particularly when electron densities vary significantly from a relative electron density of unity [6-8]. Radiation oncologists expect tight dose margins as shown in Figure 1, but can we be sure this is realistic and delivered accordingly in low density media? In this study, water based, non-homogenous medium is used in an attempt to replicate and mimic human equivalent tissue as close as possible. This study utilises the unique high resolution characteristics of gel dosimeters to encapsulate doses [9-12] in critical treatment regions such as the periphery of lung tumours. Film dosimetry techniques were also employed to provide additional validation of our findings and are reported here within this paper. Investigations continue with other dosimetry tools but are not discussed here.

## 2. Methods and Materials

### *2.1. Lung-tumour simulation*

To simulate lung tissue, a car washing sponge of similar size to human lung with a 47 mm diameter hole drilled out from its central core was used. The unladen sponge weighed 40 grams, and through experimentation it was determined that with a water retention weight of 640 grams the CT HU of the sponge was equivalent to that of average human lung tissue. To simulate the lung tumour [13], oxygen proof polyethylene terephthalate (PET) vials of 47 mm dia. x 150 mm hgt. filled with gelatine were used and placed within the sponge as per Figures 2 and 3.

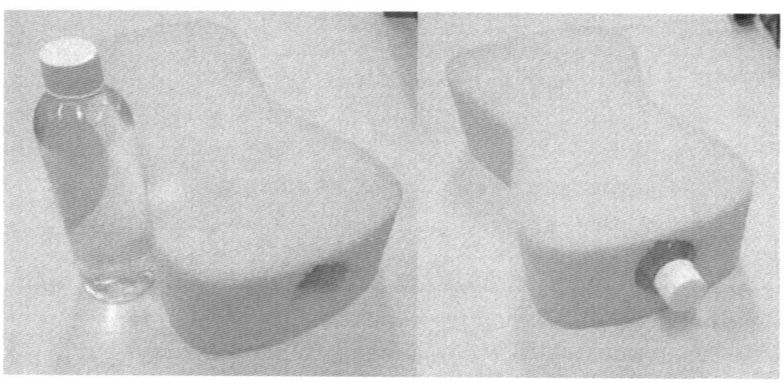

**Figure 2.** Sponge and PET vial.

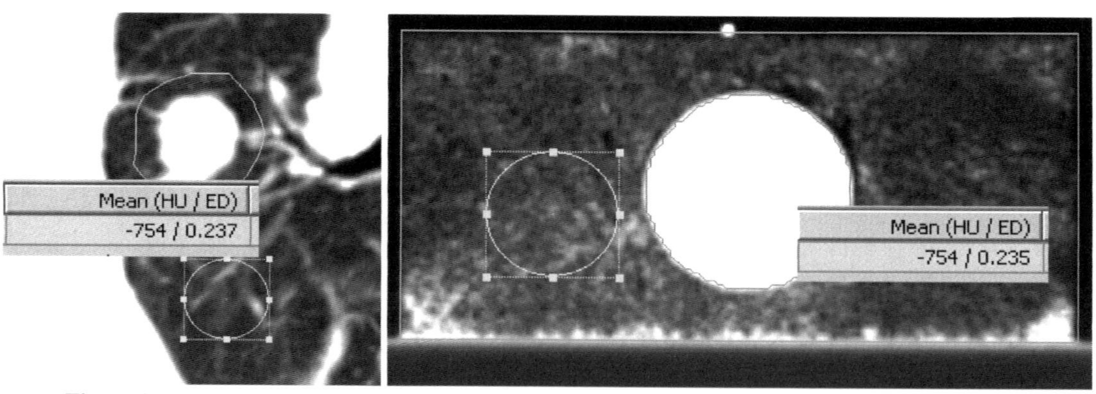


**Figure 3.** CT of lung-tumour (left) with wet sponge and gelatine filled vial simulating lung-tumour.

*2.2. Polymer gel method*

Normoxic PAG polymer gel dosimeters were manufactured using methods and techniques described elsewhere [14-20], with a composition of 5% gelatine, 3% acrylamide, 3% N, N′-Methylenebisacrylamide, 2mM hydroquinone, 5 mM tetrakis phosphonium chloride and 89% de-ionised water with a resistivity of 17MΩ. A total of 9 PET vials were filled with this gel, 4 used for the lung-tumour irradiations, 4 used for calibration and 1 used for simulation and treatment planning.

*2.3 Film method*

One PET vial was filled with a mixture of 95% $H_2O$ and 5% gelatine. A thin slit was cut halfway through the vial in the transverse plane where a square strip of EBT3 film could be slotted into the vial. With the vial in place in the sponge, a small insertion was made in the sponge to place the film inside the vial.

**Figure 4.** Film inserted in gelatine avoiding edge cutting effects in the critical dosimetric region.

*2.4 Treatment planning*

CT images of the wet sponge with the gelatine filled PET vial were exported to the Monaco TPS. Four conformal arc treatment (CAT) plans were generated for each energy, all having the same field size (4.8 cm x 4.8 cm) and calculated with 450 MU. The central axis transverse dose plane was exported for comparison with the measurements.

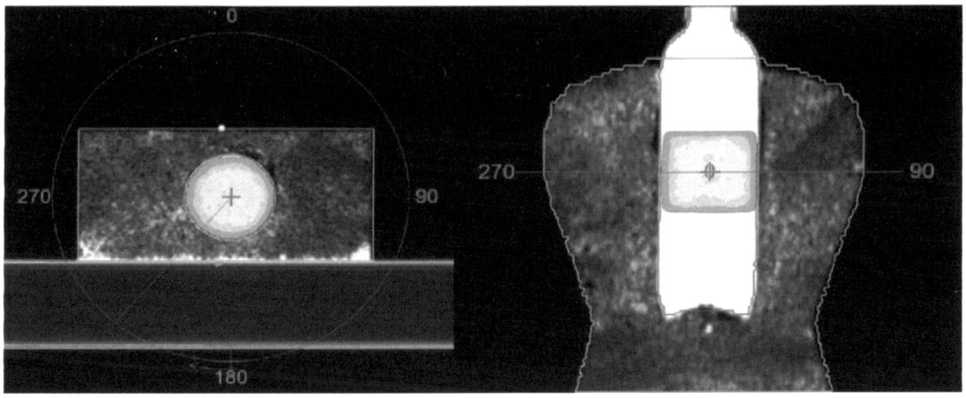


**Figure 5.** Conformal arc treatment plan in Monaco.

*2.5 Irradiation*

Before each irradiation the sponge was weighed to ensure water retention consistency. The sponge with each dosimeter were located at the isocentre of the linac and irradiated according to the conformal arc treatment plan. To generate the $R_2$-dose response calibration curve, three calibration vials were irradiated end on in a water tank using 6MV with a range of monitor units under TRS-398 conditions. The centre of each vial was placed at the reference depth of 10 cm as shown in Figure 6. One vial remained un-irradiated to acquire the background.

**Figure 6.** Polymer gels post-irradiation (left) with irradiation of calibration vials.

*2.6 Polymer gel readout*

The polymer gels were imaged 12 hours post-irradiation using a Siemens Avanto 1.5T clinical MRI scanner with head coil. A multi-echo pulse sequence with 16 echoes, echo spacing of TE = 40 ms, slice thickness of 5 mm, pixel spacing of 1 x 1 mm$^2$ and with 128 phase encoding steps was used to generate $T_2$ maps [21]. The arc distributions were imaged on the central transverse axis and the calibration vials were imaged at the centre of the vial and at ±5 cm where known doses were delivered.

*2.7 Film readout*

The EBT3 films were readout using a method developed in-house on an EPSON® V850 scanner with a scanning resolution of 75 dpi. A calibration curve was generated for each energy extending up to 10Gy.

*2.8 Analysis*

Data analysis between the TPS, polymer gel dosimeters and films were performed using MATLAB™ [22]. The film data was symmetrised by "flipping" the image over to create a full circle as only half the dose distribution was captured on the film squares. This is based on the assumption that the dose distribution is symmetric due to the delivered field being a conformal arc. Absorbed dose was used for each dosimetry method; however, profile scaling was applied to match profiles to the central region of the Monaco profile to allow a more effective analysis of the build-up region.

## 3. Results and Discussion

*3.1. $R_2$-dose response*

Figure 7 shows the $R_2$-dose response curve with an $R_2$-dose sensitivity of 0.26 s$^{-1}$Gy$^{-1}$ in the linear region extending up to 5Gy. The maximum absorbed dose in the gel phantoms was less than 5.5 Gy as delivered by the conformal arc plan.

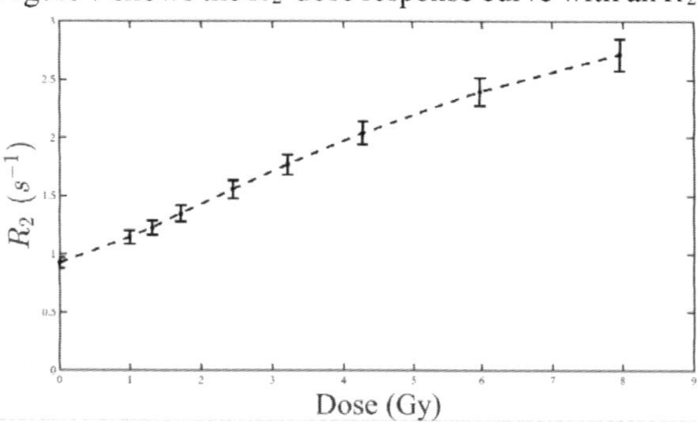


**Figure 7.** $R_2$-dose response curve.

*3.2. Gel and film measurement compared to TPS*

Figure 9 shows the profile comparisons taken in the vertical direction of the dose maps as, shown in Figure 8, for each energy studied. While the film and polymer gel measurements do not match exactly, they both indicate that Monaco is overestimating the dose in the build-up peripheral region of tumour equivalent tissue in

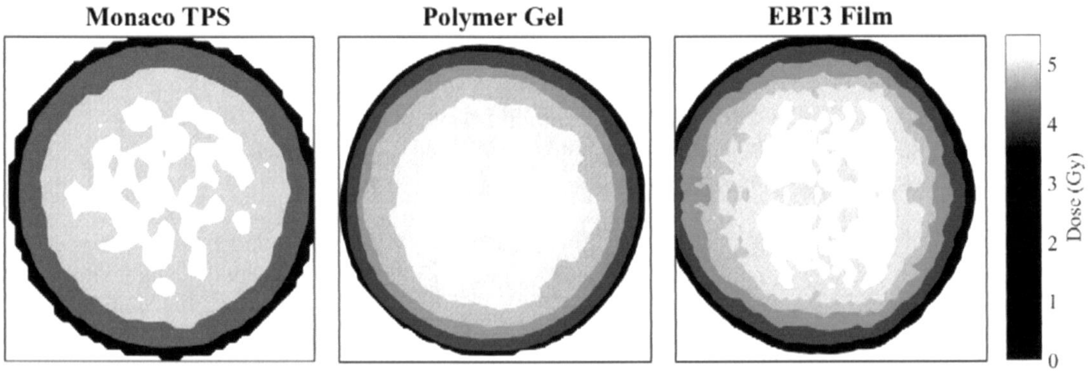


**Figure 8.** 6MV FFF dose maps showing the peripheral build-up ring.

lung. The overestimation of the TPS compared with measurement is of similar magnitude for each energy. This finding was not anticipated and is of concern, in consideration of the tight dose margins required by radiation oncologists (as per Figure 1) when using high dose fractionation regimes, especially when a tumour under-dose rapidly leads to a reduction in the tumour control probability. This indicates Monaco may not be correctly modelling regions where a state of electronic disequilibrium exists. Simulating and measuring doses accurately under such conditions, in regions of steep dose gradients is challenging and many dosimeters are not suitable for this purpose. While polymer gel dosimeters are ideal for a study of this complexity, investigations will continue with other dosimetry techniques and methods to characterise this finding in greater detail.

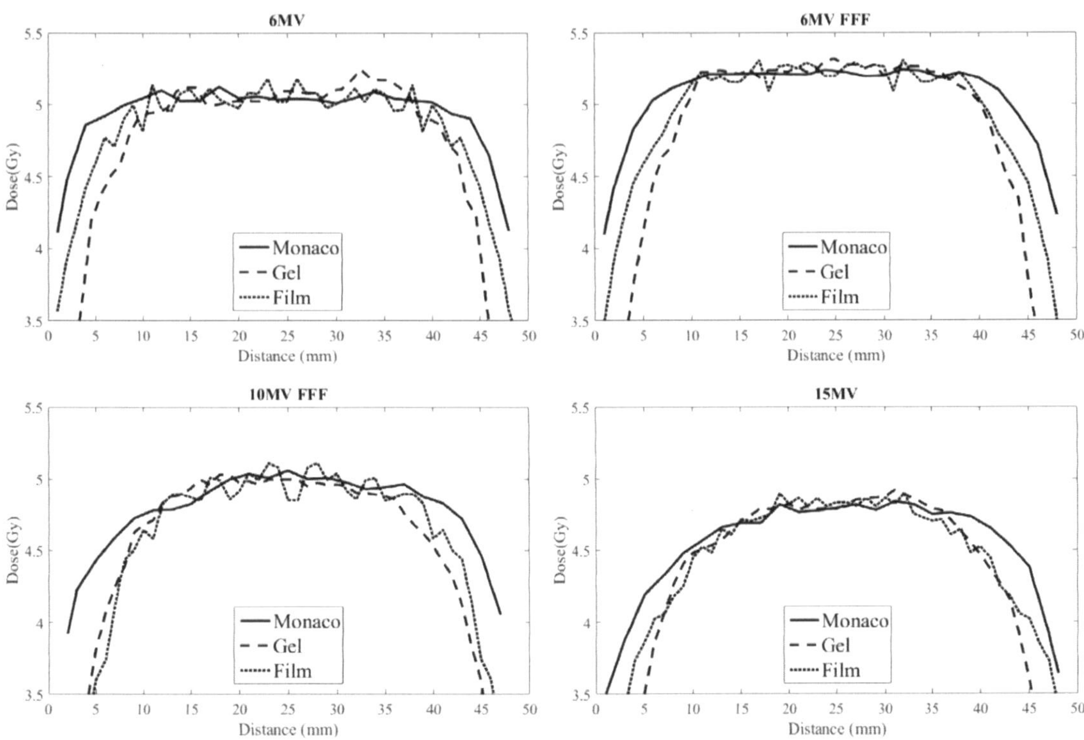


**Figure 9.** Comparison of polymer gel and film profiles against the TPS for each photon energy.

### *3.3. Gel profiles – energy evaluation in lung tumours*

Figure 10 shows a comparison of the absorbed dose profiles for each energy in the polymer gel phantoms. This shows that 6MV and 6MV FFF are comparable in terms of absorbed dose and build-up. On the Elekta system, the PDDs for 6MV and 6MV FFF are matched (at 10 cm deep in water for 10 x 10 $cm^2$ field), yielding similar beam qualities, but have different energy spectrums. The profiles in the polymer gel show that 6MV and 6MV FFF can be used inter-changeably without compromising the peripheral regions of a lung tumour. The higher energy 10MV FFF exhibits a slightly lower maximum dose for equivalent monitor units and clearly shows a lower dose contribution in the periphery as might be expected. To make up this difference and to expand the iso-dose lines, a higher total dose needs to be delivered, potentially compromising organs at risk and therefore, careful consideration should be given to its use. While 10MV FFF provides an efficiency dividend, it comes with a dosimetric expense. For equivalent monitor units, 15MV shows a lower total dose and a significantly lower dose contribution to the periphery and hence; should not be considered for use under such circumstances.

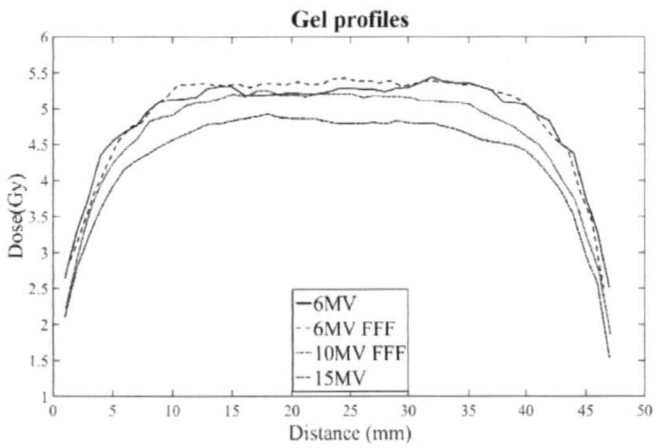


**Figure 10.** Combined gel profiles for all energies.

## 4. Conclusion

Two independent dosimetry techniques suggest the Monaco TPS is overestimating dose in the build-up peripheral region of tumour equivalent tissue in lung. Further investigations will be performed to fully characterise this finding and quantify the extent of overestimation. The use of 10MV FFF in lung tumours should be considered carefully, while there is an efficiency gain that can lead to a geometric gain, there is a dosimetric compromise compared with 6MV and 6MV FFF.

## 5. References

[1] Prendergast B *et al* 2012 *J. Appl. Clin. Med. Phys.* **14** 64-71
[2] Xing L *et al 2005 Med. Dosim.* **31** 91-112
[3] Folkert M and Timmerman R 2017 *Adv. Drug Del. Rev.* **109** 3-14
[4] Vial P *et al* 2008 *Med. Phys.* **35** 4362-74
[5] Wang L *et al* 2002 *J. Appl. Clin. Med. Phys.* **3** 51-9
[6] Andreo P 2015 *Phys. Med. Biol.* **60** 309-37
[7] Liu H and Keall P *et al* 2002 *Med. Phys.* **29** 922-4
[8] Chetty I J *et al* 2007 *Med. Phys.* **34** 4818-53
[9] Baldock C *et al* 2010 *Phys. Med. Biol.* **55** R1-R63
[10] Hurley C *et al* 2006 *Nucl. Instrum. Meth. A* **565** 801-11
[11] Mather M L *et al* 2003 *Phys. Med. Biol.* **48** N269-75
[12] Baldock C 2009 *J. Phys.: Conf. Ser.* **164** 012002
[13] Venning A J *et al* 2005 *Med. Phys.* **32** 1047-53
[14] Venning A J *et al* 2004 *J. Phys.: Conf. Ser.* **3** 155-8
[15] Venning A J *et al* 2005 *Phys. Med. Biol.* **50** 3875-88
[16] De Deene Y *et al* 2002 *Phys. Med. Biol.* **47** 2459-70
[17] De Deene Y *et al* 2002 *Phys. Med. Biol.* **47** 3441-63
[18] De Deene Y 2004 *J. Phys.: Conf. Ser.* **3** 34-57
[19] Baldock C *et al* 1998 *Phys. Med. Biol.* **43** 695-702
[20] De Deene Y *et al* 2006 *Phys. Med. Biol.* **51** 653-73
[21] De Deene Y and Baldock C 2002 *Phys. Med. Biol.* **47** 3117-41

[22] Murry P *et al* 2000 *Austral. Phys. Eng. Sci. Med.* **23** 44-51

# Feasibility of radiosurgery dosimetry using NIPAM 3D dosimeters and x-ray CT

**Justus Adamson[1], Jaclyn Carroll[2], Michael Trager[2], Paul Yoon[2], Jacob Kodra[2], Fang-Fang Yin[1], Evan Maynard[3], Michelle Hilts[4], Mark Oldham[1*] and Andrew Jirasik[5*]**
[1]Department of Radiation Oncology, Duke University Medical Center, Durham, North Carolina, USA
[2]Medical Physics Graduate Program, Duke University, Durham North Carolina, USA
[3]Department of Physics and Astronomy, University of Victoria, Victoria, BC, Canada
[4]Medical Physics, BC Cancer - Kelowna, BC, Canada
[5]Department of Physics, Irving K Barber School of Arts and Sciences, University of British Columbia – Okanagan Campus, Kelowna, BC, Canada
*Equal contribution

E-Mail: justus.adamson@duke.edu

**Abstract**. We investigated the feasibility of using N-isopropylacrylamide (NIPAM) dosimeters with x-ray CT to verify radiosurgery dose. Dosimeters were prepared at one facility and shipped to a second facility for irradiation. A simulation CT was acquired and plans prepared for a 4 field box, and a 4 arc VMAT radiosurgery plan to 6 targets with 1cm diameter. Each dosimeter was aligned via CBCT and irradiated, followed by 5 diagnostic CTs acquired after >24 hours, which were averaged for analysis. Absolute dose calibration was applied and dose evaluated for both plans. Hounsfield Units were proportional to dose above 10-12Gy. For the 4-field box, mean difference between measured and predicted dose >10Gy was -0.13Gy ±1.69Gy and gamma index was <1 for 72% and 65% of voxels using a 5% / 1mm and 3% / 2mm criteria, respectively (threshold = 15Gy, global dose criteria). For the multifocal SRS case, mean dose within each target was within -0.14Gy± 0.55Gy of the expected value, and gamma index was < 1 for 94.0% and 99.5% of voxels, respectively (threshold = 15Gy). NIPAM based 3D dosimetry with x-ray CT is well suited for validating radiosurgery spatial alignment, as well as dose distributions when dose is above 10-12Gy.

## 1. Introduction

Single isocenter volumetric-modulated arc therapy (VMAT) has recently been applied for radiosurgery of multifocal intracranial disease [1,2]. This technique has many advantages, but also requires increased attention to spatial accuracy especially for rotational errors, as small targets may be located at a distance from the isocenter [2-4]. Another major challenge is the difficulty in verifying the complex dose distribution. The radiosurgery plans include high dose gradients and dispersed small target volumes which are often similar or smaller in size than the detector spacing of many of the technologies used for typical VMAT verification. Film measurements provide high spatial resolution, but are limited to a single plane and require additional effort for digitization. A number of specialized digital detectors exist for verifying dose planes at sufficiently high resolution for radiosurgery, but for smaller field sizes than those encountered in multifocal radiosurgery [5].

3D dosimetry systems have unique advantages for these special circumstances, in that they can offer comprehensive dosimetric measurements at high spatial resolution [6-9]. Despite potential advantages, 3D dosimetry is challenged by the limited access to optical CT or MRI for reading out dose information, and lack of specialized commercial analysis [10]. Alternatively, for some 3D dosimeter materials such as N-isopropylacrylamide (NIPAM) [11,12], radiation dose invokes a change in mass density which can be read out as change in Hounsfield Unit in x-ray CT [13]. To date, there has been much previous research which has included increasing the sensitivity of dosimeter materials [14,15], and improving the dose readout through selective CT acquisition and reconstruction parameters [16], and post processing [17,18]. Despite these improvements, the main limitation to x-ray CT 3D dosimetry is still the limited dose sensitivity due to the small size of density changes (approximately 1 mg/cm$^3$ per 1 Gy of absorbed dose) [19], however this becomes much less problematic in the setting of high dose and high dose gradients for radiosurgery.

In this study, we investigate the feasibility of using a NIPAM dosimeter with x-ray CT for verification of radiosurgery.

## 2. Materials and Methods

Two NIPAM dosimeters and one "blank" dosimeter without the active agents were prepared at one facility (Department of Physics and Astronomy, University of Victoria, Victoria, BC, Canada). Each dosimeter was poured into a cylindrical plastic jar with a 9.5cm diameter and 15.3cm height. The dosimeters were then shipped commercially to a second facility (Department of Radiation Oncology, Duke University, Durham NC, USA) for irradiation, being packaged in an insulated box with dry ice, thus remaining at a refrigerated temperature. The dosimeters arrived at the second facility after 2 days, after which they were stored in a refrigerator at all times except during the X-ray CTs and irradiation. The subsequent treatment planning and irradiation were performed on the same day that the dosimeters arrived.

### *2.1. Treatment Planning & Delivery*

*2.1.1. General Details.* For this example, simulation CTs of each dosimeter were acquired for treatment planning; fiducials were placed on the superior portion of the dosimeter to allow for reproducible setup. Treatment plans were prepared within the Varian External Beam Treatment Planning software version 13.6 (Varian Medical Systems) with the Anisotropic Analytical Algorithm (version 13.6.23). We utilized a Truebeam STX linear accelerator with 6MV photons and HD-MLCs.

The first dosimeter received a treatment plan in which the top half included a 3-field irradiation pattern used for the absolute dose calibration, which has been described in detail previously [12]. The plan utilizes three rectangular (3cmx7cm) fields at oblique angles, designed so that the high dose area includes a range of dose values to aid in the absolute dose calibration. For this case, the prescribed dose was 20Gy, delivered in a single irradiation with 6MV photons, with a maximum dose of 27.0Gy. The lower half of the first dosimeter included a simple 4-field box irradiation, which served as a simple evaluation of the feasibility of using the NIPAM dosimeter for remote dosimetry, and as an example of the possibility of including the calibration and test dosimetry in the same dosimeter. The 6MV photon 4-field box also consisted of a 20Gy irradiation with open fields of size 5cm square; the maximum planned dose of the 4 field box was 25.0Gy.

A single isocenter multifocal VMAT SRS plan was prepared for the 2$^{nd}$ dosimeter with 6 targets, each with a 1cm diameter. The SRS plan utilized 4 non-coplanar VMAT arcs with a 6MV photon beam; the treatment planning technique has been described previously [20,21]. Each target was prescribed a dose of 20Gy, with the maximum dose being 31.3Gy.

IOP Conf. Series: Journal of Physics: Conf. Series **1305** (2019) 012004 doi:10.1088/1742-6596/1305/1/012004

### 2.2. *Analysis*

To minimize statistical noise, we created an average CT image for each of the NIPAM dosimeters (blank, calibration, and SRS test), which consisted of the average of all 5 individual image series. The calibration and SRS test dosimeters were processed including subtracting the background image (from the average CT of the blank dosimeter). The CT image of the dosimeters was registered to the treatment planning dose matrix, and the planning dose matrix was re-sampled to match the resolution of the dosimeter CT image. A calibration curve was created to convert Hounsfield Units from the dosimeter after background subtraction, h, to dose, d, using the fit:

$$d = d_0 + a_1(h - a_2)^{a_3} \quad (1)$$

where $d_0$ is the cutoff dose below which the Hounsfield Units within the dosimeter are not above the background signal, and $a_1$, $a_2$, and $a_3$ are variables that are fit through an iterative optimization process via in-house code in Matlab (Mathworks, Natick, MA). For this analysis, $d_0$ was set to 10Gy, as Hounsfield Units below this dose level were not above the background signal. The iterative optimization yielded values of $a_1$, $a_2$, and $a_3$ of 0.850 Gy/HU, -20.160 HU, and 0.978 (no units), respectively. This fit was then used to convert Hounsfield Units to dose for the SRS test dosimeter.

The SRS test dosimeter dose was then compared to the treatment planning system dose within Matlab, including gamma index analysis, dose profiles, and isodose comparisons. Gamma analysis was performed using various criteria with a lower dose threshold of 30% of the maximum dose in order to eliminate noise from the dosimeter and low dose regions.

## 3. Results

The planned and measured dose distribution (both when with and without subtracting the background signal) for the entire first dosimeter is shown in Figure 1. As is visible in Figure 1, while the dosimeter yielded a signal that was proportional to dose in the high dose areas (above 10Gy), this proportionality was non linear and had poor dose sensitivity for lower values. This was also the case for the second dosimeter below a threshold of around 12Gy. After fitting the HU values and dose distribution from the 3 field plan, the fit values from equation 1 to convert HU after subtracting the background signal to absolute dose was $a_1$=0.850 Gy / HU, $a_2$=-20.160 HU, and $a_3$=0.978. When the background was not subtracted, the fit values were $a_1$=0.027 Gy / HU, $a_2$=969.477 HU, and $a_3$=1.6881. For both cases the threshold dose $d_0$ below which voxels were excluded from the analysis was 10Gy. The scatter plot of HU and dose with and without subtracting the background signal, along with the dose fit from Equation 1 is shown in Figure 2(a) and 2(b).

Figure 3 shows dose profiles intersecting the four-field box plan. For this plan, the overall difference (mean ± standard deviation) between measured and predicted dose for all dose values above 10Gy was -0.20Gy ± 1.72Gy and -0.13Gy ± 1.69Gy with and without the background signal subtracted, respectively. The gamma index over the 4 field dose plan was less than 1 for 72% and 65% of voxels when using a 5% / 1mm and 3% / 2mm criteria, respectively (threshold = 15Gy, using a global dose criteria).

Figure 4 shows the dose distribution for the TPS and measurement from the SRS plan. The mean dose per target was within -0.14Gy ± 0.54Gy of the expected value from the TPS. The gamma index over the entire dosimeter was less than 1 for 94.0% and 99.5% of voxels when using a 5% / 1mm and 3% / 2mm criteria, respectively (threshold = 15Gy, using a global dose criteria).

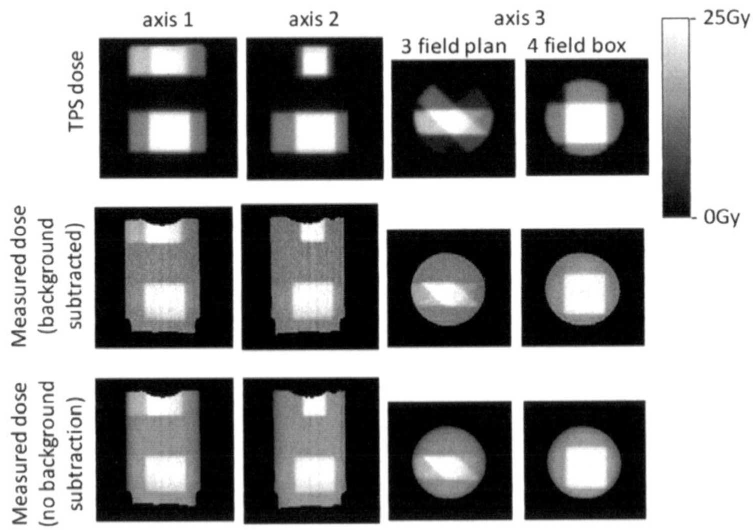


**Figure 1.** Planned and measured dose distribution for 3 and 4 field treatment plans after applying absolute dose calibration.

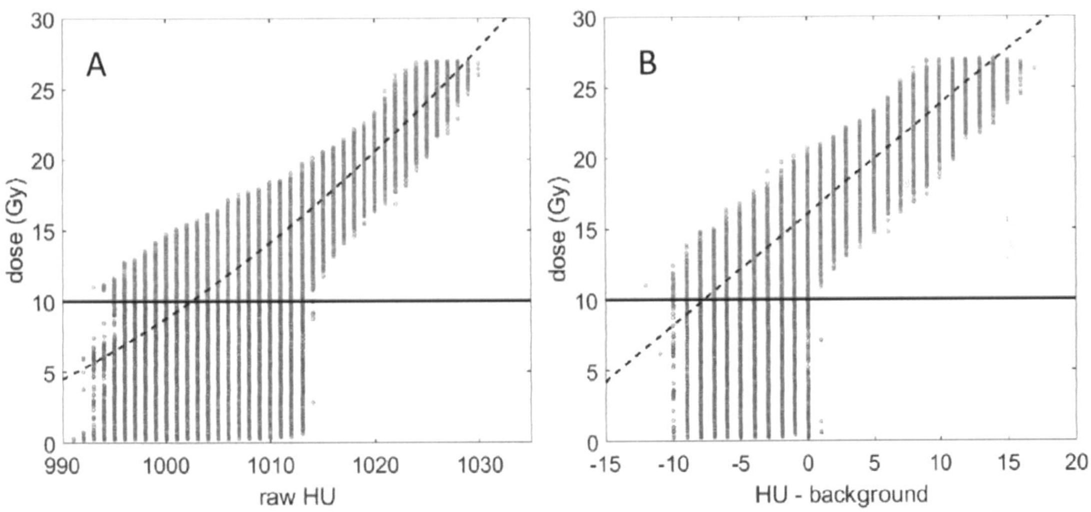


**Figure 2.** Relationship between dose and Hounsfield Unit for absolute dose calibration portion of the dosimeter after subtracting background (a) and using raw Hounsfield Units (b). Points excluded from the fit are shown below the 10Gy threshold line.

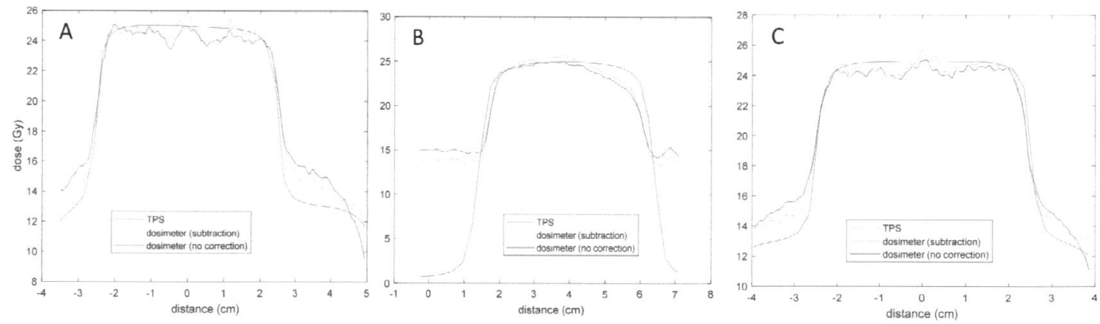


**Figure 3.** Planned and measured dose profiles across the 4-field box for each axis.

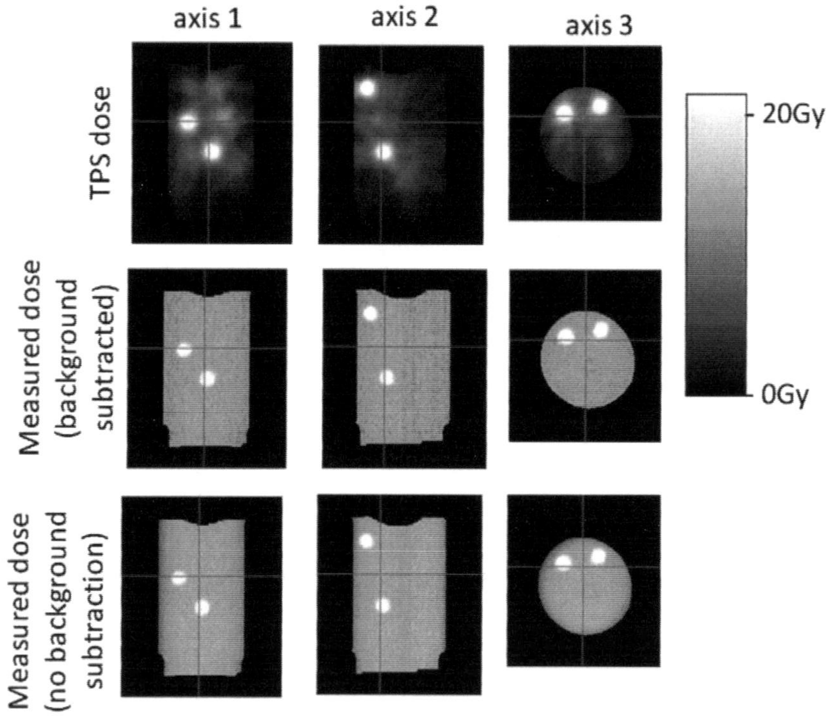


**Figure 4.** Planned and measured dose distributions for the SRS irradiation.

## 4. Discussion and Conclusion

In this study we used a polynomial fit of dose values above 10-12Gy. Earlier studies used an S-shaped dose calibration curve which has included these lower dose values; in this study, we used a simpler polynomial fit which excludes the area of lower sensitivity but allows for easier conversion back to dose. We anticipate that using a greater number of averaged CT scans would increase this sensitivity and allow for better dose resolution in the lower dose range. This represents a proof of principle of the ability to verify accurate alignment of the radiotherapy dose immediately after irradiation using a NIPAM dosimeter with x-ray CT. High dose areas (greater than 10-12Gy) were well correlated with expected dose.

## 5. References

[1] Clark G M *et al* 2010 *Int. J. Radiat. Oncol. Biol. Phys.* **76** 296-302

[2] Stanhope C *et al* 2016 *Pract. Radiat. Oncol.* **6** 207-213
[3] Roper J *et al* 2015 *Int. J. Radiat. Oncol.* **93** 540-6
[4] Faught A M *et al* 2016 *Med. Phys.* **43** 5437
[5] Sarkar V *et al* 2016 *J. Radiosurgery* **4** 213-23
[6] Sakhalkar H S *et al* 2009 *Med. Phys.* **36** 71-82
[7] Hurley C *et al* 2006 *Nucl. Instrum. Meth. A* **565** 801-11
[8] Mather M L *et al* 2003 *Phys. Med. Biol.* **48** N269-75
[9] Baldock C 2009 *J. Phys.: Conf. Ser.* **164** 012002
[10] Rowshanfarzad P *et al* 2011 *J. Appl. Clin. Med. Phys.* **12** 3645
[11] Johnston H *et al* 2012 *Phys. Med. Biol.* **57** 3155-75
[12] Jirasek A *et al* 2017 *Med. Phys.* 736-46
[13] Hilts M *et al* 2000 *Phys. Med. Biol.* **45** 2559-71
[14] Chain J N M *et al* 2011 *Phys. Med. Biol.* **56** 2091-102
[15] Jirasek A *et al* 2010 *Phys. Med. Biol.* **55** 5269-81
[16] Hilts M *et al* 2005 *Phys. Med. Biol.* **50** 1727-45
[17] Hilts M and Jirasek A 2008 *Med. Phys.* **35** 344-55
[18] Jirasek A et al 2012 *Phys. Med. Biol.* **57** 3137-53
[19] Baldock C *et al* 2010 *Phys. Med. Biol.* **55** R1-63
[20] Stanhope C *et al* 2016 *Pract. Radiat. Oncol.* **6** 207-13
[21] Morrison J *et al* 2016 *Med. Dosim.* **41** 285-9

IOP Conf. Series: Journal of Physics: Conf. Series **1305** (2019) 012005 doi:10.1088/1742-6596/1305/1/012005

# Surface dose accuracy in VMAT head and neck radiation treatment using bolus

**KM Alexander[1], J Gooding[2], LJ Schreiner[1,2] and T Olding[1,2]**
[1]Department of Physics, Queen's University, Kingston, Ontario, Canada, K7L3N6
[2]Cancer Centre of South Eastern Ontario at Kingston Health Sciences Centre, 76 Stuart Street, Kingston, ON, Canada K7L 2V7

E-Mail: kevin.alexander@kingstonhsc.ca

**Abstract**. The accuracy of Eclipse™-calculated head and neck VMAT plan dose delivery was assessed in the situation when there is a gap between placed bolus on a head and neck thermoplastic mask and the body surface. GafchromicEBT3 film dosimetry was used for dose delivery validation. The results from 3%, 3 mm 2D gamma comparison indicate better than 95% agreement between measured film dose and calculated plan dose for gap widths up to 25 mm between placed bolus and body surface.

## 1. Introduction

Radiation treatment (RT) of head and neck cancers often involves the decision of whether or not to treat the planning target volume (PTV) of the high risk disease to the skin surface. At some cancer centres, no additional effort is made to compensate for the photon dose build-up region and bring full prescription dose to the surface of the skin. At our centre, radiation oncologists occasionally request the use of bolus on the surface of the head and neck mask to ensure full surface build-up to the treated volume.

A previous study examined the deviation from calculated plan in the case of significant tumour volume shrinkage over the course of treatment [1]. The standard practice at our centre was to initiate a CT re-scan when the daily cone beam computed tomography (CBCT) imaging of a head & neck patient at time of treatment indicated a change in external body contour greater than 1 cm compared to the planning CT anatomy. The treatment approved plan was then calculated and verified for target coverage and organ-at-risk (OAR) sparing on the new CT anatomy. If the resultant dose distribution was acceptable, then the treatment would continue as before, else a re-plan was initiated. A custom-built wax head and neck phantom incorporating gel, Gafchromic EBT3 film (ISP, Wayne, NJ), ion chamber and MOSFET dosimeters was used for the evaluation. The overall conclusion from the previous study was that a 1 cm change in external body contour was an appropriate point to initiate a re-scan and potential re-plan. There was some discrepancy between the MOSFET and film surface dose results, indicating a slight overdose and significant underdose (up to 20% compared to the calculated plan dose) respectively. This reported difference was significant enough in its implications to warrant further investigation.

The purpose of this work is to assess the accuracy of the treatment planning surface dose calculations at our centre for head and neck RT in the case where there is a gap between the bolus placed on the immobilization mask and the body surface (see figure 1a). Given that this type of 3D dose delivery had been validated at depth using gel and ion chamber dosimeters in the previous study, the focus is shifted in this work to ensuring the accuracy of the dose at target (skin) surface. EBT3 film is well-suited to this type of measurement and was selected for use in this investigation.

## 2. Materials and Methods

### *2.1. Phantom Design*

A multi-purpose cylindrical acrylic phantom [2] was used for surface dose evaluations, along with Superflab (Mick RadioNuclear, Mount Vernon, NY) bolus with clinically validated dosimetric properties. This phantom has two possible coronal planes for incorporation of EBT3 film in a well-determined fixed geometry which simplifies the measurement dose-to-plan dose registration process. Film was placed in the more anterior plane (see figure 1c). Layers of bolus were affixed to the surface of the cylinder to simulate a tumour (total bolus thicknesses of 10 mm, 15 mm, 20 mm and 25 mm) and a thermoplastic mask was fitted to the phantom. A large cut piece of bolus was taped in place on the surface of the mask, sufficient in size to ensure full dose build-up to the "tumour" volume.

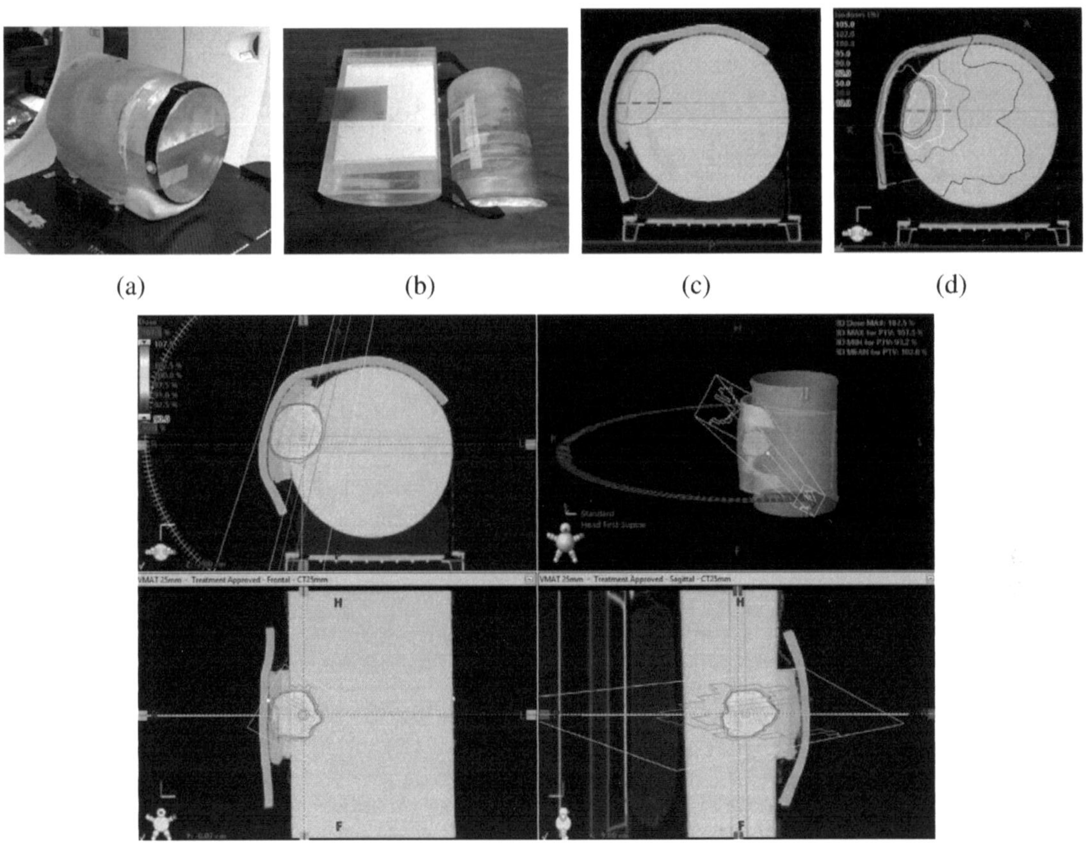

(a) (b) (c) (d)

(e)

**Figure 1.** (a) Multi-purpose cylindrical acrylic phantom, (b) phantom showing the film placement, and (c) axial CT image showing the 15 mm tumour anatomy (pink contour), where the dashed red line indicates the film position. (d) Eclipse screen capture of the "no tumour" anatomy showing overlaid plan isodose with dashed line again showing film position, and (e) Eclipse screen capture of the full 25 mm tumour plan, windowed to show target dose coverage.

### *2.2. Delivery Validation*

CT scans of the cylindrical phantom (with bolus and mask, see figure 1a) were acquired according to our centre's standard head and neck imaging protocol (120 kV, 2 mm slice thickness, 2 mm spacing) using a Philips Brilliance Big Bore CT scanner (Philips Medical Systems, Cleveland, OH). The phantom was immobilized on a Type-S Overlay head and neck board (Civco Medical Solutions, Coralville IA, USA) during CT planning and treatment delivery.

A RapidArc™ VMAT plan was optimized and calculated in Eclipse™ v.13.6 (Varian Medical Systems, Palo Alto, CA) using the Analytical Anisotropic Algorithm (AAA) for each of the phantom anatomies with various gaps. PTV targets were contoured to the phantom "tumour" surface on each CT anatomy. For VMAT planning, dual coplanar clockwise/counterclockwise 6 MV partial arcs were employed, extending from just beyond midline anteriorly to 180 degrees posteriorly (figure 1e). The collimator angle on the two arcs was set to ± 30 degrees to minimize overlap of multi-leaf collimator (MLC) leakage. The optimized plan was normalized to a prescription of 2.0 Gy. Screen captures of plans on different anatomies are shown in figures 1d and 1e, with one showing isodose lines to visualize the full dose distribution and the other showing dose color wash to visualize target dose coverage and uniformity. Full sheets of Gafchromic EBT3 film (20.3 cm x 25.4 cm) were cut into 7 cm x 12.7 cm pieces (long edge cut in half) and placed in a coronal plane, as shown in figure 1b.

All treatment plans were delivered on a Varian TrueBeam linear accelerator (Varian Medical Systems, Palo Alto, CA) with cone beam CT (CBCT) setup verification. EBT3 film dose readout measurements were acquired using an in-house built CCD-lightbox film scanner [3]. A calibration set of 6 MV 10x10 $cm^2$ films were delivered under machine reference conditions to doses spanning the range of 0-4.5 Gy. The calibrated film dose data was compared to Eclipse dose in 3D Slicer (www.slicer.org) using the Film Dosimetry Analysis extension [4].

## 3. Results

### *3.1. Dosimeter-Plan Dose Comparison*

Dosimetric results comparing film measurements against Eclipse planned are shown in figure 2 for the various stages of bolus thicknesses. Excellent agreement was found between film dosimeter measurements and Eclipse calculated doses, with gamma pass rates of > 95% within the PTV contour for each bolus thickness.

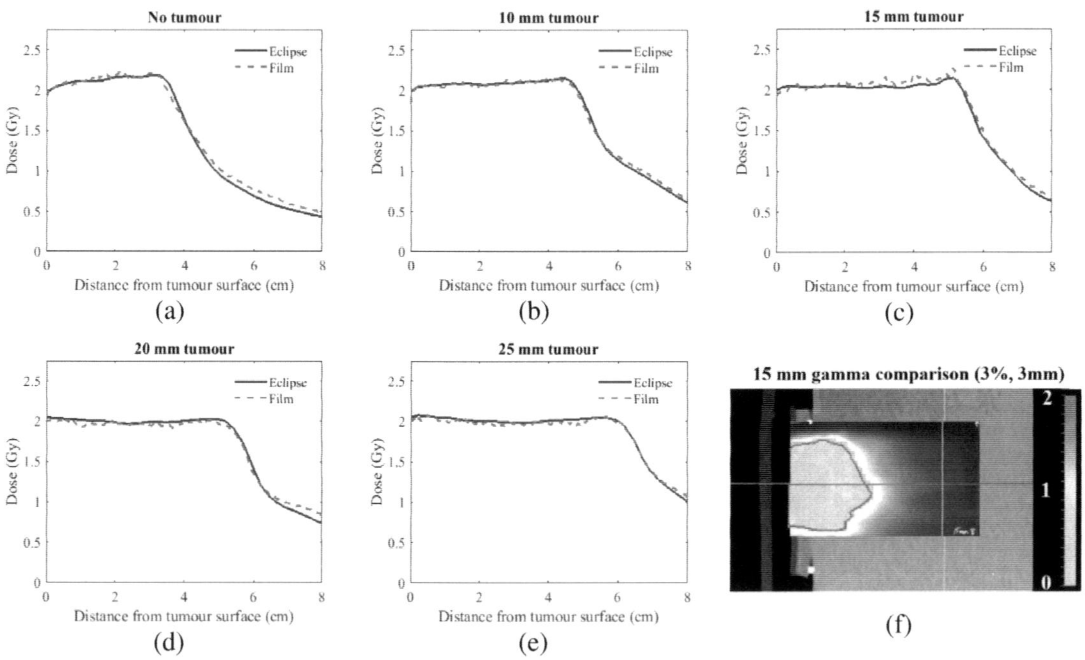


**Figure 2.** (a)-(e) Dose profiles about the central axis (profile direction shown in (f)) comparing the Eclipse dose measurement and EBT3 film measurements for various bolus thicknesses.

### *3.2. Film Edge Effects*

After initial film dosimeter analysis, a systematic discrepancy in film measurements at the tumour surface was observed for dose measurements for all cases where an original manufacturer edge (non-cut edge) was positioned at the tumour surface. The experiment was repeated by irradiating two films simultaneously (one film with a cut edge at the surface, a second film with a manufacturer edge at the surface). This repeat experiment confirmed the discrepancies (within the first 1 cm of the tumour) for films with the manufacturer edge at the surface. Up to 12 % surface dose difference was observed across all film irradiations, as shown in figure 3.

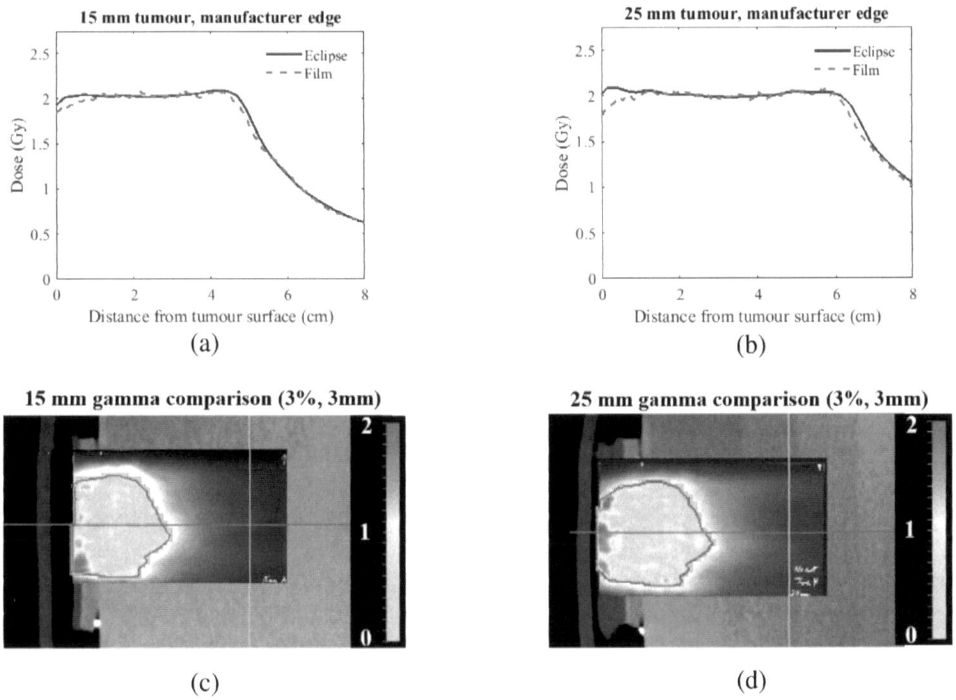


**Figure 3.** Dose profile comparing Eclipse and film measurement for a film irradiated with a manufacturer film edge at the tumour surface with a (a) 15 mm bolus, and (b) 25 mm bolus. Gamma dose comparisons for films irradiated with a manufacturer film edge at the surface for each of the (c) 15 mm bolus (gamma agreement of 86.5%), and (d) 25 mm bolus (gamma agreement of 83.7%).

## 4. Discussion & Conclusions

Accurate surface dose measurements are difficult to perform. MOSFETs are frequently used for surface dose measurements, but they are not tissue equivalent and are sensitive to positioning errors. While a 3D gel dosimeter measurement is desirable due to its tissue equivalence and high-resolution readout, optical CT measurements are degraded within 5 mm of the gel jar surface due to refraction and other optical artifacts. Based on our past experience [1] we chose to use EBT3 film, as it is tissue equivalent and enables a high-resolution 2D dose measurement.

As shown in figure 2, film dose measurements were shown to be in good agreement for air gaps of up to 25 mm between the bolus on the mask and the body surface. While the mask bolus was removed and replaced after each film irradiation, we note that the minor variations in bolus placement does not significantly affect dose measurements. The clinical practice of placing bolus on the mask to act as a beam spoiler was confirmed to accurately deliver full dose to the skin surface due to the generation of secondary electrons by the mask-placed bolus. No further bolus material is needed beneath the mask to ensure full dose buildup. Placement of bolus material under the mask can be awkward and somewhat variable on a day-to-day basis, so avoidance of this practice is preferred.

Small non-uniformity across radiochromic film is reported in the literature [5], however the drop in response at the edge of the manufacturer film edge was unexpected. As presented, a significant dose difference of up to 12% at the tumour surface was found when irradiating films with a manufacturer film edge at the tumour surface. This discrepancy is likely inherent to the manufacturer's film coating process. Disagreement within the first 1 cm may also help to explain some of the discrepancy in the dose measurements previously reported [1].

While this phantom lacks the anthropomorphic geometry used in Alexander *et al.* 2017, it was chosen to realize a more robust setup which allows for better localization of the film dosimeter plane for precise registration with the planning CT and Eclipse calculated dose volume.

Overall, this investigation reassures us that air gaps between bolus and the body surface are accurately considered in the Eclipse treatment planning algorithms.

## 5. Acknowledgements

This work was funded by the Canadian CIHR funding agency, project MOP-115101.

## 6. References

[1] Alexander K M *et al* 2017 *J. Phys.: Conf. Ser.* **847** 012038
[2] Olding T *et al* 2013 *J. Phys.: Conf. Ser.* **444** 012052
[3] Lalonde M *et al.* 2016 *Med. Phys.* **43** 4938
[4] Alexander KM *et al.* 2017 *J. Phys.: Conf. Ser.* **847** 012061
[5] Micke A *et al* 2011 *Med. Phys.* **38** 2523

IOP Conf. Series: Journal of Physics: Conf. Series **1305** (2019) 012006 doi:10.1088/1742-6596/1305/1/012006

# Radiotherapy in the presence of magnetic fields: a brief review of detector response characteristics and the contribution of 3-D measurements to the study of dose distributions at interfaces

**Simon J Doran**
CRUK Cancer Imaging Centre, Division of Radiotherapy and Imaging, Institute of Cancer Research, London, SW7 3RP, UK

E-Mail: Simon.Doran@icr.ac.uk

**Abstract.** The combination of MRI and radiotherapy on a single platform has the potential to revolutionise image-guided delivery of radiation doses. However, in order to realise these ambitions, good dosimetry must be available. The electron return effect gives rise to significant perturbations of dose at interfaces between tissue and air within the body, and this might lead to difficulties in dose compensation if air cavities move during treatment. In this article, I review briefly the ways in which the available methods of dosimetry are affected by the presence of magnetic fields and discuss the contribution that three-dimensional measurements can make to studies in this area. The methods of MRI and optical computed tomography have well known issues in imaging close to interfaces. These are described together with progress so far in providing solutions.

## 1. Introduction

The creation of treatment platforms that can acquire Magnetic Resonance Imaging (MRI) data during radiotherapy has enormous potential benefits for real-time, image-guided therapy. Improved tumour visibility in MRI, as compared with cone-beam CT, has led to significant investment over the last decade and the development of a number of systems - see the Introduction section of ref. [1] for a useful summary - some of which are already available commercially and approved for treating patients in the clinic. Much has already been written (and will not be discussed here) about the ways in which these different designs have tackled the various engineering challenges inherent in combining the two modalities. Each has consequences for the treatments that can be delivered and the quality assurance procedures required.

This article considers the general problem of how we can make reliable measurements of radiation doses to tissue in the presence of magnetic fields. What is different about the way dose is deposited? How are the various detectors at our disposal affected by magnetic fields? And what effects must we be particularly alert for as we plan our treatments? I start by reviewing briefly the different areas of dosimetry research that have been undertaken and then focus in more detail on the problems of 3-D dosimetry at the interfaces between different materials.

## 2. Background

### *2.1. Why is radiation dose deposition different in a magnetic field?*

Whilst the incident photons employed in radiotherapy are unaffected by the presence of a magnetic field, the dose deposited in tissue has large contributions from secondary electrons and these are subject to the Lorentz force, whose direction is perpendicular to both the applied magnetic field and the instantaneous direction of motion of the electron. Two effects are important:

(i) For those electrons that remain within the tissue, trajectories can be calculated using the formalism developed by Jette [2] and these lead to a modified dose deposition. Where lateral electron equilibrium exists, build-up distances are decreased; there is a small increase in the penumbra perpendicular to the magnetic field; and there is a small shift of the depth dose curve compared with the $B = 0$ case [3].
(ii) Electrons can exit the original tissue and travel through air, either externally or within a body cavity, in circular arcs relatively unimpeded by scattering. For the magnetic fields and electron energies typically found in combined MR-radiotherapy systems - see Table 1 of [4] - the radius of curvature is of order mm or cm. This means that the location in which the dose from these secondary electrons is finally deposited might be some distance from the point at which the electrons previously left the tissue. This is the so called electron return effect (ERE) [4, 5].

Magnetic field effects on dose deposition can be simulated using Monte Carlo methods [6].

*2.2. Measurement using ionisation chambers*
Typically, reference dosimetry is performed with ionisation chambers containing an air cavity. The paths of the ion pairs created are perturbed by the magnetic field and this leads to variations in the chamber response with model, strength of the magnetic field and orientation of the chamber with respect to the field [7-11]. For accurate dosimetry, it has been shown that it is necessary to place the ionisation chambers in water and that inserting them directly into solid phantoms is not advisable due to the likely presence of air gaps that may perturb the dose reading [12, 13].

*2.3. Solid-state detectors*
Reynolds *et al.* [14] simulated both a PTW60003 diamond detector and an IBA PFD diode detector and found that both required corrections for use with a magnetic field that depends on the relative orientations of magnetic field, detector long axis and incident photon beam. Gargett *et al.* [15] also found an orientation dependence in Monte Carlo simulations of a novel 2D silicon diode. Although it is generally difficult to perform identical tests *in situ* with and without a magnetic field present, several groups have compared the performance of clinical multi-diode devices mounted in an MR linac with the same tests carried out on a conventional linac at $B = 0$. Houweling *et al.* [16] investigated an MR-compatible ArcCHECK device in the presence of a 1.5 T magnetic field and found it to be similar within 1% to reference data of Li *et al.* [17]. De Vries *et al.* [18] performed a detailed comparison between the detector response characteristics of an MR-compatible Delta4 positioned in both MR- and conventional linacs and noted the same general agreement in properties, but an orientation dependence of between 5 and 10%.

*2.4. EPID-based dosimetry*
Although Raaymakers *et al.* [19] demonstrated some time ago the feasibility of using a megavoltage portal imaging device in conjunction with their MR linac, it is only recently that attempts have been made to adapt existing back-projection algorithms for quantitative portal dosimetry [1] and characterise EPID detectors in detail [20]. Issues making this difficult are:

(i) the elements of the MR scanner situated between the patient and the EPID, which comprise a non-uniform attenuating medium that also alters the photon energy spectrum;
(ii) changes in dose deposition in the patient, as described in Sec. 2.1;
(iii) possible changes in the dose-response characteristics of the EPID in a magnetic field.

Torres-Xirau *et al.* have addressed issue (i) successfully in a "mock-up" of an MR linac [1]. There is also encouraging agreement in the EPID profiles of normalised pixel intensity between data for different field sizes acquired in both the presence and absence of a magnetic field [20].

### *2.5. Response of radiochromic film*

Despite the ability to obtain excellent relative dosimetry using Gafchromic EBT2 film [21], there have been persistent concerns about the potential for low-level effects of magnetic fields on the absolute reading from radiochromic films. Raaijmakers *et al.* [22] noted that measurements in magnetic fields up to 1.3 T led to readings in the range 1 - 4% lower than at B = 0, and Reyhan *et al* [23] observed a similar decrease in optical density, whilst Reynoso *et al.* [24] observed dramatic reductions of up to 15%. Potential explanations proposed have included a change in the underlying yield of the radiochemical reaction and a localised temperature increase associated with thermal energy deposition by MR imaging sequences. Early results using the more recently introduced EBT3 film [25] suggest an effect of magnitude approximately 2%.

### *2.6. 3-D dosimeters*

For ionisation chambers, diodes and EPIDs, energy deposition and quantitative readout happen contemporaneously. By contrast, for 3-D dosimeters and films, the dose detection process has *two stages*: first, the radiation dose leaves some form of "chemical imprint" on the detector, and then, at some later juncture, we read this out with one of several possible 3-D imaging techniques (typically, Magnetic Resonance Imaging (MRI), x-ray computed tomography (CT), optical CT or ultrasound imaging) [26]. The effect of magnetic field on the dose-imprinting process can thus be separated from the readout, the main exception being where the readout device is actually the MR linac itself and a "real-time" dose readout is required [26][27].

Given the close historical links between gel dosimetry and MRI, it is perhaps surprising that (to the best of my knowledge) no studies have been performed to measure the effect of magnetic field on the radiochemical yield (G-value) of ferric ions. Similarly, it is not clear whether comparisons have been performed to determine whether there is any magnetic field dependency of the polymerisation reaction in polymer gel dosimetry.

For the PRESAGE® dosimeter, which is designed for optical readout, three sets of measurements [27-29][28-30] all found only very small effects (~1%) at the edge of detectability when comparing the optical response of dosimeters irradiated in the presence and absence of a magnetic field.

## 3. Dosimetry at interfaces

Understanding the effects that occur at interfaces between different media is of great importance in many clinical dosimetry problems. Of particular concern for the MR linac are air-containing cavities, as found in the head and neck, lung and pelvic areas, where the ERE may modify dose significantly. Although strategies exist to optimise dose to compensate for these perturbations in static situations, the cases of air in the bowel and rectum represent considerable challenges for MR linac therapy because the position of the air cavity may move during treatment delivery. Precisely those cases that might benefit most from intra-fractional image-guided therapy are those where the electron return effect may present the greatest hazards [30][31].

Since body cavities often have irregular shapes for which prediction of dose in a magnetic field requires 3-D models, it is likely to be insufficient to use 2-D radiochromic film for patient-related dose verification. Similarly, the inhomogeneities involved and potentially high spatial resolution required suggest that neither ionisation chambers nor diode arrays will be suitable. 3-D materials moulded to relevant anthropomorphic shapes are thus attractive potential solutions.

However, as detailed below, there are a number of problems that must first be addressed.

### *3.1. MRI measurements at interfaces*

The presence of interfaces is a concern in MRI because of the differences in magnetic susceptibility (or, equivalently, magnetic permeability) between materials, particularly tissue and air. The main magnetic field $B_0$ is perturbed and this leads to shifts in the Larmor frequencies of the imaged nuclei ("NMR spins"). Two effects are important:

(i) Since the Larmor frequency of a spin relates directly to the position at which it appears in the final image, images can become distorted. In general, both positional information and measured image intensities are affected, with signal "pile-up" in some voxels and signal voids in others.
(ii) Distributions of precession frequencies within voxels can lead to the cancelling out of spin contributions (additional "dephasing", which is expressed as a dramatically decreased $T_2$*). This is a second cause of "signal dropout", i.e., apparent voids in the images.

However, despite these potential issues, robust experimental design, using the correct types of sequence allows quantitative dosimetric imaging near interfaces [31, 32][32, 33] An area where MRI may have a significant role to play is in the creation of anthropomorphic phantoms, with a range of different densities [33][34].

### 3.2. *Inhibition of polymer gels*

In the early days of polymer gel dosimetry, a major confounding factor preventing accurate dosimetry at interfaces was the failure of the gels to polymerise. Oxygen free radicals have an inhibitory effect on the radiation detection process, because they "mop up" the free radicals created by the incident photons, thus preventing them taking part in the polymerisation process [34][35]. Since many commonly used phantom construction materials are permeable to oxygen, early solutions to the inhibition problem involved the use of glass vials of various sizes. However, as noted in [32][33], the use of glass containers is not suitable for interface dosimetry because its density perturbs the very dose distributions it is aiming to measure. Other solutions have involved the use of the thermoplastic Barex® to form gel containers, or, starting with the work of Fong *et al.* [35][36], the use of dosimeter gels that are unaffected by oxygen.

### 3.3. *Optical measurements at interfaces*

Perhaps the most significant difference between optical and x-ray CT is that, whilst x-rays travel in straight lines through the sample, regardless of any interfaces present, light in the visible spectrum is refracted at boundaries between materials of different refractive indices. In order to minimise this effect, samples are almost always placed within some form of "aquarium" containing a so-called "matching liquid" of similar refractive index. An imperfect match leads to a prominent image artefact at the interface, with significant perturbations of image intensity extending both into and outside the sample. The situation has been modelled both for gel samples inside containers [36][37] and solid PRESAGE® dosimeters that do not require any external support [37][38].

Historically, many of the applications of optical CT have used large cylindrical samples, for which the effect typically manifests itself as a high intensity ring surrounding the dosimeter and occupying a small fraction of the analysed region that can often be excluded. However, the spatial extent of the artefact does not scale with sample radius [36][37] and this leads to potential problems whose aim is to make accurate measurements of dose next to small moulded air cavities.

As in the case of MRI, an understanding of the physics governing the measurement process has allowed useful measurements of dose to be obtained close to boundaries, by ensuring careful optimisation of refractive index and judicious choice of interface positions in experiments [27, 38, 39][28, 39, 40].

### 3.4. *PRESAGE®*

Recently, it has become apparent that the optical edge artefact discussed above may, in fact, mask a genuine change in physical and dosimetric properties at the outer edges of PRESAGE® samples. Anecdotal reports have noted that when machining PRESAGE® cylinders, some regions of the sample are softer than others. During their work on remote dosimetry of the Viewray MR-guided IMRT platform, Rankine *et al.* [29][30] observed disagreements "in the periphery of the dosimeter" and these were similar to results seen and analysed in detail by Dekker *et al.* [40][41]. A radially non-uniform sensitivity was observed, which was batch dependent. At this IC3Ddose meeting, Costa *et al.* present studies specifically designed to measure and correct for this type of radial dependence. This is vital if credible measurements of dosimetry at interfaces are required.

## 4. Conclusion
The translation of combined MRI-radiotherapy platforms from the research laboratory to the clinic is now well underway. Together with intense technical development on the systems themselves has come a renewed focus on dosimetry in the presence of magnetic fields. Although most of the physical principles involved are now well established, issues still remain. As with other areas of radiotherapy, 3-D dosimetry has its part to play, and the study of interfaces between different tissues has the potential to be of great interest moving forward.

## 5. Acknowledgments
SJD acknowledges CRUK and EPSRC support to the Cancer Imaging Centre at ICR and RMH in association with MRC and the Department of Health C1060/A10334, C1060/A16464, and NHS funding to the NIHR Biomedical Research Centre, Clinical Research Facility in Imaging and Cancer Research Network. This report is independent research funded by the National Institute for Health Research. The views expressed in this publication are those of the author and not necessarily those of the NHS, the National Institute for Health Research or the Department of Health.

## 6. References
[1] Torres-Xirau I *et al* 2017 *Phys. Med. Biol.* **62** 6322
[2] Jette D 2000 *Med. Phys.* **27** 1705-16
[3] Raaymakers B W *et al* 2004 *Phys. Med. Biol.* **49** 4109
[4] Raaijmakers A *et al* 2008 *Phys. Med. Biol.* **53** 909
[5] Kirkby C *et al* 2008 *Med. Phys.* **35** 1019-27
[6] Bielajew A F 1993 *Med. Phys.* **20** 1171-9
[7] Meijsing I *et al* 2009 *Phys. Med. Biol.* **54** 2993
[8] Reynolds M *et al* 2013 *Med. Phys.* **40** 042102
[9] Smit K *et al* 2013 *Phys. Med. Biol.* **58** 5945
[10] Spindeldreier C *et al* 2017 *Phys. Med. Biol.* **62** 6708
[11] O'brien D *et al* 2016 *Med. Phys.* **43** 4915-27
[12] Agnew J *et al* 2017 *Phys. Med. Biol.* **62** 1731
[13] Hackett S et al 2016 *Med. Phys.* **43** 3961-8
[14] Reynolds M *et al* 2014 *Med. Phys.* **41** 092103
[15] Gargett M *et al* 2015 *Med. Phys.* **42** 856-65
[16] Houweling A *et al* 2016 *Phys. Med. Biol.* **61** N80
[17] Li G *et al Phys. Medica* **29** 295-303
[18] de Vries J *et al* 2018 *Phys. Med. Biol.* **63** 02NT02
[19] Raaymakers B *et al* 2011 *Phys. Med. Biol.* **56** N207
[20] Torres-Xirau I *et al* 2018 *Phys. Med. Biol.* **63** 025006
[21] Smit K *et al* 2014 *Phys. Med. Biol.* **59** 4099
[22] Raaijmakers A *et al* 2007 *Phys. Med. Biol.* **52** 4283
[23] Reyhan M L *et al* 2015 *J. Appl. Clin. Med. Phys.* **16** 325-32
[24] Reynoso F J *et al* 2016 *Med. Phys.* **43** 6552-6
[25] Delfs B *et al* 2018 *Phys. Med. Biol.* **63** 035028
[26] Baldock C *et al* 2010 *Phys. Med. Biol.* **55** R1-63
[27] Lee H J *et al Phys. Med. Biol.* **63** 045021

[28] Costa F *et al* 2018 *Phys. Med. Biol.* **63** 05NT01
[29] Lee H J *et al* 2017 *Radioth. Oncol.* **125** 426-32
[30] Mein S *et al* 2017 *Med. Phys.* **44** 6018-28
[31] Uilkema S *et al* 2015 *Med. Phys.* **42** 7182-9
[32] Hepworth S *et al* 1999 *Nucl. Instrum. Methods Phys. Res* **422** 756-60
[33] Vergote K *et al* 2003 *Radioth. Oncol.* **67** 119-28
[34] De Deene Y *et al* 2006 *Med. Phys.* **33** 2586-97
[35] Hepworth S *et al* 1999 *Phys. Med. Biol.* **44** 1875
[36] Fong P M *et al* 2001 *Phys. Med. Biol.* **46** 3105
[37] Doran S J *et al* 2001 *Phys. Med. Biol.* **46** 3191-213
[38] Doran S J and Yatigammana D N 2012 *Phys. Med. Biol.* **57** 665
[39] Lee H J *et al* 2017 *J. Phys.: Conf. Ser.* **847** 012057
[40] Wai P *et al* 2009 *Appl. Rad. Isotop.* **67** 419-22
[41] Dekker K H *et al* 2016 *Med. Phys.* **43** 4585-97

# Polymer gel-based measurements of the isocenter accuracy in an MR-LINAC

**S Dorsch[1,2,4], P Mann[1,4], A Elter[1,2,4], A Runz[1,4], S Klüter[3,4] and C P Karger[1,4]**
[1]Department of Medical Physics in Radiation Oncology, German Cancer Research Center (DKFZ), INF 280, 69120 Heidelberg, Germany
[2]Faculty of Physics and Astronomy, University of Heidelberg, Heidelberg, Germany
[3]Department of Radiation Oncology, University Hospital Heidelberg, INF 400, 69120 Heidelberg, Germany
[4]National Center for Radiation Research in Oncology (NCRO), Heidelberg Institute for Radiation Oncology (HIRO), Heidelberg, Germany

Email: s.dorsch@dkfz.de

**Abstract**. As magnetic resonance-guided radiotherapy (MRgRT) is becoming increasingly important in clinical applications, the development of new quality assurance (QA) methods is needed. One important aspect is the alignment of the radiation and imaging isocenter. MR-visible polymer gels offer a way to perform such measurements online and additionally may allow for 3-dimensional (3D) evaluation. We present a star shot measurement irradiated and scanned with a 0.35 T MR-LINAC device evaluating the polyacrylamide gelatin (PAGAT) gel dosimeter immediately and 48 h after irradiation. The gel was additionally scanned at a 3 T MR device 5 h and 52 h after irradiation. The evaluation revealed an isocircle radius of 0.5 mm for both imaging devices and all image resolutions and time points after irradiation. The distance between radiation and imaging isocenter varied between 0.25 mm and 1.30 mm depending on the applied image resolution. This demonstrates that evaluation of a star shot measurement in a 0.35 T MR-LINAC is feasible, even immediately after irradiation.

## 1. Introduction

Integrating magnetic resonance imaging (MRI) into radiation devices, such as linear accelerators (LINACs) or $^{60}$Co-based systems [1-4], allows for image-guided radiation therapy (IGRT) improving the accuracy of dose delivery by acquisition of real-time images with excellent soft tissue contrast and without additional exposure of the patient. However, due to the static magnetic (B) field and the Lorentz force acting on the secondary electrons, the electrons will be deflected resulting in an altered dose distribution as compared to the case without magnetic field [5]. Therefore, the magnetic field changes not only the response of ionization chambers [6,7] but generates also asymmetric beam profiles [8], which may complicate geometrical quality assurance (QA) tests [9]: Most prominently, the isocircle radius, which is the smallest circle containing the central axis of all intersecting beams of a star shot measurement [10] - a well-established method for isocenter QA on conventional LINACs - will increase with increasing magnetic field strength. As a consequence, an increased isocircle radius cannot be attributed solely to mechanical machine misalignments. Therefore the interpretation of such tests also has to be reconsidered. To compensate for the effect of the magnetic field the use of high density material has been suggested [9].

Star shot measurements at conventional LINACs are usually performed with films. Using polymer gel (PG) instead, could offer a possibility to check the alignment between the irradiation and imaging isocenter [11] of MR-LINAC devices in 3D. Polymer gels [12] have revealed the same radiation isocenter accuracy as radiochromic films and can be evaluated immediately after irradiation, if only geometrical parameters are of interest [13]. In this paper we present the first irradiation of a star shot in PG within a 0.35 T MR-LINAC as well as the image-evaluation in the same device immediately and 48 h after irradiation. Additionally the PG was scanned with a 3 T MR device 5 h and 52 h after irradiation.

## 2. Material & Methods

### 2.1. Polymer gel

As PG the PAGAT (PolyAcrylamide Gelatin gel fabricated at ATmospheric conditions) [14] polymer gel was used (Table 1). A spherical flask composed of borosilicate glass was used as gel container (outer diameter = 8.5 cm, volume = 250 ml, wall thickness = 1 mm). The spherical flask was placed in the center of a cylindrical phantom (outer diameter = 23 cm) filled with a 3.6 g/l NaCl- and 1.25 g/l $CuSO_4$-solution.

**Table 1.** Composition of the used PAGAT polymer gel in weight percent and mmol.

| Incredient | amount |
|---|---|
| Water | 89% |
| Gelatine | 6% |
| Acrylamid | 2.5% |
| N,N′-methylene-bis-acrylamide | 2.5% |
| bis[tetrakis(hydroxymethyl)phosphonium] chloride (THPC) | 5mM |

### 2.2. Irradiation

The phantom was first aligned with a clinical 0.35 T MR-LINAC (MRIdian, ViewRAY, Oakwood Village, OH, USA) by means of dedicated fiducials on the gel-filled flask using the laser-system of the device. In a second step, MR-images were acquired and an image-based correction was performed to position the phantom exactly at the image isocenter of the device. The irradiation was performed with five beams (0°, 72°, 144°, 216° and 288°), a field size of 0.4 x 2.49 cm² and 3 Gy per beam at the isocenter.

### 2.3. Evaluation

Immediately after irradiation, the phantom was scanned on the MR-LINAC using two consecutive single slice double spin echo sequence with the following imaging parameters: TR = 2000 ms, $TE_{1,1}$ = 200 ms, $TE_{1,2}$ = 600 ms, $TE_{2,1}$ = 400 ms, $TE_{2,2}$ = 800 ms, slice thickness = 20 mm, resolution = 0.5 x 0.5 mm² and 4 averages with a total acquisition time of 1 h 15 min each. Imaging was repeated 48 h after irradiation when the polymerization of the PG was known to be completed. At this time point, another measurement with a better signal to noise ratio (SNR) was acquired using TR = 4000 ms, $TE_1$ = 200 ms, $TE_2$ = 600 ms, slice thickness = 20 mm, three averages and resolution = 1.0 x 1.0 mm².

In addition, a measurement on a 3 T Biograph mMR (Siemens Healthineers, Erlangen, Germany) was performed 5 h after irradiation using a single slice multi spin echo with 32 equidistant echoes with the following imaging parameters: TR= 4000 ms, TE = 30 – 960 ms, slice thickness = 20 mm and a resolution of 0.5 x 0.5 mm² with a total acquisition time of 29 min 10 s. This measurement was also repeated 52 h after irradiation and additionally with a better SNR using a resolution of 1.0 x 1.0 mm².

The acquired images were processed by an in-house developed Matlab (The Mathworks Inc., Natick, USA)-based PG evaluation tool [15] to generate T2-maps. The images prepared in this way were evaluated with the commercial software Mephisto (Version mcc 1.8, PTW, Freiburg, Germany). Each beam was selected manually in the software and the isocircle was determined; this procedure was repeated 6 times to determine the standard deviation (SD) of the Mephisto evaluation.

## 3. Results

Figure 1 displays the evaluated $T_2$-Maps for the different imaging devices and resolutions. The corresponding isocircle radii ($IC_r$) consistently revealed values of 0.5 mm independent of the applied imaging device and image resolution (Table 2). Reproducibility as indicated by the standard deviation (SD) was better by a factor of two for the 3T MR-device. The distance ($IC_d$) of the radiation isocenter, defined by the center of the isocircle, to the imaging isocenter, defined by the fiducials on the phantom, revealed values of up to 0.65 mm for the high- and up to 1.30 mm for the low-resolution images (Table 2).

**Table 2.** Isocircle radii ($IC_r$) and distance ($IC_d$) for the different imaging devices and resolutions (mean ± 1 SD).

| | 0.35 T MR-LINAC | | | 3 T Biograph mMR | | |
|---|---|---|---|---|---|---|
| resolution [mm²] | time point [h] | $IC_r$ [mm] | $IC_d$ [mm] | time point [h] | $IC_r$ [mm] | $IC_d$ [mm] |
| 0.5 x 0.5 | 0 | 0.51 ± 0.11 | 0.25 ± 0.11 | 5 | 0.51 ± 0.05 | 0.61 ± 0.09 |
| 0.5 x 0.5 | 48 | 0.48 ± 0.06 | 0.65 ± 0.17 | 52 | 0.49 ± 0.03 | 0.44 ± 0.08 |
| 1.0 x 1.0 | 49 | 0.55 ± 0.06 | 1.30 ± 0.09 | 52 | 0.48 ± 0.03 | 1.03 ± 0.10 |

## 4. Discussion

This is the first study demonstrating the feasibility of star shot measurements in an MR-LINAC using PG. Although the images acquired with the 0.35 T MR-device of the MR-LINAC had a higher noise content than those of the 3 T device, well-comparable isocircle radii with clinically acceptable accuracy were obtained.

Since the 0.35 T MR-LINAC images with high resolution (0.5 x 0.5 mm² pixel size) showed a low SNR (Figure 1 (a) and (c)), an additional image with a resolution of 1.0 x 1.0 mm² was obtained 48 h after irradiation to analyze the effect of an improved SNR (Figure 1 (e)). As a result, the same isocircle radius was obtained indicating that the lower SNR of the high resolution images are still good enough for a quantitative evaluation. In contrast, the SNR of the images of the 3 T MR-device is much better even for the higher resolution (Figure 1 b,d,f), which however, did not lead to differences in the isocircle radii. It is thus feasible to determine geometric parameters like the isocircle radius even in images with very low SNR.

Additionally, it should also be mentioned that the automatic beam-fitting-algorithm of Mephisto was yet significantly less error prone using the MR-LINAC images with 1.0 x 1.0 pixel size in contrast to the 0.5 x 0.5 mm² pixel size.

The values determined for the isocircle distance are within the recommended tolerance limits for conventional LINACS (< 1 mm distance between the laser and the irradiation isocenter [16]) for measurements with 0.5 x 0.5 mm pixel size, for both images with 1x1 mm voxel size the values slightly exceed this recommended value. This is most likely due to the pixel size of 1 mm, which leads to an increased uncertainty in the determination of the fiducial position in the image. Since these fiducials represent the imaging isocenter, $IC_d$ may be increased.

Furthermore, it should be noted that the $T_2$-map of the images recorded on the MR LINAC was determined using only 4 echoes ($R^2 = 0.90$-$0.94$), so that it can be assumed that the fit over all echoes is not as stable as in the case with 32 echoes (3 T mMR $R^2 > 0.98$). Unfortunately, there is no multi-spin-echo sequences available on the MR-LINAC device, yet, but it is expected to be implemented soon. Since we intend to use the presented measurement technique in the future to measure the alignment accuracy of irradiation and imaging isocenter in 3D with an uncertaintly of below 1mm in axial direction, it is essential to reduce the slice thickness from 20 mm to at least 1 mm. For this, the signal intensity is expected to drop significantly (SNR~slice thickness), which would further decrease the signal to noise ratio. This may be compensated by using more numbers of signal averages/acquisitions (NSA) (SNR~√NSA), which however, will increase the measurement time. Nevertheless, we expect that the combination of extended measurement time and the use of a multi-spin-echo sequence with 32 echoes will significantly enhance the signal of the images, so that even images with a significantly reduced slice thickness may provide sufficient information to perform irradiation and imaging isocenter measurements.

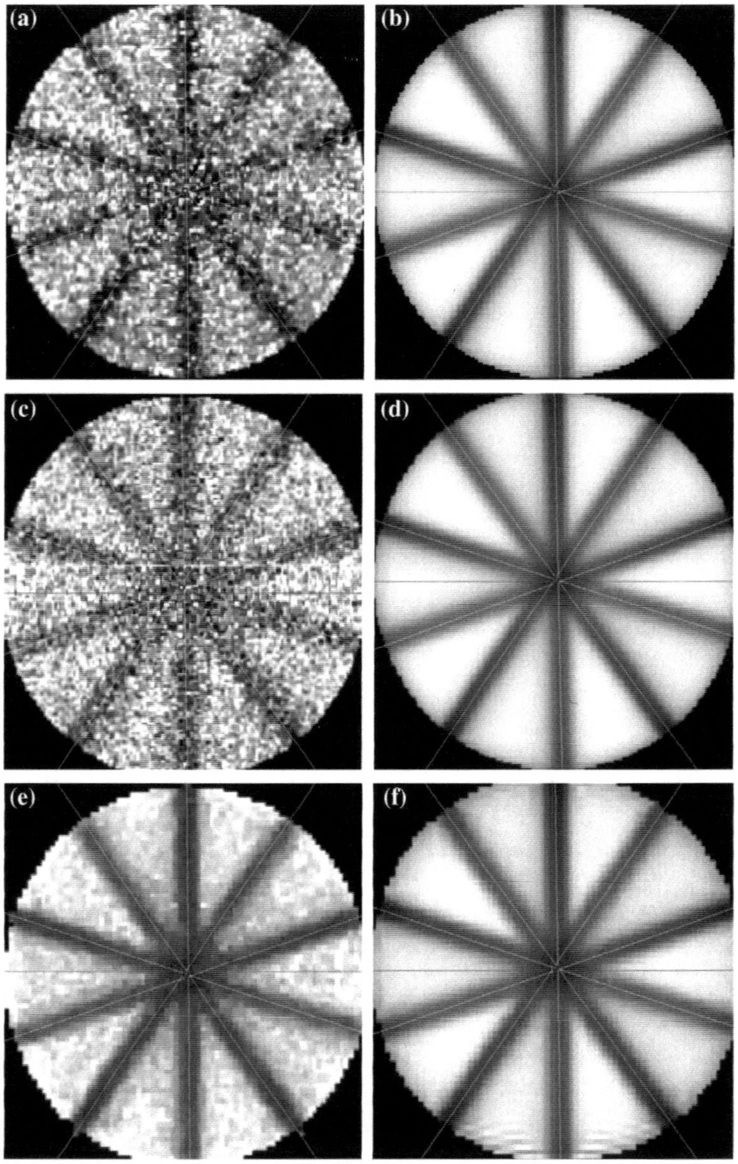


**Figure 1.** PG-based $T_2$-Maps acquired with a 0.35 T MR-LINAC (a,c,e) and a 3 T MR device (b,d,f) at different time points after irradiation: immediately (a), 5 h (b), 48 h (c), 49 h (e) and 52 h (d,f) after irradiation. The magnetic field is orientated towards the reader. Resolution is 0.5 x 0.5 mm² (a-d) and 1.0 x 1.0 mm² (e,f). The isocircle is delineated in blue.

IOP Conf. Series: Journal of Physics: Conf. Series **1305** (2019) 012007 doi:10.1088/1742-6596/1305/1/012007

## 5. Conclusion

The evaluation of a star shot measurement in a 0.35 T MR-LINAC is feasible, even immediately after irradiation. Although the SNR was significantly worse than that of the 3 T device, well-comparable isocircle radii and isocenter distances were obtained for an image resolution of 0.5x0.5 $mm^2$. With this low magnetic field strength, however, significantly longer acquisition times are needed.

## 6. Acknowledments

This work has received funding from the EMPIR programme co-financed by the Participating States and from the European Union's Horizon 2020 research and innovation programme. Additionally the authors would like to thank Professor Dr. K. Kopka for providing the lab space for the production of the polymer gel.

## 7. References

[1] Lagendijk J J W *et al* 2008 *Radiother. Oncol.* **86** 25-9
[2] Fallone B G *et a.* 2009 *Med. Phys.* **36** 2084-8
[3] Keall P J *et al* 2014 *Semin. Radiat. Oncol.* **24** 203-6
[4] Mutic S *et al* 2014 *Semin. Radiat. Oncol.* **24** 196-9
[5] Raaymakers B W *et al* 2004 *Phys. Med. Biol.* **50** 1363-76
[6] Meijsing I *et al* 2009 *Phys. Med. Biol.* **54** 2993-3002
[7] Smit K *et al* 2013 *Phys. Med. Biol.* **58** 5945-57
[8] Raaijmakers J E *et al* 2008 *Phys. Med. Biol.* **53** 909-23
[9] van Zijp H M *et al* 2016 *Phys. Med. Biol.* **61** N50-9
[10] Treuer H *et al* 2000 *Phys. Med. Biol.* **45** 2331-42
[11] Rankine L J *et al* 2013 *Phys. Med. Biol.* **58** 7791-801
[12] Baldock C *et al* 2010 *Phys. Med. Biol.* **55** R1-63
[13] Dorsch S *et al* 2018 *Phys. Med. Biol.* **63** 11NT02
[14] De Deene Y *et al* 2002 *Phys. Med. Biol.* **47** 3441-63
[15] Mann P *et al* 2017 *Phys. Med. Biol.* **62** 573-95
[16] Klein E E *et al* 2009 *Med. Phys.* **36** 4197-212

# Dose rate and fractionation dependence of methacrylic acid based polymer gels using optical and MRI techniques

**Hannah J. Lee, Yvonne Roed and Geoffrey S. Ibbott**
Department of Radiation Physics, UT MD Anderson Cancer Center, Houston, TX

E-mail: gibbott@mdanderson.org

**Abstract**. Polymer gels formulated with methacrylic acid as the main monomer source have been investigated for several decades. However, there is some discrepancy in the literature regarding their dose rate and fractionation dependence. This study investigated whether the read-out technique, optical versus MRI, affected the measured dependence for dose rate and fractionation.

## 1. Introduction

Polymer gels are a type of 3D dosimeter that react to the absorption of radiation dose with the polymerization of monomers as a result of the byproducts of radiolysis of water molecules. The linking and crosslinking of monomers produce a measurable, localized 3D dose distribution that can be measured with computed tomography (CT), optically, or with magnetic resonance imaging (MRI) techniques [1-7]. A variety of monomers have been investigated for polymer gel applications, most notably beginning with N,N'-methylene-bis-acrylamide (bis) monomers [8], acrylamide and bis monomers (BANANA, BANG®, and PAG) [1, 9-12], acrylic acid and bis monomers (BANG®-2) [13], methacrylic acid monomers (BANG®-3, BANGkit™, and MAGAT) [14], and methacrylic acid and ascorbic acid (MAGIC) [15].

Mixed reports of dose rate (DR) dependence and fractionation (fx) dependence exist for polymer gels and particularly for those containing methacrylic acid, such as BANG®-3, BANGkit™, MAGAT, and MAGIC gels. These reports use a variety of read-out methods: CT, optical, and MRI. As a polymer gel reacts due to absorption of radiation, the change in gel density is expected to be directly proportional to the change in CT number [3]. Similarly, the optical density (OD) of an irradiated polymer gel is expected to increase with radiation dose. However, the relationship of OD versus dose is dependent on the wavelength at which the measurements are taken [4, 16]. Polymer gels exhibit a stronger change and mono-exponential decay in spin-spin relaxation response (time constant $T_2$) compared to spin-lattice relaxation response (time constant $T_1$) and was first described by Zimmerman and later called the fast exchange model [17, 18]. As a result, the changes in polymer gels read-out using MRI are typically reported as $T_2$ or transverse relaxation rate $R_2$ ($=1/T_2$) values.

Reports of dose rate dependence include a threshold beginning at 6 Gy while others have indicated a range for which their dose rate dependence occurs (30 cGy/min to 400 cGy/min) [2, 19-21]. A summary of these findings is below in Table 1 with their respective read-out methodologies. The literature thus far agrees that methacrylic acid-containing gels do not exhibit any energy dependence, which could confound some of the reported dose rate dependence [21, 22].

**Table 1.** Dose rate (DR) dependence in literature for methacrylic acid containing polymer gels.

| Reference | Gel | Read-out Method | DR Dependence | Range |
|---|---|---|---|---|
| Sathiyaraj *et al* 2018 [23] | MAGAT | CT | Y | 250-1500 cGy/min |
| Massillon-JL *et al* 2010 [24] | BANG3®-Pro-1 | Optical (635 nm) | N | 3.7-15.2 cGy/min |
| Sathiyaraj *et al* 2018 [23] | MAGAT | Optical (400 nm) | Y | 250-1500 cGy/min |
| De Deene 2004 [2] | nMAG | MRI | Y | 25-1000 cGy/min |
| Bayreder *et al* 2006 [20] | MAGAT | MRI | Y | 60-500 cGy/min |
| De Deene *et al* 2006 [21] | nMAG | MRI | Y | 30-400 cGy/min |
| Murakami *et al* 2007 25] | BANGkit™ | MRI | Y | 68-408 cGy/min > 6 Gy |
| Marques *et al* 2010 [26] | MAGIC-*f* | MRI | N | 44-1000 cGy/min |
| Pavoni *et al* 2012 [27] | MAGIC-*f* | MRI | N | 200-500 cGy/min |
| Roed 2017 [22] | BANG3®-Pro | MRI | Y | 0.095-0.682, 60-425 cGy/min |
| Khan *et al* 2018 [28] | MAGIC | MRI | Y | 60-1800 cGy/min |

Similarly for fractionation dependence, mixed reports exist with both optical and MRI read-out methodologies. A summary of this literature can be found below in Table 2.

**Table 2.** Fractionation (fx) dependence in literature for methacrylic acid containing polymer gels.

| Reference | Gel | Read-out Method | fx Dependence |
|---|---|---|---|
| Massillon-JL *et al* 2010 [24] | BANG3®-Pro-1 | Optical (635 nm) | N |
| Vredevoogd 2012 [29] | BANG3®-Pro-2 | Optical (785 nm) | Y |
| Karlsson *et al* 2007 [19] | nMAG | MRI | Y |
| Pavoni *et al* 2012 [27] | MAGIC-*f* | MRI | N |
| Roed 2017 [22] | BANG3®-Pro | MRI | Y |

For the above reasons, this work used a methacrylic acid-based polymer gel (MAGAT) formulation for comparisons of dose rate and fractionation dependence when read-out with optical and MRI techniques.

## 2. Materials and Methods

### *2.1. MAGAT gel fabrication and read-out methodologies*

The MAGAT formulation used to make the polymer gels in-house was as follows: 8 wt% 300 bloom gelatin (Sigma-Aldrich), 6% methacrylic acid (MAA) (ACROS Organics), 5mM tetrakis (hydroxymethyl) phosphonium chloride (THPC) (Sigma-Aldrich), and 86 wt% deionized water. Gelatin served as the gel matrix; MAA was the source of monomers; THPC was an oxygen scavenger; and water was the source of radicals to initiate the polymerization process during irradiation. Gelatin was soaked in water until fully dissolved then heated gradually to 50°C. After maintaining 50°C for an hour, the temperature was lowered to 40°C and MAA and THPC were added. The mixture was stirred while the temperature was further gradually lowered to 35°C, poured into cuvettes or jars, and moved to 4°C for complete gel solidification. After at least 10 hours at 4°C, the gels were then acclimated to room temperature (~21°C) over at least another 10 hours. All MAGAT gels in this study were irradiated within 24 hours of creation in order to reduce the risk of oxygen contamination, especially for the cuvettes.

For the optical read-out method of this study, the gels were poured into standard size cuvettes with an optical path length of 1 cm (Thermo Fisher Scientific) for optical density (OD) read-out in the visible range using a GENESYS™ 10S UV-VIS spectrophotometer (Thermo Fisher Scientific). The cuvettes were topped with plastic paraffin film to reduce oxygen contamination. The absorption spectra (350-700 nm, 1 nm wavelength steps) of all cuvettes were measured prior to irradiation and 22 hours post-irradiation. The xenon lamp light beam dimensions were 2 mm (wide) by 7 mm (tall), and the central point of measurement in the cuvette was 8.5 mm from the bottom of the cuvette (away from the region of possible oxygen contamination at the top of the cuvettes). The net OD was calculated by subtracting the OD of un-irradiated samples from that of irradiated samples. Three intra-batch cuvettes were averaged for each given data point. A single batch was used for all of the optical measurements and was the same as one of the MRI read-out method batches.

For the MRI read-out method of this study, the gels were poured into cylindrical polyethylene terephthalate (PET) plastic containers that were 45 mm in diameter and filled to a height of 90 mm. Similar to the optical read-out samples, all MRI read-out gels were imaged prior to irradiation and 22 hours post-irradiation using two different sequences: T2-Weighted (T2W) and $T_2$ map series. The T2W images were acquired with the following parameters: acquired voxels 1x1x3 $mm^3$, reconstructed voxels 0.63x0.63x1.50 $mm^3$, turbo spin echo, 5 averages (NSA or NEX), TR = 2100 ms, and TE = 439 ms. The $T_2$ map was calculated with images acquired with the following parameters: acquired voxels 1x1x3 $mm^3$, reconstructed voxels 0.42x0.42x3.00 $mm^3$, turbo spin echo, 5 averages, TR = 240 ms, TE = 5.5 + n*11 (8 echoes = 5.5, 16.5, 27.5, 38.5, 49.5, 60.5, 71.5, and 82.5 ms). The first echo of each $T_2$ map series (TE = 5.5 ms) was not used for the $T_2$ map calculation since it may not have been acquired in a steady state. The $T_2$ map of each sample was calculated in ImageJ using the QuickVol II plugin, where $T_2$ is calculated by a linear regression model [30]. The goodness of fit for calculating $T_2$ using this model measured by $R^2$ was on average 0.914±0.004. The final reported $R_2$ values were calculated as $1/T_2$. The temperature of these samples was also measured immediately prior to irradiation to ensure room temperature equilibration using an infrared thermometer (Cen-Tech™). Three batches were used for the MRI read-out comparison and with one gel per batch per data point (three inter-batch gels averaged for each given data point).

### *2.2. MR-linac irradiation set-up*

An integrated 1.5 T magnetic resonance imaging and 7 MV linear accelerator system (pre-clinical MR-linac, Elekta AB, Stockholm, Sweden) was used for all irradiations and MR imaging of the gels in this study. The radiation isocenter and MR isocenter of the pre-clinical "2ATL" MR-linac system are located 143.5 cm from the linac target and approximately 14.2 cm above the surface of the couch. The cuvettes were centered within a 1.3 cm thick acrylic slab with 4.4 cm of solid water on top for build-up to reach a total of 5 cm depth at the center of the cuvettes (Figure 1a). The MRI gels were set-up on top of solid water such that 4 cm of gel was below the isocenter height and 5 cm of gel was above the isocenter height (with the region of interest and measurement at 5 cm depth) (Figure 1b). All samples received a total dose of 3 Gy at 5 cm depth using a 20x20 $cm^2$ field with gantry angle 0°.

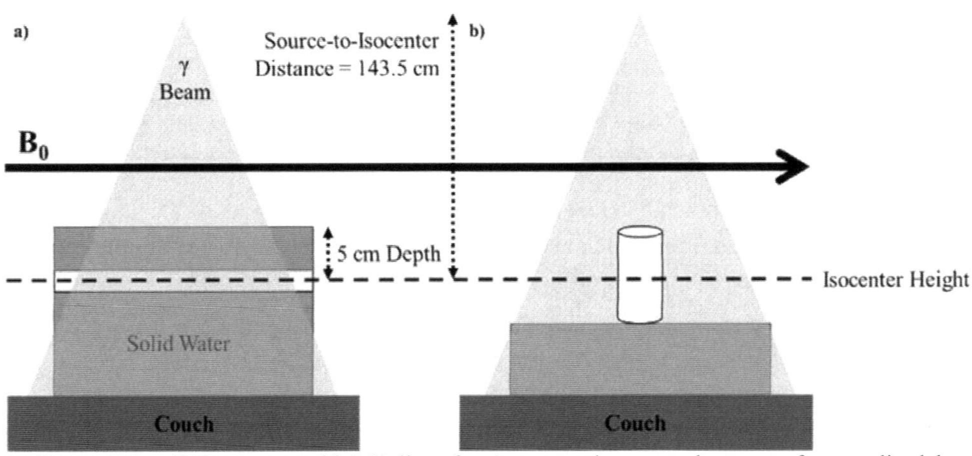


**Figure 1.** Irradiation set-up with MR-linac for a) cuvettes in a central cut-out of an acrylic slab and b) cylindrical PET gels.

The dose rate dependence was investigated using two methods: changing the gun duty cycle (GDC) (changing the photon fluence rate) and changing the distance of the samples relative to the source. The first method (GDC) changes the average dose rate but not the instantaneous dose rate. For the first method, GDC of 100% (400 MU/min), 87% (348 MU/min), 50% (200 MU/min), and 14% (56 MU/min) were used. The second method (by distance) changes both the average and instantaneous dose rates. For this method, dose rate was changed using the distance from the source-to-center-of-sample from 100% (143.5 cm) to 87% (153.7 cm) and to 130% (125.9 cm). Both the minimum and maximum dose rates using the second method were limited by the size of the MR-linac bore.

The fractionation dependence used the same set-up as for the 100% dose rate (and used 400 MU/min). The total dose of 3 Gy was divided into three different fractionation schemes: 1 fx (3 Gy), 2 fx (1.5 Gy + 1.5 Gy), and 3 fx (1 Gy + 1 Gy + 1 Gy). Each of these fractions had a 5 minute interval in between.

## 3. Results and Discussions

### *3.1. Optical absorption spectra and results*

The optical absorption spectra of the MAGAT samples were measured prior to and 22 hour post-irradiation between 350 and 700 nm. Wavelengths of 400, 635, and 785 nm have been reported in the literature using spectrophotometers or optical CT methods for similar methacrylic acid containing polymer gels. Figure 2 demonstrates that the apparent relationship between OD or net OD and radiation dose can vary depending on the wavelength selected. In order to compare with existing literature, wavelengths of 400, 500, and 650 nm were selected for further analysis. A larger separation between the OD of the irradiated and un-irradiated samples existed for 500 and 650 nm compared to 400 nm. 400 nm was also closer to the ultraviolet region where the signal became noisier.

When changing the average dose rate by changing the gun duty cycle, measurements taken at 400 nm were increased by up to 7%, at 500 nm by up to 33%, and at 650 nm by up to 51% at 14% dose rate when compared to 100% dose rate (Figure 3). When changing the average and instantaneous dose rate by changing the distance, measurements taken at 400 nm did not change, at 500 nm up to 4%, and at 650 nm up to 5% at 130% dose rate when compared to 100% dose rate (Figure 3). When changing the number of fractions, measurements taken at 400 nm were increased by up to 11%, at 500 nm up to 46%, and at 650 nm up to 86% (Figure 3).

### *3.2. Optical versus MRI read-out measurements*

When changing the average dose rate by changing the gun duty cycle, T2W measurements were decreased by up to 15% (50% DR), and $R_2$ values were increased by up to 6% (14% DR) (Figure 3). When changing the average and instantaneous dose rate by distance, T2W measurements were decreased by up to 6% (130% DR), and $R_2$ values were decreased by up to 1% (130% DR) (Figure 3). When changing the number of fractions, T2W measurements were decreased by up to 9% (2 fx), and $R_2$ values were increased by up to 3% (3 fx) (Figure 3). The temperature of all MRI read-out gel samples increased on average by 2.6±0.7°C after absorption of 3 Gy. This temperature change is solely attributed to the polymerization process occurring immediately after irradiation began. For this reason and to allow for a stable polymerization state, all samples were imaged 22 hours post-irradiation.

Comparing the optical to MRI read-out results, the $R_2$ values correlated more closely with the 400 nm optical results. However, the T2W measurements had a larger standard deviation and did not follow any trends of change with dose rate or fractionation, which could be due to differences in signal gain factor during image acquisition. Changing the average dose rate by changing the GDC appeared to have a slightly greater effect on the MAGAT response compared to changing both the average and instantaneous dose rate by changing the distance, especially when comparing both 87% dose rates. Radicals formed during the radiolysis of water contributed to both the polymerization of monomers (contributing to increased response of gel) and to the termination of polymer chains (contributing to decreased response of gel). The overall trends in the changes in response agreed with the literature, where the response of MAGAT was expected to be increased (greater OD, smaller T2W and $T_2$ values, greater $R_2$ values) with decreasing dose rates, due to the production of fewer radicals and fewer polymer chains being terminated by radicals. Similarly, with multiple fractions delivering the same total dose (and therefore the same total number of radicals generated), an increased number of fractions resulted in fewer radicals contributing to polymer chain terminations (versus polymerization) due to the time in between each fraction and therefore an increased response of MAGAT.

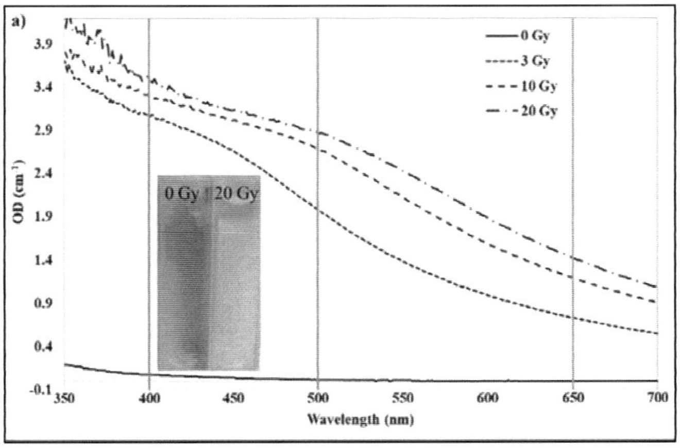


(a)

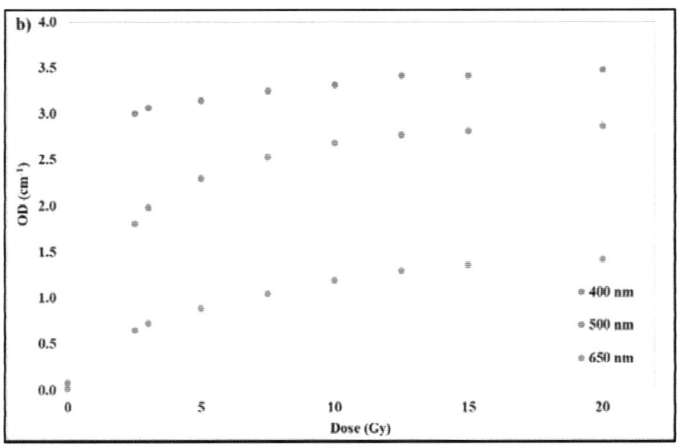


(b)

**Figure 2.** a) Spectra of un-irradiated and irradiated cuvettes of MAGAT gel and b) dose-response curves for 400 nm, 500 nm, and 650 nm.

## 4. Conclusions

The results of this study demonstrated that $R_2$ MRI measurements can be correlated to optical measurements but depend highly on the wavelength selected for optical measurements. As vendors continue to provide optical CT scanners dedicated to 3D dosimetry, this study emphasizes that regardless of the dosimeter, careful assessment of the characterization steps should be taken dependent on the wavelength used and compared to other conventional methods in order to make accurate dose comparisons between the dosimeter read-out measurements and treatment plan calculated doses.

Future work should investigate repeating this study at different dose levels since this specific MAGAT formulation may have a different dose threshold for dose rate and fractionation dependence for optical and MRI read-out techniques. Energy dependence should also be characterized to ensure that it is not confounding the dose rate dependence results.

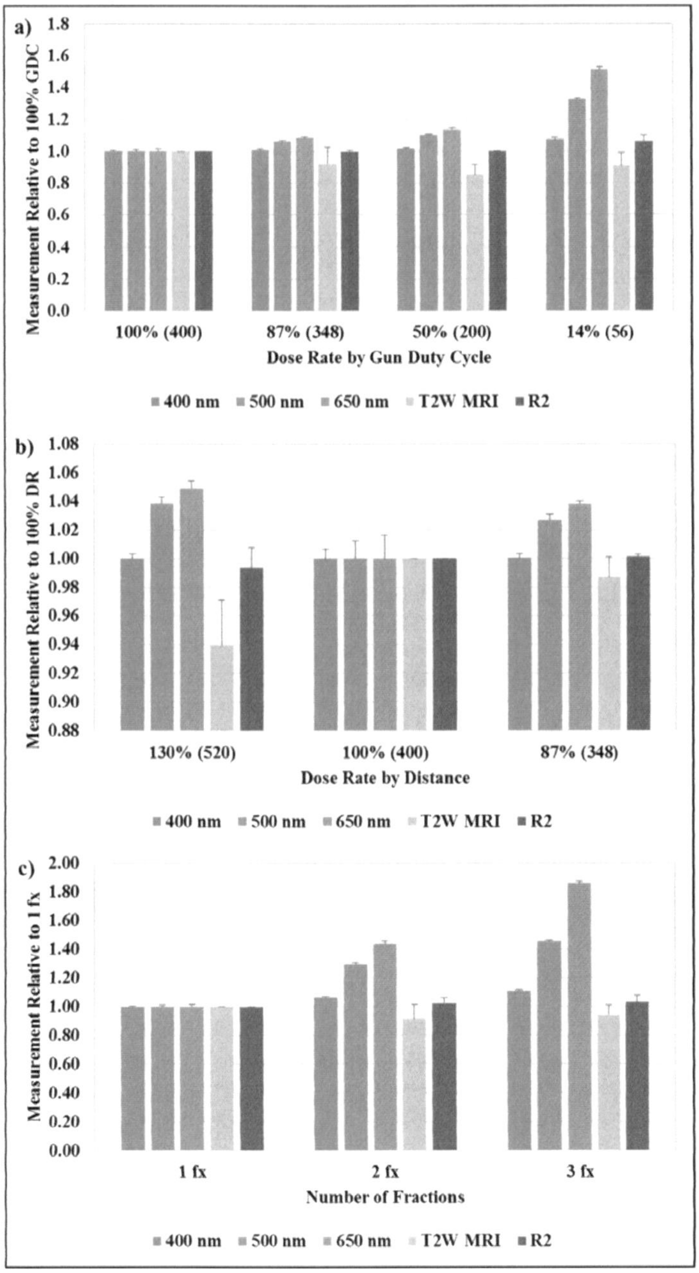


**Figure 3.** a) Normalized measurements for dose rates modulated by the gun duty cycle (GDC) relative to 100% (# MU/min in parenthesis), b) normalized measurements for dose rates modulated by distance relative to 100%, and c) Normalized measurements for fractionation relative to 1 fraction (fx).

## 5. References

[1] Baldock C *et al* 2010 *Phys. Med. Biol.* **55** R1-63
[2] De Deene Y 2004 *J. Phys.: Conf. Ser.* **3** 34-57

[3] Hilts M 2006 *J. Phys.: Conf. Ser.* **56** 95-107
[4] Oldham M 2004 *J. Phys.: Conf. Ser.* **3** 122-35
[5] Hurley C et al 2006 Nucl. Instrum. Meth. A 565 801-11
[6] Mather M L et al 2003 Phys. Med. Biol. 48 N269-75
[7] Baldock C 2009 J. Phys.: Conf. Ser. 164 012002
[8] Kennan R P *et al* 1992 *Proc. Int. Soc. for Magnetic Resonance in Medicine (New York)*
[9] Baldock C *et al* 1998 *Phys. Med. Biol.* **43** 695-702
[10] Maryanski M J *et al* 1994 *US Patent* 5321357
[11] Maryanski M J *et al* 1994 *Phys. Med. Biol.* **39** 1437-55
[12] Maryanski M J *et al* 1993 *Magn. Reson. Imaging* **11** 253-8
[13] Maryanski M J *et al* 1996 *Med. Phys.* **23** 699-705
[14] Maryanski M and Barry M 1998 *Med. Phys* **25** A178
[15] Fong P M *et al* 2001 *Phys. Med. Biol.* **46** 3105-13
[16] Maryañski M J *et al* 1996 *Phys. Med. Biol.* **41** 2705-17
[17] Zimmerman J R *et al* 1957 *J. Phys. Chem.* **61** 1328-33
[18] Deene Y De 2010 *J. Phys.: Conf. Ser.* **250** 12015
[19] Karlsson A 2007 *Phys. Med. Biol.* **52** 4697-706
[20] Bayreder C *et al* 2006 *Med Phys* 2006 **33** 2506-18
[21] De Deene Y *et al* 2006 *Phys. Med. Biol.* **51** 653-73
[22] Roed Y 2017 *Polymer Gel Dosimetry in Magnetic Resonance-Image Guided Radiation Therapy (University of Houston)*
[23] Sathiyaraj P and Samuel J J 2018 *J. Cancer. Res. Ther.* **14** 287-91
[24] Massillon-JL G *et al* 2010 *Appl. Radiat. Isotop.* **68** 144-54
[25] Murakami Y *et al* 2007 *Phys. Med. Biol.* **52** 2301-11
[26] Marques T *et al* 2010 *J. Phys. Conf. Ser.* **250** 12012
[27] Pavoni J F *et al* 2012 *Radiat. Meas.* **47** 1074-82
[28] Khan M *et al* 2018 *Phys. Med. Biol.* **63** 06NT01
[29] Vredevoogd K 2012 *UT GSBS Diss Theses*
[30] Schmidt K 2006 QuickVol II Users Guide

# Characterization of small PRESAGE® samples for measurements near the dosimeter edges

**Filipa Costa[1], Simon Doran[2], John Adamovics[3], Simeon Nill[1], Ian M Hanson[1] and Uwe Oelfke[1]**

[1]Joint Department of Physics, The Institute of Cancer Research and The Royal Marsden NHS Foundation Trust, London SM2 5NG, UK

[2]CRUK and EPSRC Cancer Imaging Centre, Institute of Cancer Research, London SM2 5NG, UK

[3]Department of Chemistry and Biology, Rider University, Lawrenceville, NJ 08648, USA

E-mail: filipa.costa@icr.ac.uk

**Abstract**. Measurements near the edges of PRESAGE® 3D dosimeters will be important for validating the electron return effect (ERE) that can occur at tissue-air interfaces during radiotherapy treatment delivered with the Elekta MR-linac. We investigate and characterize the spatially non-uniform response of small samples of PRESAGE® to radiation in a conventional linac. We develop a correction to compensate for these non-uniformities and obtain dose values near the dosimeter edges. Five samples from the same batch were uniformly irradiated in a water tank with a broad beam. The non-uniform response of the samples to radiation was investigated and a radial dose-correction function was generated from each sample to obtain a correction image. We then applied these correction images to another sample from the same batch, irradiated with four beams in an inhomogeneous medium, and compared this with the relevant simulated data. Additionally, we irradiated samples after physically removed their edges (axially and top and bottom edges). Higher sensitivity to radiation was observed at the edges (~6mm) of the samples in comparison with the central region. Applying the dose correction function improved agreement between simulations and measurements, but only partial correction was possible. A uniform response was observed on the samples with the edges removed, which we propose as the best option to measure dose at the edges of PRESAGE® samples.

## 1. Introduction

Radiotherapy for lung cancer can benefit greatly from real-time plan adaptation as motion is always present and high doses of radiation are given [1]. In the MR-linac the constant B0 magnetic field can create a strong dose increase at the proximal side of air cavities and a decrease at the distal side. This is known as electron return effect (ERE) and is a consequence of the Lorentz force acting perpendicularly to the motion of the electrons [2]. In previous work [3] we have shown that small diameter PRESAGE® 3D dosimeters [4] can be used to measure dose at dosimeter-air interfaces and to detect accurately the ERE using the flat end of a cylindrical dosimeter. Non-uniform radiation sensitivities within the samples were observed in that work. Different radiation sensitivities at the center and the edges have also previously been reported for large PRESAGE® samples [5, 6] by other groups, and are thought to arise as a result of the manufacturing process. In addition, measurements near the interfaces between the dosimeter and surrounding optical matching liquid have historically been regarded as problematic because of the potential for refractive index mismatches. Thus, the outermost 5mm of data are often disregarded in studies using PRESAGE®. This is normally of little consequence either because the samples used are large and so the discarded fraction of data is insignificant, or because these outer shells of data contain only very low doses (dose penumbra region).

In this work we investigate and characterize the non-uniform response of small samples of PRESAGE® to radiation in a conventional linac, using samples from the same batch in order to develop a correction image to compensate for those non-uniformities. We tested the correction on a dosimeter which was irradiated with a simple dose distribution, but in a clinical relevant scenario, inside a lung phantom and mimicking a tumour. Alternative ways to measure doses near the PRESAGE® samples edges were also investigated.

## 2. Material and Methods

### *2.1. Presage® 3D dosimeters*

A PRESAGE® (Heuris Pharma, Skillman, NJ) 3D dosimeter consists of a solid-state polyurethane matrix with a free radical initiator and a leuco-dye. When irradiated, the leuco-dye oxidises from leuco malachite green to malachite green showing an optical density (OD) change whose peak absorption maximum is at around 633 nm. A batch of cylindrical PRESAGE® samples was manufactured using silicone moulds developed in house, in order to obtain samples with a smooth finish and consistent diameter throughout the length of the sample. Each sample was 3.5 cm in diameter and 5.5 cm (±0.2 cm) long, had a flat side (region corresponding to the bottom of the mould) and a curved side (meniscus region corresponding to the top of the mould). Five samples were randomly selected from the batch and their flat side glued to a 3D printed cap. The cap shape was created so each sample could be positioned in a reproducible way when being scanned and irradiated.

### *2.2. Optical-CT scan*

To measure changes in optical-CT pixel value, PRESAGE® samples were always scanned before and after being irradiated (within the same day). The scanner used in this work was a modified version of the one described by McErlean et al 2016 [7], which includes a new telecentric lens (TC2MHR096-C, Opto-Engineering, Italy) to allow to image samples up to 10 cm diameter and a new rotation stage (CR1/M-Z7K, Thorlabs Ltd., Ely, UK). Each sample is placed inside a tank with a matching liquid, which was carefully adjusted to the same refractive index as PRESAGE® in order to minimize edge artefacts originating from index mismatch. Raw data consisting of 1000 projections each of matrix 320 × 320 pixels were obtained over 180° rotation to create reconstructed images with a voxel size of $0.2mm^3$.

### *2.3. Non-uniform dose response characterization and correction*

The existence of a non-uniform dose response of the dosimeter to radiation was initially investigated using only two samples. Irradiations were carried out at our research Elekta Synergy Linac. The sample holder was placed inside a water tank (MP1, PTW, Germany) at 90 cm source-to-surface (SSD) distance and 10 cm depth (Figure 1a). To deliver a uniform dose of 2Gy, the dosimeter was irradiated with a 6 MV beam, a 10 x $10cm^2$ field size and gantry at 0° with four equal beams. The sample was rotated by 90° between each beam irradiation. This process was repeated to irradiate uniformly the samples with accumulated doses of 4Gy, 6Gy, 8Gy and 10Gy. An optical-CT scan was performed between irradiations, always 60 minutes after the delivery of the last beam. This experiment was performed for both Sample 1 and 2. Two months later, three further samples (3 to 5) were irradiated under the same experimental conditions but with different doses (2Gy + 4Gy + 4Gy), thus obtaining accumulated doses of 2Gy, 6Gy and 10Gy. A pre-scan was taken for each sample. All samples were kept in a fridge (~8°C) and taken out 90 minutes prior to the pre-scan to allow for equilibration with the ambient temperature.

A possible correction to a dosimeter non-uniform response to dose at the edges was investigated by taking the radial average of each sample (averaged in lengths thought the middle region of the sample (from slices 100-250 in Figure 1c), for each dose level). The relation (always found to be linear) between dose and optical-CT pixel value was characterised separately for each radial distance, leading to both a gradient *m* and intercept value *c* that varied as a function of radial distance from the central axis of the sample. Thus, each of our correction image datasets was reduced to a calibration of form:

$$\Delta I(D,r) = c(r) + m(r)D \qquad (1)$$

where $\Delta I$ is the change in optical-CT pixel value between the pre-scan and the post-irradiation scan. D corresponds to the dose and $r$ is the radial distance of each sample which is given by $r^2 = x^2 + y^2$. Since $\Delta I$ corresponds to the variation in optical-CT pixel value, $c \approx 0$, and we can ignore it. Note that, although the calibration is obtained via a radial average, the correction is applied on a voxel-by-voxel basis i.e.:

$$D(x,y,z) = [I_{\text{post}}(x,y,z) - [I_{\text{pre}}(x,y,z)]/m(r) \qquad (2)$$

where z corresponds to the direction along the sample length. The calibration from each correction image series was tested on a dose distribution obtained with another sample from the same batch. Note that this only corrects axial edge effect, and not the top and bottom edges. The research version of Monaco (5.19.03, Elekta AB, Stockholm, Sweden) treatment planning system (TPS) was used to simulate a plan consisting of four equidistant 6 MV beams delivered to a sample of PRESAGE® placed within a lung phantom (QUASAR™ MRI4D motion phantom, Modus Medical, London, Ontario) to a maximum 9.7Gy. The sample was cut at both ends to be 5 cm long. For comparison, EBT3 Gafchromic films (ISP Corp, Wayne NJ, USA) were irradiated in the same conditions as the PRESAGE® samples. A pieces of film was placed between two sets of two half Perspex cylinders of 5 cm length and oriented in the sagittal direction, while a second piece of film was oriented in the coronal direction. Films were scanned 24 hours after irradiation and data analysis was performed, in the red and green channels, using Film QA Pro software (Ashland Inc., Wayne, NJ).

### 2.4. *Physical removal of the dosimeters edge*

On the assumption that the varying sensitivities to radiation were introduced into the samples during the manufacturing processes the edges of the samples were removed. Sample 1 was reduced in diameter from 3.5 cm to 2.5 cm using a metal lathe, and Sample 2 was reduced in length by removing 6mm from each end. Both samples were scanned after being cut and then uniformity irradiated with additional 6Gy. The samples were then scanned 60 minutes after the irradiation.

## 3. Results and discussion

### 3.1. *Non-uniform dose response characterization and correction*

Results obtained with all five samples showed that the samples responded linearly with dose as expected, but seem to have higher sensitivity to radiation in the edges when compared to the middle. This effect is clearly visible for both axial edges and also top and bottom edges of the samples. This is shown in Figure 1b and c for Sample 1 with accumulated dose of 10Gy. In Figure 1d, the radial average of optical-CT pixel value is shown for each accumulated dose. Here we can see that this effect occurs on the last 25-30 pixels (corresponding to 5 to 6mm), and the remaining dosimeter shows a relatively flat profile (i.e., uniform dose sensitivity). Absolute differences between sample central region and edges can go up to 20%. The magnitude of the edge effect is dependent on the accumulated dose and increases with dose. In Figure 1e the relationship between optical-CT pixel value and dose is shown for Sample 1 for different radial distances. For larger radial distances (near the edges), the gradient of the fitted lines increase, but the linearity between optical-CT pixel value and dose is kept. This happens for all five samples. As the absolute gradient values for each sample were different (data not shown), the gradient values were normalized to the central region of each sample as shown in Figure 2a. Figure 2a shows how the profiles are broadly comparable between samples, but sample 3, 4 and 5 have a longer edge effect, which could be due to their age in relation to Sample 1 and 2. A 2D gradient map correction image was calculated for each sample based on the normalized gradient for each radial position as shown in Figure 2b for Sample 1.

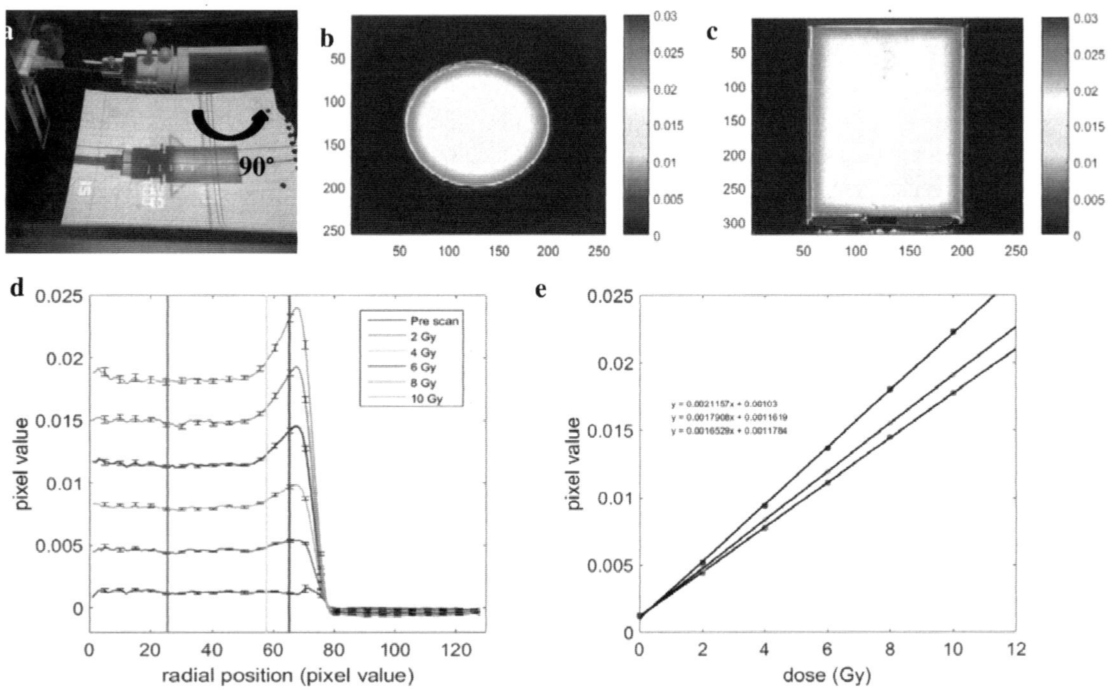


**Figure 1.** (a) Experimental set-up with a sample positioned in the water tank at 10 cm depth. The holder allows rotating the sample by 90° and locking it in position. 2D reconstructed optical-CT pixel image of a central (b) axial and (c) sagittal slices taken from Sample 1, with 10Gy accumulated dose. (d) Radial average for each accumulated dose obtained by averaging central region (from pixel 100-250) of the sample shown in c. One standard deviation is also shown. (e) Dose versus optical-CT pixel value obtained for three different radial positions taken from Sample 1 and given by the vertical lines in d.

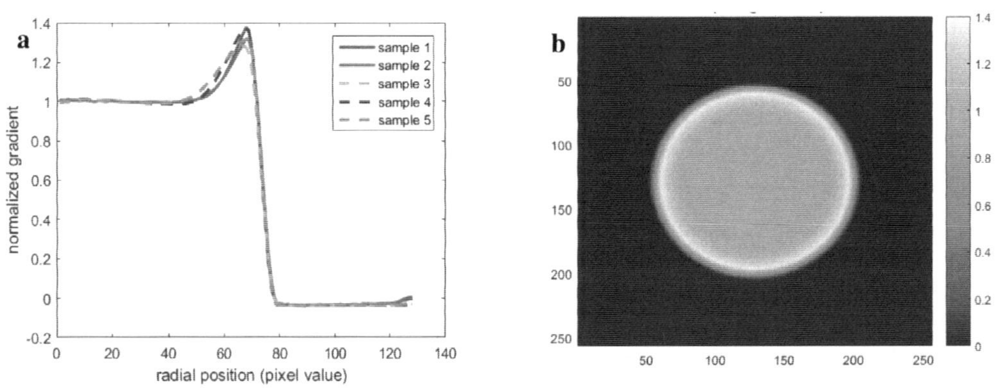


**Figure 2.** (a) Normalized gradient values versus radial position for all the 5 samples, obtained by plotting dose in relation to optical-CT pixel value, for different radial distances. Radial position > 80 correspond to the background. (b) 2D gradient map correction image obtained from Sample 1, based on the normalized gradient for each radial position.

In Figure 3, 2D reconstructed normalized optical-CT image and profiles obtained with a PRESAGE® sample irradiated with a lung phantom are shown. Measured profiles, taken in the middle of the target in several directions, are compared with the research Monaco simulated data, after normalizing both to the region of maximum dose. Measured profiles are shown with and without applying a 2D normalized gradient correction image (Presage corrected and Presage not corrected respectively). Despite the differences between all five samples, the correction images are comparable. For this reason in Figure 3, we applied a correction image that corresponds to an average of all five corrected images obtained from each sample. When a correction image is applied to the PRESAGE® data, in the regions with worse agreement, the difference between simulated and measured doses can be reduced by 12%. However, by comparison with simulations we can say that only partial correction is observed. A "bump" seems to be visible on two of the sides of the samples (at the left-right (LR) and anterior-posterior (AP) directions) even after correction, which could be a residual optical artefact at the boundary. As expected from the results in Figure 1d, which shows a very small edge effect for lower doses, the edge effect only occurs when the doses are high (here ≥ 30% of the maximum dose used). No effects are visible in the diagonal and sagittal profiles as, at the edges, only very low doses are present.

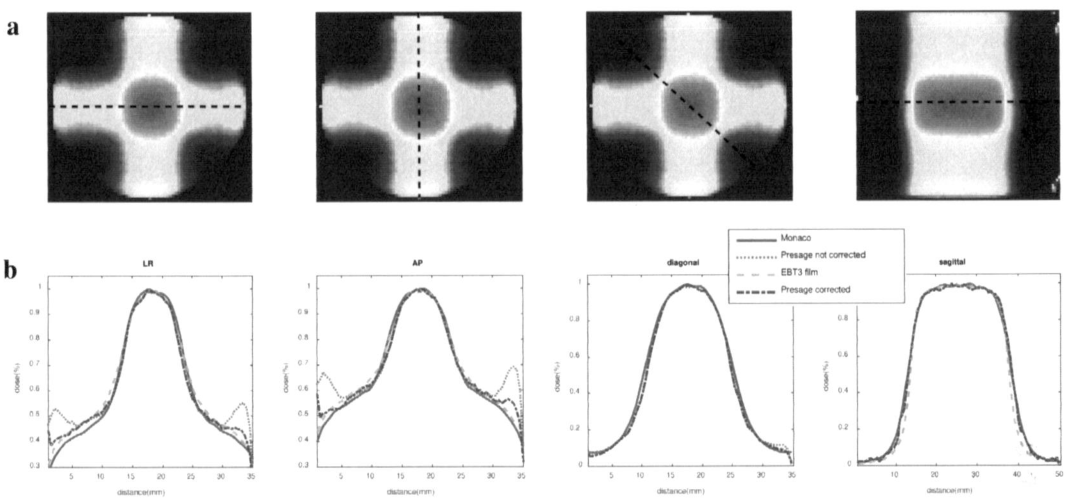


**Figure 3.** (a) Central axial slice 2D color-coded normalized optical-CT pixel value image showing results from a sample of PRESAGE® irradiated inside a lung phantom with no correction applied. Central sagittal slice is also shown. (b) Normalized simulated profiles taken at the central axial slice at left-right (LR), anterior-posterior (AP), diagonal and sagittal directions are plotted against measured data with and without correction. Green channel EBT3 film results are also shown.

Film results taken from the green channel are also shown in Figure 3 and were normalized to the maximum value of the profile. Higher doses at the edges are visible agreeing well with PRESAGE® when a "bump" is not present. Normalized results on the red channel showed values at the edges higher than the green (up to 3%). Further measurements are needed to obtain a better understanding of these differences.

### *3.2. Physical removal of the dosimeters edge*

In Figure 4, we observe the disappearance of the edge effect in the regions where the sample edges are removed, and the non-reappearance of the edge effect after each sample is irradiated with an additional 6Gy. A long vertical line is visible in the profile (~1mm) which we attribute to an optical-CT edge effect. This could be due to the roughness of the sample surface after being cut and a possible slight mismatch between the samples and matching liquid. The dose increase at the sample edges on the non-cut regions is still visible. This is clearly visible on the profiles but also looking at the 2D sagittal reconstructed images.

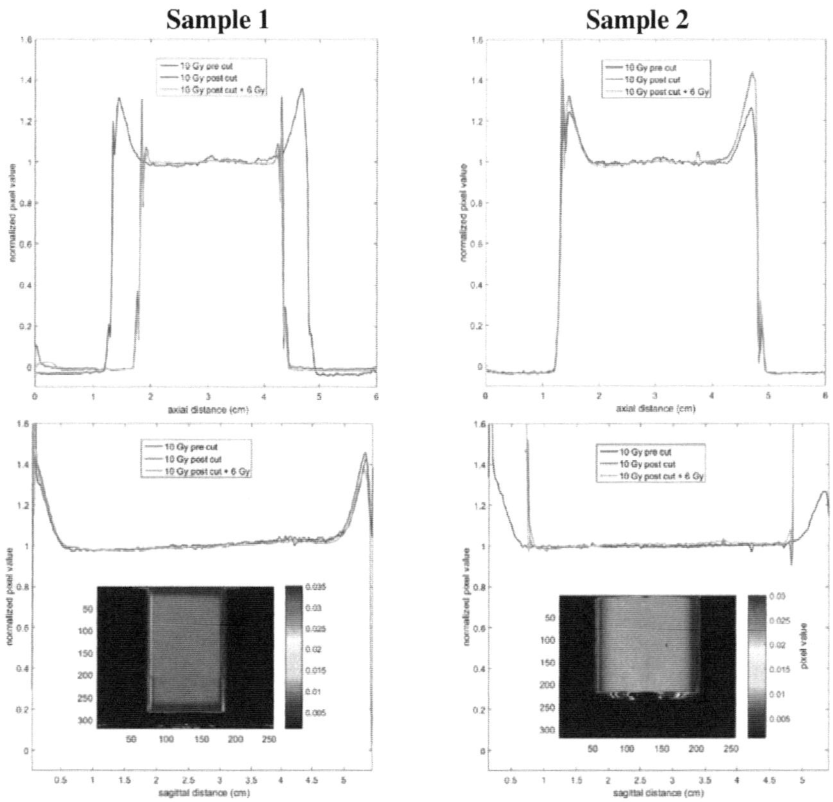


**Figure 4.** Normalized axial (top) and sagittal (bottom) 1D central optical-CT pixel values taken from Sample 1 (cut in diameter) and Sample 2 (cut in lengths) before being cut (10Gy pre cut), after being cut (10Gy post cut) and after irradiating with 6Gy more (10Gy post cut + 6Gy). A sagittal 2D pixel value image of the samples after being cut and irradiated with 6Gy more (10Gy post cut + 6Gy) are displayed.

## 4. Conclusion

In this work we show that small cylindrical PRESAGE® samples (3.5 cm diameter) have a different response to radiation in the central region in comparison to the edges. We propose a correction to compensate for this effect and investigate alternatives to obtain a uniform response throughout the samples, so that the dose can be measured right up to the boundary of the sample. Partial correction is possible and was applied here using data from homogeneous irradiated samples, but further work is still needed to develop a robust correction. However, we believe that physically removing the edges of the samples is a good alternative for measurements near dosimeter-air interfaces, provided that bigger samples are obtained. This will be important for studies on the MR-linac to measure accurately the ERE and will be tested in the future.

## 5. Acknowledgements

SJD acknowledges receipt of CRUK and EPSRC support to the Cancer Imaging Centre at ICR and RMH in association with MRC and Department of Health C1060/A10334, C1060/A16464 and the NHS funding to the NIHR Biomedical Research Centre and the Clinical Research Facility in Imaging. The ICR is supported by Cancer Research UK under Programme C33589/A19727. The authors acknowledge support from Elekta AB under a research agreement.

## 6. References

[1] Bainbridge H E *et al* 2017 *Radiother. Oncol.* **125** 280-5
[2] Raaijmakers A J E *et al* 2005 *Phys. Med. Biol.* **50** 1363-76
[3] Costa F *et al* 2018 *Phys. Med. Biol.* **63** 05NT01

IOP Conf. Series: Journal of Physics: Conf. Series **1305** (2019) 012009 doi:10.1088/1742-6596/1305/1/012009

[4] Doran S *et al* 2015 *Phys. Med. Biol.* **60** 709-26
[5] Dekker K H *et al* 2016 *Med. Phys.* **43** 4585-97
[6] Mein S *et al* 2017 *Med. Phys.* **44** 6018-28
[7] McErlean C M *et al* 2016 *Phys. Med. Biol.* **61** 320-37

IOP Conf. Series: Journal of Physics: Conf. Series **1305** (2019) 012010 doi:10.1088/1742-6596/1305/1/012010

# Dosimetry needs for MRI-linacs

**U Jelen[1] and J Begg[1,2,3]**
[1]Department of Medical Physics, Ingham Institute for Applied Medical Research, Liverpool, Australia
[2]Department of Medical Physics, Liverpool and Macarthur Cancer Therapy Centre, Liverpool, Australia
[3]South Western Sydney Clinical School, School of Medicine, University of New South Wales, Sydney, Australia

E-mail: Urszula.Jelen@health.nsw.gov.au

**Abstract**. The pursuit of real-time image-guided radiotherapy has prompted the development of hybrid devices coupling MRI scanners with radiotherapy treatment units, usually linear accelerators (linacs). One of the challenges in MRI-linac technology is the magnetic field impact on the dose deposition. Dose deposition effects have to be considered in the radiation therapy chain as they alter the dose distribution in patients and influence the response of many commonly used radiation detectors. In this presentation specific issues of dosimetry for MRI-linacs will be reviewed and illustrated with examples from the Australian MRI-linac commissioning process.

## 1. Introduction

The pursuit of real-time image-guided radiotherapy has prompted the development of hybrid devices coupling MRI scanners with radiotherapy treatment units, usually linear accelerators (linacs) [1]. One of the challenges in MRI-linac technology is the magnetic field impact on charged particles, which affects both the radiation beam generation in the linear accelerator as well as the dose deposition in the irradiated volume. While the former can be remedied with magnetic shielding and is not discussed further, the latter alters the dose distribution in patients and in radiation detectors and is the focus of this article: specific issues of dosimetry for MRI-linacs will be reviewed and illustrated with examples from the Australian MRI-linac commissioning process.

## 2. MRI-Linac Systems

Four MRI-linac designs exist to date, covering a range of magnetic field strengths and employing two beam-to-magnetic-field orientations [2]: *perpendicular* (also referred to in the literature as *transverse*) or *parallel* (also referred to in the literature as *inline*). In terms of the interference between the magnetic field and the dose deposition, each system therefore has different characteristics, which are reviewed below and summarised in Table 1 (starting from lowest field strength warranting least interference). Discussion of the imaging properties of these systems is beyond the scope of this manuscript, however, it should be mentioned that all systems have been demonstrated to provide image quality sufficient for MR guided radiotherapy (MRgRT) [3-6].

The MRIdian (ViewRay, Oakwood Village, OH, USA) system uses a 0.35 T split superconducting magnet in a perpendicular configuration with a radiation beam from, originally three cobalt sources, or more recently a 6 MV linear accelerator. The accelerator rotates around the patient couch on a gantry installed in the magnet split. The split design ensures an unobstructed beam path [7] and the relatively low magnetic field of this system results in minimal influence on dose deposition.

The Aurora-RT system (MagnetTx Oncology Solutions, Edmonton, Alberta, Canada) was developed at the University of Alberta and consists of a biplanar 0.5 T magnet and a 6 MV linac, which are both installed on a gantry and rotate around the patient couch. Whilst both beam-to-magnetic-field orientations are possible, in the current design the radiation beam axis and the magnetic field are parallel. Two features of this system should be highlighted from the dosimetry point of view. Firstly, the high temperature superconducting magnet allows the system to be turned on and off relatively easily allowing zero field measurements [2]. Secondly, an iron yoke used within the bore to restrict the extent of the fringe field reduces electron contamination [8]) (see section *3.1.2.*).

The Australian MRI-Linac consists of an open bore 1.0 T magnet and a linac which can produce photon beams of two energies 4 & 6 MV. While the versatile design permits radiation beam entry (and patient positioning) in either orientation, the beam is parallel to the magnetic field in the current system. Uniquely, the linac is mounted on rails, allowing variation of the source-to-isocentre distance (SID) and therefore enabling dosimetric measurements at different magnetic field strengths [2,6].

The system commercialised as Elekta Unity, has been developed at the University Medical Centre Utrecht (Utrecht, the Netherlands) in collaboration with Elekta AB (Stockholm, Sweden) and Philips Medical Systems (Best, the Netherlands). It consists of a modified closed bore 1.5 T Ingenia MR scanner and an Elekta 6 MV linac, which rotates around the patient couch. The radiation beam is perpendicular to the magnetic field and passes through the magnet cryostat. Although the magnet has been modified to minimise radiation beam perturbation, it attenuates the beam, introduces additional scatter [1,9] and prohibits the use of a (small) range of beam angles [9].

**Table 1.** Properties of current MRI-Linac systems.

| System (vendor) | Magnet properties | | | Orientation | Radiation beam properties | | |
|---|---|---|---|---|---|---|---|
| | Type | Strength | Opening | | Energy | SID | Field size / leaf width |
| MRIdian (Viewray) [7] | split, super-conducting | 0.35 T | 70 cm bore | perpendicular | 6 MV | 0.9 m | 27.4×24.1 cm / 4.15 mm |
| Aurora-RT (MagnetTx) [10,11] | biplanar, high temp. super-conducting | 0.5 T | 110 (W) × 60 (H) cm | parallel (perpen-dicular possible) | 6 MV | 1.22 m | 30 cm / 5 mm |
| Australian MRI-linac | open, super-conducting | 1.0 T | 62 cm bore, 50 cm gap | parallel (perpen-dicular possible) | 4 & 6 MV | 1.9–3.3 m | 30–50 cm / 9.5–16.5 mm |
| Unity (Elekta) [9,12] | closed, super-conducting | 1.5 T | 70 cm bore | perpendicular | 6 MV | 1.435 m | 22×57.4 cm / 7.2 mm |

SID – source-to-isocentre

## 3. Dose deposition in presence of the magnetic field

While the photons produced in the accelerator are not affected, both the contaminant electrons (originating in the accelerator and in other materials/air upstream from the patient or phantom) as well as the secondary electrons released in patient (or phantom) are affected by the magnetic field. The Lorenz force acts orthogonally to the magnetic field and to the direction of the electron motion, therefore its effects on the dose distribution depend on the orientation of the radiation beam relative to the magnetic field. It should be emphasised that, both the dose deposition in matter as well as the response of the dosimeters are affected by the magnetic field.

### *3.1. Influence on dose deposition by radiation beams*

*3.1.1. Transverse configuration.* In MRI-linacs employing a transverse configuration, the radiation beam axis, and hence to the predominant component of electron motion direction, is perpendicular to the magnetic field. The Lorentz force causes the electron paths between the collisions to become curved, resulting with trajectories with a preferential direction change and a shorter overall range. Macroscopically, this results in a shifted and asymmetric beam penumbra and in a decreased build-up distance [13]. Furthermore, upon entering low-density materials (such as lung or air cavity), depending on the radiological thickness of this region relative to their trajectory radius, electrons may be curved back, towards the interface, in a phenomenon termed the electron-return effect (ERE) [14]. ERE leads to localised dose increases at such tissue interfaces, including at the beam exit surface [14]. Additionally, in transverse MRI-linac systems, the Lorentz force sweeps contaminant electrons away from the incident beam axis. This leads to the reduction of the skin dose within the primary beam [8], however these electrons can potentially reach the patient surfaces outside the primary beam [15].

*3.1.2. Inline configuration.* In inline MRI-linacs the radiation beam axis is parallel to the magnetic field, however, due to scattering, the electrons have motion components orthogonal to the magnetic field. The Lorentz force causes the electrons to spiral around the magnetic field direction and successive energy losses in collisions lead the shrinkage of their helical orbits [16]. Macroscopically, this results in the reduction of the beam penumbra and in dose enhancement on the beam central axis, especially pronounced in low density materials. Furthermore, it reduces the dose deposition perturbations due to density heterogeneities [16]. On the other hand, strong inline magnetic field reduces the divergence of the contaminate electrons and concentrates them around radiation beam axis, which can lead to increased skin doses if not counteracted [17].

*3.2. Influence on treatment plan dose distributions.*

Non-negligible modifications of dose distributions, especially at density interfaces, have been reported for clinical treatment plans for various tumour locations, if the influence of magnetic field was not accounted for [18-20]. Conversely, Oborn *et al* [21] showed a case of tumour surrounded by low density tissue (lung) in inline field, where these effects could be beneficial and lead to target dose enhancement. A full review of studies investigating the effects of magnetic fields on treatment plans is beyond the scope of this manuscript. However, the need for dose computation algorithms that model magnetic field influence accurately, allowing to compensate or to even exploit these effects in plan optimisation [22], should be emphasised. This in turn requires high resolution 3D dose measurement systems, such as radiochromic or polymer materials, for the benchmarking of the dose computation algorithms as well as for the treatment plan verification.

*3.3. Influence on detector response*

The response of radiation detectors is related to the paths of the electrons traversing their active volume. In the magnetic field the electron trajectories change, however this change may be different in the materials constituting the detector (e.g. air cavities in ion chambers, silicon wafers in diode detectors etc.) than the trajectory change in the surrounding medium (phantom). As a result, the reading of the detector may not represent the dose that would be deposited in the medium in its absence. Furthermore, many detectors are not symmetric, therefore the change in their response is dependent on their orientation in the magnetic field.

## 4. Dosimetry for MRI-linacs: current status and open issues

The general dosimetry framework for MRI-linac systems, similarly to other radiotherapy modalities, aims at ensuring that the therapeutic dose distributions are delivered with sufficient accuracy. In this section, dosimetry issues for MRI-linacs and related technical and clinical aspects different to conventional radiotherapy, will be discussed and illustrated with examples from the Australian MRI-linac commissioning process.

IOP Conf. Series: Journal of Physics: Conf. Series **1305** (2019) 012010 doi:10.1088/1742-6596/1305/1/012010

### *4.1. Dose measurements*

*4.1.1. Detector response.* Detector response to radiation may be modified by different factors depending on the principle of operation of the dosimeter. Ideally, detectors used in MRI-linac dosimetry should have no or a well-characterised dependency of response on these factors and their linearity and reproducibility should not be affected. While the most appropriate detectors for dosimetry in high magnetic fields are still subject of research, investigations of the characteristics of standard detectors show that the change in reading can be in the order of several percent. The current state of knowledge on the use of various types of detectors in the magnetic field, grouped by their principle of operation, is reviewed below.

Several groups investigated the responses of common types of air-filled ion chambers in the presence of magnetic fields either experimentally or by employing Monte Carlo (MC) simulations. Negligible response changes of about 1–2% have been reported in inline magnetic fields of up to 2 T [23,24]. Both over- and under-responses, peaking at ca. 10% at field strengths of about 1 T, have been observed in transverse fields [23,25-26]. It should be noted that in these studies long axis of the ion chambers was typically oriented perpendicular to the magnetic field. O'Brien *et al* [27] described a difference between two possible perpendicular orientations, resulting from the electrons generated in the chamber stem being curved either towards or away from the sensitive volume. Additionally, they showed that the response change for chambers positioned parallel to the field is less than 1%. These observations were confirmed in a systematic MC study by Reynolds *et al* [28].

Recognising the fact that these effects should be included in the formalism, several authors, experimentally or using MC, derived correction factors ($k_B$), rather than of response changes reported earlier, for a number of standard cylindrical [27,29-32] and parallel-plate chambers [31]. Parallel-plate chambers were shown to require larger $k_B$ values (up to 8.9%) in both the parallel and perpendicular orientations to the magnetic field and to have larger angular sensitivity [31]. At the Australian MRI-linac a unique approach for $k_B$ measurements is being pursued, which exploits the linac-on-rail system enabling dosimetric measurements at different magnetic field strengths including ~0 T [33]. It should be noted that the $k_B$ factors obtained in these studies are contingent on the definitions introduced by the formalism and that they were obtained mostly for transverse MRI-linacs. Finally, it should be mentioned that recently a mobile water calorimeter has been developed and characterised for use in MRI-linacs allowing the direct determination of the correction factors [34].

Reynolds *et al* [35] simulated the response of two detectors relying on ionisation in solids, diamond detector PTW60003 (PTW, Freiburg, Germany) and PFD diode (IBA Dosimetry, Schwarzenbruck, Germany), in various orientations in transverse and inline magnetic fields up to 1.5 T. Similar to the case of ion chambers, the response change was more pronounced in transverse (up to ~20% at 1.5 T) as compared to in inline magnetic field (<2% at 1.5 T). In transverse fields however, unlike for ion chambers, the change in response was monotonically increasing in the examined range field strengths. The effect was strongly dependent on the detector rotation around the beam and magnetic field directions and it was most pronounced with the detector axis perpendicular to the beam. In a follow up paper the same group confirmed some of these results experimentally in inline magnetic fields up to 0.2 T [24].

Among the dosimeters relying on chemical transitions induced by radiation, two common types are radiochromic and polymer materials. Dose under-responses were reported for Gafchromic EBT2 films (Ashland Advanced Materials, Bridgewater NJ, USA) exposed to 0.35 T [36] and 1.5 T fields [37,38]. These effects have been attributed to the occurrence of magnetokinetic changes in polymer orientation observed microscopically [36, 38]. Additional changes in response were observed for EBT2 films that underwent concurrent imaging and correlated with the radiofrequency exposure and associated temperature increase [38]. However, two conference contributions on similar experiments with EBT3 films reported no changes in dose response in the presence of 0.35 T [39] and 1.5 T fields [40]. Only minimal effect of magnetic field on dose response (up to 5%) and its linearity (up to 1.6%) were observed for three types of 3D chemical dosimeters, each using a different response mechanism: radiochromic plastic (PRESAGE, in-house made) in 1.5 T field, radiochromic gel (FOX, in-house made) in 1.5 T field and polymer gel (BANG, MGS Research, Madison, CT, USA) in 1 T field [41].

Wen *et al* [40] and Wang *et al* [42] reported no response modification for thermoluminescence detectors (TLDs) in a 1.5 T transverse field. Spindeldreier *et al* [43] observed a relatively small dose response decrease with increasing magnetic (-1.3%/T) and no influence on the angular dependence for optically stimulated luminescent dosimeters (OSLDs). Therefore, luminescence based detectors are considered promising candidate for dosimetry in magnetic fields.

Plastic scintillation detectors (PSDs) are non-magnetic and have density similar to water, which potentially renders their dose response independent of magnetic field. Stefanowicz *et al* [44] investigated this in fields up to 1 T and observed unexplained response increase by up to 7%. More recently, Therriault-Proulx *et al* [45] reported 2.4% dose response increase in fields up to 1.5 T, which they interpreted as the change in the inherent dose deposition within the phantom, and concluded that PSDs show great promise for use in magnetic fields. However, other works suggest that optimal noise-subtraction methods for PSDs are still to be established [46].

*4.1.2. Air gaps.* Air gaps might be present between the dosimeter and the surrounding material e.g. when using plastic phantoms or even in water phantoms when using non-waterproof ion chambers with sleeves. It has been demonstrated experimentally [47] and through MC simulations [48] that in a 1.5 T transverse magnetic field this may lead to the reduction of a Farmer-type chamber response by 0.7–1.2% in a standard solid water phantom. This effect was attributed to a reduction in ionisation produced in the immediate vicinity of the chamber wall rather than to ERE [48]. Agnew *et al* [49] investigated this systematically in a phantom with a recess manufactured within the chamber holder allowing the placement of the air gap at a controlled angle relative to the radiation beam. They observed variability of the Farmer-type chamber response, depending on the air bubble position, of 0.25% at 0.25 T and 3.8% at 1.5 T. To the best of the authors' knowledge, these effects have not been investigated for inline MRI-linacs. Filling the chamber holder with water has been proposed as a method to mitigate air gap effects. However, it is difficult to apply in transverse MRI-linacs, where chamber orientation parallel to the magnetic field is preferred in order to minimise the dose response perturbation [27,31].

Small air volumes may also be present in detector arrays. This was investigated for a monolithic silicon array detector, Magic Plate (M512), developed at the University of Wollongong. A small air gap above the silicon chip in this device compensates for different scattering and electron spectra in silicon compared to water and was optimised in absence of magnetic field. Measured response changes of 2% in an inline and of 10% in a transverse ~1 T field as well as profile shift and penumbra width overestimation in MC simulations have been attributed to the presence of this gap [50].

*4.1.3. Absolute dosimetry.* For the MRgRT systems employing a low magnetic field strength, reference dosimetry based on TG-51 code of practice has been shown feasible [51]. However, given the significant response changes in higher magnetic fields for air-filled ion chambers [23-32] commonly used for reference dosimetry, the need to introduce a new correction factor and formalism has been recognised [32]. First approaches defined it as a ratio of the measured signal of the chamber without the magnetic field to that with the magnetic field present under the same conditions [26], however, several aspects hinder such a simple expansion of existing codes of practice.

Due to the construction characteristics of MRI-linacs (see Table 1) the reference conditions (e.g. specific SSD or field size) cannot be always fulfilled, requiring extension of the formalism [27]. The effect of the magnetic field on the dose distribution may change the relationship between the beam quality specifiers and the Spencer-Attix water-to-air restricted stopping power ratios for which they are surrogates [27,32]. Finally, it had been suggested that the correction for the change in reading of the ion chamber should be separated from the change in dose distribution [30].

Under these considerations, several methods have been proposed to include the response modification in the formalism for reference dosimetry [27,30-31], however the establishment of the code of practice, in order to warrant the same level of uncertainty as for conventional radiotherapy modalities, is still among the open issues in MRgRT. It should be complemented by a systematic characterisation of various detectors in terms of their response in the magnetic field and by recommendations on their suitability for different dosimetry tasks. To this end, a standardised experimental methodology and/or benchmarks of the MC codes for radiation transport simulations used in detector response modelling are required.

*4.1.4. Relative dosimetry.* Constancy of detector response with the off-axis position, depth or field size is a prerequisite in relative dosimetry.

Many ion chambers were shown to exhibit stronger angular response dependence in the magnetic field [23-31]. Through MC simulations Reynolds *et al* [35] showed response changes of 10-20% for the PTW0003 diamond and the PFD diode detector in a water tank at the periphery of the beam, when oriented perpendicular to a transverse magnetic field of 0.5 T. The response variation was much smaller with detectors positioned parallel to a transvers magnetic field and in an inline magnetic field. Woodings *et al* [52] observed increase of the angular dependence for the diamond detector (PTW60019) in a transverse 1.5 T field, which was deemed relevant for relative dosimetry at distant off-axis positions or at different gantry angles. At the Australian MRI-linac the PTW60019 detector has been recently used for percentage depth dose (PDDs) profiles measurements at near zero and 1.0 T fields [53]. Furthermore, MO*Skin*, MOSFET type dosimeters with the effective point of measurement at 70 μm (as recommended by ICRP for skin dosimetry) developed at the University of Wollongong, are currently employed for acquisition of high-resolution PDDs.

Different orientation of individual detectors in detector arrays leads to non-negligible artefacts in profile measurements [54] for transverse MRI-linacs. These effects have not been observed for the Australian MRI-linac employing an inline configuration [55].

The suitability of various detectors for specific relative dosimetry tasks in a transverse magnetic field of 1.5 T has been investigated recently by O'Brien *et al* [56]. They observed discrepancies between the results obtained with ion chambers of various volumes and with solid state detectors for PDD and output factor measurements for small fields and in the estimation of the lateral shift in the dose distribution. The latter has been simulated using MC and attributed to the detector density; the detectors containing dense materials dampen the magnetic field effects while the low density detectors accentuate it [57]. Finally, the effective point of measurement (EPOM) for air-filled ion chambers has been shown to change in the magnetic field, whereas for solid state detectors (diamond, diode) it remains unaffected [56].

Last but not least, the risk of detector response perturbation during beam scanning, due to motion in the magnetic field, has been considered, however no such effects have been observed for ion chambers [58] and for diamond detector [52].

*4.2. Hybrid tests*

The hybrid character of the MRI-linac treatment unit requires the assessment of its concurrent functionalities, for instance dose deposition during imaging, congruence of the imaging and radiation isocentres and end-to-end testing. The issues relevant to the verification of the dose deposition during imaging [6] have been addressed in other sections. In the context of non-dosimetric hybrid tests, there is a need for development and/or characterisation of dual MR/CT-visible synthetic tissue equivalent materials [59]. It is also interesting to note that even alignment tests pose additional challenges when executed on an MRI-linac. For example, in transverse MRI-linacs the result of the star shot test is altered by the magnetic field, due to the lateral shift of the deposited dose. To overcome this, QA phantoms using electron dense materials that reduce the magnetic field effect on the dose deposition [60] and the use of polymer gels to visualize the intersecting beams with MR [61] have been proposed.

### *4.3. Technical aspects*

*4.3.1. Magnetic field compatibility.* The foremost technical requirement for the dosimetric equipment (e.g. phantoms, detectors) to be used in MRI-linacs is compatibility with the magnetic field. The two main compatibility issues are the safety and image quality aspects due to the presence of ferrous materials or unscreened mechanical or electrical components respectively.

Many authors reported on the use of standard point dose detectors (ion chambers, diodes, etc.) in the magnetic field [23,25-27,30,32,33,52], however some vendors introduced additional MR safety certification for these models. Other types of equipment exist now in MR-conditional versions, modified for use in MRI-linacs, perhaps most notably the detector arrays like Starcheck maxi MR and Octavius Detector 1500 MR (PTW, Freiburg, Germany), ArcCHECK-MR and IC PROFILER-MR (Sun Nuclear Corporation, Melbourne, FL USA) or MR-Delta4 (Scandidos AB, Uppsala Sweden). It should be underlined that these devices still require characterisation of their dose response in the magnetic field (discussed in section *4.1.1*) [54,62,63].

In the context of magnetic field compatibility other equipment should also be mentioned. In-house designed or manual 1D tanks MR-compatible water tanks are currently used, while a motorised 3D plotting tank PTW-MP3T (PTW, Freiburg, Germany), developed in collaboration with the University Medical Centre Utrecht (Utrecht, the Netherlands) [58], is scheduled to be released in Q4/2018. It's worth mentioning that the scanning speed of this tank is lower than that of the conventional ones due to the use of different motors. Finally, MR-compatible motion phantoms e.g. MRI-LINAC Dynamic Phantom (CIRS, Norfolk, VA, USA) or QUASAR MRI-Compatible Respiratory Motion Phantom (Oncology Systems Limited, Shrewsbury, UK) are available, addressing the fact that treatment of moving targets is seen as one of the most promising applications of MR-linacs.

*4.3.2. Other compatibility aspects.* MRI-linacs have restrictions in terms of accessibility and space around isocentre when compared to conventional linacs. Most notably, the standard full-size water tanks cannot be used in the magnet bore. This necessitates the development of dedicated water tanks, like the above mentioned PTW-MP3T scheduled to be released with two vendor specific sizes and scanning ranges. Furthermore, not all MRI-systems feature a full set of isocentric lasers for positioning. Fiducial markers for alignment based on portal imaging or, ideally, MR-visibility are desirable in the design of dosimeters and of auxiliary dosimetric equipment [59,64]. Finally, concurrent dose measurement and MR image acquisition scenarios require detectors to cause minimal RF interference (e.g. from the power supplies and readout system) and image distortion.

### *4.4. Clinical aspects*

The greater targeting accuracy provided by MRI-Linacs lends their use in the most challenging cases involving small fields, steep dose gradients, non-uniform target doses, high fraction doses and employing dynamic delivery techniques, ultimately, with concurrent real-time tracking. Additionally, planned dose distributions must account for the effects of the magnetic field, like the dose perturbations at tissue interfaces, which may add to their complexity. These properties determine the requirements for dosimetry systems with special emphasis on high spatial resolution and volumetric information. Ion chambers, detector arrays or even films, commonly used for dose distribution verification in radiation therapy, may be insufficient for verification of dose distributions in MRgRT. One promising alternative are 3D gel or plastic dosimeters [65,66]. Some of these dosimeters are read-out using MR, which enables their use at MRI-linacs in the treatment set-up and even for real-time dosimetry if required [67].

Another aspect of MRgRT, which sets new requirements for dosimetry, is the online adaptive therapy workflow, where the patient remains on the table during the plan adjustment process. This has two consequences for the patient specific QA process: the original treatment plan may never be applied and the adapted plan cannot be measured before application. This may render obsolete the patient specific pre-treatment QA, but at the same time it increases the requirements for dose computation accuracy and prompts the need for online verification tools e.g. based on transient dosimetry using electronic portal

IOP Conf. Series: Journal of Physics: Conf. Series **1305** (2019) 012010 doi:10.1088/1742-6596/1305/1/012010

imaging devices (EPIDs). For application on MRI-linacs, EPID dosimetry requires some modifications to account for: the influence of the magnetic field on the functioning of EPID and on the dose deposition in the material traversed by the beam (patient or phantom) and, in some designs, for the presence of additional material in the beam. A back projection algorithm accounting for the latter has been recently developed for the Unity system [68]. Implementation of EPID dosimetry is ongoing within the Australian MRI-linac program.

## 5. Summary

Several areas have been reviewed, where improvements or new developments in dosimetry are needed to facilitate safe clinical implementation and to support further development of MRgRT. With the expected rapid adoption of MRI-linacs, the need for a new formalism for dosimetry protocols and for recommendations on the suitability of various detectors for different dosimetry tasks was discussed. The limitations of current dosimeters were indicated and several possible developments that will benefit MRg were highlighted, such as validation and further optimisation of 3D dosimeters, dosimetric benchmarking of MC radiation transport codes for independent dose calculation and implementation of EPID dosimetry in the presence of magnetic fields. Finally, the need for the development of phantoms specific to the hybrid nature of the MRI-linac system was underlined.

## 6. Acknowledgments

This work has been funded by the Australian NHMRC Research Council Program Grant APP1132471. JB has received a SWSLHD Early Career Researchers Program grant and an Australian Government Research Training Program Scholarship.

## 7. References

[1] Lagendijk JJW *et al* 2008 *Radiother. Oncol.* **86** 25-9
[2] Liney GP *et al* 2018 *Clin. Oncol.* [in press]
[3] Fallone G *et al* 2009 *Med. Phys.* **36** 2084-8
[4] Mutic S *et al* 2014 *Semin. Radiat. Oncol.* **24** 196-9
[5] Raaymakers BW *et al* 2017 *Phys. Med. Biol.* **62** L41-50
[6] Liney GP *et al* 2018 *Phys. Med. Biol.* **63** 135005
[7] MRIdian Linac Advantage brochure L-0083 Rev E, retrieved from https://viewray.worldsecuresystems.com/brochure on 3-08-2018
[8] Keyvanloo A *et al* 2012 *Med. Phys.* **39** 6509-21
[9] Woodings SJ *et al* 2018 *Phys. Med. Biol.* **63** 085015
[10] http://www.magnettx.com retrieved on 3-08-2018
[11] Private communication Prof. G.Fallone
[12] https://mrrt.elekta.com retrieved on 3-08-2018
[13] Raaymakers BW *et al* 2004 *Phys. Med. Biol.* **49** 4109-18
[14] Raaijmakers AJE *et al* 2005 *Phys. Med. Biol.* **50** 1363-76
[15] Hackett SL *et al* 2018 *Phys. Med. Biol.* **63** 095001
[16] Bielajew A 1993 *Med. Phys.* **20** 1171-9
[17] Oborn B *et al* 2012 *Med. Phys.* **39** 874-90
[18] Kirkby *et al* 2008 *Med. Phys.* **35** 1019-27
[19] van Heijst TC *et al* 2013 *Phys. Med. Biol.* **58** 5917-30
[20] Chen X *et al* 2016 *Med. Phys.* **43** 4797-802
[21] Oborn B *et al* 2016 *Med. Phys.* **43** 368-77
[22] Raaijmakers AJE *et al* 2007 *Phys. Med. Biol.* **52** 7045-54
[23] Reynolds M *et al* 2013 *Med. Phys.* **40** 042102-1 042102-7
[24] Reynolds M *et al* 2015 *Med. Phys.* **42** 2837-40
[25] Meijsing I *et al* 2009 *Phys. Med. Biol.* **54** 2993-3002
[26] K Smit *et al* 2013 Phys. *Med. Biol.* **58** 5945-57

[27] O'Brien DJ *et al* 2016 *Med. Phys.* **43** 4915-2
[28] Reynolds M *et al* 2017 *Med. Phys.* **44** 4322-8
[29] Spindeldreier CK *et al* 2017 *Phys. Med. Biol.* **62** 6708-28
[30] van Asselen B *et al* 2018 *Phys. Med. Biol.* **63** 125008
[31] Malkov VN and Rogers DWO 2018 *Med. Phys.* **45** 908-25
[32] Pojtinger S *et al* 2018 *Phys. Med. Biol.* **63** 11NT03
[33] Begg J *et al* 2018 *6th MR in RT Symposium, Nicolai Church, Utrecht, The Netherlands, 30 Jun - 3 Jul 2018*
[34] de Prez L *et al* 2016 *Phys. Med. Biol.* **61** 5051
[35] Reynolds M *et al* 2014 *Med. Phys.* **41** 092103-1
[36] Reynoso FJ *et al* 2016 *Med. Phys.* **43** 6552-6
[37] Raaijmakers AJE *et al* 2007 *Phys. Med. Biol.* **52** 4283-91
[38] Reyhan ML *et al* 2015 *J Appl Clin Med Phys* **8** 325-32
[39] Barten DLJ *et al* 2017 *Radiother. Oncol.* **123** S115-6
[40] Wen Z *et al* 2016 *Med. Phys.* **43** 3632
[41] Lee H *et al* 2017 *Radiother. Oncol.* **125** 426-32
[42] Wang J *et al* 2016 *Med. Phys.* **43** 3873-4
[43] Spindeldreier CK *et al* 2017 *Radiat. Meas.* **106** 346-51
[44] Stefanowicz S *et al* 2013 *Radiat. Meas.* **56** 357-60
[45] Therriault-Proulx F *et al* 2018 *Radiat. Meas.* **116** 10-3
[46] Simiele E *et al* 2018 *6th MR in RT Symposium, Nicolai Church, Utrecht, The Netherlands, 30 Jun - 3 Jul 2018*
[47] Hackett SL *et al* 2016 *Med. Phys.* **43** 3961-8
[48] O'Brien DJ and Sawakuchi GO 2017 *Med. Phys.* **44** 3830-8
[49] Agnew J *et al* 2017 Phys. *Med. Biol.* **62** 1731-43
[50] Gargett M *et al* 2018 *Biomed. Phys. Eng. Express* **4** 035041
[51] Goddu S *et al* 2012 Med. Phys. **39** 3968
[52] Woodings S *et al* 2018 *Phys. Med. Biol.* **63** 05NT04
[53] Roberts N *et al* 2018 *6th MR in RT Symposium, Nicolai Church, Utrecht, The Netherlands, 30 Jun - 3 Jul 2018*
[54] Smit K *et al* 2014 *Phys. Med. Biol.* **59** 1845-1855
[55] Jelen U *et al* 2018 *6th MR in RT Symposium, Nicolai Church, Utrecht, The Netherlands, 30 Jun - 3 Jul 2018*
[56] O'Brien DJ *el al* 2018 *Med. Phys.* **45** 884-97
[57] Looe HK *et al* 2017 *Phys. Med. Biol.* **62** 5131-48
[58] Smit K *et al* 2014 *Phys. Med. Biol.* **59** 4099-109
[59] Steinmann A *et al* 2018 *Med Phys.* **45** 773-82
[60] van Zijp HM *at al* 2016 *Phys. Med. Biol.* **61** N50–9
[61] Dorsch S *et al* 2018 *Phys. Med. Biol.* **63** 11NT02
[62] Houweling A *et al* 2014 *Phys. Med. Biol.* **61** N80
[63] de Vries JHW *et al* 2018 *Phys. Med. Biol.* **63** 02NT02
[64] Alnaghy SJ *et al* 2016 *Australas. Phys. Eng. Sci. Med.* **39** 921-32
[65] Lee HJ *et al* 2017 *J. Phys.: Conf. Ser.* **847** 012057
[66] Costa F *et al* 2018 *Phys. Med. Biol.* **63** 05NT01
[67] Lee HJ *et al* 2018 *Phys. Med. Biol.* **63** 045021
[68] Torres-Xirau I *et al* 2017 *Phys. Med. Biol.* **62** 6322-40

IOP Conf. Series: Journal of Physics: Conf. Series **1305** (2019) 012011 doi:10.1088/1742-6596/1305/1/012011

# The MD Anderson experience with 3D dosimetry and an MR-linac

**Geoffrey S. Ibbott, Hannah J. Le and Yvonne Roe**
Department of Radiation Physics, UT MD Anderson Cancer Center, Houston, TX

E-mail: gibbott@mdanderson.org

**Abstract**. A recent extension of image guidance in radiation therapy has been brought about by the introduction of the MR-linac; a hybrid system comprising an MR imager and a medical linear accelerator. The University of Texas MD Anderson Cancer Center is one of seven institutes around the world that have collaborated with the manufacturer to develop this system and bring it into clinical use. In the process, a great deal has been learned about the influence of the magnetic field on radiation dose deposition and upon dosimetry systems. A number of dosimetry systems have been evaluated and issues affecting their performance have been investigated. The potential value of three-dimensional dosimetry systems has been explored.

## 1. Introduction

The benefits of visualizing a patient's anatomy while in the treatment position led to the development of image-guided radiation therapy (IGRT) [1]. As an improvement over contemporary x-ray based IGRT, the feasibility of using magnetic resonance imaging [MRI) for radiation therapy position verification is being investigated, leading to the development of MR-image guided radiation therapy (MR-IGRT) [2]. Compared to x-ray imaging, MRI does not use ionizing radiation to generate images, and thus removes the concern for harmful effects from the imaging dose. MRI also offers superior soft tissue contrast, allowing the physician to delineate the target volume and surrounding structures with less uncertainty and potentially smaller margins. MR imaging protocols exist that acquire images rapidly, requiring less time than a CBCT image, and offering the potential to acquire 4D images in multiple planes. Because there appears to be no biological interaction between MR imaging and the treatment radiation, it is possible to image the patient during treatment and to monitor the location of the target volume in real-time. Functional imaging capabilities of MRI also allow the assessment of treatment response, during and post-treatment. Such assessment can allow treatment to be adapted according to the radiation response, providing a more personalized treatment for each patient.

Several MR-IGRT treatment machines have been designed; one system is currently commercially available while a second has received the CE Mark and is being marketed in Europe and elsewhere. The Unity MR-linac which combines a 7 MV linear accelerator with a diagnostic-quality MRI scanner was developed as a collaboration between The University Medical Center Utrecht (The Netherlands) and comprises a 7-MV linac (Elekta AB, Stockholm, Sweden). mounted on a ring around a 1.5 T MRI system (Philips, Best, The Netherlands) MRI system [3]. This system is presently installed at several sites and has received the CE Mark; at this writing it continues to undergo pre-clinical evaluation in the US. The MRIdian system based on a 0.3 T MRI and cobalt sources, which have subsequently replaced by a linear accelerator, also is commercially available (ViewRay, Oakwood, OH) [4]. Other systems are in development in Canada by the Cross Cancer Institute [5] and the Ingham Institute in Australia [6].

Despite the substantial benefits, MR-IGRT faces several challenges before it is likely to reach its full potential [2]. A major concern is the change in the delivered dose distribution when strong (≥1 T) magnetic fields are used for MR imaging. In photon beam radiation therapy, dose is delivered by secondary electrons that carry the energy transferred from the incoming photons. Secondary electrons are mostly forward peaked and deposit energy downstream, as they travel deeper into the patient, in segments of straight lines. In a magnetic field, however, the trajectories of secondary electrons are curved by the Lorentz force, shortening the path. When a secondary electron is generated in a dense medium and travels into a less dense medium, the electron may be able to follow a circular path and return into the original medium. Due to the lack of electron equilibrium, dose is increased in the upstream medium near the interface. This phenomenon is referred to as the electron return effect (ERE) and has been shown to pose clinical concerns [7]. Figure 1 shows the trajectory of an electron traveling from one medium to the next; the path changes in the presence of the magnetic field. The figure clearly demonstrates the ERE. [8] In measurements with a PRESAGE® dosimeter and radiochromic film (Figure 2) the ERE is observed to enhance dose at the interface by a factor of 1.3 ~ 1.4, over about 1 cm range [9].

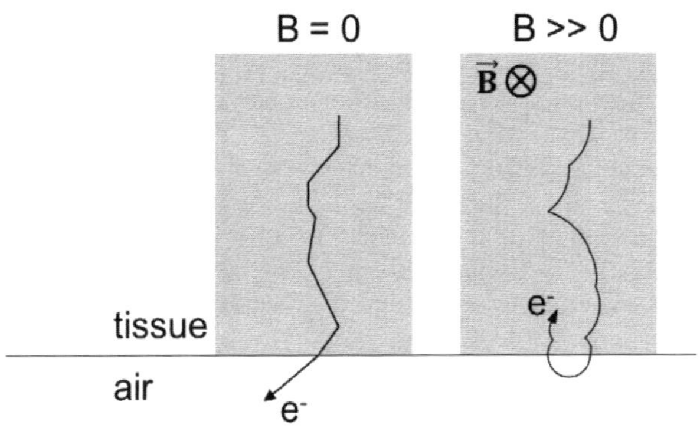


| E (MeV) | r (mm) |
|---|---|
| 0.5 | 2.2 |
| 1.0 | 3.4 |
| 2.0 | 5.6 |
| 4.0 | 10.0 |
| 6.0 | 14.5 |

**Figure 1.** (Left) Examples of the trajectory of a secondary electron traveling in zero (B=0) and strong (B >> 0) magnetic field. In a magnetic field, the electron may travel back to the original medium (redrawn after Raaijmakers, 2008) [8]. (Right) Radius of helical electron trajectory (r) as a function of electron energy (E) with B = 1.5 T (reproduced by permission) [8].

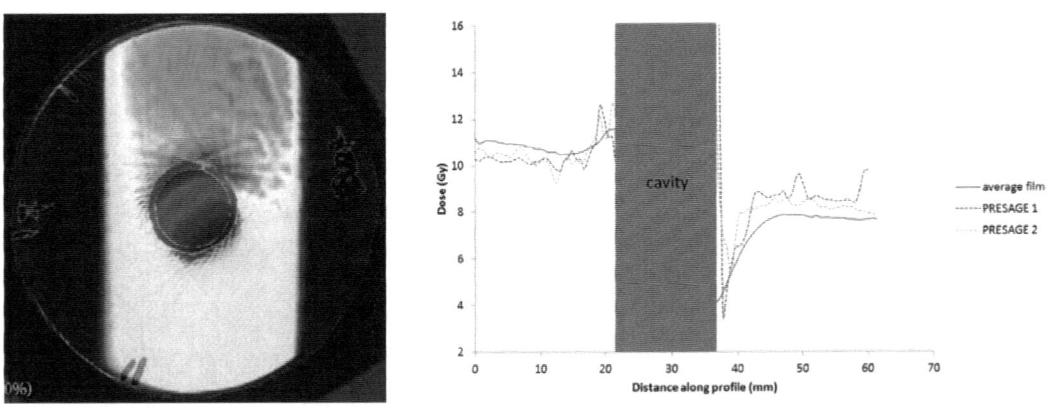


**Figure 2.** (Left) A PRESAGE®-type dosimeter containing an air cavity that has been irradiated with a single field. The ERE is demonstrated clearly. (Right) A comparison between profiles taken from two measurements with PRESAGE® and several with radiochromic film (reproduced by permission) [9].

The ERE can increase the dose at interfaces but the amount of increase depends on several factors. These include the media comprising the interface, the shape and orientation of the interface, the strength and direction of the magnetic field, and the energy of the beam. Some measures such as using opposing beams or multiple beam angles have been shown to reduce the dose enhancement due to the ERE in simple geometries [7, 10]. However, it remains a concern for the treatment of heterogeneous treatment sites and regions containing airways and air-filled cavities.

## 2. Inadequacies of point and planar detectors

Conventional quality assurance (QA) procedures in radiation therapy generally rely on point and planar dosimeters to compare the measured dose to the planned dose at a few points of interest and in a few planes. For example, the Imaging and Radiation Oncology Core-Houston (IROC-Houston, formerly the Radiological Physics Center) QA Center uses thermoluminescence dosimeters (TLDs) and radiochromic film to evaluate the dosimetric performance of institutions wishing to participate in national cooperative group clinical trials [11]. In 2010, IROC-Houston reported that 20-30 % of institutions participating in credentialing procedures had failed to deliver treatments that matched their own treatment plans within the IROC criteria. These data were measured using phantoms representing sites such as the head and neck, pelvis, spine, and lung [12]. A more recent report shows that even for irradiations performed since 2012, about 10% of irradiations did not meet the passing criteria [13]. It is of note that the IROC analysis evaluated the delivered dose in only two or three planes. Consequently, it is conceivable that a more comprehensive investigation would reveal either better or worse agreement between the plan and the delivered dose. Work in this area has demonstrated that, in some cases, a 3D evaluation identifies regions of disagreement that, because they appeared outside the plane of the 2D dosimeter, were not detected with film analysis [14, 15]. It is apparent that a more thorough means of measuring dose distributions is needed for MR-IGRT treatments, where the magnetic field changes the delivered dose distribution in a complicated fashion.

## 3. Opportunities offered by 3D dosimetry

In an effort to improve upon the sparse sampling offered by 0D and 2D dosimetry, three dimensional (3D) dosimeters were developed to provide volumetric dose information with high spatial resolution [16, 17]. Materials that have been investigated recently include polymerizing gels, radiochromic gels, and radiochromic plastic materials such as PRESAGE® (Heuris Pharma, Skillman, NJ, USA) [18]. Gel dosimeters can be analyzed with one of several read-out methods including MRI, x-ray CT, and optical-CT [19, 20]. Radiochromic plastics do not generate a signal that is visible with MR or x-ray CT, and therefore are customarily analyzed with optical CT methods [21]. The capacity of 3D dosimetry to measure and compare volumetric dose distributions offers this modality the potential to perform as a complement to, or a substitute for 0D and 2D dosimetry.

## 4. Dosimeters in magnetic fields

### *4.1. General concerns*

Before 3D dosimeter systems could be employed as viable QA tools in magnetic field environments, it was essential to first determine whether a magnetic field could affect the response of the dosimeters. There have been few investigations on how a strong magnetic field affects conventional dosimeters. A 2009 report described the magnetic field effects on a Farmer NE2571 ion chamber [22]. Two experiments were described: a GEANT4 Monte Carlo simulation and measurements with the Farmer NE2571 chamber in the magnetic field produced by an electromagnet. Depending on the orientation of the chamber and the strength of the magnetic field, the response of the ion chamber varied 10-15 %. More recent data have been reported that characterize a number of cylindrical ion chamber models in a 1.5 T magnetic field [23]. These authors showed that chamber design and construction played important roles in the magnitude of the effect. For some of the instruments evaluated, the orientation of the ion chamber axis to the magnetic field contributed a significant effect; the influence of the magnetic field was considerably greater when the chamber was perpendicular to the magnetic field. O'Brien and colleagues also have shown that even small air gaps around an ionization chamber altered the reading of the instrument, suggesting that the use of water-equivalent plastic phantoms could lead to measurement errors [23, 24].

The complications experienced with ion chambers largely can be minimized through selection of an appropriate chamber design, by avoiding air gaps and by taking care to align the chamber parallel to the magnetic field. Perhaps the best solution, at least for reference dosimetry, involves the use of a suitable ion chamber in a water phantom. However, the design of most MR-guided treatment units makes this inconvenient and potentially dangerous due to the high voltages in use.

For making relative measurements, motor-driven water phantoms are desirable, but most are unsuitable for use in a magnetic field. Consequently, a 3D volumetric dosimeter is attractive to avoid some of these complications. In addition, some 3D dosimeters such as Fricke gels exhibit a response that is measurable with MR imaging. This offers the very unusual opportunity to analyze the dosimeter with the MR component of the MR-IGRT system, immediately following irradiation with the treatment component of the system. A further benefit is the avoidance of moving the dosimeter from the treatment device to an imaging system. In fact, with the MR-linac, imaging is possible during irradiation, allowing for a real-time display of accumulated dose.

### *4.2. Influence of magnetic fields on dosimeter sensitivity*

The effect of a magnetic field on the sensitivity of a polyurethane-based radiochromic dosimeter similar to PRESAGE® was investigated using cuvettes that fit into a PMMA phantom that was designed to fit between the pole pieces of an electromagnet [25]. Doses of approximately 1, 4, 7 and 10 Gy were delivered, with either B = 0 or B = 1.5 T. The net OD change due to irradiation was calculated by taking the difference between the average OD of the irradiated cuvettes and that of unirradiated cuvettes that were handled identically. OD was determined using a GENESYS™ 10S UV-VIS spectrophotometer (Thermo Fisher Scientific, Waltham, MA, USA). ANOVA was performed in R-statistical software to test whether the magnetic field had a significant influence on the dose response curve. The measurements suggested an under-response of approximately 2% in the presence of the magnetic field. However, it was subsequently recognized that the dose delivered in a 1.5 T magnetic field per photon (and consequently, per monitor unit) is approximately 0.5% lower than in the absence of a magnetic field [23]. It is likely that what appeared to be an under-response of the dosimeter was actually a reflection of the lower delivered dose in the magnetic field. Despite the small decrease in response compared to measurements at B = 0 T, the response of the polyurethane radiochromic dosimeter at B = 1.5 T was strictly linear, with $R^2 > 0.99$.

### *4.3. The benefits of Fricke gel dosimeters in a magnetic field*

Preliminary work has been conducted by Lee et al to investigate the feasibility of using 3D Fricke-type gel dosimeters both for analysis of dose distributions and for real-time dose observations [26]. The Elekta Unity MR-Linac with a 1.5 T magnetic field was used for these studies. Fricke-type dosimeters were prepared in 97% w/w Milli-Q water with 3% w/w gelatin (300 Bloom), 1 mM ferrous ion, 0.05 mM xylenol orange, 50 mM sulfuric acid, and 1 mM sodium chloride. The dosimeters were prepared in a plastic flask approximately 8.5 cm in diameter and 6 cm in height. The dosimeters were stored at 4° C prior to irradiation and imaging.

To demonstrate the response post-irradiation as well as in real time, the dosimeters were irradiated in air, with a part of each dosimeter outside the treatment field to act as a reference. A pair of perpendicular fields was used, an "anterior" field delivered 10 Gy to the center of the dosimeter through the top of the dosimeter and a "lateral" field delivered 20 Gy through the side of the dosimeter. This arrangement constructed an overlapping region of high dose with regions of lower dose on either side. MR imaging was performed with the MR-Linac to observe the change in paramagnetic properties pre- and post-irradiation using a T1-weighted sequence of TR = 500 ms and TE = 20 ms. MRI during irradiation was done in the MRL using a balanced fast field echo sequence with TR = 5 ms and TE = 1.7 ms.

Lee et al. observed a significant increase in pixel value between un-irradiated and irradiated regions. The increase in pixel value and corresponding dose was also visible during irradiation. Figure 3 demonstrates this increase in pixel value with dose, indicating that the signal increases in a linear fashion. Visibly, the dosimeter underwent a color change from yellow to purple with the formation of the xylenol orange – ferric complex. Following irradiation, the dosimeter demonstrated the 3D dose distribution as indicated in Figure 4.

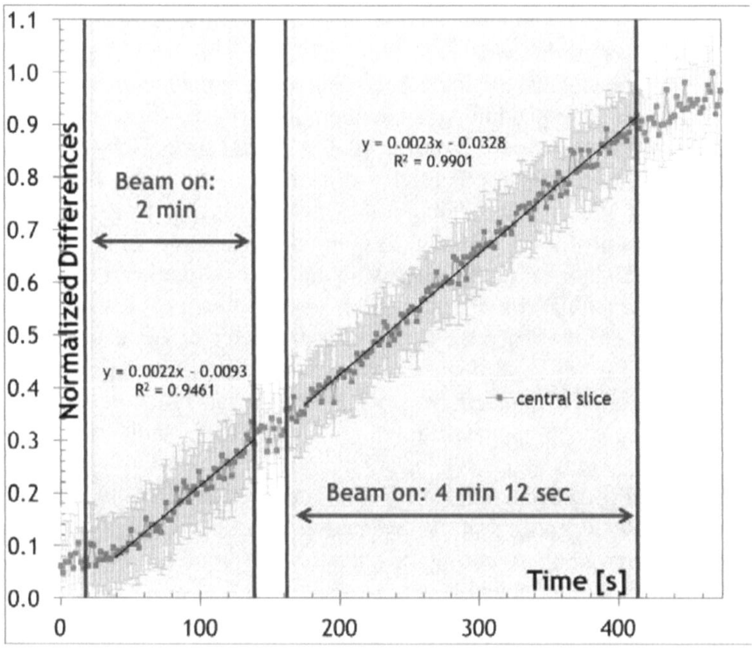


**Figure 3.** An example of the increase in MR signal of a gel dosimeter during irradiation [26].

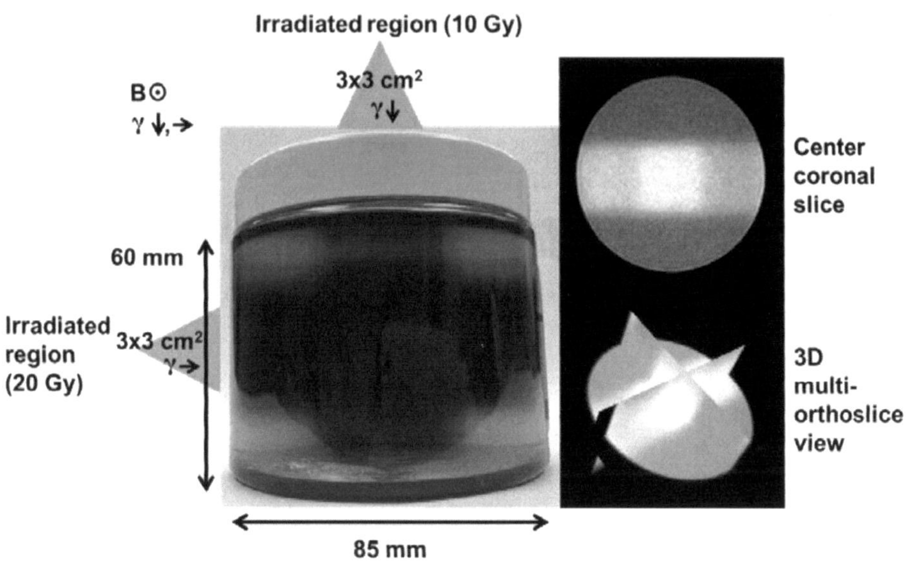


**Figure 4.** A gel dosimeter irradiated with two perpendicular beams. The MR images on the right demonstrate the 3D nature of the measurement [26].

Perhaps the most comprehensive investigation of 3D dosimetry for measurement of dose distributions from an MR-guided treatment unit was published by Rankine et al [27]. This study involved the use of a PRESAGE® dosimeter to evaluate the dose distributions from a ViewRay MR-cobalt system. The authors evaluated several simple dose distributions, and also several IMRT distributions selected from those recommended by the AAPM [28]. As has been shown above, the 3D dosimetry system demonstrated excellent agreement with other methods including treatment planning system calculations and ionization chamber measurements. This test was admittedly less challenging, given the low field strength of the magnetic field, but is still a good demonstration of the value of 3D dosimetry.

## 5. Array dosimeters in magnetic fields

Several 2D and quasi 3D array detectors have been evaluated in magnetic fields. Such detector arrays have benefits for routine QA procedures due to their large dimensions and the ability to provide information about the radiation output and beam profile in real time. However, such devices exhibit differences in their behavior in magnetic fields.

The Starcheck maxi® MR (PTW, Freiburg, Germany) was evaluated for its ability to display machine output and beam profiles. This device contains ion chambers arranged in orthogonal and diagonal linear rows, with 3 mm spacing. Measurements of output were generally consistent with measurements using conventional ion chambers in a water phantom. Measurements of beam profile demonstrated the influence of the magnetic field due to the detector geometry. Most of the chambers along the array's Y axis are surrounded by phantom material, except for the chambers close to the central axis which are surrounded by other chambers. When adjacent to another chamber, a chamber's response appears to be increased approximately 2% over that of chambers surrounded by phantom material.

Similarly, the IC Profiler-MR (Sun Nuclear, Melbourne, Florida, USA) likewise has been evaluated. It shows a similar response depending on the orientation of the ionization chambers. The response of the IC Profiler-MR is improved by executing the manufacturer's calibration procedure in the MR-linac beam, with the magnetic field energized.

Finally, the ArcCheck-MR (Sun Nuclear, Melbourne, Florida, USA) has been evaluated in a magnetic field. Unlike the two devices described above, the ArcCheck uses semiconductor detectors. No direct influence of the magnetic field on the detectors was observed, but issues arose with the calibration procedure. Sun Nuclear recommends irradiating the ArcCheck from a number of different orientations, several of which are either not achievable with the Elekta MR-linac or cause the beam to pass through high-atomic number structures, interfering with the measurement. A solution for this issue is presently being investigated.

## 6. Summary

The complex dose distributions produced by today's treatment equipment and delivery techniques require more advanced dosimetry systems to provide confidence that the delivered distribution is consistent with the planned distribution. 3D dosimetry techniques are valuable to enable acquisition of volumetric information with a single irradiation. The emerging field of MR image-guided radiotherapy requires the presence of strong magnetic fields that can affect the performance of most conventional dosimetry systems. However, several novel 3D dosimeters have been shown to perform well in the presence of magnetic fields and are shown to provide quantitative dose distributions in volumetric fashion. While the available data are preliminary, these results indicate the potential for 3D dosimeters, including both gels and radiochromic polyurethane, to provide reliable measurements in clinically relevant circumstances.

## 7. References

[1] Jaffray D A 2007 *Semin. Radiat. Oncol.* 17 243-4
[2] Ibbott G S *et al* 2018 Clinical 3D Dosimetry in Modern Radiation Therapy (Mijnheer B Ed). Boca Raton, FL: Taylor & Francis Group
[3] Raaymakers B W *et al* 2009 *Phys. Med. Biol.* 54 N229-37
[4] Mutic S and Dempsey J F 2014 *Semin. Radiat. Oncol.* 24 196-9
[5] Fallone B G *et al* 2009 *Med. Phys.* **36** 2084-8
[6] Keall P J *et al* 2014 *Semin. Radiat. Oncol.* **24** 203-6

[7] Raaijmakers A J *et al* 2005 *Phys. Med. Biol.* **50** 1363-76
[8] Raaijmakers A J E 2008 MR-guided radiotherapy: Magnetic field dose effects: The University of Utrecht
[9] Choi G *et al* 2016 *Med. Phys.* **43** 3856
[10] van Heijst T C *et al* 2013 *Phys. Med. Biol.* **58** 5917-30
[11] Ibbott G S et al 2008 *Int. J. Radiat. Oncol. Biol. Phys.* **71** S71-5
[12] Ibbott G 2010 *Med. Phys.* **37** 3334
[13] Molineu A *et al* 2014 *Med. Phys.* **41** 352
[14] Oldham M *et al* 2012 **84** 540-6
[15] Lafratta R *et al* 2015 *J. Phys.: Conf. Ser.* **573** 012055
[16] Mather M L *et al* 2003 *Phys. Med. Biol.* **48** N269-75
[17] Hurley C *et al* 2006 *Nucl. Instrum. Meth. A* **565** 801-11
[18] Guo P Y *et al* 2006 *Med. Phys.* **33** 1338-45
[19] Baldock C *et al* 2010 *Phys. Med. Biol.* **55** R1-63
[20] Baldock C 2009 *J. Phys.: Conf. Ser.* **164** 012002
[21] Jackson J *et al* 2015 *Phys. Med. Biol.* **60** 2217-30
[22] Meijsing I *et al* 2009 *Phys. Med. Biol.* **54** 2993-3002
[23] O'Brien D J *et al* 2016 *Med. Phys.* **43** 4915-27
[24] O'Brien DJ *et al* 2015 *Med. Phys.* **42** 3724
[25] Wen Z *et al* 2016 Med. Phys. **43** 3632
[26] Lee H *et al* 2016 *Med. Phys.* **43** 3660
[27] Rankine L J et al 2017 *Int. J. Radiat. Oncol. Biol. Phys.* **97** 1095-104
[28] Ezzell G A *et al* 2009 *Med. Phys.* **36** 5359-73

# Evaluation of a lung-equivalent gel dosimeter for MR image-guided radiation therapy

**BA McDonald[1,2], HJ Lee[1] and GS Ibbott[1]**
[1]Department of Radiation Physics, UT MD Anderson Cancer Center, Houston, TX
[2]The University of Texas MD Anderson Cancer Center UTHealth Graduate School of Biomedical Sciences, Houston, TX

Email: bmcdonald@mdanderson.org

**Abstract.** With recent advances in magnetic resonance image-guided radiation therapy (MR-IGRT), Fricke gel dosimetry has demonstrated value for its ability to measure three-dimensional dose distributions in the presence of a strong magnetic field. This strong magnetic field causes hot and cold spots in dose distributions at the interfaces of lung and normal tissue due to a phenomenon known as the electron return effect (ERE). In this paper, we report the development of lung-equivalent gel dosimeters to better measure dose to lung tissue caused by the ERE. Small polystyrene beads of variable sizes were mixed into Fricke xylenol orange gelatin (FXG) and ferrous oxide xylenol orange (FOX) gels. Lung-equivalence was confirmed by measuring the average CT number of each gel. The effects of gel type, bead size, and voxel size on uniformity and signal intensity were investigated. The smallest beads (<1 mm) exhibited the best uniformity, with values comparable to conventional gel with 2 mm voxels. Signal intensity followed an inverse relationship with uniformity, but FXG low-density dosimeters generated enough signal to produce acceptable quality images. The spin-lattice relaxation rate (R1 = 1/T1) increased with dose, which enabled us to measure dose to both soft tissue and lung due to the ERE using a phantom simulating the soft tissue-lung interface.

## 1. Introduction

In recent years, a 1.5 T magnetic resonance imaging (MRI) system has been integrated with a 7 MV linear accelerator to produce an MR-image guided radiation therapy (MR-IGRT) system known as the MR-linac (Elekta AB, Stockholm, Sweden) [1-2]. MRI offers several advantages compared to conventional on-board imaging techniques for IGRT such as orthogonal x-ray planar imaging and computed tomography (CT). These advantages include superior soft tissue contrast, improved verification of patient setup, reduced ionizing radiation dose to the patient, and the ability to generate highly conformal treatment plans through adaptive replanning.

In the MR-linac, the strong magnetic field ($B_0$-field) presents several dosimetric challenges due to the Lorentz forces acting on the secondary electrons. At interfaces between high- and low-density regions, a phenomenon known as the electron return effect (ERE) causes hot and cold spots in the dose distribution [3-4]. Secondary electrons entering a low-density region can be re-directed back toward the high-density region, causing more dose to be deposited in the denser tissue. The ERE is of particular concern in the thoracic region where there are interfaces between lung, airways, soft tissue, and bone.

As a result of the ERE and the highly conformal treatment plans that can be delivered on the MR-linac, there is a demand for a dosimetry system that can measure steep dose gradients in three dimensions. Conventional quality assurance (QA) tools can provide, at best, discrete point

IOP Conf. Series: Journal of Physics: Conf. Series **1305** (2019) 012012 doi:10.1088/1742-6596/1305/1/012012

measurements in a pseudo-3D array [5]. Gel dosimetry has been used as an alternative QA method due to its ability to measure continuous volumetric dose distributions [6]. Fricke-based gels are particularly useful for the MR-linac because the radiation-induced oxidation of $Fe^{2+}$ to $Fe^{3+}$ and the corresponding change in spin-lattice relaxation rate (R1 = 1/T1) is linear with dose and can be measured using MRI [7-9].

Conventional gel formulations are water-equivalent, so they are optimized for measuring dose to soft tissue but not to lung tissue. Previous studies have explored the development of lung-equivalent gels using two different methods to reduce the density of the gel [10-15]. One technique involves beating the gel to a foam-like consistency before it sets, and the second involves mixing gel with small polystyrene beads to simulate air pockets. In this study, we used the latter method with two Fricke-type gels to optimize lung-equivalent gels for use with MR-IGRT systems.

Two concerns with the low-density gel dosimeters are the reduced MRI signal intensity and the heterogeneity caused by the presence of the polystyrene beads. Polystyrene generates no MR signal [16], so the total MR signal in the low-density gel is reduced compared to conventional gel. Further, the heterogeneous composition of the low-density gels may interfere with the resolution at which absorbed dose can be measured. This investigation was an effort to mitigate these issues by changing the sizes of the polystyrene beads and imaging voxels.

## 2. Materials and Methods

### *2.1. Gel Preparation*

The two radiochromic gel formulations used in this study were Fricke xylenol orange gelatin (FXG) and ferrous oxide xylenol orange (FOX) [17-18]. FXG and FOX gels were formulated in-house using the following chemical components and concentrations: ~96% w/w deionized water, 4% w/w 300 bloom gelatin, 0.05 mM xylenol orange disodium salt, 50 mM sulfuric acid, and 1 mM ammonium iron(II) sulfate hexahydrate (FXG only) or 1 mM iron(II) oxide (FOX only). All chemicals were obtained from Sigma-Aldrich. For each batch, the gelatin was first mixed with water at 40°C to ensure complete dissolution, then cooled to 25°C prior to the addition of other components. The gels were refrigerated at 4°C for about 24 hours as they solidified, then they were stored at room temperature for about 3 hours prior to irradiation.

Each lung-equivalent gel was created by adding the liquid gel formulation into a cylindrical container (5.5 cm diameter, 11.5 cm height) filled with expanded polystyrene foam beads. Each container was filled completely with polystyrene beads prior to the addition of gel to ensure uniform mixing. Four different sized polystyrene beads were used (<1 mm, 2-3 mm, 2-4 mm, and 6-10 mm), and the conventional gel with no polystyrene beads (referred to as "gel-only") was used for comparison. Lung-equivalent dosimeters were created using both FXG and FOX.

### *2.2. CT Number and Spin-Lattice Relaxation (T1) Characteristics*

The average CT number (Hounsfield units (HU)) of each dosimeter was measured to determine lung tissue equivalence. The dosimeters were imaged using CT (Philips Brilliance Big Bore, 120 kVp, 300 mAs, $1\times1\times1$ mm$^3$ voxel size).

Gels were irradiated with a uniform $20\times20$ cm$^2$ field on a pre-clinical MR-linac such that the dose at the center of each gel was approximately 31 Gy. T1 maps were generated using fifteen inversion recovery images (MR-linac, Philips Turbo Spin Echo pulse sequence, TR/TE = 2500/20 ms, pixel size = $0.833\times0.833$ mm$^2$, slice thickness = 5 mm, number of signal averages (NSA) = 1, and inversion times ranging from 150–1400 ms). A pixel-wise T1 calculation was performed using in-house software, and the average T1 value of each gel was determined. Gels were imaged both pre- and immediately post-irradiation to avoid degradation of the dose map due to diffusion.

### *2.3. MR Signal Intensity and Uniformity*

The gels were imaged with a T1-weighted spin echo (SE) pulse sequence (TR/TE = 500/20 ms, NSA = 1) with 1×1×1 $mm^3$, 2×2×2 $mm^3$, 3×3×3 $mm^3$, and 5×5×5 $mm^3$ voxel sizes (henceforth referred to as 1 mm, 2 mm, 3 mm, and 5 mm voxel sizes). Uniformity was calculated from these T1-weighted images using the Uniformity Normalized Absolute Average Deviation (UNAAD) method [19-20], which is a pixel-wise measurement of the difference in each signal intensity value from the average. This method defines uniformity $U$ as

$$U = 100\left(1 - \frac{1}{N \cdot \bar{Y}}\sum_{i=1}^{N}|Y_i - \bar{Y}|\right)$$

where $\bar{Y}$ is the average pixel value within a region of interest (ROI), each $Y_i$ is the pixel value of a single pixel, and $N$ is the number of pixels in the ROI. The higher the value, the more uniform the image (maximum value of 100).

### *2.4. Electron Return Effect*

A heterogeneous phantom containing low-density gel surrounded by conventional gel was created to simulate the lung-soft tissue interface. The inner cylinder (5.5 cm diameter) contained FXG with 2-4 mm polystyrene beads. The outer cylinder (10 cm diameter) contained conventional FXG gel. The phantom was oriented in the MR-linac so that the axis of the cylinder was parallel to the $B_0$-field and perpendicular to the radiation beam. A uniform 16×22 $cm^2$ half-blocked field (gantry 90°) was delivered to the phantom such that the absorbed dose to water at the center of the phantom was approximately 23 Gy. The phantom was imaged using an inversion recovery pulse sequence to produce a T1 map as described in section 2.2.

## 3. Results and Discussion

### *3.1. CT Number and Spin-Lattice Relaxation (T1) Characteristics*

The CT numbers for the gel-only dosimeters were extremely close to that of water (0 HU) (Table 1). For each low-density dosimeter, CT numbers were comparable to those of lung tissue (~ -700 HU), which indicated lung tissue equivalence. FOX exhibited slightly lower CT numbers than FXG for each bead size. Standard deviations in CT number increased as bead size increased.

**Table 1.** Average CT numbers (HU) and average T1 values (ms) (pre- and post-irradiation) for the FXG and FOX low-density and conventional gel dosimeters.

| | FXG | | | FOX | | |
|---|---|---|---|---|---|---|
| Bead size | CT # | T1 (pre) | T1 (post) | CT # | T1 (pre) | T1 (post) |
| <1 mm | -596 ± 23 | 835 ± 10 | 578 ± 10 | -611 ± 31 | 999 ± 14 | 710 ± 11 |
| 2-3 mm | -603 ± 167 | 839 ± 6 | 590 ± 12 | -655 ± 127 | 1009 ± 10 | 714 ± 12 |
| 2-4 mm | -606 ± 228 | 845 ± 7 | 592 ± 13 | -635 ± 184 | 1014 ± 12 | 722 ± 15 |
| 6-10 mm | -559 ± 368 | 822 ± 109 | 584 ± 78 | -591 ± 354 | 970 ± 190 | 697 ± 126 |
| Gel-only | 15 ± 5 | 848 ± 2 | 608 ± 8 | 13 ± 6 | 1026 ± 6 | 769 ± 14 |

For both types of gel and both pre- and post-irradiation, the gel-only dosimeter had a slightly higher T1 value than each of the low-density dosimeters (Table 1). Polystyrene is not MR-visible, so it was expected that the presence of polystyrene would lower the average T1 values compared to gel alone [16]. All gels showed a decrease in T1 when irradiated, and the T1 value of each low-density dosimeter decreased by roughly the same amount as the gel-only dosimeter for each type of gel. The standard deviations in T1 were substantially higher for the 6-10 mm bead gels than for smaller beads and gel-only, which may translate to greater uncertainty in measured dose in future studies.

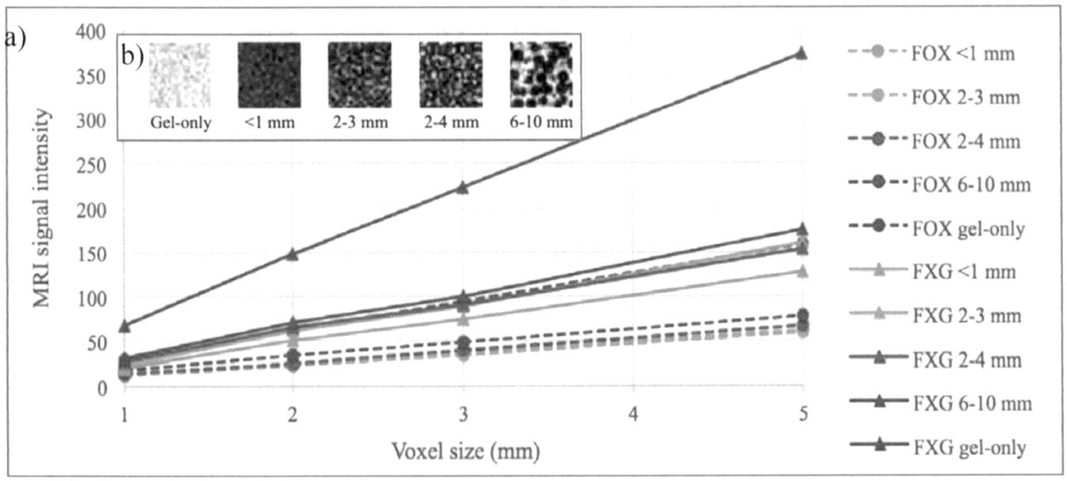


**Figure 1.** a) MRI signal intensity of each dosimeter as a function of voxel size. b) 40×40 mm$^2$ ROIs (T1-weighted SE, 1 mm voxels) of each FXG dosimeter are shown to illustrate signal intensity and uniformity.

### *3.2. MR Signal Intensity and Uniformity*

MR signal intensity increased linearly with voxel size for all gels, but the slopes of the signal intensities varied based on gel type and composition (Figure 1a). For each type of gel, the MR signal intensities of the low-density dosimeters were lower than that of the corresponding gel-only dosimeter. This is because the addition of the polystyrene beads reduces the amount of gel per unit volume, and the polystyrene produced no visible MR signal.

All FXG dosimeters produced a higher signal than their FOX counterparts, with the FXG low-density gels and the FOX low-density gels forming two distinct clusters on the graph (Figure 1a). This clustering demonstrated that the bead size had minimal impact on the signal intensity of the low-density dosimeters. Interestingly, the FOX gel-only curve fell within the low-density FXG cluster; the addition of polystyrene beads to FXG reduced the signal intensities to levels comparable with the FOX conventional gel. Considering that conventional FOX gel has been used successfully in dosimetry studies before [17-18], the signal intensity of low-density FXG is sufficient to produce acceptable quality images for dose analysis.

In general, uniformity improved with smaller bead size and larger voxel size (Figures 1b and 2). The signal uniformity for gels with 6-10 mm beads did not increase with voxel size because the beads were larger than all the voxel sizes used for imaging. For the three smallest bead sizes, uniformity increased most sharply between 1 mm and 3 mm voxel sizes, but there was little change beyond 3 mm. The smallest beads (<1 mm) performed the best, with uniformity values of FXG <1 mm comparable to those of FXG and FOX conventional gels at 2 mm voxel sizes and above. There was no consistent trend between uniformity and gel type, which suggested that bead size was the primary factor affecting uniformity.

### *3.3. Electron Return Effect*

A line profile was drawn through a cross-sectional slice of the dose distribution of the lung-soft tissue phantom (Figure 3). Within the low-density inner cylinder, dose generally decreased as the beam penetrated deeper into the phantom. Small fluctuations were observed that might have been reduced if a smaller bead size had been used. There was a region of enhanced dose just to the left of the inner cylinder and a region of reduced dose just to the right of the inner cylinder. The hot and cold spots were offset slightly with respect to the direction of the beam due to the curved path of the electrons in the magnetic field. These results are qualitatively consistent with previous Monte Carlo calculations and 3D dosimetry measurements of the ERE for phantoms with similar geometries [3, 17, 21]. Future work relating R1 to absorbed dose for the low-density gel would be required to determine the absolute percent change in dose.

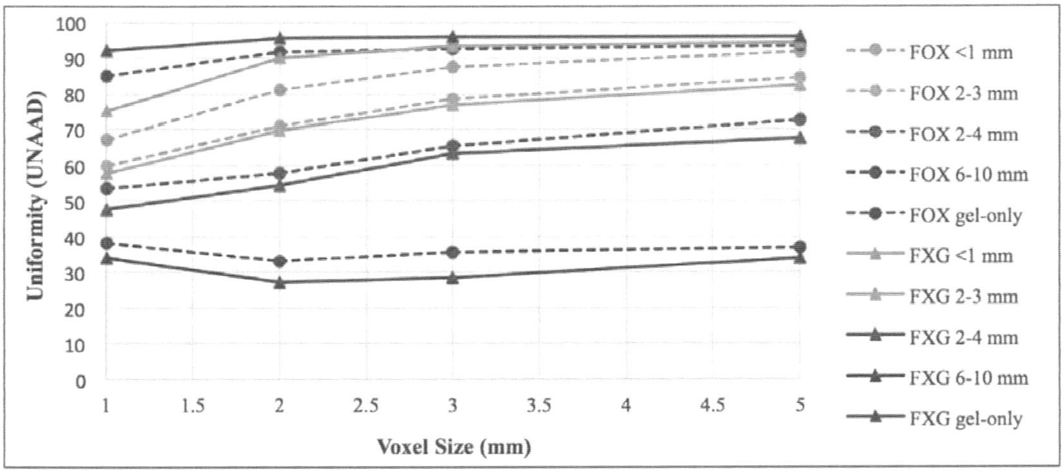


**Figure 2.** Signal uniformity of each dosimeter as a function of voxel size.

Previous phantom measurements of the ERE have used dosimeters composed of normal tissue-equivalent materials containing air cavities [17, 22], allowing only dose to normal tissue to be measured. Another study used a lung-equivalent material inside a cavity but used film and ion chambers to generate two-dimensional and point measurements, respectively [23]. Our phantom design allowed us to measure the volumetric dose distribution to both normal tissue and lung tissue, which presented a significant advantage for simulating clinically relevant scenarios.

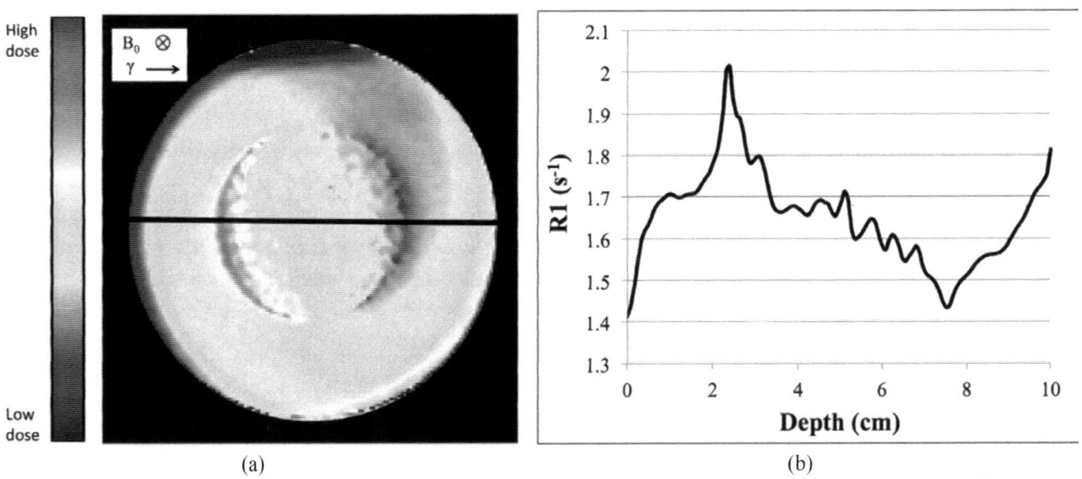


**Figure 3.** a) Cross-sectional slice of the R1 map of the lung-soft tissue phantom and b) line profile.

## 4. Conclusion

We demonstrated that low-density, lung-equivalent dosimeters can be created by adding small polystyrene spheres to FXG and FOX radiochromic gels. All low-density gels had lung-equivalent CT numbers, regardless of bead size or gel type. The addition of polystyrene beads did reduce the MR signal intensity compared to conventional gel, but FXG low-density gels produced signal intensities very similar to that of FOX conventional gel. Uniformity improved with smaller bead sizes and larger voxel sizes, with the FXG <1 mm bead dosimeters reaching the same uniformity levels as conventional gel with 2 mm voxels. Based on these results, the low-density formulation containing FXG with <1 mm beads will most likely produce the highest resolution dose maps in future work.

This study also demonstrated that irradiation caused the T1 values of the low-density gels to decrease by roughly the same amount as the T1 values of the corresponding conventional gels. Consequently, we were able to demonstrate that a volumetric dose distribution could be measured in a phantom containing both soft tissue- and lung-equivalent gels as a proof of concept. (Future studies would include calibrating the T1 values to absorbed dose for the low-density gels to extend our work to absolute dosimetry applications.) These results indicate that our low-density gel formulation can be used with conventional gel to measure the effects of the ERE on dose to lung and normal tissue.

## 5. References

[1] Lagendijk J J W *et al* 2008 *Radiother. Oncol.* **86** 25-9
[2] Raaymakers B W *et al* 2008 *Phys. Med. Biol.* **54** N229-37
[3] Raaijmakers A J E *et al* 2005 *Phys. Med. Biol.* **50** 1363-76
[4] Raaijmakers A J E *et al* 2008 *Phys. Med. Biol.* **53** 909-23
[5] Houweling A C *et al* 2016 *Phys. Med. Biol.* **61** N80-9
[6] Baldock C *et al* 2010 *Phys. Med. Biol.* **55** R1-63
[7] Fricke H and Morse M 1928 *J. Roentgenol. Radium Ther. Nucl. Med.* **18** 430-2
[8] Gore J C *et al* 1984 *Phys. Med. Biol.* **29** 1189-97
[9] Ibbott G S *et al* 2016 *AIP Conference Proceedings* **1747** 0 4002
[10] Ølberg S *et al* 2000 *Phys. Med. Biol.* **45** 2761-70
[11] Haraldsson P *et al* 2006 *Phys. Med. Biol.* **51** 919-28
[12] Claeys C *et al* 2006 *J. Phys.: Conf. Ser.* **56** 149-52
[13] De Deene Y *et al* 2006 *Med. Phys.* **33** 2586-97
[14] De Deene Y *et al* 2013 *J. Phys.: Conf. Ser.* **444** 012026
[15] Pak F et al 2016 *Int. J. Radiat. Res.* **14** 349-53
[16] Steinmann A *et al* 2017 *Med. Phys.* **45** 773-82
[17] Lee H J *et al* 2017 *J. Phys.: Conf. Ser.* **847** 012057
[18] Lee H J *et al* 2018 *Phys. Med. Biol.* **63** 045021
[19] National Electrical Manufacturers Association 2008 *NEMA Standards Publication MS 3-2008*
[20] Goerner F L *et al* 2013 *Med. Phys.* **40** 082302
[21] Lee H *et al* 2016 *Med. Phys.* **42** 3312
[22] Costa F *et al* 2018 *Phys. Med. Biol.* **63** 05NT01
[23] Wooten H *et al* 2014 *Med. Phys.* **41** 524-5

IOP Conf. Series: Journal of Physics: Conf. Series **1305** (2019) 012013 doi:10.1088/1742-6596/1305/1/012013

# MRI-based iPAGAT polymer gel dosimetry using fast recovery spin echo sequences

**K Fujino[1], K Ono[1], S Hayashi[2], K Sasaki[1], K Hioki[3], M Miyazawa[4], Y Akagi[1] and Y Hirokawa[1]**

[1]High-precision Radiotherapy Center, Hiroshima Heiwa Clinic, Hiroshima, 730-0856 Japan

[2]Department of Clinical Radiology, Hiroshima International University, Hiroshima 739-2695, Japan

[3]Division of Radiation Therapy, Hiroshima University Hospital, Hiroshima, 734-8551 Japan

[4]R-TECH. INC, Tokyo, 175-0092, Japan

E-mail: kfujino.2018@gmail.com

**Abstract.** In an estimation of iPAGAT polymer gel dosimeters using a magnetic resonance imaging (MRI) system, the dosimetric characteristics were investigated with two echo sequences: a spin echo (SE) and a fast recovery fast spin echo (FRFSE) sequence. FRFSE can shorten total scan time compared to SE because of its short repetition time (TR) and long echo train length (ETL). Although the R2-dose response measured from FRFSE was decreased compared to SE, both responses were fitted quadratic curve with a high correlation coefficient. In addition, their calibrated dose distributions showed high conformity to the planned data in the three-dimensional (3D) gamma index, dose volume histogram (DVH) and surface rendering analyses; however, results of SE were a little superior to FRFSE. In conclusion, it is suggested that FRFSE can present the accurate characteristics of polymer gel dosimeters under the optimum parameters.

## 1. Introduction

Polymer gel dosimeters consist of radiosensitive materials and form chemical polymerizations by irradiation [1-4]. A major advantage of these gel dosimeters is their ability to record three-dimensional (3D) dose distribution when compared to two-dimensional (2D) dose distribution [5] such as films. In the readout and dosimetric analysis, magnetic resonance imaging (MRI), especially a single spin echo (SE) or multi-echo spin echo (MSE) sequence, is regarded as the first choice [6]. However, because the SE sequence generally takes a long time to scan and often limits the use of gel dosimeters to a clinical setting, other sequences that are capable of taking images in a shorter time compared to the SE sequence, while also maintaining image quality, seem to be required. In contrast, a fast recovery fast spin echo (FRFSE) sequence can shorten total scan time because of its short repetition time (TR) and long echo train length (ETL). The purpose of this study is to compare the dosimetric characteristics between two MRI sequences, SE and FRFSE, using a customized PAGAT polymer gel dosimeter with improved dose sensitivity (iPAGAT).

## 2. Materials and methods

### *2.1. Preparation of iPAGAT*

A polyacrylamide-based gel dosimeter with improved dose sensitivity (iPAGAT) was prepared that we had developed in a previous study, containing 4.9% w/w gelatin, 2.9% w/w Acrylamide, 2.9% w/w *N,N'*-methylenebisacrylamide, 10 mM THPC, and 0.3 M magnesium chloride (2.5% w/w) as a sensitizer [7-10]. First, the gelatin was added to distilled water and the solution was heated to 50°C in a water bath. Next, the other reagents were added and mixed using a hot magnetic stirrer while maintaining at 40°C. Finally, the solutions were poured into two cubic acrylic containers (10 × 10 × 10 $cm^3$), cooled down, and stored in an incubator at 20°C until irradiation.

### *2.2. Irradiation*

A volumetric-modulated arc therapy (VMAT) plan with 6 MV photon beam and a single fraction of 2 Gy was created for the C-shape structure, in addition to a step-and-shoot intensity-modulated radiotherapy (IMRT) plan to obtain the dose calibration curve using the Eclipse treatment planning system (TPS). Both the VMAT and IMRT plans were irradiated to the iPAGAT dosimeters inserted into I'mRT Phantom (IBA Dosimetry, Germany) by the Novalis Tx linear accelerator (Varian/BrainLAB, Palo Alto, CA/Germany).

### *2.3. MRI readout*

Two days after irradiation, the iPAGAT dosimeters were read out by the 3.0 T MRI system (Discovery MR750, GE Healthcare, Waukesha, WI). To enable dosimeters to adjust to the MRI scanning room temperature (approximately 21.5°C), they were stored in the MRI scanning room for a period of several hours prior to the readout. R2 measurements were performed employing three different MR pulse sequences as dual-echo SE (SE2), dual-echo FRFSE (FRFSE2) and quad-echo FRFSE (FRFSE4). The SE2 parameters were: TR = 7000 ms, TE = 20 and 250 ms, slice thickness = 3 mm, FOV = 256 × 256 $mm^2$, matrix size = 256 × 256, NEX = 1, and scan time = 104 min. While the FRFSE2 parameters were: TR = 3000 ms, TE = 20 and 350 ms (the effective echo times were 22.176 and 349.272 ms), ETL = 128, slice thickness = 3 mm, FOV = 256 × 256 $mm^2$, matrix size = 256 × 256, NEX = 4, and scan time = 22 min. And the FRFSE4 parameters were: TR = 3000 ms, TE = 100-500 ms (the effective echo times were 99.568, 199.136, 298.704 and 497.84 ms), ETL = 128, slice thickness = 3 mm, FOV = 256 × 256 $mm^2$, matrix size = 256 × 256, NEX = 2.

### *2.4. 3D dose analysis*

First, an R2 map images with each sequence was created on both phantoms, where the R2 value is known as the spin-spin relaxation rate and is the inverse of T2 relaxation time. Regarding SE2 and FRFSE2, R2 values were calculated using signals measured with two different echoes. On the other hand, regarding FRFSE4, R2 values were calculated from linear regression with four different echoes.

Secondly, an R2-dose response curve with each sequence was investigated on the basis of the R2 map image of the calibration phantom. And then, for the purpose of adjusting its dose levels to the C-shape phantom, this curve was calibrated by two different R2 values which were selected manually within the lowest and highest irradiated areas of the C-shape phantom.

Finally, all R2 values of the C-shape phantom with each sequence were translated to suitable dose values with the calibrated dose-response curve, thereby a dose map of the C-shape phantom was created. After that, the dose map of each sequence was compared to the planned theoretical one, and a conformity of them was investigated using the three-dimensional (3D) gamma index, dose volume histogram (DVH), and surface rendering analyses. These analyses were all performed using in-house-developed software.

## 3. Results

The irradiated iPAGAT dosimeters clearly precipitated (Fig. 1). Within the calibration phantom, the theoretical dose values of 9 divisions were 16.2, 66.0, 111.2, 161.4, 212.6, 256.5, 300.6, 352.2 and 394.5 cGy in ascending order. On all MRI sequences, the R2-dose responses of the calibration phantom were fitted quadratic curve with a high correlation coefficient (SE2: 0.999, FRFSE2: 0.999 and FRFSE4: 0.999), and they were calibrated in an approximate dose range of the C-shape phantom and fitted likewise (SE2: 0.999, FRFSE2: 0.999 and FRFSE4: 0.999) (Fig. 2).

(a) (b)

**Figure 1.** Irradiated iPAGAT dosimeters: step-and-shoot IMRT for the dose-response curve (a) whose doses were 16.2~394.5cGy and VMAT for the C-shape while avoiding core structure (b).

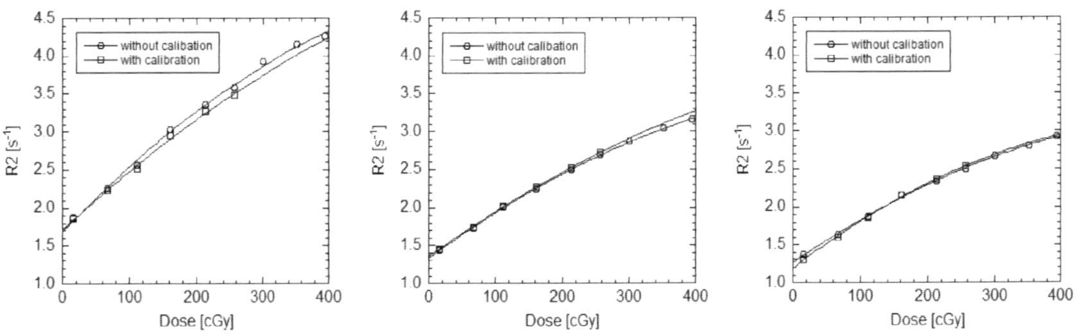


(a) (b) (c)

**Figure 2.** R2-dose response curves of the iPAGAT calibration dosimeter on each MRI sequence: SE2 (a), FRFSE2 (b) and FRFSE4 (c).

In regard to the dose maps of the C-shape phantom, a considerable conformity to the planned one was observed on all sequences, namely the pass rates of the 3D gamma index analysis using 3%/3 mm criteria in the axial plane achieved more than 99% (SE2), 93% (FRFSE2) and 99% (FRFSE4) (Fig. 3). With respect to similarity of the DVH and surface rendering analyses, there were no significant difference visually among three sequences (Fig. 4).

IOP Conf. Series: Journal of Physics: Conf. Series **1305** (2019) 012013 doi:10.1088/1742-6596/1305/1/012013

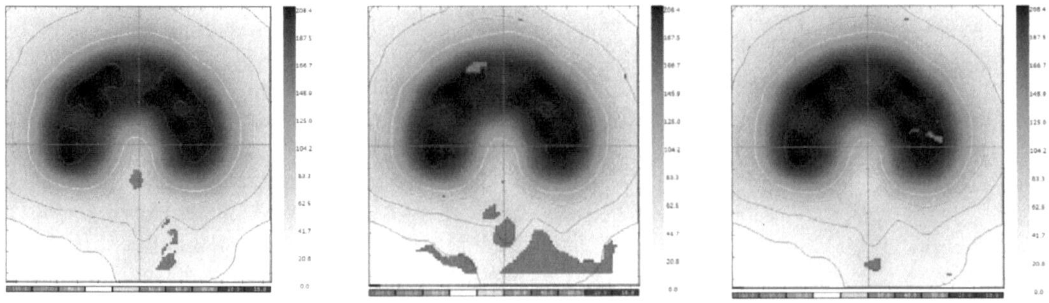

(a) (b) (c)

**Figure 3.** Axial images of the 3D gamma index analysis compared between the planned dose map and the measured one with each MRI sequence: SE2 (a), FRFSE2 (b) and FRFSE4 (c).

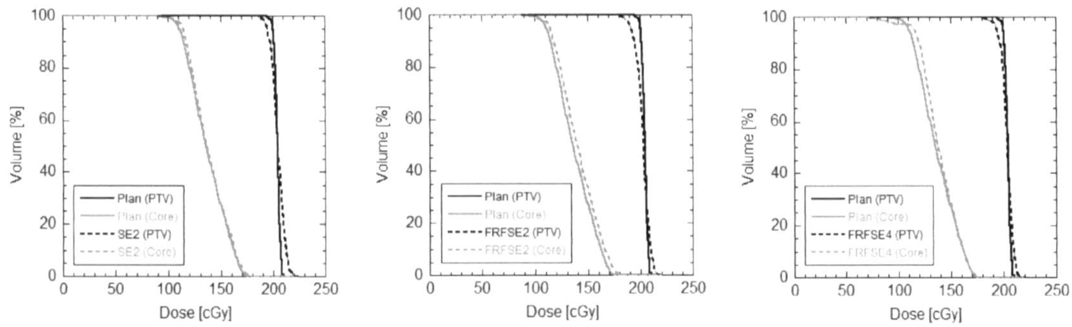


(a) (b) (c)

**Figure 4.** DVH curves for the C-shape using the planned data and the measured one with each MRI sequence: SE2 (a), FRFSE2 (b) and FRFSE4 (c).

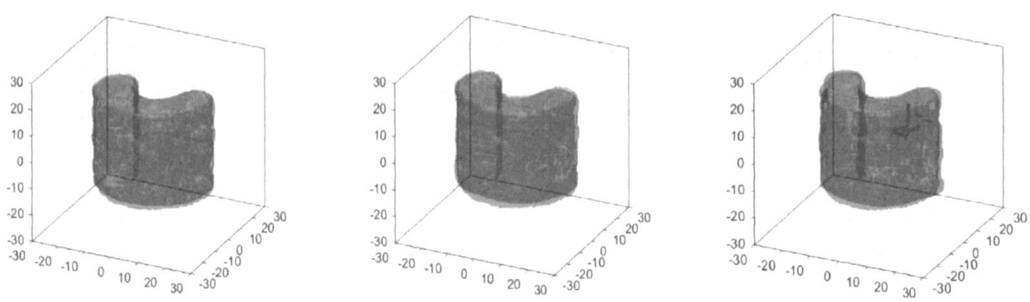

(a) (b) (c)

**Figure 5.** Surface rendering images of the 95% isodose volume for the C-shape using the planned data (red) and the measured one (blue): SE (a), FRFSE2 (b) and FRFSE4 (c).

## 4. Discussion

The readout technique of the uniquely recorded radiation dose in a polymer gel dosimeter using MRI was explored and well established in the previous study, and De Deene suggested that medical physicists should have an understanding of the basic principles of MRI, the optimization of quantitative imaging sequences, and protocols [11]. In this study, three MRI pulse sequences, namely dual-echo SE, dual-echo FRFSE and quad-echo FRFSE, were compared with the dosimetric characteristics of iPAGAT dosimeters. A foremost feature of a fast spin echo (FSE) sequence is that it can record plural different areas of k-space simultaneously because of its rather long ETL. In other words, the k-space is divided into several sections and the scan time is shorter depending on ETL compared to standard SE sequences. Furthermore, the FRFSE sequence is an enforced recovery of transverse magnetization that cannot be absolutely relaxed to longitudinal magnetization, carried out by a negative 90-degree pulse following FSE data acquisition. By means of the above process, the image contrast can be acquired by much shorter TR compared to FSE. Papoutsaki researched dosimetric characteristics using two pulse MESE sequences and a new multi-echo single-shot turbo spin echo pulse sequence (MEHASTE) for VIPET polymer gel dosimetry, and both techniques showed a linear R2-dose response. And also, MEHASTE sequences were significantly faster than MESE sequences due to their inherent design (higher ETL factor) [12].

Although the above-mentioned FRFSE sequence techniques are very useful in clinical due to their short scanning time, there are also several problems in relation to performing T2 quantitative evaluation. One of such problems is the MT (Magnetization Transfer) effect that leads to a low image contrast caused by the excitation of wide-range of frequency protons including bound water and off resonance by the 180-degree pulse of other slices while multi-slices are acquiring. Also, the T2 filter effect, that is caused by the existing different TE signal data in k-space depending on the increase of ETL, leads to a blurred image. Furthermore, though several factors may contribute to the bright fat phenomenon with FSE, the dominant mechanism is thought to be T2 prolongation secondary to the disruption of J-coupling interactions that normally take place between adjacent fat protons; therefore, the signal of the non-irradiated region in the polymer gel dosimeter is not decreased with the FSE sequence due to the shorter 180-degree pulse interval compared to factor J.

Actually, the measured absolute R2 values of FRFSE sequence are not adequate against the SE sequence by less than approximately 30%, and the contrast of FRFSE images are visually lower than that of SE. On the 3D gamma index analysis of the C-shape phantom, remarkable disagreement between the FRFSE2 dose distribution and the planned one is observed on the posterior region to the PTV of the C-shape structure in the axial plane. This disagreement is significantly improved on the FRFSE4 dose distribution and its pass rate reaches nearly equal to SE, while those in the coronal and sagittal plane is still a little inferior to SE. Considering the above limitations, these results should be improved by optimum MRI conditions in further studies. Regarding the results of the DVH and surface rendering analyses on FRFSE4 sequence, some irregularities in their shapes stand out against other sequences, so an improvement of the in-house software is thought to be required. In addition, De Deene also reported that a difference in the temperature history between the calibration vials and verification phantom pre-irradiation has an effect on R2-dose sensitivity [13]. In this study, we used the same calibration phantom and a simple normalizing method as the "two-point" calibration to overcome the difficulties. Eventually, though an analyzation of MRI with the FRFSE sequence is worthful in shortening its total time significantly, it seems to require more consideration on the comparison of 3D dosimetric characteristics against planned theoretical data.

## 5. Conclusion

In the MRI based iPAGAT polymer gel dosimetry with dose responses and distributions, the FRFSE sequence can shorten total scan time greatly compared with SE, so it has some possibility of rendering a clinical usage of polymer gel dosimeters more user-friendly. However, it has several issues of assuring image quality, and further improvement is required.

## 6. References

[1] Baldock C *et al* 2010 *Phys. Med. Biol.* **55** R1-63
[2] Hurley C *et al* 2006 *Nucl. Instrum. Meth. A* **565** 801-11
[3] Mather M L *et al* 2003 *Phys. Med. Biol.* **48** N269-75
[4] Baldock C 2009 *J. Phys.: Conf. Ser.* **164** 012002
[5] Vial P *et al* 2008 *Med. Phys.* **35** 4362-74

[6] Martin L 2006 *J. Phys: Conf Series* **56** 008, 86-94
[7] Hayashi S *et al* 2012 *Radiat. Phys. Chem.* **81** 884-888
[8] Hayashi *S et al* 2013 *J. Phys: Conf Series* **444** 012094 1-4
[9] Hayashi S *et al* 2018 *Radiol. Phys. Technol.* **11** (in press)
[10] Ono K *et al* 2015 *Med. Phys.* **42** 3406
[11] De Deene Y 2013 *J. Phys: Conf Series* **444** 012033, 1-9
[12] Papoutsaki M V *et al* 2013 *J. Phys: Conf Series* **444** 012088, 1-4
[13] De Deene Y *et al* 2007 *Phys. Med. Biol.* **52** 2719-2728

# Polymer gel dosimetry in the presence of a strong magnetic field

**Y Roed[1,2], L Pinsky[1] and G Ibbott[2]**
[1]Department of Physics, University of Houston, Science and Research Building 1, 3507 Cullen Blvd. Houston, TX 77204 USA
[2]Department of Radiation Physics, The University of Texas MD Anderson Cancer Center, 1400 Pressler St., Houston, TX 77030 USA

Email: gibbott@mdanderson.org

**Abstract**. The integration of magnetic resonance (MR) imagers with radiotherapy units provided a new opportunity to demonstrate the value of polymer gels as volumetric dosimeters. The purpose of this work was to investigate the use of methacrylic-acid based polymer gels for quality assurance of patient-specific treatment plans delivered with these novel treatment machines. The characterization of the gel was performed while the gel was subjected to a strong magnetic field and in the absence of the magnetic field. Additionally, an end-to-end phantom study was conducted using an MR image-guided radiotherapy (MR-IGRT) unit. This data will be used to support the implementation of volumetric dosimeters in MR-IGRT.

## 1. Background

Novel treatment modalities integrate magnetic resonance imaging (MRI) with a linear accelerator or with $^{60}Co$ sources for volumetric image guidance during radiation therapy. Several different designs of radiotherapy units integrated with magnetic resonance imaging have been introduced recently [1-4].

Using MRI for image guidance has advantages over conventional computed tomography (CT) or cone-beam CT: The patient is not exposed to any additional imaging dose during the course of the treatment; MRI provides superior image contrast allowing for a more accurate delineation of the tumor and adjacent organs; and it enables more accurate adaptation of the patient's treatment plan through the use of MR image acquisition for assessment of inter- and intra-fraction changes.

Lorentz forces are exerted on secondary electrons produced during photon interactions in the presence of a magnetic field and lead to a three-dimensional shift in delivered dose distributions in phantoms and patients. The magnetic field limits the use of, or requires modifications of electronic detectors used for quality assurance (QA) of the machines and for QA of patient-specific treatment plans. Several vendors provide MR-compatible electronic dosimeters that measure dose distributions at best in quasi-3D. Hence, 3D detectors such radiochromic plastics, radiochromic gels, and polymer gels have been suggested for MR image-guided radiotherapy (MR-IGRT) treatment plan verification [5].

Polymer gels have been used to measure complex dose distributions with steep dose gradients using an MR scanner for readout. The spin-spin relaxation rates ($R_2$) of the irradiated polymer gel change due to radiation-induced polymerization and increase monotonically with an increase in dose. These changes in $R_2$ can be measured with MRI [6-8] making polymer gels favorable detectors for QA of MR-IGRT units.

This paper presents two sets of data: the characterization of a polymer gel in the presence of a strong magnetic field and an end-to-end phantom study performed with an MR-IGRT treatment unit. Characterizing a polymer gel shed light into potential effects of the magnetic field during irradiation and the end-to-end phantom study was used to advocate the implementation of polymer gel dosimeters in MR-IGRT.

Experiments initially performed with methacrylic-acid based polymer gels (MAGAT) on an orthovoltage unit (Philips RT-250) found a dose-rate dependence of up to 30 % a low dose rates [9]. Both De Deene *et al* [10, 11] and Bayreder *et al* [12] performed studies that concluded a dose-rate dependence when methacrylic acid was the monomer source in the polymer gel dosimeters. Hence, before commencing studies with a different methacrylic-acid based polymer gel (BANG3-Pro) information was requested from the manufacturer (MGS Research, Madison, CT) who stated that this particular formulation was neither dose-rate dependent nor energy dependent [15]. This statement was based on the investigations performed by Massillon-JL *et al* [13] and Massillon *et al* [14] whose findings supported a dose-rate independence and a fractionation independence of this gel formulation.

## 2. Methods

Methacrylic-acid based polymer gel (BANG3-Pro) dosimeters were provided by the manufacturer (MGS Research, Madison, CT) in customized glass vials of 5 cm diameter and 4 cm height. The design enabled the initial characterization of the polymer gels using an electromagnet (GMW Dipole Electromagnet 3472-70, GMW Associates, San Carlos, California) installed in a vault that housed a cobalt unit (Theratron 780C, AECL, Chalk River, Canada) and subsequently in a vault that housed a Versa HD linear accelerator (Elekta AB, Stockholm, Sweden). These measurements were performed prior to the installation of an MR-linac (Elekta in collaboration with Philips Medical Systems, Best, The Netherlands); a 7 MV linear accelerator integrated with a 1.5 T MR scanner.

The dose response, dose-rate dependence, and fractionation dependence of the polymer gel were investigated in the cobalt beam and in the 6 MV photon beam from the Versa HD linac in the absence of a magnetic field. Dose response measurements were repeated inside a strong magnetic field using the electromagnet inside the linac vault. Following the installation of the MR-linac, the dose-rate dependence and fractionation dependence of the gel were evaluated with the MR-linac.

To assure full scatter conditions all glass vials were irradiated one at a time inside a full water-equivalent phantom at 10 cm depth. Twenty-four hours after exposure the irradiated gels were scanned with a GE 3 T MR scanner (GE Discovery MR750, Waukesha, Wisconsin). $R_2$ values were determined from the MR images. For end-to-end testing of the MR-image guided treatment process, a large, spherical dosimeter filled with the same polymer gel was positioned and irradiated on the MR-linac.

### *2.1. Dose response*

A dose response curve was acquired by placing five dosimeters inside a full phantom and irradiating them to 1, 3, 5, 7, and 9 Gy using a cobalt irradiator at a dose rate of approximately 64 cGy/min. The dose response measurements were repeated by delivering the same dose levels with Versa HD linac to five additional dosimeters while they were exposed to a 1.04 T magnetic field that was generated with an electromagnet. The electromagnet was positioned at a distance of 300 cm away from the radiation source while the gantry was rotated to 90 degrees to avoid any interference of the magnetic field with the linear accelerator.

Due to the differences in setup the dose response curves acquired with the cobalt irradiator ($B = 0$ T) and the Versa HD linac ($B = 1.04$ T) could not be directly compared. To confirm the response of the gel to 6 MV radiation two additional dosimeters were irradiated to 5 Gy and to 9 Gy after the magnetic field of the electromagnet was turned off.

$\Delta R_2$ values were obtained from MR images for all irradiated dosimeters one day after exposure and were plotted against the respective dose values. A linear fit was applied to the two dose response data sets when $B = 0$ T and the data set when $B = 1.04$ T. The slope of the linear fit obtained from two data

IOP Conf. Series: Journal of Physics: Conf. Series **1305** (2019) 012014 doi:10.1088/1742-6596/1305/1/012014

points with B = 0 T was compared to the slope of the linear fit obtained from five data points with B = 1.04 T.

### 2.2. *Dose-rate dependence*
A change in dose rate was achieved with the Versa HD by extending the distance between the dosimeter and the radiation source. Ten Gy were delivered to a dosimeter inside a full phantom at a source-to-surface distance (SSD) of 100 cm. A second dosimeter was exposed to the same dose after the setup was moved to an extended SSD of 290 cm. The dose rate at the extended SSD was 14% of the rate at the standard treatment distance. $\Delta R_2$ values were compared between the standard and extended SSDs.

Due to the fixed source-to-isocenter distance of the MR-linac the dose rate was changed at the console. Ten gray were delivered to a dosimeter at a dose rate of approximately 450 cGy/min (100 %) and to a second dosimeter at a dose rate of approximately 63 cGy/min (14 %). The lower dose rate was chosen for consistency with the irradiations with the Versa HD linac. The ΔR2 values from the irradiations with high and low dose rates were compared.

### 2.3. *Fractionation dependence*
A dose of 10 Gy was delivered to one gel dosimeter in a single fraction using the Elekta Versa HD. The same dose was delivered to another dosimeter in three fractions of 3.33 Gy each. The time between each fraction measured 9 minutes, to yield a total irradiation time of 21 minutes, for consistency with the low-dose rate irradiations. $\Delta R_2$ values were determined from MR scans a day after irradiation and compared between single fraction and multiple fractions.

The same dose delivery pattern was repeated using the MR-linac with an equal time between each fraction. Again, $\Delta R_2$ values from the single fraction dose delivery were compared to the multiple fraction dose delivery.

### 2.4. *End-to-end phantom study*
The workflow used in MR-IGRT was followed with the spherical gel dosimeter to perform a full end-to-end phantom study with the MR-linac. CT images of the dosimeter were used to create a reference treatment plan. The CT images were fused with MR images acquired with the MR-linac prior to irradiation to generate an adapted plan with the Monaco treatment planning system (research version 5.19.02). A dose of 7.5 Gy was delivered according to the AAPM TG-119 head and neck treatment plan [16]. The head and neck plan consisted of nine IMRT beams that were delivered in a step-and-shoot fashion. Dose constraints were set for structures drawn to represent a head and neck PTV, a spinal cord, and parotid glands. The dosimeter was MR imaged again for dose analysis 24 hours post-irradiation. A 3D gamma analysis was performed on dose maps using 3D Slicer with the SlicerRT extension [17]. IROC Houston has determined that a minimum of 85 % of all analysed pixels must pass 7 %/4 mm gamma criteria with a 10 % dose threshold for a head and neck phantom to pass the QA test [18]. The results of the 3D gamma analysis were compared with the IROC passing criteria.

## 3. Results and Discussion

### 3.1. *Dose response*
$R_2$ values exhibited a linear increase with dose when dose response measurements were performed with the cobalt irradiator. A straight-line fit resulted in a slope of 0.0104 $s^{-1}cGy^{-1}$ and an intercept of 6.830 $s^{-1}$. The standard deviation was calculated for each dose level and displayed as error bars in Figure 1 (A).

When irradiated with 6 MV photons inside a 1.04 T magnetic field, the dose response curve of the polymer gels appeared linear up to a dose of 9 Gy as the black diamonds indicate in Figure 1 (B). A linear fit provided a slope of 0.0144 $s^{-1}cGy^{-1}$ and an intercept of 9.612 $s^{-1}$. The red squares in Figure 1 (B) show the $R_2$ values measured at 5 Gy and at 9 Gy. Connecting the two points with a straight line was supported by the linearity of measurements performed with the cobalt irradiator. A linear fit generated a slope of 0.0144 $s^{-1}cGy^{-1}$ and an intercept of 9.514 $s^{-1}$.

$R_2$ values obtained at 5 Gy and at 9 Gy differed by 0.4 % and by 0.5 %, respectively, when comparing measurements with and without a magnetic field. This difference was smaller than the 2 % to 3 % uncertainty of $R_2$ values. Both dose response curves exhibited the same slope therefore, no apparent difference between the two dose response curves could be determined as. Any influence of the magnetic field on the dose response was smaller than the uncertainty in the measurements.

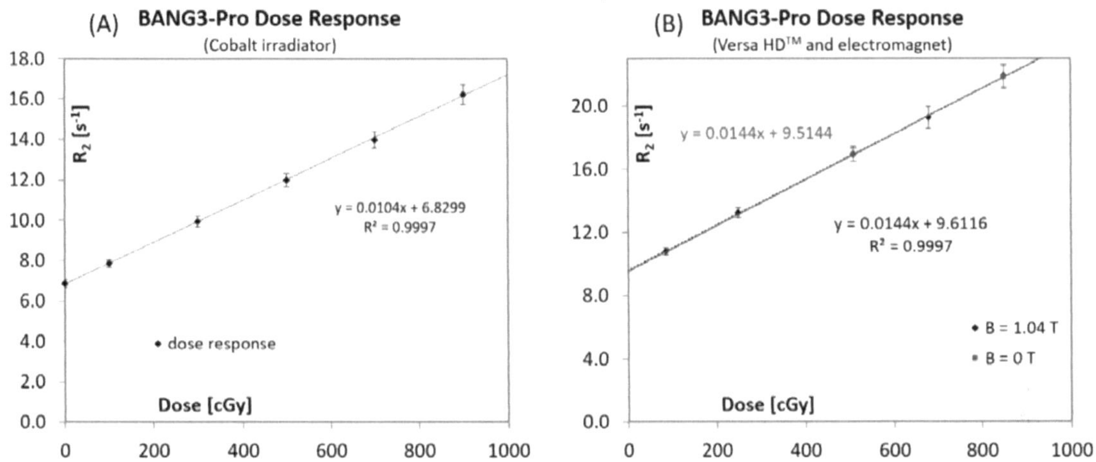


**Figure 1**. Dose response curves using (A) a cobalt irradiator and (B) a Versa HD linear accelerator and an electromagnet. Error bars represent one standard deviation.

### *3.2. Dose-rate dependence and fractionation dependence*

Table 1 shows that $\Delta R_2$ values measured at the reduced dose rate at extended distance were 26 % higher than the values at 100 cm SSD. A corresponding 86 % reduction in dose rate on the MR-linac resulted in a 24 % increase in $\Delta R_2$ values compared to delivering the same dose at 100 % dose rate. The difference of 2 % in measurements with and without a magnetic field fell within the uncertainty of $\Delta R_2$ measurements of 2 % to 3 %. No apparent effect of the magnetic field on the dose-rate dependence was detected.

The $\Delta R_2$ value at 10 Gy determined from the linear fit of the dose response in Figure 1 (A) was lower than a $\Delta R_2$ of 10.80 $s^{-1}$ measured with the 6 MV photon beam at 100 % dose rate. However, the reported dose-rate dependence suggests that the response measurements using the cobalt irradiator would yield higher $\Delta R_2$ values. This difference reflects the use of different bathes of BANG3-Pro gel through the course of this study.

**Table 1.** Dose-rate dependence measurements.

| | B = 0 T | | B = 1.5 T | |
|---|---|---|---|---|
| | dD/dt = 14 % | dD/dt = 100 % | dD/dt = 14 % | dD/dt = 100 % |
| $\Delta R_2$ [$s^{-1}$] | 13.63 | 10.80 | 12.30 | 9.93 |
| ratio $\Delta R_2$ | 1.26 | | 1.24 | |

Table 2 shows 8 % higher $\Delta R_2$ values when the dose was delivered in multiple fractions compared to when the same dose was delivered in a single fraction. When irradiated in a magnetic field, 8 % higher $\Delta R_2$ values were obtained when delivering the same dose in multiple fractions compared to a single fraction. No difference in ratios of $\Delta R_2$ values was determined when investigating fractionation dependence of the polymer gel in the presence and absence of the magnetic field. These results lead to the conclusion that the fractionation dependence was independent of magnetic field.

**Table 2.** Fractionation dependence measurements.

| | B = 0 T | | B = 1.5 T | |
|---|---|---|---|---|
| | multiple fx | single fx | multiple fx | single fx |
| $\Delta R_2$ [$s^{-1}$] | 11.72 | 10.80 | 10.72 | 9.93 |
| ratio $\Delta R_2$ | 1.08 | | 1.08 | |

Studies have indicated that dose-rate dependence and fractionation dependence were correlated [19]. Hence, it was not surprising that both a high dose-rate dependence and a significant fractionation dependence were observed with this polymer gel.

### *3.3. End-to-end phantom study*

Figure 2 (A) (left) shows the scale of the planned dose distribution that ranged from 0 Gy (shown in green) to 7.5 Gy (shown in red). The same scale was used for the gamma maps in Figure 2 (B) (right) ranging from 0 in green up to 1 in yellow and >1 in red. A black line was included on the gamma map to indicate the horseshoe-shaped PTV. The sparing of the spinal cord was visible from the horseshoe shape of the dose distribution seen on the axial views in Figure 2. The indentations in the dose distribution indicate the sparing of the parotid glands which was seen on coronal views (not shown). A gamma pass rate of 85.6 % was determined using the IROC head and neck 7 %/4 mm gamma criteria and a 10 % dose threshold.

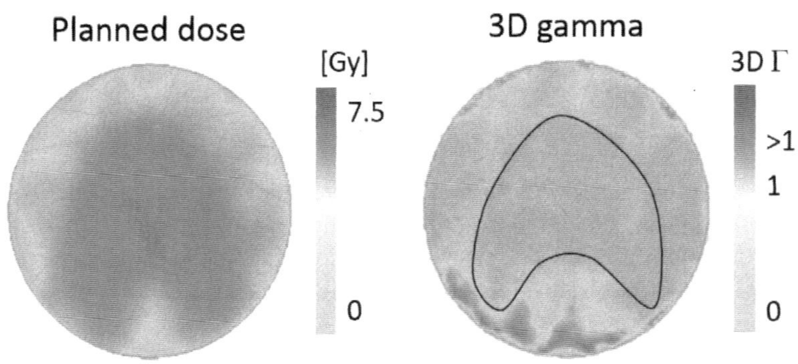


**Figure 2.** Representative axial slice of AAPM TG-119 head and neck plan: (A) planned dose on the left and (B) 3D gamma map on the right. Horseshoe-shaped black line added to indicate the PTV.

Several reasons were considered for the less than expected gamma pass rates. The gamma maps for each slice indicated that the gamma analysis failed in regions of beam entrance and exit inside the dosimeter. This could be due to the high dose-rate dependence. Also, the MR signal was scaled to the prescription dose using $T_2$-weighted MR images instead of $R_2$ maps.

## 4. Conclusion

For the first time, an MR-IGRT treatment system was used to deliver and analyze a dose distribution in a clinically relevant volume with a polymer gel dosimeter. The polymer gel with its current chemical composition worked well for visualizing the complex 3D dose distributions delivered with the MR-linac.

However, the strong dose-rate dependence and the non-negligible fractionation dependence rendered the gel unsuitable for clinical dosimetry and for quality assurance of patient-specific treatment plans.

The strong dose-rate dependence is of concern as IMRT and VMAT dose deliveries exhibit steep dose gradients and varying dose rates due to beam modulation and to differences in the depths of tissue penetrated during delivery. This dependence can lead to an overestimation of the signal in regions of lower dose rates. Furthermore, due to the fractionation dependence a further overestimation of the measured dose can occur when the treatment is heavily modulated or has to be interrupted. Therefore, minimizing these dependences is essential to perform accurate multi-beam or modulated treatment plan validation.

These results warrant the investigation of non-methacrylic-acid based polymer gels so that they can be used as clinical, 3D dosimeters for MR-IGRT.

## 5. References

[1] Raaymakers *et al* 2009 *Phys. Med. Biol.* **54** N229-237
[2] Mutic S and Dempsey J F 2014 *Seminars in Radiation Oncology* **24** 196-199
[3] Fallone B G *et al* 2009 *Med. Phys.* **36** 2084-88
[4] Keall P J *et al* 2014 *Seminars in Radiation Oncology* **24** 203-206
[5] Ibbott G *et al* 2017 *Clinical 3D Dosimetry in Modern Radiaton Therapy* ed B Mijnheer (Bova Raton: CRC Press) chapter 26
[6] Baldock C 2009 *J. Phys.: Conf. Ser.* **164** 012002
[7] Baldock C *et al* 2010 *Phys. Med. Biol.* **55** R1-63
[8] Venning A J *et al* 2005 *Med. Phys.* **32** 1047-53
[9] Roed Y *et al* 2015 *Med. Phys.* **42** 3501
[10] De Deene Y 2004 *J. Phys.: Conf. Series* **3** 34-57
[11] De Deene Y *et al* 2006 *Phys. Med. Biol.* **51** 653-673
[12] Bayreder C *et al* 2006 *Med. Phys.* **33** 2506-18
[13] Massillon-J L G et al 2010 *Appl. Radiat. Isotop.* **68** 144-154
[14] Massillon G *et al* 2008 *Med. Phys.* **35** 2920
[15] Maryanski M 2015 *Personal Communication*
[16] Ezzell G *et al* 2009 *Med. Phys.* **36** 5359-73
[17] SlicerRT: 3D Slicer 2017 *www.slicer.org*
[18] Molineu A *et al* 2014 *Med. Phys.* **41** 352
[19] Karlsson A *et al* 2007 *Phys. Med. Biol.* **52** 4697-706

# 4D scintillation dosimetry for the MRI-linac: proof of concept

**P Brůža[1], D Gladstone[1,2,3], J Cammin[4], O Green[4] and B W Pogue[1,2,5]**
[1]Thayer School of Engineering, Dartmouth College, Hanover, NH 03755, USA
[2]Department of Medicine, Geisel School of Medicine, Dartmouth College, Hanover, NH 03755, USA
[3]Norris Cotton Cancer Center, Dartmouth Hitchcock Medical Center, Lebanon, NH 03756, USA
[4]Department of Radiation Oncology, Washington University School of Medicine, St. Louis, MO 63110, USA
[5]DoseOptics LLC, Lebanon, NH 03756, USA

E-mail: Petr.Bruza@dartmouth.edu

**Abstract.** A new method of time-resolved volumetric (4D) dosimetry combining transversal projected view scintillation imaging with the multi-leaf collimator (MLC) geometry information is presented and demonstrated in a magnetic resonance (MRI) guided linear accelerator (linac). The setup consisted of a time gated intensified camera and a cylindrical plastic scintillator phantom. Positioning the camera outside the 0.35 T magnetic field suppresses the interference between the MRI-linac and dosimeter camera. Transversal view images of the scintillation light were recorded at 20 Hz framerate and the light distribution along optical axis was decoded from the MLC data by Fourier algorithm. Considering scintillation light as dose surrogate, the dose volume was reconstructed with sub-millimeter resolution, and this was tested on an intensity modulated delivery of a TG119 C-shape plan. 3D gamma analysis of the recorded cumulative dose volume as compared to a Monte-Carlo simulation reported 95% pass rate at 3%/3mm criteria. By enabling the use of measurement-based 3D beam comparison metrics, the presented method may provide a comprehensive solution for volumetric end-to-end dosimetry and fast machine performance checks in this challenging environment of an MRI-linac.

## 1. Introduction

Novel magnetic resonance (MRI) guided radiotherapy systems offer very sharp dose gradients (<2 mm 20%-80%) and the possibility of adaptive re-planning [1,2], allowing a highly conformal beam delivery. This approach may enable not only an unprecedented tissue localization, but for certain cases also the use of hypofractionation, leading to fewer patient visits and improved clinical outcomes [2]. Unsurprisingly, the fusion of an MRI scanner with a linear accelerator has brought unique challenges to dosimetry practice [3]. The strong magnetic field impedes quality assurance (QA) efforts by inducing angular-dependent corrections in ionization chambers and semiconductor dosimeters, as well as by introducing MRI-compatibility requirement on the used equipment. The latter issue basically rules out actuator-based point scanning systems, leaving hand-actuated systems [4] and film dosimetry as the remaining options. A large effort has been invested into alternative techniques, including gel dosimeters [5] and MRI-compatible 2.5D diode arrays [4,6,7]. Non-trivial response calibration and handling of the gel dosimeters or additional calculation tools in the case of 2.5D diode array may, however, introduce other sources of significant uncertainties. Spatial resolution of the 2.5D diode array is also generally low. Both end-to-end dosimetry as well as fast machine performance checks would benefit from a stand-alone dosimeter that is MRI-compatible and minimally intrusive to the scanning sequence, MRI tissue equivalent (e.g. water-based), sensitive (equivalent noise level <3 cGy), able to record or reconstruct a 3D dose distribution, and with spatial resolution high enough to resolve steep dose gradients (<1 $mm^3$). Especially for end-to-end dosimetry, 3D imaging capability would allow analysis of dose volume histograms (DVH) in addition to standard gamma analysis [8] thereby enabling direct assessment of the clinical impact of dosimetric errors [9]. Based on earlier research of Cherenkov [10] and scintillation dosimetry [11,12] imaging, we developed a 3D dose imaging system that combines projection view scintillation imaging of a transverse dose distribution with orthogonal (cross-beam profile) geometry information. In [10] we demonstrated this technique using a portal image for decoding the 3D dose distribution. As the portal imagers are not available on all MRI-linac systems, our presented approach uses multi-leaf collimator (MLC) geometry information obtained from machine log files. In combination with "traditional" transverse projectional imaging of dose-induced visible scintillation in a cylindrical phantom, we demonstrate for the first time a 4D (3D + time) reconstruction of dose delivered by MRI-linac using a complex intensity modulated radiation therapy (IMRT) plan.

## 2. Materials and methods

### 2.1. Experimental setup

Our dosimetry technique consists of three basic components: a) a scintillating phantom, b) an intensified camera with long focal distance lens; and c) a reconstruction algorithm that used camera data stream and MRI-linac log files to reconstruct 3D dose. The initial experiment was performed on a 0.35 T MRI-linac (MRIdian, ViewRay, Mountain View, CA), and the actual setup is shown in the Fig. 1. A cylindrical plastic scintillator (Ø5" ×2.5", Bicron BC412, Saint-Gobain Crystals, OH) was aligned such that the center of gravity coincided with the isocenter, and the base was parallel with the rotational axis of the linac gantry. An intensified complementary metal oxide semiconductor (iCMOS) camera (C-Dose, DoseOptics, Lebanon, NH) was positioned at the distal couch base, optical axis coinciding with bore axis of the linac. The intensifier of the iCMOS camera was gated to the linac pulses using a stray X-ray point detector with fast transient response time [13]. A total of 5220 scintillation image frames at 20 Hz framerate were captured upon delivery of a C-shape plan for TG119 using a 6 MV flattening-filter free beam. Linac log files recorded MLC positions, beam on-off time, fractional dose, and gantry angle, with temporal resolution of 10 μs.

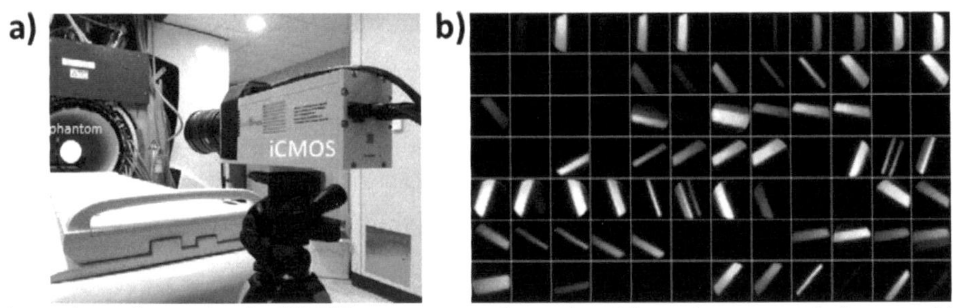


**Figure 1.** a) experimental setup showing iCMOS camera setup at bore axis and focused on target at the isocenter; b) montage of selected beamlets as imaged by the camera at 20 fps, 5220 frames total (800cGy delivery).

### 2.2. Reconstruction algorithm

The reconstruction algorithm relies on an approach developed earlier for 3D electronic portal imaging device (EPID)-Cherenkov dosimetry [14]. We establish our analysis in assumption that the scintillation image is a surrogate of the dose distribution, projected along optical axis and scaled by calibration constant

$$d_T = a\, I_T \tag{1}$$

For clarity, it is assumed that the scintillation image is free of any optical aberrations and scatter due to the phantom re-absorption.

After extracting necessary beam information from linac log files, the reconstruction begins with creating a geometrical projection of the MLC shape across the phantom volume, yielding beam volumes $m$ for each individual field. A transverse projection MLC image $m_T$, similar to the transverse scintillation image $I_T$ but lacking any information about depth dose distribution and penumbra, is created by summing the voxel intensities of the MLC shape volume in the direction of the optical axis of the camera:

$$m_T = \sum_y m(x, y, z, t) \tag{2}$$

In the case of MRI-linac, the coordinate space used here ($x,y,z$ for real space, $u,v,w$ for corresponding spatial frequency space, and $t$ for time) corresponds to IEC61217 gantry coordinate system. The dissimilarity between projection MLC image $m_T$ and scintillation image $I_T$ can be expressed for each cross-beam profile across axis of the beam $z$ as a projectional correction kernel $\hat{s}_T$:

$$\mathcal{F}_x\{I_T\} = \hat{s}_T(u, z)\,\mathcal{F}_x\{m_T\} \tag{3}$$

Where $\mathcal{F}$ represents fast Fourier transform and the subscript $_x$ indicates the independent variable over which the transform is applied. We follow this notation in the rest of this manuscript. Equation 3 also indicates that the depth dose distribution is captured by the magnitude of correction kernel, and specific penumbra shape is represented by the spectral shape of $\hat{s}_T$.

Until now, we were operating exclusively with the transverse projection images. Assuming circular symmetry of the dose distribution kernel (isotropic, homogeneous phantom), we can expand the projectional correction kernel $\hat{s}_T$ to the third dimension (along optical axis of the camera, $y$-coordinate) by exploiting principles of the projection-slice theorem and Fourier-Hankel-Abel (FHA) cycle. As $\hat{s}_T$ represents a projection of a circularly symmetric 2D kernel for each slice along beam axis, we can reconstruct the 3D correction kernel from $\hat{s}_T$ by applying the FHA cycle [14]:

$$\hat{s}(u, v, z, t) = \mathcal{H}_{x,y}\,\mathcal{A}_x^{-1}\mathcal{F}_u^{-1}\left\{\frac{\mathcal{F}_x\{I_T\}}{\mathcal{F}_x\{m_T\}}\right\} \tag{4}$$

Where $\mathcal{H}\mathcal{A}^{-1}\mathcal{F}^{-1}$ represents a series of zeroth order Hankel, inverse Abel, and inverse Fourier transforms, respectively. Again, subscripts of the transform symbols indicate independent variables over which the transforms are applied. The final dose volume is then calculated as partial convolution of the 3D correction kernel with the MLC-based beam volumes $m$:

$$d(x, y, z, t) = a\mathcal{F}_{u,v}^{-1}\{\hat{s}\,\hat{m}\} \tag{5}$$

### *2.3. Image post-processing*

The image post-processing and dose reconstruction was implemented in Matlab (Mathworks, Natick, MA) environment. First the image dataset was corrected for the optical scatter and divided into image sub-stacks for each beam field. Timestamps of all frames were shifted to match the MRI-linac log file data. Based on gantry angle and MLC shape information from the log files, an array of MLC-based beam volumes was created with resolution of 0.5 mm. Coordinates of MLC data and scintillation data were matched and scaled according to the preceding optical calibration, followed by the 3D analysis. Resulting 3D dose distribution movie was summed in time to get cumulative dose distribution. Finally, we performed gamma analysis-based validation of the measured data with Monte-Carlo based treatment planning system (TPS), and the results were rendered using The Visualization Toolkit (VTK, Kitware, NY) software system.

## 3. Results and Discussion

The imaged cumulative dose distribution (represented by cumulative scintillation image, Fig. 2a) matches well the transverse projection view distribution calculated by the Monte-Carlo based treatment planning system (ViewRay, Mountain View, CA) (Fig. 2b). 3%/3 mm gamma test success rate for the scintillation image as compared against projected TPS image data was 96% for the pixels with intensity over 10% of the maximum intensity value. Importantly, comparing the reconstructed 3D dose to the TPS calculated dose volume, 3%/3 mm gamma test success rate was 95% for the voxels receiving over 10% of the maximum dose.

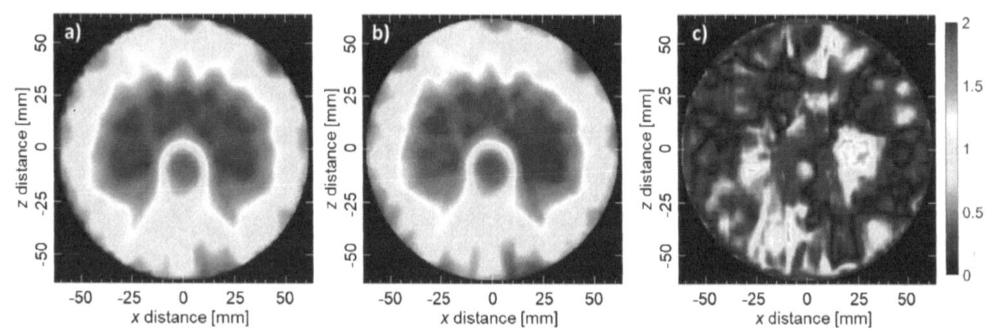


**Figure 2.** a) Cumulative dose distribution calculated by ViewRay TPS, summed across the 57 mm depth of the scintillator; b) measured cumulative dose distribution. Both dose distribution profiles are normalized. c) Result of gamma analysis, indicating 96% pass rate at 3%/3mm criteria and 10% threshold.

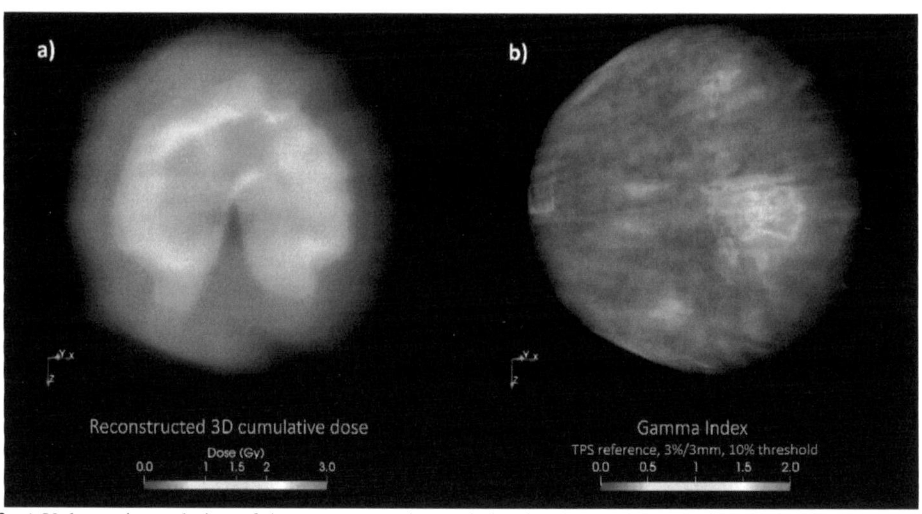


**Figure 3.** a) Volumetric rendering of the reconstructed cumulative 3D dose volume; b) 3D distribution of gamma values comparing reconstructed cumulative dose and TPS dose volumes.

## 4. Conclusion

We presented a novel 4D dosimetric system that is capable of reconstructing time-resolved dose volumes from projections of dose (scintillation images) and cross-beam profile information based on logged MLC configuration data. The innovation of this study lies in enabling fully volumetric dose imaging with high spatial (0.5 ×0.5 ×0.5 mm$^3$) and temporal (50 ms) resolution. The system is MRI compatible, as demonstrated on a proof-of-concept measurement of a dose delivered by 0.35 T MRI-guided radiation therapy linac. Gamma analysis revealed a good match between standard C-shape plan prepared for TG119-based commissioning. 2D and 3D 3%/3mm gamma test rates were 96% and 95%, respectively, indicating an elevated region in one of the lobes of the planning target volume. The source of discrepancy will be studied further. Following the current proof-of-concept study, an investigation of the accuracy of the presented technique is warranted. Similarly, the phantom composition will be optimized for minimal optical scatter, water equivalency, and MRI detectability. The presented system potentially offers a comprehensive solution for volumetric end-to-end dosimetry as well as a rapid machine performance assessment not only for MRI guided therapy, but also for standard C-arm and tomotherapy systems.

## 5. Acknowledgements

This research was supported by National Institutes of Health research grants R44 CA199681 and P30 CA023108.

## 6. References

[1] Green O *et al* 2018 *Med. Phys.* 45 3728-3740
[2] Pathmanathan A U *et al* 2018 *Int. J. Radiat. Oncol. Biol. Phys.* **100** 361-373
[3] Cai B *et al* 2018 *Z. Med. Phys.* **28** 211-223
[4] Woodings S J *et al* 2018 *Phys. Med. Biol.* **63** 085015
[5] Lee H J *et al* 2018 *Phys. Med. Biol.* **63** 045021
[6] Houweling A C *et al* 2016 *Phys. Med. Biol.* **61** N80
[7] de Vries J H W *et al* 2018 *Phys. Med. Biol.* **63** 02NT02
[8] Nelms D E *et al* 2013 *Med. Phys.* **40** 111722
[9] Pasler M *et al* 2018 *Phys. Imaging in Rad. Onc.* **5** 76-84
[10] Andreozzi *et al* 2018 *Med. Phys.* **45** 2647-59
[11] Goulet M *et al* 2014 *Med. Phys.* **41** 082101
[12] Robertson D *et al* 2014 *Phys. Med. Biol.* **59** 23
[13] Ashraf M R *et al* 2018 *Med. Phys. (in press)*
[14] Bruza P *et al* 2017 *IEEE T. Med. Imaging* **36** 2099-103

# Cherenkov imaging of total skin electron irradiation (TSEI)

**Timothy C. Zhu**
Dept. of Radiation Oncology, University of Pennsylvania, Philadelphia, PA 19104, USA

E-mail: tzhu@pennmedicine.upenn.edu

**Abstract.** Total Skin Electron Irradiation (TSEI) utilizes high-energy electrons to treat skin cancers, mostly mycosis fungoides. The otherwise invisible radiation beam can be visualized using the optical Cherenkov light emitted from interaction between incident electron beams and tissue during radiation therapy. Using a gated camera system with built-in intensifier, the Cherenkov emission can be used to evaluate the dose uniformity on the surface of the patient in real-time. Each patient was also monitored on the skin surface using *in-vivo* diode or OSLD dosimeters.

## 1. Introduction

Total Skin Electron Irradiation (TSEI) utilizes high-energy electrons, usually 6 MeV, to treat skin cancer of the entire body. At University of Pennsylvania, we utilize a Stanford Technique [1] to deliver the TSEI at a source-to-surface distance of 500 cm from a Varian Truebeam accelerator with a dose rate of 2500 MU/min in HDSE mode (Fig. 1a). It is usually required that the dose uniformity to the entire body surface to be less than 10%. For the optimized gantry angles (90°±16°), we found that the variation of percentage dose of the dual field, normalized to incident onface electron beam (gantry angle (90°), is well within 5% (Fig. 1b). The small bump at 103 cm is due to the electron backscattering from a lead backscatter sheet used to boost scalp dose [2].

We used a gated camera provided by Dartmouth College to measure the Cherenkov emission generated by the TSEI [3]. The Cherenkov light signal is generated when the speed of high-energy electrons in the medium is higher than the speed of light in the medium [4]. Experiments have shown that, for beams with the same energy, there is a linear relationship between the Cherenkov signal intensity and dose deposited on the surface of the phantom with uniform optical property [3].

## 2. Methods and Materials

Based on prior experience from Dartmouth for Cherenkov imaging of TSEI [5], we have started a clinical trial for TSET [6]. The setup of Cherenkov imaging is shown in Fig 2, where the Cherenkov imaging camera mounted on a tripod is at about 400 cm from the TSE stand. Patient anatomic positions are obtained via a 3D scanner.

IOP Conf. Series: Journal of Physics: Conf. Series **1305** (2019) 012016 doi:10.1088/1742-6596/1305/1/012016

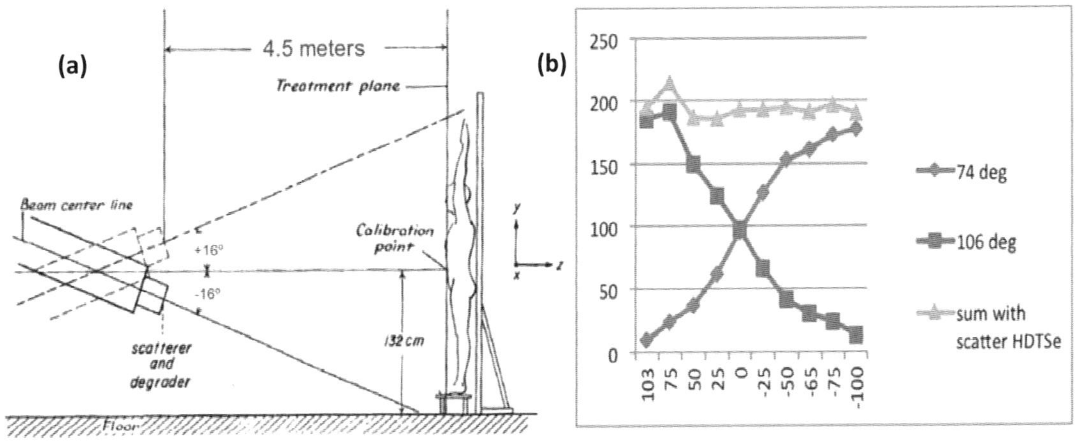


**Figure 1.** (a) Schematics of TSEI using a Stanford technique with a dual electron fields with gantry tilted up and down 16 degrees. (b) The profile of dual electron beams. The y-axis is the percentage value assuming normalized to a single on-face electron beam as 100%. The x-axis is the distance at the Stand in cm.

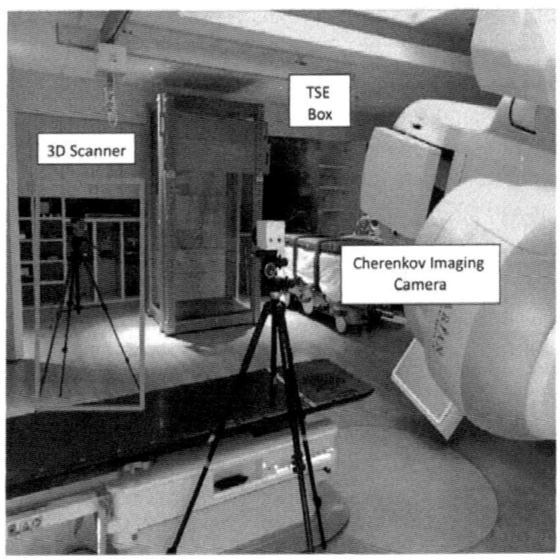


**Figure 2.** Experimental setup for the Cherenkov imaging of TSEI. The patient is placed in the TSE box at 500 cm away from isocenter. The Cherenkov imaging camera is placed on a tripod at about 400 cm from the patient. A 3D scanner is used to imaging the anatomic position of the patient in 3D.

### *2.1. Cherenkov signal dependence on tissue optical properties*

There is a difference between dose IVD dose profile and Cherenkov intensity map. A perspective correction factor can be identified as:

$$CF\theta = (D(x,y)/D_{max})/(Ch(x,y)/Ch_{max}), \quad (1)$$

Where $Ch(x,y)$ is the Cherenkov intensity and $D(x,y)$ is the measured dose value at off-axis location (x, y) on a plane of TSE phantom (e.g., a PVC panel) at SSD = 500 cm. Figure 3 shows the correction factors in the vertical direction [6]. There is no need for correction in the horizontal direction (data shown elsewhere) [6]. The correction factor is substantially larger for the dual field than for the single field in the vertical direction. Thus, a Cherenkov correction factor is applied to the acquired patient data in that direction.

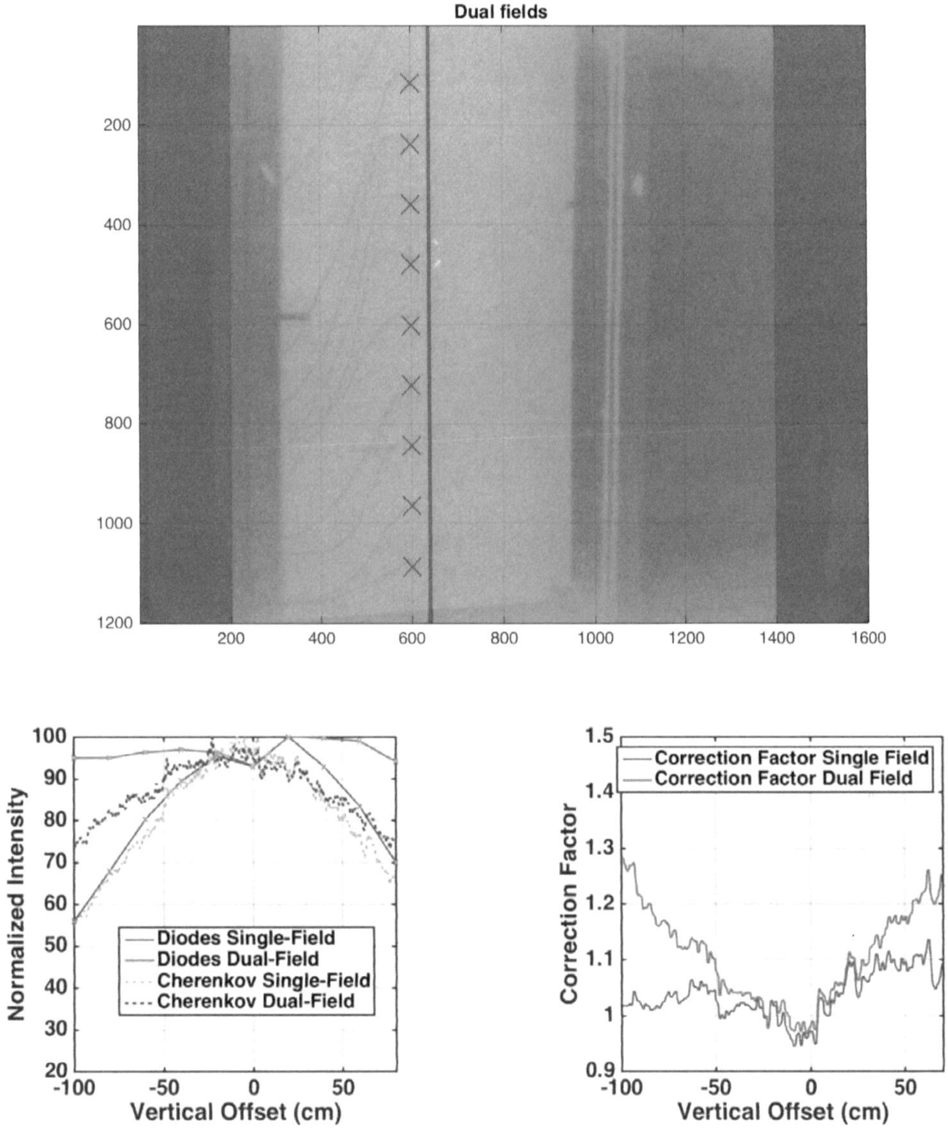


**Figure 3.** (a) Cherenkov raw image for dual field in 1200x1600 pixel map. The isocenter of the radiation field is located at pixel location (600, 500). The distance between pixels is 0.1647 cm projected at SSD = 500 cm. The "x" marked the diode locations, which is spaced at 20 cm apart on the PVC board. The red line is where the Cherenkov intensity is taken for comparison. (b) Comparison of Cherenkov and Diode dose profiles for dual fields composed of gantry angles 74° and 106° and a single field of gantry angle 90°. The vertical offset is in cm on the PVC panel. (c) The correction factor is determined using Eq. (1) in the same coordinates as Fig. 3b for single and dual fields, respectively.

## *2.2. Cherenkov signal dependence on incident angles*

The Cherenkov intensity is linear to the delivered dose from a 6 MeV electron beam at depth of 6 mm. However, when tissues with different optical properties are used, the slope of this linear dependence will change because of the absorption and scattering of the Cherenkov photons by the tissue. Figure 4 shows the relationship between Cherenkov intensity and dose on the central axis on the PVC board for different tissue-simulating solid phantoms, with optical properties ($\mu_a$, $\mu_s$') = (0.130, 11.2), (0.156, 1.9), (0.0425, 14.1), and (0.4, 5.5) cm$^{-1}$. It seems to be proportional to the value of diffuse reflectance at 630 nm [6].

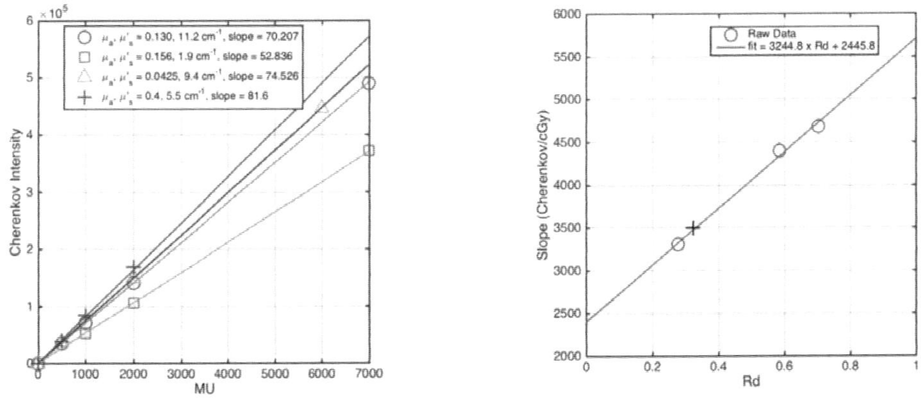


**Figure 4.** (a) Cherenkov intensity vs. dose with three different optical properties ($\mu_a$, $\mu_s$'). (b) Slope of Cherenkov response vs. value of diffuse reflectance at 630 nm of the tissue simulating phantoms.

### *2.3. Conversion of Cherenkov signal to dose*

The conversion of Cherenkov signal to dose follows two steps: First, the perspective correction factor is applied in the vertical direction; Second, the corrected Cherenkov intensity is normalized to the value at the umbilicus point, where the dose is measured using a diode. The corrected Cherenkov intensity is then converted to dose.

## 3. Results

Figure 5 showed the Cherenkov converted dose for a patient for all 6 postures during TSET.

## 4. Discussions

The conversion of Cherenkov intensity to dose for the upper row of Fig. 5 is done by renormalizing the corrected Cherenkov intensity to umbilicus point, where the dose is measured. For the bottom row in Fig. 5, the diode reading at umbilicus point cannot be used because the body blocked it. We made the conversion by multiplying the same Dose-to-Cherenkov ratio for the corresponding day for the patient. This way, the conversion can be made without worrying about the effect of tissue optical properties.

Figure 5 shows that the dose on the legs is lower than the dose in other parts of the body even though the optimal angles (74 and 106) gives uniform profile in the phantom conditions (see Fig. 1b and Fig. 2a). The reasons for the low dose are unknown but we can speculate two possible reasons: (1) Dose conversion error due to approximation made to the perspective correction factor. The perspective correction factor (Eq. (1)) should be a two-dimensional correction covering (x,y) on the patient plan, however, we have approximated this as a one-dimensional correction factor (Fig. 3b). In addition, for oblique fields (LAO, LPO, RAO, RPO) the plan where the correction should be obtained is tilted by 30 degree relative to the PVC board used in Fig. 3a, that may also introduce errors in the conversion. In addition, our current conversion factor did not take into account of the camera response distortion. (2) The dose is lower near the floor due to floor scattering as pointed out by a recent study, where the dose near the floor is lower on a solid-water material compared to that on the PVC material [7]. Thus on the patient, the dose distribution may indeed be lower near the feet. A selection of asymmetrical gantry angles for TSET may provide an appreciable more uniform dose distribution [7]. Further studies using a human shaped phantom coupled with in-vivo dosimetry on the patient feet should help us understand the cause of the lower dose.

We have studied 10 patients so far, the ratio of dose reading on Chest point is compared to that at umbilicus. They agree to within 5% and is presented elsewhere [6].

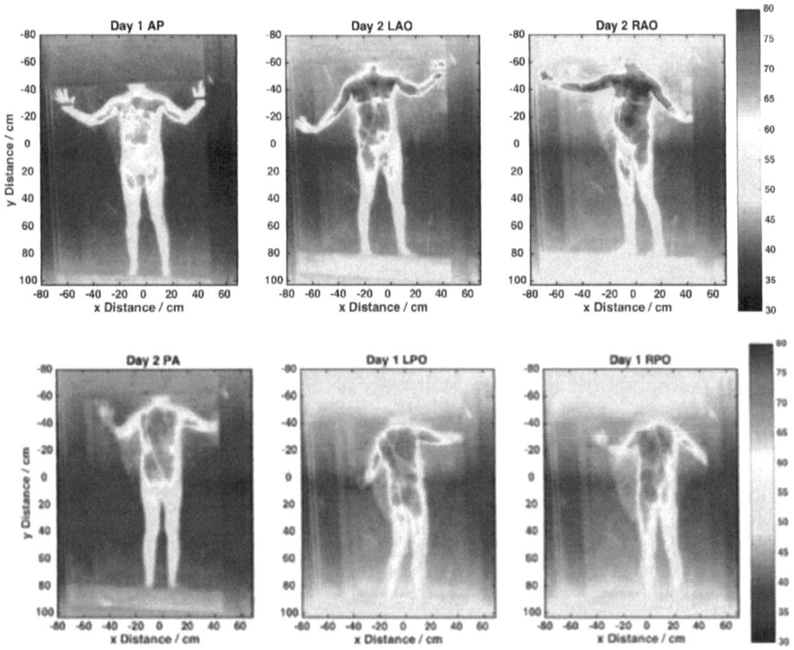


**Figure 5.** Cherenkov converted dose distribution for the 6 positions of a TSET patient in Day 1 and Day 2 for a total of 200 cGy/2-day cycles. X and Y (cm) is at patient location (SSD=500 cm).

## 5. Conclusions

We conclude that Cherenkov imaging provide a mean to study the distribution of surface dose for patient undergoing TSET.

## 6. Acknowledgments

The Author wishes to thank Yunhe Xie, PhD and Heather Petroccia, Ph.D. for taking the Cherenkov emission patient data, and Yihua Zhu, Tianshun Miao, and Yihong Ong, PhD for phantom and tissue optical properties measurements to obtain correction factors. The Author wishes to acknowledge technical assistant and providing the Cherenkov Camera from Brian Pogue, PhD and Petr Pruza, PhD of Dartmouth College and DoseOptics LLC.

## 7. References

[1] Cox R S et al 1990 *Int. J. Radiat. Oncol. Biol. Phys.* **18** 659-69
[2] Chen Z *et al* 2004 *Int. J. Radiat. Oncol. Biol. Phys.* **59** 872-85
[3] Rongxiao Z *et al* 2013 *Phys. Med. Biol.* **58** 5477
[4] Cerenkov P A 1934 *Doklady Akad. Nauk SSSR* **2** 451
[5] Andreozzi J M 2016 *Med. Phys.* **43** 993-1002
[6] Xie Y *et al* 2018 *Proc. SPIE* **10478** 1047816
[7] Andreozzi J M 2018 *Med. Phys.* **45** 2639-46

# Measurement of build-up region dose with optical cone-beam computed tomography scanner

**Sarah Garisto[1] and Kevin Jordan[1,2]**
[1]London Regional Cancer Program, London Health Sciences Centre, London, Ontario, Canada
[2]Dept of Medical Biophysics, Western University, London, Ontario, Canada

E-mail: kevin.jordan@lhsc.on.ca

**Abstract**. Measurement of the dose gradients from the entrance surface to depth is a standard task for characterizing an ionizing radiation beam. Most gel dosimeters provide spurious results near an air interface, limiting their value for this geometry. In this study, a 3D dosimeter system consisting of a low-diffusion, radiochromic hydrogel cast in a custom polyethylene terephthalate (PETE) vessel and imaged with a modified commercial optical cone-beam computed tomography (CBCT) scanner was employed. The cylindrical vessel wall and flat ends were constructed from a 0.025 cm thick PETE sheet. The optical CBCT scanner was modified to place the entire vessel in the centre of the field of view or to have the vessel base at the optical axis. Pre-irradiation and post-irradiation scans were acquired with the sample mounted in the standard and elevated positions. The sample was irradiated with a 2x2 cm square, 6 MV x-ray beam. Normalized attenuation coefficients from the central quarter of the reconstructed beam images were compared to a Monte Carlo depth dose calculation. Placing the vessel base at the optical axis allowed accurate dose measurements to within 0.2 cm of the entrance face and for the standard position to within 0.4 cm. These measurements validated the Monte Carlo calculation and provide an alternative to parallel plate ion chambers for dose measurement in the build-up region.

## 1. Introduction

Measurement of dose from the entrance surface to depth of maximum dose, the build-up region, is required for characterizing a megavoltage radiation beam. Typically, near surface measurements are obtained with a parallel plate ion chamber and merged with 1D scans with diodes or small cylindrical ion chambers in a water tank. The dose gradients could also be obtained from Monte Carlo calculations once the output has been validated for selected beams. Water-equivalent, 3D gel dosimeters [1] could also be used to measure dose in the build-up region but previous attempts to measure near a gel-air interface have often resulted in spurious results unless the gel has been cast within hours of irradiation. Recently, our group has prepared custom 1 cm square cuvettes for near surface 1D gel dosimetry and recorded 3D doses in custom 15 cm diameter vessels with a planar laser CT scanner. The small cuvettes allowed measurements to within 0.1 cm of the surface and the 3D large vessel system to within 0.3 cm of surface.

A feature of cone-beam geometry is incomplete sampling near the ends of the field of view (FOV). As the horizontal and vertical acceptance angles increase, the fraction of FOV length under-sampled also increases. In this study, custom cylindrical vessels, 10 cm in diameter, having thin flat ends were scanned using a modified commercial optical CBCT scanner. The goal is to determine how near the entrance planar interface accurate 3D data could be measured.

## 2. Methods

Low diffusion, radiochromic hydrogel in a custom cylindrical vessel with thin, flat ends was irradiated with a 2x2 cm, jaw-defined, 6 MV photon beam perpendicular to the flat ends with source to surface distance of 90 cm. The sample was scanned with vessel centre and entrance end face positioned at the scanner's optical axis, analogous to a previous study [2]. Reconstructed 3D, normalized, attenuation coefficients were compared to previous Monte Carlo 3D dose calculations.

### 2.1. *Vessel construction*

PETE sheet (Dura-Lar®) was cut to form a cylinder, 10 cm OD x 11 cm height x 0.025 cm thick. The ends were welded to form an opaque seam <0.1 cm wide. The uniform thickness of the sheet provided much higher optical quality than selected samples of blow molded vessels. The ends were flat discs of the same material, welded into the cylinder ends. A small ink dot was placed 5.0 cm above the base on the vessel wall to serve as a height fiducial.

### 2.2. *Optical CBCT scanner*

A Vista16 optical CBCT scanner from Modus QA was modified by placing a 1.5 cm diameter aperture at the 530 nm LED source diffuser, similar to a previous study [3]. The smaller area source reduced stray light artifacts, manuscript submitted. Custom mounts were developed to couple the custom vessel to the rotating scanner lid and position the vessel either in the centre of the FOV or with base coplanar with the optic axis, see figure 1.

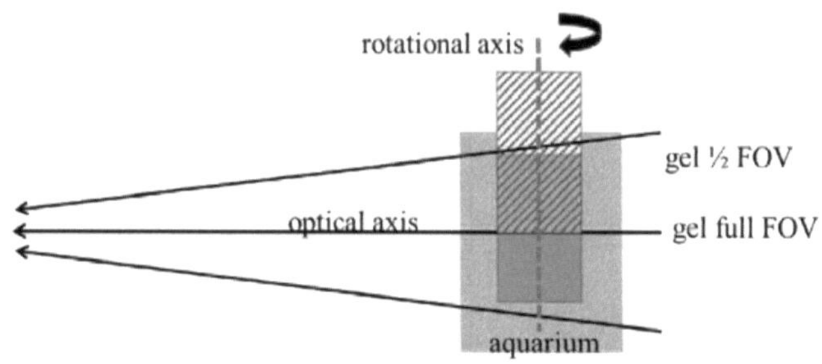


**Figure 1.** Drawing of side cross-section of optical CBCT scanner indicating gel centered on optical axis (full FOV) or base of gel at optical axis.

One thousand projections over 360 degrees were acquired in 4 minutes. The natural logarithm of ratios of reference and data image sets formed the projection images. Reconstructions were calculated with VistaScan program: 0.5 mm isotropic voxels, iterative reconstruction algorithm with default parameters (ordered subset convex (initial/final subset count 100/10, iteration=10, power=0.5, ordered subset angular distance) with total variation minimization (iteration count =10, update constant=0.020). Reconstructions were analyzed in Matlab.

### 2.3. *Solution and gel samples*

A uniform coloured solution was prepared by adding nigrosin dye to a 3.5% propylene glycol solution to achieve a transmission of ~70% along vessel diameter. For this solution, water was used as a refractive index optimized solution in the aquarium. The radiochromic gel consisted of 5% gelatin (porcine, 300 Bloom, type A), 32 mM trichloroacetic acid, 1 mM leuco crystal violet and 0.7 mM sodium dodecyl sulfate [4]. The gelatin was first dissolved in deionized water with 1 mM $H_2O_2$ at 55 °C for 16 hours prior to lowering temperature to 32 °C to add radiochromic chemicals. The sample was placed in a 20 °C water bath for ~3 hours prior to scanning and irradiation. A 7.5% propylene glycol solution was found to optimize refractive index by maximizing the transmitted signal just inside the vessel wall.

IOP Conf. Series: Journal of Physics: Conf. Series **1305** (2019) 012017 doi:10.1088/1742-6596/1305/1/012017

## 3. Results

Inspection of the reconstructed transverse slices from the solution scans, with the vessel base at optical axis, revealed the 4th slice from the end was unaffected by the vessel end. This corresponds to 0.15 to 0.2 cm from the end face. For the gel scans, mean attenuation coefficients per 0.05 cm slice were averaged over the central 0.5x0.5 cm of the beam, normalized and plotted against a Monte Carlo calculation, see figure 2. The gel depths were scaled to water equivalent depths using a mass density of 1.005 g $cm^{-3}$. Note, no additional spatial filtering was applied to remove spurious reconstruction artifacts such as the spikes at maximum dose for the full FOV scan.

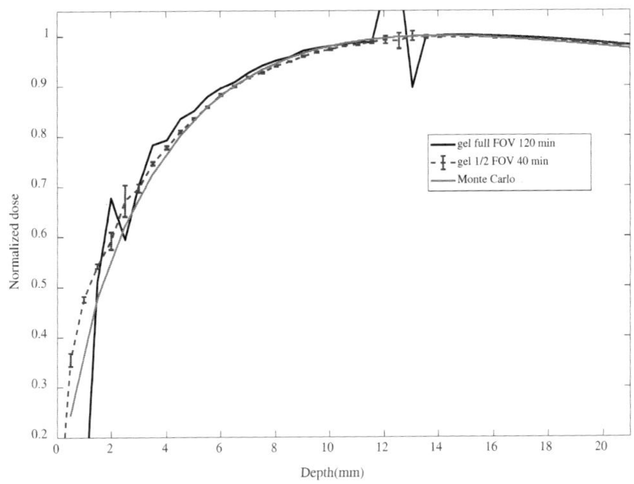


**Figure 2.** Plot of normalized depth doses with gel scanned in standard, full field of view position (solid line), gel bottom elevated to the optic axis (dashed line, error bars equal twice standard deviation of mean) and Monte Carlo calculation (dotted line, statistical uncertainty 1%).

## 4. Discussion

The agreement of gel reconstruction with Monte Carlo indicates that accurate, 3D dose measurements within 0.2 cm of the planar, entrance surface are possible with optical CBCT readout. Because small field dosimetry is challenging, larger fields sizes such as 3x3 cm or greater can be used to independently validate the calculations against a Markus parallel plate ion chamber. For the small fields, alternate dosimeters such as radiochromic film, optically stimulated luminescence detectors, thermoluminescent chips and plastic fibre scintillators could provide independent measurements. While a single beam was measured in this feasibility experiment, the capacity to record several small beams in the same 10 cm diameter gel represents a more appropriate use of the 3D dosimeter. Also, analysis of a set of different field sizes in the same experiment versus single beam per dosimeter should provide more robust measurements.

## 5. Summary

The combination, low diffusion radiochromic hydrogel in a cylindrical vessel with a thin flat end and placing the gel end at the scanner's optical axis allows accurate 3D dosimetry to within 0.2 cm of the entrance, planar surface with optical CBCT readout. These 3D measurements can then validate Monte Carlo calculations for small field sizes.

## 6. Acknowledgements
Conflict of interest: author K Jordan has a licensing agreement with ModusQA concerning optical CT for radiation dosimetry. S Garisto is recipient of a H E Johns student scholarship and was partially funded by a Department of Oncology research award. ModusQA provided the Vista16 scanner in support of a Collaborative Health Research Projects grant. M Mulligan performed the Monte Carlo calculations. G DeWale machined custom components for the modified scanner mounts.

## 7. References
[1] Baldock C *et al* 2010 *Phys. Med. Biol.* **55** R1-R63
[2] Babic S *et al* 2009 *Phys. Med. Biol.* **54** 2463-81
[3] Dekker K *et al* 2016 *Phys. Med. Biol.* **61** 2910-25
[4] Jordan K *et al* 2017 *J. Phys.: Conf. Series* **847** 012009

IOP Conf. Series: Journal of Physics: Conf. Series **1305** (2019) 012018 doi:10.1088/1742-6596/1305/1/012018

# Feasibility study of a dry optical CT scanner using aspherical lenses

**Yves De Deene**
School of Engineering, Macquarie University, North Ryde, Sydney, Australia

E-mail: Yves.DeDeene@mq.edu.au

**Abstract**. Dry scanners have been proposed as alternatives to scanners in which a refractive index matching fluid is used. Previous dry scanners either have a small field-of-view or employ a thick cast with a higher refractive index than the phantom. This may not always be feasible. A new design is proposed in this study where two aspherical meniscus lenses are employed to direct the laser beams in parallel lines through the phantom.

## 1. Introduction

Different optical CT scanners have been proposed in recent years to scan three dimensional dosimeters. The first generation of optical CT scanners are based on a scanning laser beam and a photodetector to measure the transmitted light. The most rapid and less vibration prone optical laser scanners consist of a Galvano-mirror to move the laser beam across the lateral direction of the phantom. In order to avoid refraction of the laser beam on the curved surface of a cylindrical phantom, a tank filled with a refractive index matching fluid is mostly used. The refractive index matching fluid consists mostly of a mixture of glycerol and water for gel based dosimeters and silicone based dosimeters, while a silicone oil is used for polyurethane (PRESAGE) based dosimeters.

For the end-user of 3D radiation dosimeters, the use of a fluid bath with refractive index matching fluid has some disadvantages in terms of practicality: The fluid bath and the fluid need additional maintenance as floating dust particles can create image artifacts in the dose maps. Regular cleaning of the optical scanner and around the scanner is needed as the employed liquids are mostly relatively greasy. Unequal evaporation of the components of the matching fluid may also cause a shift in the refractive index, hence causing a mismatch in refractive index over time. Over time, the fluid bath needs refilling which requires mostly an iterative process to obtain the correct mixture. Motivated by these drawbacks, some research groups have proposed a 'dry' approach where no refractive index matching fluid is applied.

The challenge in the design of a dry scanner is to compensate for the refraction of the light by the phantom. A first attempt was undertaken by Maryanski et al using the gel dosimeter phantom as focusing lens [1]. With this design, only an effective field-of-view of 1/3 of the phantom size was obtained. Ramm et al [2] investigated the possibility to extend the field-of-view by adding a thick cast of high refractive index material around the gel dosimeter. The latter approach is limited by the availability of a casting material with a refractive index higher than that of the dosimeter system. Doran and Yatigammana [3] investigated the possibility of removing the refractive index matching fluid tank and capturing the non-parallel light ray trajectories with a CCD camera. Reconstruction of the captured data is performed using an iterative SART reconstruction algorithm. As the light rays are travelling along non-parallel directions, weighting factors need to be applied in the reconstruction that take the divergence of the ray thickness into account. The field-of-view of the reconstructed image is restricted to a physical limit which depends on the refractive index of the dosimeter phantom.

In the current study, the feasibility of using lenses to direct as many parallel rays as possible through the dosimeter phantom is investigated. While the proposed method does not allow a complete coverage of the entire cross-section of the phantom, it provides a solution for configurations where the earlier method fails as a result of high refractive index dosimeter systems such as PMMA based (Flexydos3D) and polyurethane based dosimeters (PRESAGE™).

## 2. Methods and Materials

### 2.1. Theory

Spherical lenses have a shape which is part of the surface of a sphere. Spherical lenses do not have a constant focal distance at all points across the lens. As a result, spherical aberration occurs. To compensate for spherical aberration, aspherical lenses have been proposed as early as the 10th century by the Persian scholar Ibn Sahl [3]. The lens shape of an aspherical lens is described by the equation:

$$z(r) = \frac{r^2}{R_l\left(1 + \sqrt{1 - (1+k)\frac{r^2}{R_l^2}}\right)} + \sum_{n=2}^{m} A_{2n} r^{2n} \tag{1}$$

where r is the radial coordinate, $R_l$ is the vertex radius, $k$ the conic constant and $A_{2n}$ the coefficients of a correction polynomial for higher-order aspheric optical components. Depending on the value for the conic constant $k$, different types of conic sections can be identified: $k > 0$ defines an oblate elliptic section; $k = 0$ a spherical section; $-1 < k < 0$ a prolate elliptic section; $k = -1$ a parabolic section; and $k < -1$ a hyperpolic section.

The optical design for the dry laser scanner consists of two aspherical lenses on either side of the cylindrical dosimeter phantom (Figure 1). With two sections on each side of either lens, this results in 4 aspherical sections that need to be optimized in order to direct parallel lines through the dosimeter phantom. When no higher order terms are considered in equation 1, this still results in an 8-dimensional parametric space for which exist an infinite number of possible solutions. Some additional constraints are needed for the optimization problem. The following geometric constraints were added: (1) a predefined distance between each lens and the phantom; (2) a predefined thickness of each lens and (3) laser light rays follow parallel lines in the first lens. The optimization problem is solved sequentially in four stages each with their own objective function (Figure 1):

(1) For the entrance section of the first lens, optimize the lens shape ($k$ and $R_l$) so that all light rays originating from a source at the same focal distance $F$. The objective function here is the standard deviation of all light ray travelling through lens 1 (equation 2).

$$\min_{\left(R_l^{(I)}, k^{(I)}\right)} \sum_{\forall i} (\theta_{i1})^2 \tag{2}$$

(2) The light rays leave the first lens under an optimal angle so that light rays will travel along parallel lines within the phantom. The objective function that is to be minimized is the sum of the squared angle of all light rays traveling through the phantom (equation 3).

$$\min_{\left(R_l^{(II)}, k^{(II)}\right)} \sum_{\forall i} (\theta_{ip})^2 \tag{3}$$

(3) The light rays entering the second lens are expected to travel in parallel lines. The objective function to be minimized is the sum of the squared angle of all light rays traveling through lens 2 (equation 4).

$$\min_{\left(R_l^{(III)},k^{(III)}\right)} \sum_{\forall i} (\theta_{i2})^2 \tag{4}$$

(4) The parameters of the third lens are optimized by targeting convergence of all light rays in a focal spot behind the lens. The objective function is the quadratic difference between the focal spot of the light rays and the target focal distance (equation 5).

$$\min_{\left(R_l^{(IV)},k^{(IV)}\right)} \sum_{\forall i} (\Delta F_i')^2 \tag{5}$$

where $\Delta F_i'$ is the difference between the target focal length and the point where light ray $i$ crosses the main axis ($\Delta F_i' = F_i' - F$) ).

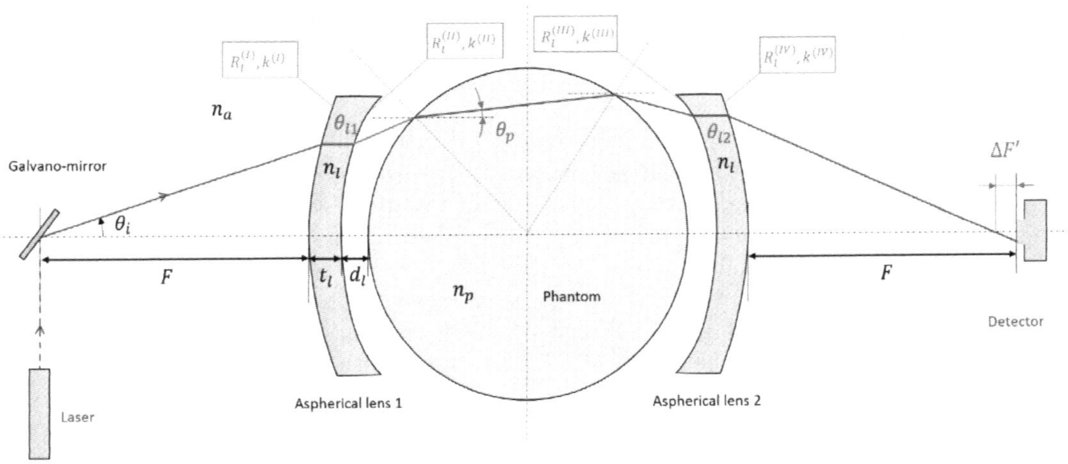


**Figure 1.** Geometry of the simulations. Fixed parameters are labeled in black, while objective functions are labeled in red. The optimization variables are indicated in green.

### 2.2. Implementation

The software was written in Matlab code. While Matlab provides built-in routines for minimization of cost functions, the minimum was sought using a brute-force technique where discrete steps were taken through both $R_l$ and $k$ in order to avoid convergence to local minima. All dimensions are considered relative to the radius of the phantom R. The values were chosen arbitrary as: $F_1/R = -2.5$ ; $t_l/R = 0.2$; $d_l/R = 0.2$; $F_2/R = -2.5$ ; where $F_1$ and $F_2$ are the focal distances for the first and second lens, $t_l$ is the thickness of the lens on the main axis and $d_l$ is the distance from the lens to the phantom (i.e. air gap) on the main axis. While not necessary, $t_l$ and $d_l$ were taken the same for both lenses.

With equidistant angular increments in the incident light rays, small deviations in distance of the parallel light rays in the phantom occur as a result of the divergence of beams after the first lens, the varying lens thickness and the curvature of the phantom. The distance in light rays is plotted as a function of incident (Galvano-mirror) angle.

Three different types of dosimeter phantoms with different refractive indices are considered. In order to test the feasibility of using a dual-wave length scanning approach [5, 6], the chromatic aberration was studied for both red light ($\lambda = 630\ nm$) and blue light ($\lambda = 480\ nm$). The aspherical lens parameters were first optimized for the red light source and ray tracing was applied using the refractive index of lenses and phantom for the blue light source.

**Table 1.** Refractive index of different materials used in the simulations

| | 480 nm | 630 nm |
|---|---|---|
| Gelatin gel (polymer gel) | 1.347 | 1.340 |
| Polyurethane (PRESAGE™) | 1.515 | 1.508 |
| PDMS (Flexydos3D) | 1.419 | 1.412 |
| Glass (Lens) | 1.5228 | 1.5152 |

## 3. Results and Discussion

### *3.1. Ray tracing and distortions*

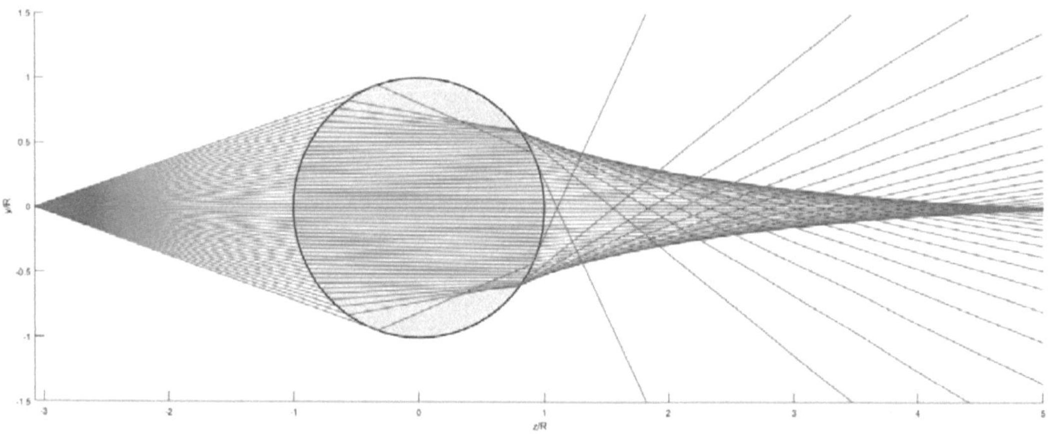

**(a)**

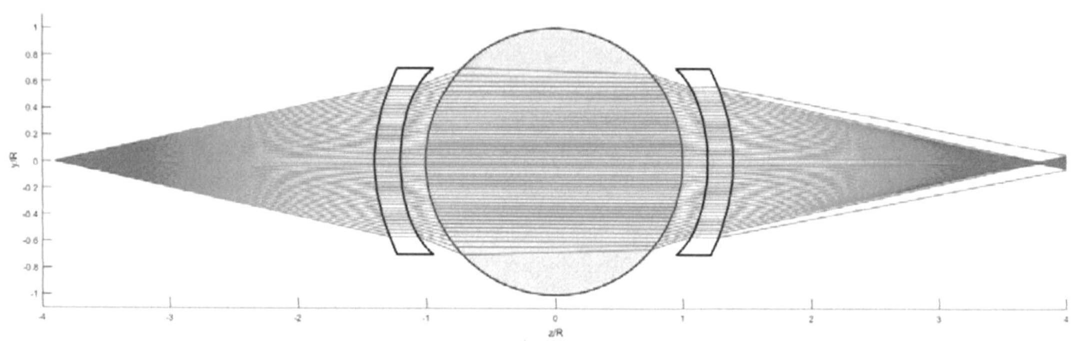

**(b)**

**Figure 2.** Simulations of ray paths for light rays originating from a point source for a configuration containing a gelatin hydrogel phantom without lenses (a) and with two aspherical lenses after optimization (b).

Light ray trajectories through a cylindrical phantom with the fluid tank removed are shown in figure 2a. In this simulation, the focal distance was optimized in order to minimize beam divergence in the phantom, i.e. equation 3 as objective function.

It is clear from figure 2a, that the rays within the phantom are not parallel, nor do the light ray trajectories in the phantom converge to a single point so that nor a radon transform or fan beam reconstruction can be used. The only valid approach to reconstruct an image from the transmitted rays is by using an iterative reconstruction algorithm [3].

Figure 2b shows the light ray trajectories for the same phantom with optimized aspherical lenses. Only a small deviation from the parallel direction is visible in the outermost simulated light rays within the phantom. However, a plot of the spatial distribution of light rays in the phantom shows that they are closer packed in the centre than at the periphery (figure 3). It was found that a tangent function fits the spatial distribution precisely. This function can be used to correct for the non-uniform distribution.

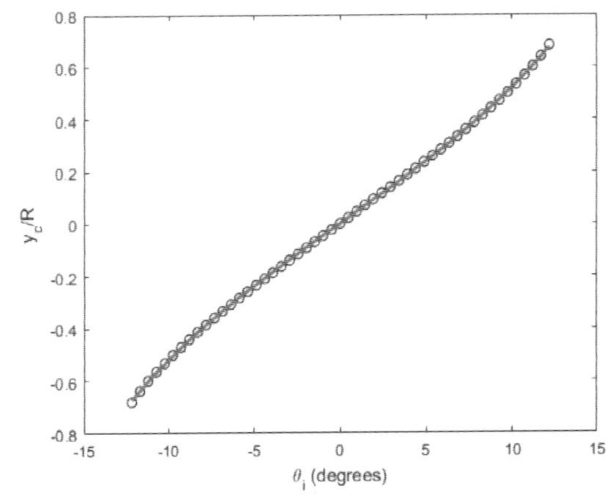


**Figure 3.** Correlation between the light ray position ($y_c/R$) measured along the central axis of the phantom and the light ray incident angle ($\theta_i$). A fit of a tangent function on the data points is shown as a red solid line.

## *3.2. Chromatic aberration (dual wavelength scanning)*

In earlier work [5, 6], we suggested the use of a blue light scan as a reference scan. As the refractive index of different materials is frequency dependent, it is important to investigate its effect on the light ray trajectories. Figure 4 shows the trajectories of both red light ($\lambda = 630\ nm$) and blue light ($\lambda = 480\ nm$) as these are the worst-case central wavelengths of the light sources used in our dual wavelength scanner.

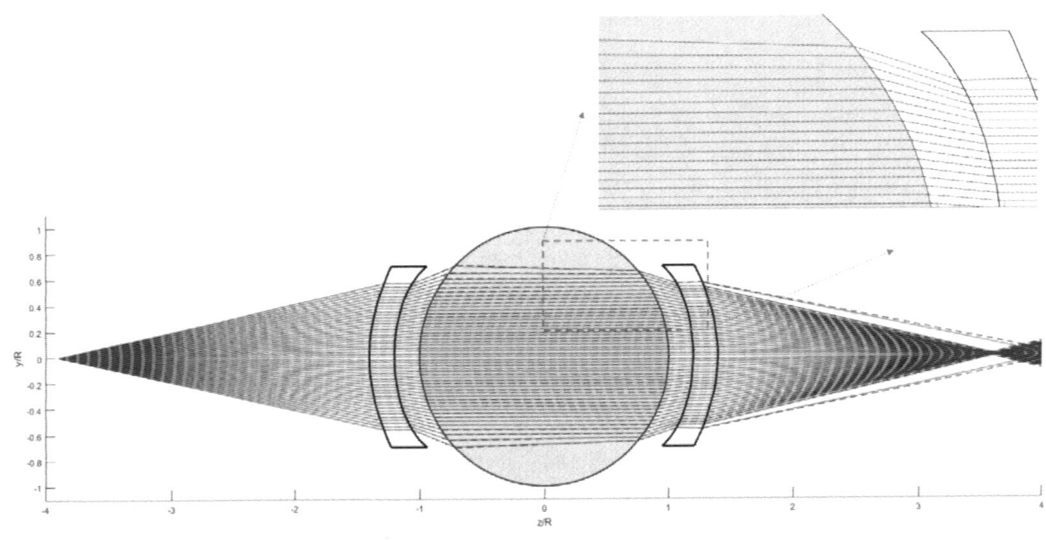


**Figure 4.** Ray tracing simulation demonstrating a negligible effect of chromatic aberration for light rays inside the phantom. The solid red lines correspond to light rays with a wavelength of 630 nm, while the dashed blue lines correspond with a wavelength of 480 nm.

While several methods have been proposed to reconstruct CT images with reduced field-of-detection (FOD), a fast alternative reconstruction method has been developed that uses the a priori information that in the dry scanner, the phantoms have a cylindrical shape with a predefined radius. In this method, missing lines in the sinogram are extrapolated prior to reconstruction using a functional relation based on attenuation and continuity between acquired and missing lines in the sinogram (figure 5).

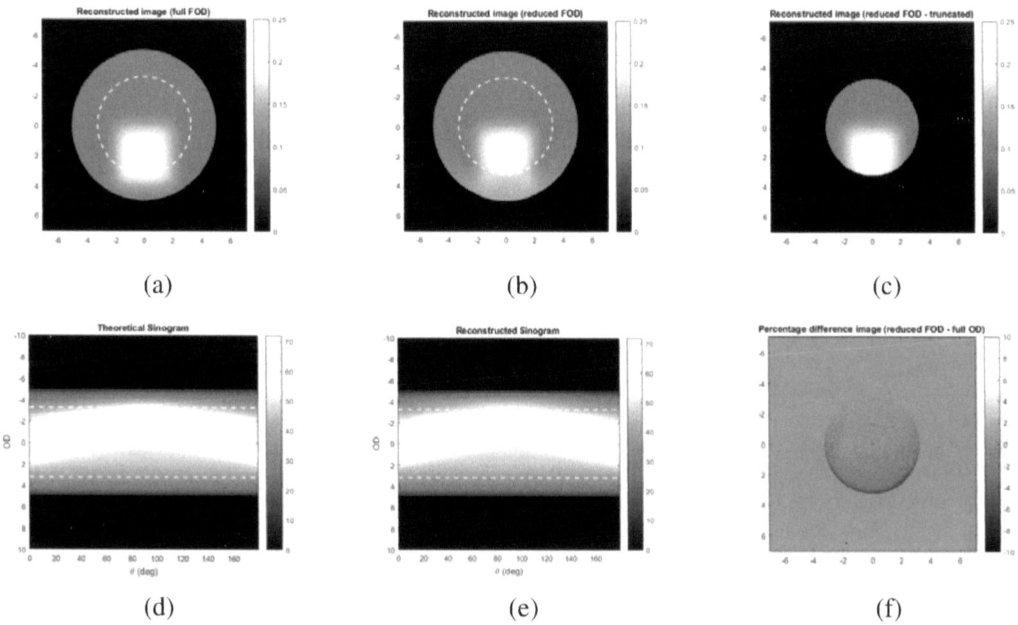


**Figure 5.** Simulation of the reconstruction of an optical CT image with off-centric square field with a reduced field-of-detection (65%). The missing lines in the sinogram are filled using an extrapolation based on the optical attenuation in a cylinder and with preservation of continuity in the projection direction. Reconstructed OD image (a) and sinogram (d) with full detection; Reconstructed image with reduced FOD (b) but where missing sinogram lines (e) have been obtained by extrapolation. Reconstructed image but reduced to acquired FOV and difference image (f).

## 4. Conclusions

In this computational simulation study, the feasibility of dry scanning using two aspherical lenses is demonstrated. It is shown how parallel light ray trajectories can be obtained in a cylindrical phantom by use of two optimized aspherical lenses. The theoretical concept can be applied to different kinds of scanner designs, such as laser scanners, fan beam scanners or cone beam scanners. The field-of-detection is physically limited by the refractive index of the phantom which is similar to another approach where the transmission of parallel lines is abandoned. However, in previous approaches, an iterative reconstruction with high computational power is required, where in the proposed method using aspherical lenses, a simple filtered back projection in combination with extrapolation of missing sinogram data can be applied.

## 5. References

[1] Maryanski M J and Ranade M K 2001 *Proc. SPIE* **4320** 764-7
[2] Ramm D *et al* 2012 *Phys. Med. Biol.* **87** 3853-68
[3] Doran S J and Yatigammana D N B 2012, *Phys. Med. Biol.* **57** 665-83
[4] Rashed R 1990 *Isis* **81** 464-91
[5] De Deene Y *et al* 2015 *Phys. Med. Biol.* **60** 1543-63
[6] De Deene Y 2015 *J. Phys.: Conf. Ser.* **573** 012058
[7] Meichner C *et al* 2015 *AIP Advances* **5** 087135

IOP Conf. Series: Journal of Physics: Conf. Series **1305** (2019) 012019 doi:10.1088/1742-6596/1305/1/012019

# Preliminary characterization of the Duke Integrated-Lens Optical-CT scanner (DIOS)

**Cielle Collins[1*], Suk Whan Yoon[1*], Jacob Kodra[1], John Adamovics[2] and Mark Oldham[1]**
[1]Duke University, Durham, North Carolina
[2]Rider University, Lawrenceville, NJ.
[*]Co-first authors.

E-mail: mark.oldham@duke.edu

**Abstract**. The present study investigates a cost effective and practical non-telecentric optical-CT scanner developed for 3D dosimetry, the Duke Integrated-Lens Optical-CT scanner (DIOS). The DIOS system includes an upgraded light-collimating tank (the LC-tank) made of solid polyurethane with precision curved ends (with lensing functionality) and a precision cylindrical central hollow for the dosimeter. The LC-tank thus collimates light from a small area light source (~2cm diameter) into parallel rays through the dosimeter, with refocusing of emergent light onto a CCD camera with a focusing lens with an aperture. The solid nature of the LC-Tank dramatically reduces the amount of required refractive matching fluid compared to earlier scanning systems. The aim of this work was to perform preliminary characterization studies of DIOS in comparison to earlier systems, particularly telecentric systems. The preliminary results indicate promising performance for the DIOS approach.

## 1. Introduction

In recent years there has been a push to develop commercial optical-CT readout 3D dosimetry systems for BANG gels, polymer gels, radiochromic gels, and radiochromic plastics [1-8]. A standard optical-CT scanner model, the Duke Large-FOV Optical Scanner (DLOS), achieved high-accuracy, high-resolution 3D dosimetry using large telecentric lenses and a large refractive matching fluid tank [7-10]. While high in image quality, this model was (1) prone to mess and hassle due to the excessive amount of refractive matching fluid needed to immerse the dosimeter, and (2) presented a significant expense exceeding $20,000 per scanner. The Duke Integrated-Lens Optical Scanner (DIOS) is a significantly cost-effective (<$2,000) and mass-producible model that seeks to match DLOS image quality standards, using a curved light-collimating tank (LC-tank). Previous works have introduced a prototype for the DIOS system [11], presenting preliminary construction and testing of the dry tank and scanner setup. The DIOS system design is illustrated in Fig. 1, which includes a 3W red LED (633nm ± 10), a solid polyurethane light collimating tank on a rotating stage, a Presage 3D dosimeter, and a 12-bit monochromatic CCD camera. A focusing lens with an aperture on the camera is included to help to reject scattered light and reduce depth-of-field blurring. The tank and Presage are both made from clear polyurethane, so that if a thin layer of RI-matched fluid is present, light passes through the tank un-impeded. The rotating stage allows a full 360° rotation so that projections can be collected at various angles for 3D reconstruction. The current work has improved the manufacturing and performance of the original DIOS scanner, mitigating spatial resolution loss and image distortions that plagued the initial version. The most significant change from the original to the upgraded system is the light source, where a small, uniform opal diffuser replaced the film diffuser. The second change is the streamlined fabrication of the spherical surfaces of the LC-tank using spherical Teflon molds of precisely known radius of curvature. The source is placed at the focal point calculated from the radius of curvature of the spherical LC-tank surface so that the tank is able to collimate the diverging LED light field into the desired parallel-CT geometry as it transmits through the Presage dosimeter. We present a comparison of the new DIOS to DLOS and demonstrate that the upgraded LC-tank has an improved flood field and improved modulation transfer function (MTF). We also discuss a method to compensate for image deformation stemming from LC-tank imperfections using grids. Preliminary findings indicate that the upgraded DIOS system has resolution and image quality sufficient for clinical implementation, while making improvements in cost and practicality compared to DLOS.

## 2. Methods

Evaluations focused on characterizing the improved MTF, the spatial consistency of the point-spread function, the accuracy of geometrical reconstruction, and the stray-light effects of this upgraded system.

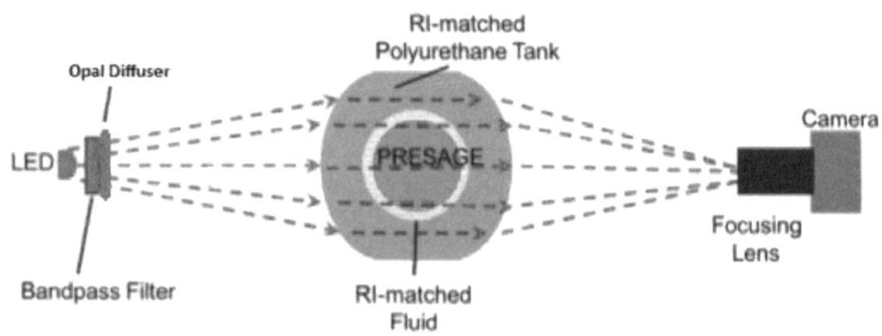


**Figure 1.** Upgraded DIOS system schematic. Novel features include the more precise tank and the large-area source with opal diffuser.

### *2.1 Upgraded LC-Tank*

A new light collimating (LC) tank has been developed with significantly less distortion and high throughput via improved manufacturing procedures, involving two spherical cap Teflon molds that "sandwich" the LC-tank from both sides. Previous methods required destruction of the mold after casting the LC-tank to extract the tank from the mold. The current method allows the Teflon molds to be separated from the LC-tank easily and reused for another tank. In addition, the process to mount the tank onto the rotating stage has been significantly simplified, requiring only drilling a single hole and pouring in slow-curing adhesives to attach the tank to the rotating stage.

### *2.2 Light Field and Flood Field Uniformity*

In addition to the new tank, a small and uniform area light source, created using an LED source and an opal diffuser, replaced the large and heterogeneous diffuser used in the previous DIOS system. This allowed for a more uniform flood field and improved the MTF. To test uniformity of the upgraded tank compared to the previous version, a flood field image was collected. A blank dosimeter was placed in the tank on the center rotating stage. Refractive matching fluid was pipetted into the tank to fill the air gap surrounding the blank dosimeter so that the field appeared uniform. A single image was taken of the field to produce a flood field image. Coefficient of variation (COV) was determined from this flood image by taking a small central area from old and new DIOS flood images and then dividing the pixel-by-pixel standard deviation by mean flood value (adjusted for dark background from thermal noise).

### *2.3 Edge and Pin-hole Image Resolution*

An optical edge and pin-hole were used to evaluate the resolution of the system. A thick piece of black aluminum foil with a straight edge was placed at the center of the tank, immersed in refractive index fluid. The foil covered half of the light field. An image of the edge was used to calculate the MTF using the built in ImageJ MTF calculator. Resolution of the system was estimated from the MTF using the 10% threshold.

A pin-hole was created using an iris immersed in refractive index fluid. This pin-hole was moved to various points around the tank for acquisition. The acquired pin-hole images were used to test the spatial invariance of the point spread function (PSF) of the system. Line profiles were collected through the center of each pinhole and the PSFs were compared.

*2.4 Tank Distortion*

The extent of tank deformation was estimated from a 3D-printed grid with 3.5 mm wide holes and 2.35 mm wide gridlines immersed within the LC-tank. This grid was placed 3.5 cm from the center of the tank towards the camera, the closest location to the camera within the tank. This was the 0 degree rotation stage position. A single image taken at this point was compared with an image taken at 3.5cm from the center of the tank *away* from the camera, the 180 degree rotation stage position. The two images were registered and compared. Two horizontal line profiles taken through the same grid position on each of the two images were compared as were two vertical line profiles. The profiles were analyzed to determine whether there was any position dependent distortion.

*2.5 Stray Light Comparison with DLOS System*

Light may reach the camera via multiple scattering events, called the stray light effect. A thick piece of light-blocking metal was placed at the center of the tank on the rotation stage. A single projection image was acquired of the partially blocked field. An additional projection was acquired of a completely blocked field, with a sheet of metal blocking the entire camera lens. The photon counts of the blocked field and the partially blocked field were compared to determine the amount of stray light present in the system. The stray light analysis for DIOS was compared with the data collected for the same experiment with the DLOS system. The DLOS system version of the experiment used a slightly wider metal piece. The line profile of the intensity of the two images compared with the dark image was analyzed to characterize the stray light.

## 3. Results and Discussion

*3.1 Light Field and Flood Field Uniformity*

The purpose of the diffuser is to compensate for the deformation of the LC-tank causing the focal point of the lens to shift and to blur. A uniform area-source is thus preferred to compensate for this imperfection. The original DIOS flood used a coarse diffuser film and consequently suffered from a textured flood field causing a high coefficient of variation. The upgraded DIOS flood field has significantly improved uniformity in the central area owing to the uniform opal diffuser ($COV_{FloodUpgraded} = 0.0082$ compared to $COV_{FloodOriginal} = 0.151$). This is demonstrated in the consistent image field in figure 2. In the upgraded DIOS image in figure 2B, the two vertical light-colored lines in the image represent the edges of the dosimeter in the tank. Several imperfections remain such as the uneven field in the bottom left side of the image in 2B where the image field is dimmer and appears to curve in at the corner. Several small speckled imperfections remain on the lens face and a thin strip of bubbles where the polyurethane set imperfectly remains at the top of the lens. The image quality is a clear step up from the previous DIOS version that lacked a uniform diffuser.

IOP Conf. Series: Journal of Physics: Conf. Series **1305** (2019) 012019 doi:10.1088/1742-6596/1305/1/012019

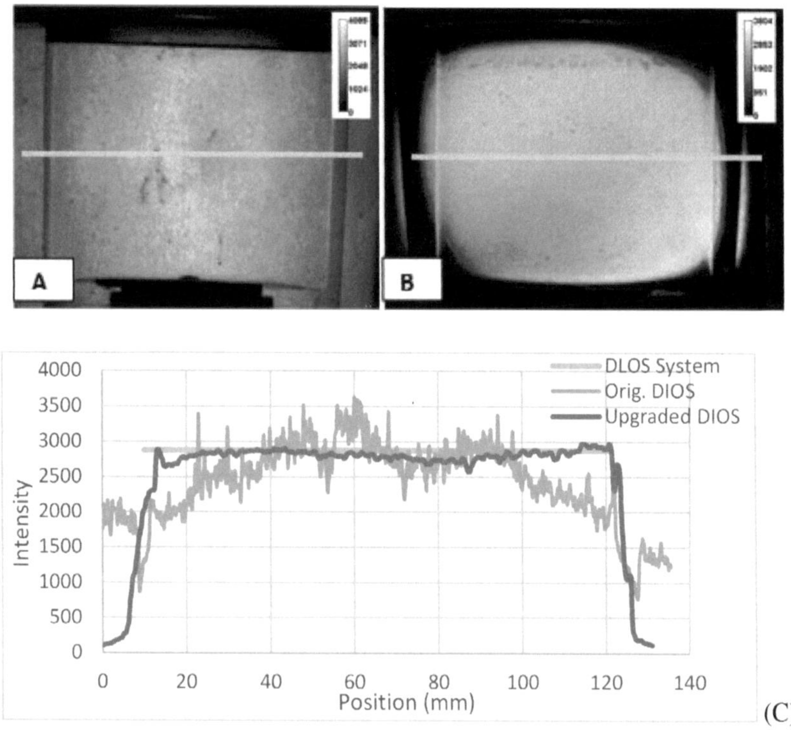


**Figure 2.** Flood field image comparison between the upgraded and the original DIOS systems and the DLOS system. (A) Flood field image from original DIOS system. (B) Flood field image from the upgraded DIOS system. (C) Center horizontal line profile comparison between the original DIOS, upgraded DIOS, and DLOS imaging systems.

### *3.2 Edge and Pin-hole Image Resolution*

Figure 3 is a comparison between the MTF measured with the original DIOS scanner and the current paper's upgraded DIOS scanner. When the MTF is 0.1, the original scanner's cut-off frequency was 1.4 $mm^{-1}$, and able to resolve 0.71 mm. The upgraded DIOS system's cut-off is 3.7 $mm^{-1}$, and able to resolve 0.27 mm. This leads to significantly improved image resolution.

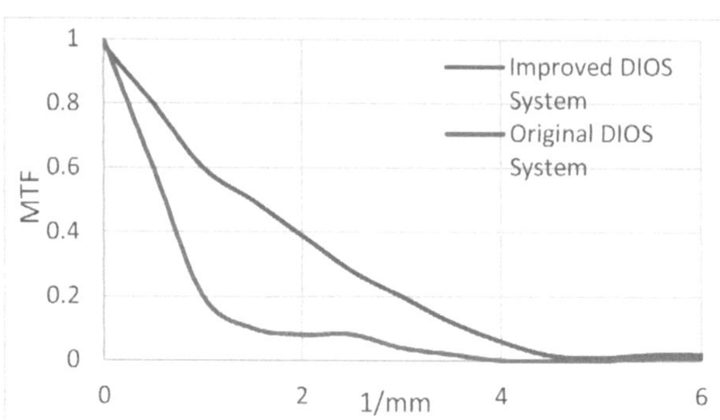


**Figure 3.** MTF comparison between the original DIOS system and the current upgraded DIOS system.

Additionally, through the collection and line profile analysis of a variety of pinhole measurements scattered throughout the tank, consistent profiles were found at various different locations in the field. This indicates that this system has the capability for deconvolution, which would allow for a full-field deconvolution correction for the PSF.

IOP Conf. Series: Journal of Physics: Conf. Series **1305** (2019) 012019 doi:10.1088/1742-6596/1305/1/012019

### *3.3 Tank Distortion*

The images in figures 4 indicate that spatial deformation is minimal with respect to position on the tank. The image of the grid placed at +3.5 cm from the center of the LC-tank (rotation stage 0° position) does not deviate significantly when compared to image of the grid placed at -3.5 cm (rotation stage 180° position), meaning light is traveling relatively collimated within the ±3.5cm with respect to the scanner center of rotation. However, qualitatively, we note there is a systemic barrel distortion causing a bulged appearance of the grids in figure 4A and B. This is likely due to the spherical surface of the lens where our flat-lens approximations fail. A deformation field correction could be applied to the images before reconstruction. Methods to correct deformation of the images will be explored in future works.

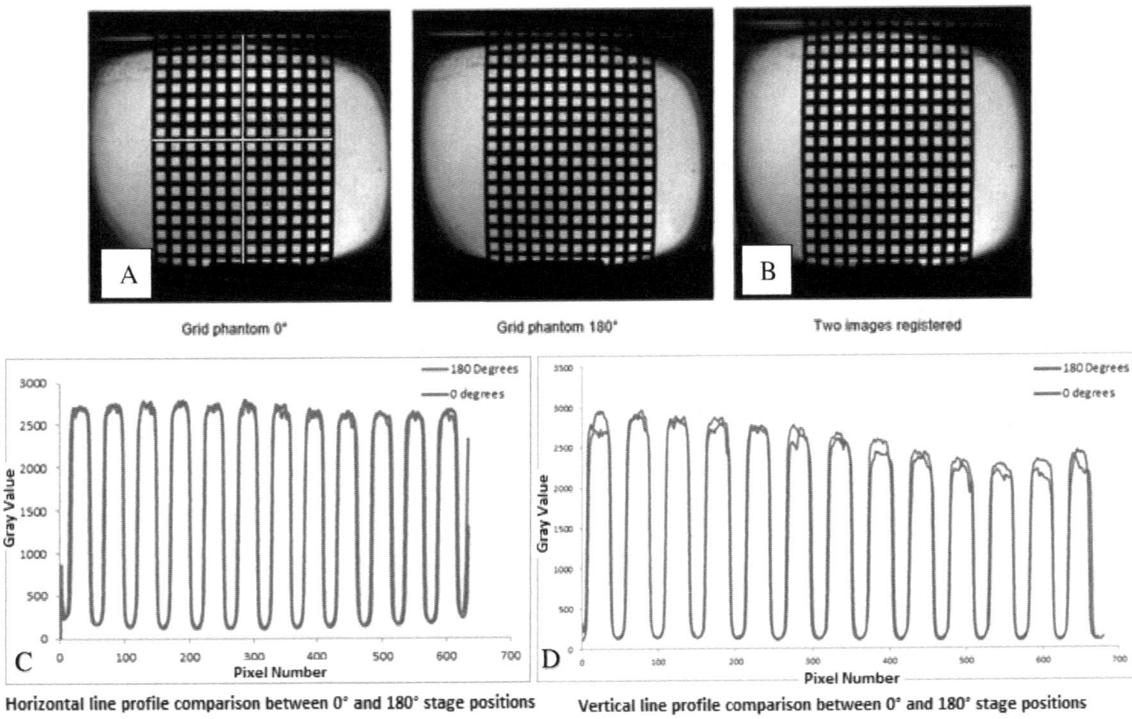


**Figure 4.** Depth of focus analysis using a grid phantom attached to the rotating stage. (A) Images taken with the grid phantom at the 0 degree rotating stage position and the 180 degree position. This test was to determine if there are any depth of focus distortions depending on location of the object in the tank. (B) The two images in A were registered for comparison. (C) Horizontal line profile depth of focus comparison, according to the yellow profile drawn in (A), between a grid phantom positioned at 0 degrees and a grid phantom positioned at 180 degrees in the tank. (D) Vertical line profile depth of focus comparison according to the vertical yellow profile drawn in (A).

### *3.4 Stray Light Comparison with DLOS System*

The stray light analysis concluded that due to the telecentric lenses used in the DLOS system, the elimination of stray light is still superior in the DLOS system compared with the DIOS system. However, both systems reduced the stray light by at least 2 orders of magnitude. Deconvolution by a PSF for the DIOS system could eliminate the stray light [4].

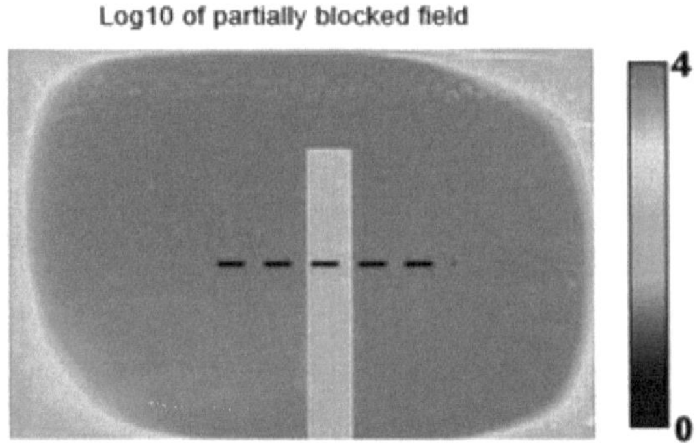


(A)

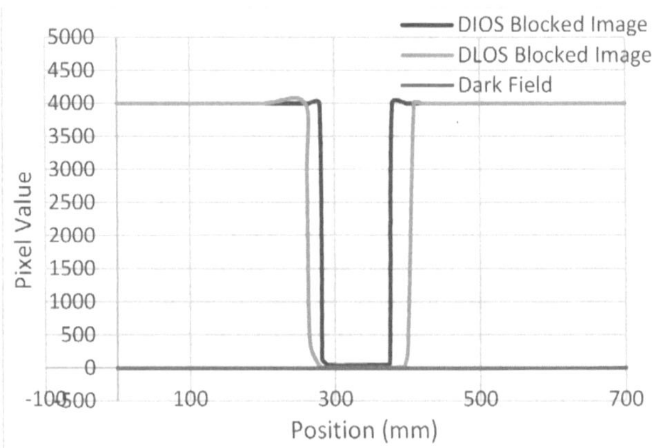


(B)

**Figure 5.** Comparison of stray light between the DLOS and DIOS systems. (A) DIOS flood projection image of an opaque light block placed in the center of the field of view. (B) Line profile, the black line shown in (A), through the DIOS system compared with the same experiment and line projection on the DLOS system and compared with a line profile through a dark field image.

## 4. Conclusions

The basic characterizations presented here demonstrate that the DIOS system establishes a parallel ray geometry with good geometric accuracy approaching that of a telecentric system (i.e. DLOS). A notable increase in stray-light under a light block was observed in the DIOS when compared to DLOS. This is to be expected as a result of the price paid for the lack of telecentricity. The exact implications of this increased scatter are unclear at this point, and will need further study in the context of 3D dosimetry verification measurements. However, the fact that the point spread function was observed to be largely spatially invariant may offer an approach to partially correct for any loss of accuracy through point-spread-deconvolution scatter correction.

## 5. Acknowledgement

The research reported was supported by the National Institutes of Health (NIH) under award number 1R41CA217421-01 A1.

## 6. References

[1] Gore J C *et al* 1996 *Phys. Med. Biol.* **41** 2695
[2] Maryañski M J *et al* 1996 *Phys. Med. Biol.* **41** 2705
[3] Doran S J *et al* 2001 *Phys. Med. Biol.* **46** 3191
[4] Krstaji N and Doran S J 2007 *Phys. Med. Biol.* **52** 3693
[5] Olding T *et al* 2010 *Phys. Med. Biol.* **55** 2819

[6] Baldock C 2009 *J. Phys.: Conf. Ser.* **164** 012002
[7] Babic S *et al* 2009 *Phys. Med. Biol.* **54** 2463
[8] Thomas A *et al* 2011. *Med. Phys.* **38** 4846-57
[9] Sakhalkar H S *et al* 2009 M*ed. Phys.* **36** 3371-7
[10] Thomas A *et al* 2011 *Phys. Med. Biol.* **56** 4433-51
[11] Miles D *et al* 2017 *J. Phys.: Conf. Ser.* **847** 012026

IOP Conf. Series: Journal of Physics: Conf. Series **1305** (2019) 012020 doi:10.1088/1742-6596/1305/1/012020

# Cause of cupping artifacts from radiochromic micelle gel dosimeters used in optical CT scanner measurement

**Takaoki Takanashi[1], Kazuya Hayashi[2], Mikio Nemoto[1,3], Hiraku Kawamura[1,4], Shin-ichiro Hayashi[5] and Hiroaki Gotoh[2]**

[1]3D Gel Dosimeter Research Laboratory, Cluster for Science, Technology and Innovation Hub, RIKEN, 2-1, Hirosawa, Wako, Saitama 351-0198, Japan
[2]Department of Chemistry and Life Science, Graduate School of Engineering Science, Yokohama National University, 79-1 Tokiwadai, Hodogaya-ku, Yokohama, Kanagawa, 240-8501, Japan
[3]Department of Radiology, Jichi Medical University Hospital, 3311-1, Yakushiji, Shimotsuke, Tochigi 329-0498, Japan
[4]Department of Radiological Technology, Faculty of Health Sciences, Tsukuba International University, 6-20-1, Manabe, Tsuchiura, Ibaraki, 300-0051, Japan
[5]Department of Clinical Radiology, Faculty of Health Sciences, Hiroshima International University, Higashi-Hiroshima, Hiroshima, 739-2695, Japan

E-Mail: gotoh-hiroaki-yw@ynu.ac.jp

**Abstract**. When a radiochromic micelle gel dosimeter is employed for optical computed tomography (CT) measurement, cupping (or dishing) artifacts appear at areas irradiated with a high dose. Anti-scatter polarizer correction is employed to remove scatter signals from optical CT data, but cupping remains. Here, measurement conditions for reducing cupping artifacts are investigated. A change in observation wavelength is found to suppress the cupping influence. Measurements involving aqueous dye solutions with varying jar sizes and dye concentrations reveal cupping artifact behavior under various conditions.

## 1. Introduction

Radiochromic gel dosimeters, i.e., micelle, polymer, and Fricke gel dosimeters, are promising tools for verification of 3D dose distribution. These 3D dose distributions can be read using magnetic resonance imaging (MRI) X-ray computed tomography (CT) and optical CT [1, 2]. In optical CT measurement using a polymer gel dosimeter [3-5], the 3D dose distribution is verified based on changes in the gel's light attenuation with increasing irradiation dose. However, the gel optical behavior features light scattering by the polymer. In optical CT measurement, the opacity caused by absorption, scattering, reflection, and refraction is evaluated as the optical density (OD). In particular, cupping artifacts are observed as a result of this scatter signal [6, 7].

For a polymer gel dosimeter, application of beam-stop array (BSA) scatter corrections yields good results [8]. For a micelle gel dosimeter, however, the 3D dose distribution is verified based on the dye attenuation. A scattering effect has little influence on the optical CT measurement in the micelle gel dosimeter. The micelle gel dosimeter has several advantages over other types of dosimeter. For example, this dosimeter has no or low diffusion, insensitivity to oxygen, lower toxicity compared to the alternatives, and is suitable for optical CT read-out. Leuco crystal violet (LCV) and leuco malachite green (LMG) micelle gel dosimeters have been reported [9, 10]. However, these gels have low dose sensitivity. To resolve this problem, addition of clay to radiochromic micelle gel has been investigated.

In this study, measurement conditions for reducing the influence of cupping artifacts are investigated, for LCV and LMG radiochromic micelle gel dosimetry samples and using the Vista 15™ Optical CT Scanner (Modus Medical Devices, Inc., Canada) [11]. Focus is placed on high-dose (>1000 MU) irradiated areas in which cupping artifacts are exhibited. As cupping artifacts appear, the dose response from the center of jar in the uniformly irradiated area is underestimated. Removal of the cupping artifact influence is important for accurate evaluation of the 3D dose distribution.

## 2. Materials and methods

### 2.1. *Gel fabrication*

In this study, LCV (Wako Pure Chemical Inc., Japan) and LMG (Wako Pure Chemical Inc., Japan) radiochromic micelle gel dyes were used. The gel formulation was 5% gelatin, 0.16% Laponite XLG (BYK Japan KK), 0.3-mM dye, 50-mM dichloromethane, and 50-mM Triton X-100. The Laponite XLG was added to water and then stirred to form a clear colorless dispersion in an oil bath at 50°C. The gelatin (Nacalai Tesque, guaranteed reagent, Japan) was slowly added to the Laponite dispersion and stirred under the same conditions for 1 h. After the gelatin dissolved, it was cooled at 35°C. The dye was added to dichloromethane and mixed under light shielding. The dye solution was then poured into the Triton X-100 (Alfa Aesar, USA) and stirred at room temperature. The gelatin solution was then added slowly to the surfactant solution and poured into polyethylene terephthalate jars (900 mL) supplied by Modus Medical Devices. The samples were stored at 4°C for gelation.

### 2.2. *Irradiation*

To achieve uniformly distributed dose irradiation, the jar (900 mL) was fixed sideways on foam (for which the radiation absorption could be neglected), and placed on the center of the cylinder to the isocenter of a medical linear accelerator (Varian CL-iX). Rotational irradiation was performed at 360° using a 10-MV X-ray beam from a direction perpendicular to the long axis of the cylinder. The irradiation doses were 250–2000MU (44–353 cGy) and 500–3000MU (88–530 cGy) for the LCV and LMG samples, respectively. To investigate the influence of the cupping artifacts in the case of a half-moon distribution, another LCV sample was subjected to 2000 cGy irradiation with the parallel-opposed half-beam irradiation fields, with equal weight in the oppositely aligned upper and lower irradiation fields.

### 2.3. *Dye solution preparation*

To study the effect of the jar size, crystal violet (CV, Wako) solution was prepared. Three types of jar supplied by Modus Medical Devices (large, L: 2.6 L; medium, M: 900 mL; small, S: 450 mL) were evaluated. The dye concentration was 6.0 μM.

### 2.4. *Optical CT scanning*

Optical CT scanning was conducted with a commercial optical cone beam scanner (Vista 15™, Modus Medical Devices). In all measurements, reference and data scans were taken using 633- and 590-nm light-emitting diode (LED) illuminations and a 1024 × 768-pixel charge-coupled display camera. For measurement without the scatter signal, polarizing films were placed at the front and back faces of the scanner aquarium and samples were measured. A total of 342 image sets were reconstructed using software provided with the Vista 15™ scanner.

## 3. Results and discussion

For the 590-nm LED illumination measurement of the LCV samples, cupping artifacts were exhibited for the samples subjected to irradiation exceeding 1000 MU (176 cGy). On the other hand, for the 633-nm illumination measurement, the gray values of all conditions were decreased and cupping artifacts were not exhibited (Figure 1). For the 633-nm illumination measurement of the LMG samples, cupping artifacts were exhibited for the samples subjected to 3000 MU (530 cGy) irradiation. For the 590-nm measurement, as for the LCV samples, the gray values of all conditions decreased and no cupping artifacts were exhibited. The results were identical with and without use of an anti-scatter polarizer. Cupping artifacts were exhibited for the samples subjected to irradiation exceeding 1000 MU (176 cGy). The dose response from the averages of the gray values in the center of each sample are shown below (Figure 2). Of the results exhibiting cupping artifacts, for the 590-nm LCV case, saturation occurred at 1500 MU (265 cGy), with underestimation at 2000 MU (353 cGy). Similarly, some underestimation at 2000 MU (353 cGy) was observed for the 633-nm LMG measurement.

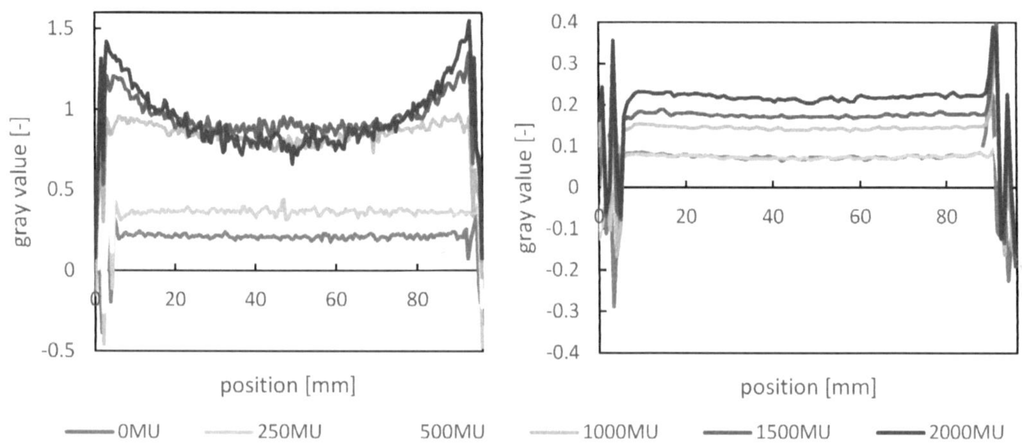


**Figure 1.** The results of LCV samples from optical CT scanning. left:590-nm, right:633-nm.

For jar L (2.6 L), the CV samples could not be measured under 590-nm illumination because of the strong light absorption when the CV concentration exceeded 3 μM. Cupping artifacts were exhibited for samples with some concentrations exceeding 3.75 and 4.5 μM for jars M and S, respectively. However, for the 633-nm illumination measurement of the CV samples, no cupping artifacts were exhibited under any conditions.

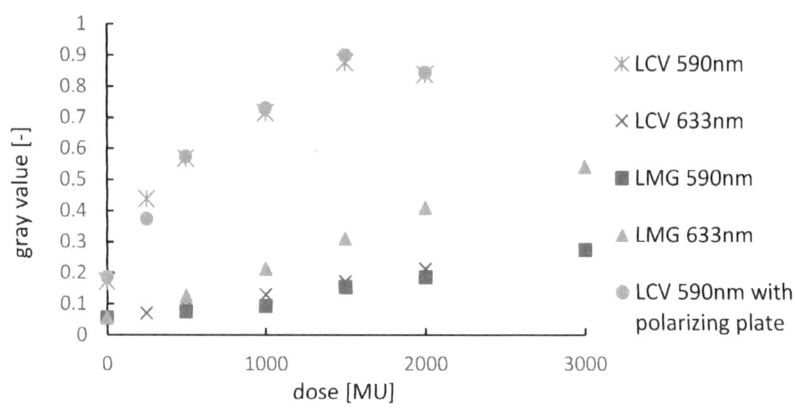


**Figure 2.** The averages of gray value in each sample. The position using calculation was the central axis and 3 mm around it.

For the half-moon distribution irradiated sample measured by optical CT, the gray values were approximately 1.0 and 0.14 for the 590- and 633-nm measurements, respectively. Figure 3 shows comparative dose profiles based on the two measured dose distributions and calculated profiles obtained from simulation with a treatment planning system (Eclipse, VARIAN Medical Systems). The calculated profile is closer to the 633-nm measured profile than the 590-nm profile. The 590-nm measured profile exhibits lower values than the calculation. The source of error in the 633-nm case may be low gray values, which render the effect of noise relatively large.

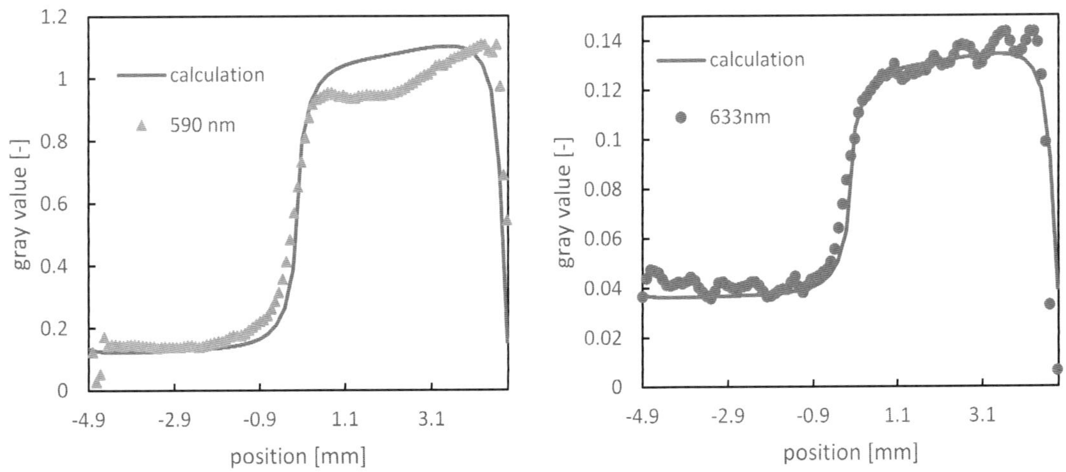


**Figure 3.** The results of half irradiated samples. The solid line shows calculation value of absorbed dose. left: 590-nm measurement, right: 633-nm measurement.

It was predicted that the scatter signal would have a weak effect on the measurement for the CV solution; however, cupping artifacts were exhibited. This indicates that the cupping artifacts are caused by factors other than the scatter signal, or the scattering from the micelles may be larger than expected. Additionally, changing the jar size causes cupping artifacts for constant CV concentration in the solution. That is, cupping artifacts are observed for the areas containing large amounts of dye and having strong absorption. When the same areas of the same samples are measured using a wavelength different from the maximum absorption wavelength of the dye, cupping artifacts are not exhibited. In those areas, the transmitted light intensity is strong. Therefore, the appearance of cupping artifacts is thought to be related to low light intensity through the sample. Strong light intensity can be achieved for measurement using wavelengths other than maximum absorption wavelength of the dye, because of the decreasing dye absorption. Therefore, by switching the observation wavelength, the transmitted light intensity is strengthened, and the influence of cupping can be suppressed.

However, for measurement using wavelengths other than the maximum absorption wavelength of the dye, it is impossible to differentiate the gray values of samples with low irradiation dose (500M (88 cGy) or less). Previously, Olding, Holmes and Schreiner reported that the influence of the scatter signal caused by a polymer gel dosimeter can be corrected through measurement using BSA [8]. Therefore, it is necessary to investigate the measurement results for the dye gel to remove the scatter signal, to confirm the possibility of suppressing the influence of cupping by changing the observation wavelength. Furthermore, cupping artifacts were also exhibited for half irradiation of the gel samples in this study. Based on this result, when a three-dimensional dose distribution is evaluated using a micelle gel dosimeter, the influence of the cupping artifacts should be considered. It is necessary to increase the dose distribution measurement accuracy using optical CT by establishing an appropriate relationship between the dose (OD) and the intensity of the traversing light.

IOP Conf. Series: Journal of Physics: Conf. Series **1305** (2019) 012020 doi:10.1088/1742-6596/1305/1/012020

## 4. Conclusion

In this work, different dye gel samples for radiochromic micelle gel dosimeters applied in optical CT were examined, to establish measurement conditions for cupping artifact reduction. Areas irradiated with a high dose exhibited cupping artifacts, but it was possible to suppress the influence of cupping by changing the observation wavelength. The source of this behavior is thought to be light scattering caused by dye, gelatin and/or micelles.


## 5. Acknowledgement

We thank A. Oe (Jichi Medical University Hospital) for medical linear accelerator operation, and BYK-Chemie Japan and Wilbur-Ellis Japan for providing nanoclay samples. This work was supported by Nuclear Technology Co., Ltd. and The Wakasa wan Energy Research Center.


## 6. References


[1] Baldock C 2009 *J. Phys.: Conf. Ser.* **164** 012002
[2] Baldock C et al 2010 Phys. Med. Biol. 55 R1-63
[3] De Jean P et al 2001 *J. Phys.: Conf. Ser.* **56** 179-86
[4] Sarabipour S *et al* 2006 *J. Phys.: Conf. Ser.* **56** 280-2
[5] Bosi *et al* 2009 *Applied Optics* **48** 2427-34
[6] Bosi S G *et al* 2007 *Phys. Med. Biol.* **52** 2893-903
[7] Bosi S G et al 2009 *J. Phys.: Conf. Ser.* **164** 012021
[8] Olding T *et al* 2009 *J. Phys.: Conf. Ser.* **164** 012031
[9] Jordan K and Avvakumov N 2009 *Phys. Med. Biol.* **54** 6773-89
[10] Babic S *et al* 2009 *Phys. Med. Biol.* **54** 6791-808
[11] Miller J *et al* 2005 *Med. Phys.* **32** 2138

IOP Conf. Series: Journal of Physics: Conf. Series **1305** (2019) 012021 doi:10.1088/1742-6596/1305/1/012021

# Influence of angular mismatch between pre- and post-irradiation scans for optical-CT gel readout: simulation based study

**Yi Du[1, 2, 3], Xiangang Wang [2*], Xincheng Xiang[2], Hao Wu [1**] and Yves De Deene**
[1]Key laboratory of Carcinogenesis and Translational Research (Ministry of Education/Beijing), Department of Radiotherapy, Peking University Cancer Hospital & Institute, 52 Fucheng Road, Beijing, 100142, China
[2] nstitute of Nuclear & New Energy Technology, Tsinghua University, Beijing, 100084, China
[3]School of Engineering, Faculty of Science and Engineering, Macquarie University, Sydney, NSW 2109, Australia

E-mail: *wangxiangang@tsinghua.edu.cn; **13552661030@139.com

**Abstract**. The pre- and post-irradiation scan strategy for optical-CT gel readout often turns out to be corrupted by angular mismatch between these two scans. In this study, we used computational simulations to investigate the influence of angular mismatch. Two phantoms are constructed: one cylindrical phantom with synthetic impurities and one elliptical phantom. The reconstructed results of angular mismatched pre- and post- data show that the dual-scan method is very sensitive to repositioning error, and *positive-negative pair errors* can be easily identified around impurities and phantom edges. From the simulation results, we believe that the angular mismatch should be less than 0.1 degree.

## 1. Introduction

Since the gel dosimeter was first proposed by Gore in 1984 [1], optical-CT has become a major readout modality for gel-based 3D radiotherapy dose distribution measurement, utilizing the property of quantitative optical density changes (ΔOD) induced by localized absorbed dose [2]. Several system designs have been proposed over the past decades [3, 4] and two compact scanners are now commercially available (Vista™ by Modus Medical Devices Inc., and OCTOPUS™ by MGS Research Inc.). Despite variations in system design, all optical-CT scanners utilize the common physical principle [5]: tomographic OD maps of gel dosimeters are reconstructed from traverse projection data captured at each angle over 360 degrees, and then absolute dose maps can be derived by calibration of ΔOD to dose.

An ideal gel dosimeter is expected to be homogeneous. However, in the process of fabrication and measurement, gel dosimeter phantoms often appear to be *imperfect* with impurities, such as dust or air bubbles inside gel matrix, stains or scratches on phantom surfaces, and seams of phantom holders, all of which degrade signal quality and reconstructed OD maps by noise and artifacts. In order to improve image quality and only obtain ΔOD maps in gel dosimeters, the strategy of using pre- scan and post-irradiation scans is commonly used [6-8]. However, although phantoms are repositioned very carefully for post scan, all too often resultant ΔOD maps turn out corrupted by phantom imperfections, which nevertheless leads us to question the reproducibility of the pre- and post- scan method. In this study, we used computational simulations to investigate the influence of angular mismatch between pre- and post-irradiation scans and discussed the required accuracy for phantom repositioning.

## 2. Methods and materials

### *2.1. Synthetic Phantoms*

In this study, two synthetic phantoms were designed with ΔOD maps to mimic dose distributions irradiated by a 6-MV Linac with 100-cm SSD.

*2.1.1. Cylindrical Phantom with Impurities.* Since most gel dosimeters are prepared in cylindrical holders, a cylindrical phantom similar to the one in [9] was used, which was 10-cm in height and 10-cm in diameter shown in Figure 1(a). The irradiation field was 2-cm × 2-cm and 2-cm offset from the central axis(CAX). It is noted that, to investigate the behaviour of phantom imperfections in angular mismatch, a pair of two-pixel patches along the z-direction on the phantom surface were constructed to simulate holder seams; in the meantime, two bright seven-pixel particles were constructed to simulate strains: one in the irradiation field and the other in the gel matrix.

*2.1.2. Elliptical Phantom.* With the advent of non-water dosimeters (Presage [10, 11] and FlexyDos3D [12, 13]), dosimeters can be mechanically strong enough to be cast in anthropomorphic shapes. To study the influence of angular mismatch to non-cylindrical phantoms, an elliptical phantom was built as shown in Figure 1(c), which was 10-cm in height 10-cm in major-axis and 5-cm in minor axis. The irradiation field was 1-cm × 1-cm and 2-cm off CAX along x-direction.

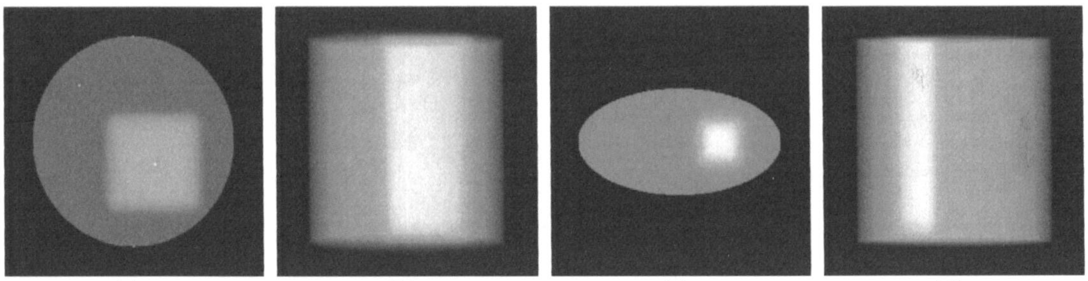

(a) (b) (c) (d)

**Figure 1.** Defected cylindrical phantom: (a) central axial slice and (b) projection image; Elliptical phantom: (c) central axial slice and (d) projection image.

### *2.2. Modelling of Optical-CT Imaging and Angular Mismatch*

We used the specifications of our Vista optical-CT scanner for geometry definition. 360 logarithmic projections from 0-degree to 359-degree of 1-degree interval were generated using the GPU accelerated pixel-driven algorithm [14]. To model the dual-scan method, the synthetic phantoms without ΔOD patches were first utilized to generate projections as pre-irradiation logarithmic data (*Ref*); then the phantoms with ΔODs were used to generate post-irradiation logarithmic data (*Data*), as shown in Figure 1(b)&(d).

To model the angular mismatch between *Ref* and *Data*, during *Data* generation we shifted all the projection angles by a small constant value ($\delta$), which means *Ref* and *Data* are angularly mismatched by $\delta$. Three mismatch scenarios were performed for both phantoms with $\delta$ = 0.1, 0.5 and 1.0 degree respectively. The projections by subtracting *Ref* from *Data* were used for tomographic image reconstruction using FDK (Hann filter) algorithm [15].

IOP Conf. Series: Journal of Physics: Conf. Series **1305** (2019) 012021 doi:10.1088/1742-6596/1305/1/012021

## 3. Results & Discussion

### *3.1. Cylindrical Phantom with Impurities*

The reconstructed images with zoomed details on the synthetic impurities are given in Figure 2, along with difference maps between the $\delta = 0$ image. From the difference maps, we can see that when pre- and post- data are not perfectly matched, the impact of phantom impurities can be easily identified. What's worse, the impurities impair image quality by forming *positive-negative pulse pairs* in the reconstructed images, and as δ increases pulse pairs become obvious.

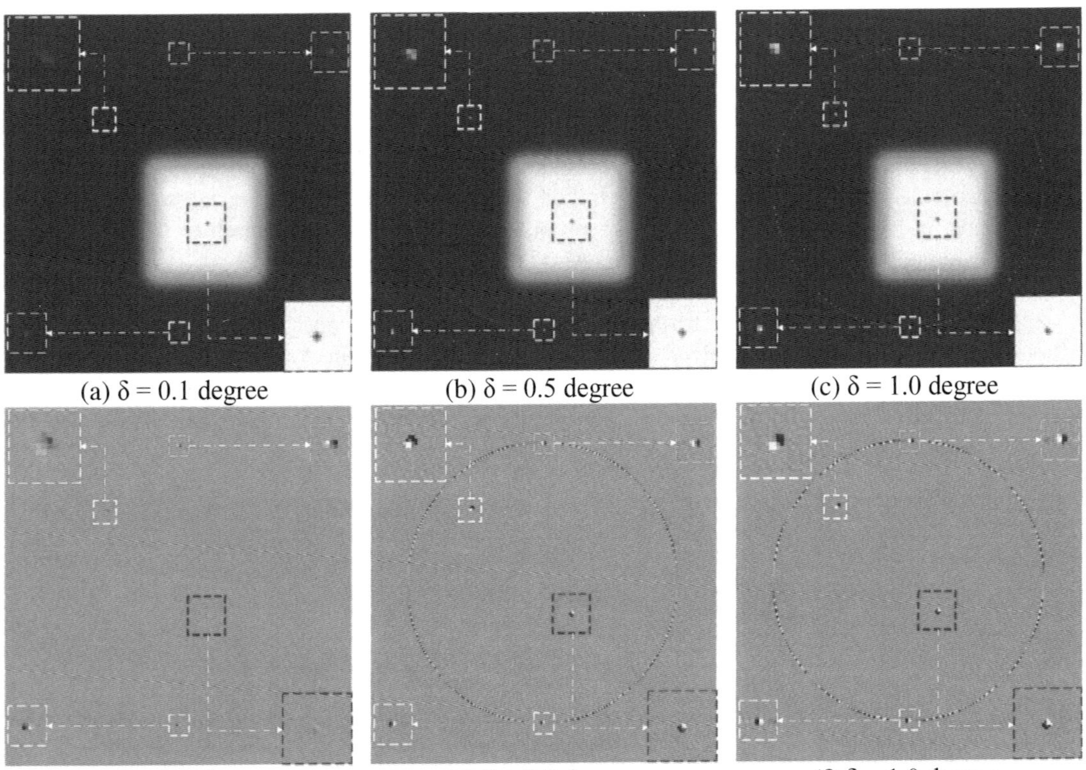

(a) δ = 0.1 degree (b) δ = 0.5 degree (c) δ = 1.0 degree

(d) δ = 0.1 degree (e) δ = 0.5 degree (f) δ = 1.0 degree

**Figure 2.** Reconstructed images with zoomed details when the angles are mismatched by (a) 0.1 degree, (b) 0.5 degree and (c) 1.0 deg. Difference maps with zoomed details between the perfect matching resultant image and (a) (b) and (c) are shown in (d) (e) and (f) respectively.

### *3.2. Elliptical Phantom*

The reconstructed images of the elliptical phantom are given in Figure 2, with difference maps between $\delta = 0$ image as well. Since the traverse distance over the phantom at each angle is not the same, we can see that the angular mismatch leads to a *positive-negative ring errors* around the phantom edge, and the ring becomes thicker as δ increases.

IOP Conf. Series: Journal of Physics: Conf. Series **1305** (2019) 012021 doi:10.1088/1742-6596/1305/1/012021

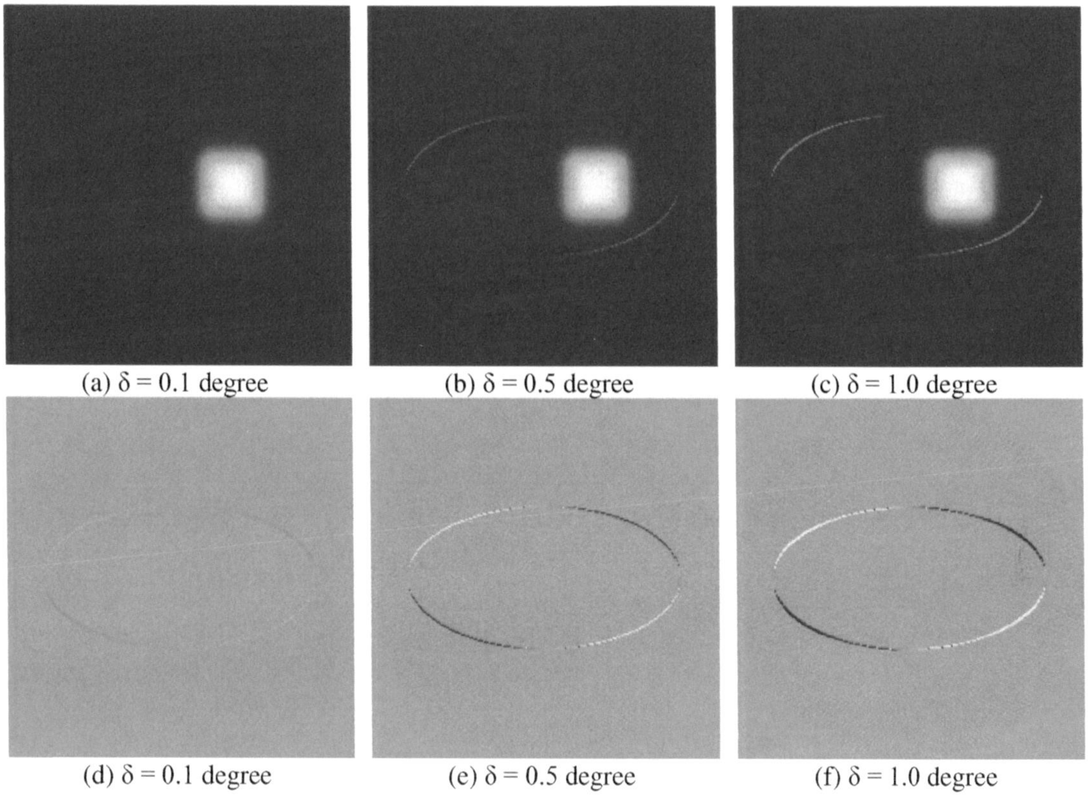


**Figure 3.** Reconstructed images with zoomed details when the angles are mismatched by (a) 0.1 degree, (b) 0.5 degree and (c) 1.0 degree. Difference maps with zoomed details between the perfect matching resultant image and (a) (b) and (c) are shown in (d) (e) and (f) respectively.

## 4. Conclusions

In this study, we can see that the pre- and post-irradiation scan method for optical-CT gel readout is very sensitive to repositioning error, and even when angular mismatch is as small as 0.1 degree, *positive-negative pair errors* can be identified around impurities and phantom edges. It is noted that the processes of scattering, reflection and refraction are not taken into account in the simulation, which all deflect light direction during light propagation. As a result, the influence of angular mismatch would be much more complicated in real cases, which implies that to obtain proper image quality the angular matching accuracy between pre- and post- scan data should be less than 0.1 degree. Considering the accuracy requirement, rather than using a precisely designed mechanical registration stage, we believe algorithm-based data registration or image processing would be a feasible cost-effective choice.

## 5. Acknowledgement

Jointly Supported by Beijing Natural Science Foundation (No. 1184014, 1174016), National Science Foundation of China (No. 61571262, 11575095), China Postdoctoral Science Foundation (No. 2017M620542) and National Key Research and Development Program (No. 2016YFC0105406).

## 6. References

[1] Gore J C and Kang Y S 1984 *Phys. Med. Biol.* **29** 1189-97
[2] Baldock C 2009 *J. Phys.: Conf. Ser.* **164** 012002
[3] Baldock C et al 2010 *Phys. Med. Biol.* **55** R1-63
[4] Doran S J 2009 *J. Phys.: Conf. Ser.* **164** 012020

[5] Jordan K J 2017 *J. Phys.: Conf. Ser.* **847** 012019
[6] Wuu C-S and Xu Y 2010 *J. Phys.: Conf. Ser.* **250** 012044
[7] Doran S J 2013 *J. Phys.: Conf. Ser.* **444** 012004
[8] Doran S J 2010 *J. Phys.: Conf. Ser.* **250** 012086
[9] Du Y *et al* 2017 *J. Phys.: Conf. Ser.* **847** 012069
[10] Schreiner L J 2015 *J. Phys. Conf. Ser.* **573** 012003
[11] Khezerloo D *et al* 2017 *Radiat. Phys. Chem.* **141** 88-97
[12] De Deene Y *et al* 2015 *Phys. Med. Biol.* **60** 1543-63
[13] Høye E M *et al* 2015 *J. Phys.: Conf. Ser.* **573** 012067
[14] Du Y *et al* 2017 *Biomed. Eng. Online* **16** 2
[15] De Deene Y 2015 *J. Phys.: Conf. Ser.* **573** 012076

IOP Conf. Series: Journal of Physics: Conf. Series **1305** (2019) 012022 doi:10.1088/1742-6596/1305/1/012022

# Fundamentals of 3D dosimetry

**L J Schreiner**
Department of Medical Physics, Cancer Centre of Southeastern Ontario at the Kingston General Hospital, Kingston, Ontario, Canada, K7L 5P9
Departments of Oncology and Physics, Queen's University, Kingston, Ontario, Canada K7L 3N6

E-mail: john.schreiner@krcc.on.ca

**Abstract**. Since 1999 it has been convention at the IC3Ddose conferences to provide an opportunity for interested attendees to come together for an early morning session to review the basic properties of one family of 3D dosimeters: volumetric chemical dosimeters. Seasoned workers in the field benefit from a refreshed perspective, while those new to the field get an overview of this one modality for 3D dose measurement early in the conference. The intent of the session is to describe the basic science and characteristics of these true 3D systems, how they work and how they are read out. This provides some basic foundation to all attendees before the conference moves to the more detailed presentations directed to various specific issues and applications (e.g., looking at specific gel or plastic dosimeter, its readout and its clinical utilization). This talk is that session.

## 1. Introduction

Modern radiation therapy has advanced considerably with the development of external beam techniques such as Intensity Modulated Radiation Therapy (IMRT), Volumetric Modulated Arc Therapy (VMAT), Stereotactic Ablative Radiation Therapy (SABR), Fractionated Stereotactic Radiotherapy (FSRT), etc.; and through improved high-dose rate brachytherapy. The radiation delivery associated with these techniques provides shaped conformal dose distributions that must be correctly registered to the patient's anatomy to achieve the treatment intent. There is a commensurate requirement for increased quality assurance (QA) associated with these complex deliveries. This extended QA includes the validation of the absorbed dose pattern in three-dimensional (3D) space: in common parlance, the validation of the dose distribution.

Various approaches are available to provide this assurance including, for example, the use of 2D and 3D arrays of detectors or electronic portal imaging detectors (EPIDs) for patient specific validation of the dose delivery. While these approaches are clinically very useful, they typically provide sparse 3D data and only surrogate validation of 3D dose delivery. A unique supplementary methodology for conformal delivery QA is offered by 3D dosimetry using volumetric chemical dosimeters probed by 3D imaging systems. Together the materials and the appropriate imaging can provide dosimetric measurements throughout an irradiated volume which can be compared, for example, to the desired dose distribution generated during treatment planning.

In this talk I will review the basic principles by which 3D chemical dosimeter systems function: presenting the main classes of dosimeters, the basic properties that mediate their radiation sensitive response, and how they are read out. Their role in clinical QA will be described.

The material is not new and many of the points addressed have been discussed previously. Rather than write another review article in these proceedings, I refer readers to some of the good existing didactic reviews [1-7]. This list, along with the many comprehensive papers that have been made freely available in the open source conference proceedings from the last seven 3D dose conferences [8-14], should provide attendees a good foundation for their continued work.

## 2. References

[1] Schreiner L J 2009 *J. Phys.: Conf. Ser.* **164** 012001.
[2] Schreiner L J and Olding T 2009 In *AAPM Medical Physics Monograph: Clinical Dosimetry Measurements in Radiotherapy* Roger D and Cygler J (ed) Madison, WI: Medical Physics Publishing.
[3] Baldock C *et al.* 2010 *Phys. Med.Biol.* **55** R1-63
[4] Oldham M 2014 In *Advances in Medical Physics* Devon J G et al (eds) Madison, WI: Medical Physics Publishing.
[5] Schreiner L J 2015 *J. Phys.: Conf. Ser.* **573** 012003
[6] De Deene Y 2018 In *Clinical 3D Dosimetry in Modern Radiation Therapy* Mijnheer B (ed) Boca Raton: CRC Press.
[7] Oldham M, Juang T and Yoon SW 2018 *In Clinical 3D Dosimetry in Modern Radiation Therapy* Mijnheer B (ed) Boca Raton: CRC Press.
[8] 2004 Proc. DosGel 2004: 3rd Int. Conf. on Radiotherapy Gel Dosimetry. C Baldock and Y De Deene (eds) *J. Phys.: Conf. Ser.* **3**
[9] 2006 Proc. DosGel 2006: 4th Int. Conf. on Radiotherapy Gel Dosimetry. Lepage M (ed) *J. Phys.: Conf. Ser.* **56**
[10] 2009 Proc. of DosGel 2008: 5th Int. Conf. on 3D Radiotherapy Dosimetry. Maris TG and Pappas E (eds) *J. Phys.: Conf. Ser.* **164**
[11] 2010 Proc. of IC3DDose 2010: 6th Int. Conf. on 3D Radiation Dosimetry. Oldham M and Newton J (eds) *J. Phys.: Conf. Ser.* **250**
[12] 2013 Proc. of IC3DDose 2012: 7th Int. Conf. on 3D Radiation Dosimetry. Thwaites D and Baldock C (eds) *J. Phys.: Conf. Ser.* **444**
[13] 2015 Proc. of IC3DDose 2014: 8th Int. Conf. on 3D Radiation Dosimetry. Bäck S and Olsson L E (eds) *J. Phys.: Conf. Ser.* **573**
[14] 2017 Proc. of IC3DDose 2016: 9th Int. Conf. on 3D Radiation Dosimetry. Baldock C and Ibbot G (eds) *J. Phys.: Conf. Ser.* **847**

# Pre-clinical and small field dosimetry

**Cheng-Shie Wuu[1], Yi-Fang Wang[1], Andy Y. Xu[1] and John Adamovics[2]**
[1]Department of Radiation Oncology, Columbia University, NY, NY 10032, USA
[2]Department of Chemistry, Rider University, Lawrenceville, NJ 08648, USA

E-mail: csw6@columbia.edu

**Abstract**. Preclinical in vivo studies have drastically improved over the past decade with the development of cone beam computed tomography (CBCT) image-guided small animal irradiation systems. Such systems produce 220 or 225 kV x-rays with square and circular field sizes ranging from 0.5 to 10 mm. The dosimetry of such equipment involving kilovoltage small-field dosimetry has not received as much attention as the megavoltage small-field dosimetry. The dosimetry of megavoltage small fields can be challenging due to lateral charged particle disequilibrium, detector volume averaging effect, and high dose gradients. Clinically there has been a rapid increase in the use of small fields in modern radiotherapy techniques such as stereotactic radiosurgery (SRS) and stereotactic radiotherapy (SBRT). This study presents dosimetric properties of image-guided small animal irradiation systems. Both EBT Gafchromic films and 3-D PRESAGE measured beam data were presented and compared with the calculated dose distribution from a commissioned planning system. For megavoltage small-field dosimetry, EBT3 films and PRESAGE dosimeters were used to measure the dose distributions for MLC-delimited 6 x 6 to 20 x 20 $mm^2$ square fields, and for selected IMRT and VMAT plans with small field sizes or segments of 1-4 cm. A single-beam optical CT scanner was used as the readout mechanism of the radiation-induced 3-D information in the PRESAGE phantoms. Measured data sets were compared with calculated results from Eclipse Acuros XB. The results for kilovoltage small animal irradiator showed that PDD data measured from EBT films and PRESAGE dosimeters are in agreement within 3%; profiles and 2-D dose distributions measured from PRESAGE present a larger penumbra compared with those from EBT films. Discrepancies between measured beam data and treatment planning data were identified. For megavoltage small fields, the measured data percent depth dose, beam profiles, and dose distributions were found to agree within experimental uncertainties for EBT films and PRESAGE dosimeters. Beam profiles for MLC-delimited field sizes less than 10 mm reveal some discrepancies in the penumbra region between measured data and calculated results. Dose distributions from EBT film and PRESAGE measurements demonstrate that 3-D dosimetry measurement for small-field IMRT and VMAT QA may be necessary to ensure a complete verification of dose delivery and accuracy.

## 1. Introduction

The development of modern radiotherapy techniques in the past 20 years have been driven largely by the advances in technology and physics. In cancer research, small animal studies are commonly used to assess the effectiveness of radiotherapy, test hypotheses concerning radiation-induced effects, and define the genetic underpinnings of specific cancers. The recent development of image-guided small animal irradiation platforms, the Xstrahl Life Sciences (Camberly, UK), Small Animal Radiation Research Platform (SARRP) and the Precision X-ray Inc. (North Branford, CT), X-RAD 225 Cx, has enabled the translation of animal-based in vivo radiation research into clinic. Both systems are offering CBCT image-guided precision irradiation, bioluminescence tomography, and treatment planning systems with dose calculation tools based on superposition convolution model (Muriplan from Xstrahl) or Monte Carlo calculation (SmART-Plan from Precision).

While this system is still under improvement to better simulate the capability and functions of modern radiotherapy equipment, a precise commissioning procedure is essential in order to maintain an accurate imaging and beam delivery. However, there are two major physical properties that are inherently challenging: (a) small radiation field dosimetry (0.5 mm to 10 mm) and (b) kilo-voltage x-ray energies (220 kVp for SARRP and 225 kVp for X-RAD 225Cx). Small field dosimetry is challenging due to the lack of charged particle equilibrium and the size of detectors commercially available, often resulting in partial chamber volume exposures, causing an underreporting of beam output. Additionally, kilo-voltage energy presents a dosimetric challenge due to the significant dose gradient over the volume of measurement device in addition to variance in the dose response of the materials used [1]. To overcome these challenges multiple sets of EBT film, 3-D PRESAGE dosimeters, and Monte Carlo approach have been used for commissioning dosimetry data by several groups [2-5]. These research studies demonstrate that kilovoltage small field dosimetry is more susceptible to measurement uncertainties than megavoltage beams. Recently, Na et al described a process of evaluating the mechanical and dosimetric characteristics of SARP utilizing a Farmer type ionization chamber, EBT3 films, and PRESAGE dosimeters. In addition, Wang et al have demonstrated that EBT3 films, with optimized protocols, can be used to acquire accurate and precise beam data of small animal kilovoltage x-ray irradiator with fields down to 0.5 mm. They have also verify the calculated beam data and dose calculation from the Muriplan with measured data and Monte Carlo simulated data. This allows investigators to image an animal by CBCT, plan the treatment and complete dose delivery in one session of anesthesia.

Modern radiotherapy such as stereotactic radiosurgery (SRS) and stereotactic body radiotherapy (SBRT) using intensity modulated radiation therapy (IMRT) or volumetric modulated radiotherapy (VMAT) may have relatively small fields (5 mm to 3 cm) to achieve desired dose distribution. An accurate dose verification of small-field treatment delivery is therefore crucial. The measurement of small-field beam data, including output factors, percentage depth dose (PDD), and lateral beam profile, is required for the commissioning of treatment planning systems. However, the dosimetric characterization of small fields is difficult due to the potentials for lateral charged-particle disequilibrium, partial source occlusion with the consequence of reduced output, and partial volume averaging effects [6]. Typical dose verification and end-to-end test for clinical treatment units uses a combination of point dose verification (ion chambers and solid state detectors) and planar isodose distribution comparison (radiochromic films) [7-9]. International atomic energy commission (IAEA) Report No. 398 provided a technical approach to deal with the issues in small field dosimetry. AAPM also formed a task group (TG-155) to set up recommendations for relative dosimetry in small fields [10]. However there is no guidance provided for small-field dosimetric measurement using 3-D dosimeters. 3-D gel and PRESAGE dosimeters can be used for accurate dose verification for modern radiotherapy techniques with high precision and spatial resolution [11-12], but require an extensive setup and specialized handling to maintain a sub-percent reproducibility. As such, they still need to be thoroughly investigated, characterized and benchmarked against other detectors [6].

3-D gel and radiochromic plastic dosimeters using MRI imaging or optical CT scanners as readout systems [13-15] have been utilized to measure dosimetric parameters from small-field megavoltage x-rays produced by linear accelerators. However, these studies were focused on measurements of output factors, percentage depth dose curves, and lateral beam profiles of small circular or square fields generated via cones or collimators [16-19]. Currently IMRT and VMAT QA were performed using ion chamber or diode arrays because of their ease of use and reliability. Comparison of the dose distribution measured in a phantom with the dose distribution calculated by the treatment planning system for the same experimental conditions is one of the routines for patient specific IMRT/VMAT QA. For the specific problems encountered in treatment planning of complex radiotherapy techniques such as the use of small fields/segments and of dynamic delivery systems, additional tests are required to verify the accuracy of dose calculations for those conditions specific for small fields. For instance, for the high dose gradient in a small irradiated volume, should the dose distribution be verified in one plane, in several planes or the total 3D dose distribution? The main disadvantage of a 2-D detector arrays is that they do not offer full 3-D dose information and the detector spacing is too large (7 mm) for small field dosimetry [20]. Gel and radiochromic PRESAGE dosimetry is the most commonly reported technique for full 3-D IMRT/VMAT QA [11, 21]. However, detailed full 3-D dose verification for small-field IMRT and VMAT radiotherapy has not been reported in the literature.

In this study, we demonstrate that an accurate and precise measurement of small-field 3-D dose distribution can be obtained using radiochromic PRESAGE dosimeters in conjunction with a single-beam optical CT scanner. Dosimetric data of PDD, lateral beam profiles, and isodose distributions for small field sizes of 6 mm to 2 cm, delimited by MLC, were obtained from PRESAGE measurements and compared with those calculated from Eclipse Acuros XB. Comparison of dose distributions for selected IMRT and VMAT plans with small fields (1 - 4 cm) were performed between PRESAGE measurements and calculations from Acuros. Radiochromic film has probably the highest intrinsic resolution of detectors which are relatively water equivalent [22]. However, film requires specific protocols to obtain acceptable accuracy for reference dosimetry. It is very suitable for lateral profile and 2-D dose distribution measurements. For comparative dosimetry, we also present corresponding measurements with EBT3 radiochromic films.

## 2. Materials and methods

### *2.1. Image-guided small animal irradiation*

All irradiations presented in this study were performed using the SARRP system, produced by Xstrahl. This platform incorporates CBCT imaging with precise radiation delivery, enabling researchers to target anatomical structures and confidently deliver photon beams with small field sizes to a specific point. Through different setups, this machine can provide two modes of operation: CBCT imaging and therapeutic irradiation. In the mode of CBCT imaging, the unit operates at 60 kVp at 0.5 mA with 1 mm Al filtration without collimation. For therapeutic irradiation, the setup is 220 kVp at 13 mA current with 0.15 mm Cu filtration. The SARRP has a source-to-isocenter distance of 35 cm and a number of fixed cone field sizes available for use. Like most of the treatment machines used clinically, a laser alignment system with millimeter accuracy is provided to support visual setup.

#### *2.1.1. Dosimetric commissioning for SARRP*

Irradiations to EBT3 films and PRESAGE phantoms were conducted for preclinical small-field dosimetry. Since EBT3 films came into clinical use, a large amount of research has been conducted to determine film characteristics and provide optimal protocols for dosimetry in the megavoltage X-ray energy range [23-25]. This is mainly due to increased demand in radiation oncology for IMRT and SRS treatment modalities. In contrast, only a small number of studies have been published aiming at providing guidelines for the dosimetry in the kilovolt energy range [3]. To acquire beam data of the small animal irradiator, PDD curves, beam profiles, and dose distribution of five field sizes were measured by films with 33 cm source to surface distance (SSD).

This study presents two methods of film measurement, axial film measurement and vertical film measurement. For axial film measurements, the film was stacked axially between 6 x 6 x 0.5 $cm^3$ plastic water sections and then fixed to a commissioning jig to be irradiated using fixed collimators. The film was oriented perpendicularly to the beam direction at various depths. In contrast, for vertical measurements, film PDD curves were obtained through vertical parallel alignment to the beam central axis, held in the center of a 6 x 6 x 2 $cm^3$ block of plastic water. A commercial Epson Expression 11000XL flatbed scanner was used for readout of irradiated EBT3 radiochromic films at 300 dpi resolution.

In the whole process of film measurement, the film digitization was the largest source of uncertainty. The error comes from (1) inappropriate scanning resolution, (2) scanning location dependence, and (3) developing time (time period between beam delivery and film scan). Therefore, it is important to optimize the protocol for film scanning to minimize these sources of error. For film measurements, a scanning resolution of 300 dpi was used to measure all the beam data while 150 dpi was used for the HD Curve generation. Resolution higher or lower than 300 dpi provided noisier or inaccurate results of small field measurements. To further minimize scanning error, the location of films on the scanner should be consistent for all measurements. The analysis of location dependence of our scanner, an EPSON 11000XL, shows a 5% difference between two films separated by 4 cm on the scanning bed. Finally, developing time of the films can significantly influence results. Therefore, an experiment was performed to find the optimal time for scanning. Films were irradiated by 8 different doses, 1, 3, 5, 7, 9, 11, 13and 15 Gy and each of the films were scanned at 6 different times, 0.5, 1, 2, 5, 24 and 48 hours after irradiation [3].

In addition to EBT3 film, cylindrical PRESAGE dosimeters of 6 cm height and diameter were used for the evaluation of beam characteristics. Commissioning data for 3 x 3 and 10 x 10 $mm^2$ fixed field sizes are used for this purpose. Each of the PRESAGE dosimeters was placed along the central axis of the beam, which is perpendicular to the gantry rotation axis, with guidance of crossed lasers.

The readouts of irradiated PRESAGE dosimeters were performed with a single laser beam optical CT scanner. For scanning the irradiated PRESAGE dosimeter is mounted on the base of the scanner rotation stage. The tank is filled with optical refractive index matching fluid to match that of the PRESAGE dosimeter. A total of 1200 projection images per axial plane were used for 3D dose reconstruction. A filtered back projection algorithm was used to reconstruct the 3D dose map. The scanner can reconstruct a transverse image to a spatial resolution of 0.1-1 mm depending on the phantom size. Although the entire readout takes 6-10 hours, as a fundamental standard of OCT in 3D dosimetry, the single laser scanner produces consistent high quality results. From the reconstructed images, PDD curves, dose distributions and transverse profiles were obtained.

The PRESAGE-measured beam data was validated using both axial and vertical film measurements. The density of the PRESAGE used in this study is 1.07 $g/cm^3$. Since the relative electron density of PRESAGE to water is 1.05, we have corrected the PDD curves acquired in the PRESAGE by a factor of 1.05, following methodology described in Newton *et al* [5].

2.2. Clinical small-field IMRT and VMAT

In this study, radiochromic plastic dosimeters, PRESAGE®, were used for clinical small-field dose measurements. Six small cylindrical phantoms (6 cm diameter × 7 cm length) were irradiated with 6 MV MLC-delimited square fields: 6 x 6 $mm^2$, 8 x 8 $mm^2$, 1 x 1 $cm^2$, 1.2 x 1.2 $cm^2$, 1.6 x 1.6 $cm^2$, and 2 x 2 $cm^2$. Five large cylindrical phantoms (10 cm diameter x 11 cm length) were used for clinical small-field IMRT and VMAT dose verification. Five IMRT or VMAT plans, with field sizes ranging 1-4 cm, were selected for dose verification. These plans were selected based on their complexity as to if they passed the patient-specific QA process using MapCheck and ArcCheck. For comparative dosimetry radiochromic EBT3 films, positioned in a 30 x 30 x 10 $cm^3$ solid water phantom, were used for PDD, lateral beam profile, and 2-D dose distribution measurements. EBT3 films were positioned at 1.5 cm depth for MLC-delimited small square field measurements, and at 5 cm depth for all IMRT/VMAT dose measurements. A commercial Epson Expression 11000XL flatbed scanner was used for readout of irradiated EBT3 radiochromic films at 300 dpi resolution, 24 hours after the irradiation.

## 3. Results

### *3.1. Preclinical small field dosimetry*

The EBT3 film reading protocol, for kV dosimetry, was carefully studied by Wang et al. In this study, films were irradiated at 8 different doses between 1-15 Gy. Each of the films was scanned at 6 different times. For doses between 1-5 Gy, at least 5 hours of developing time is necessary, while for doses

IOP Conf. Series: Journal of Physics: Conf. Series **1305** (2019) 012023 doi:10.1088/1742-6596/1305/1/012023

between 7-9 Gy, the developing time should be at least 24 hours. Finally, when the dose is 10 Gy or higher, 48 hours of developing time is suggested to achieve stable results. Percent depth dose curves of 3 different field sizes are shown in Fig. 1. Dose at different depths was normalized to the dose at 5 mm depth instead of the surface to calculate the PDD. This was done because the measurements in the buildup region are less stable than at the deeper depths. For film measurements, two thicknesses of solid water phantom, 2cm and 6 cm were placed underneath to evaluate the difference of backscattering between couch and solid water phantom.

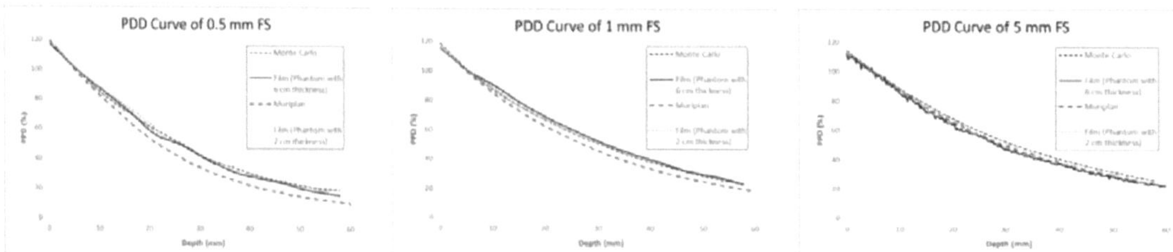


**Figure 1.** Comparison of PDD curves from Muriplan, Monte Carlo and film measurement with 2cm, cm thickness of solid water phantom.

Figures 2 and 3 show PDD and beam profile results using EBT3 radiochromic film and PRESAGE for 3 x 3 mm$^2$ and 10 x 10 mm$^2$ fields. All curves are normalized to 10 mm depth. The measured beam profiles are compared for depths of 10, 20, and 30 mm in Fig. 2(b) and Fig. 3(b). For the 3 x 3 mm$^2$ field size, the differences in PDD among PRESAGE, film axial, and film vertical measurements are within ±1% at depths less than 3 cm. At depths between 3 and 5 cm, PRESAGE has a higher depth dose than film vertical by up to 3%, but agrees with film axial measurements within ±1%.

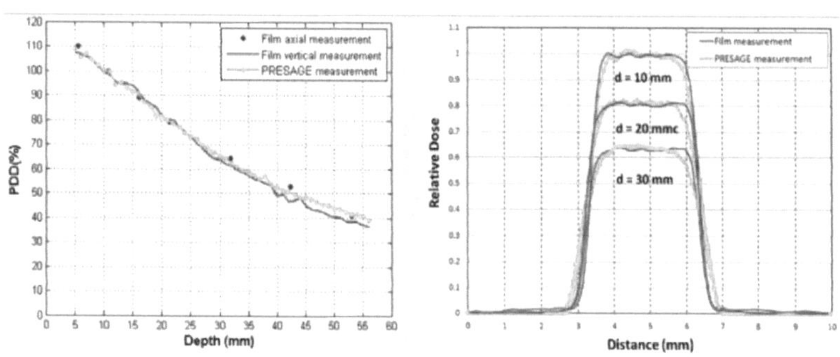


**Figure 2**. PDD and beam profile measurements for the 3 x 3 mm$^2$ field using EBT3 film and PRESAGE. Depth-dose curves were plotted from 5 to 55 mm, and the beam profiles were compared at depths of 10, 20, and 30 mm.

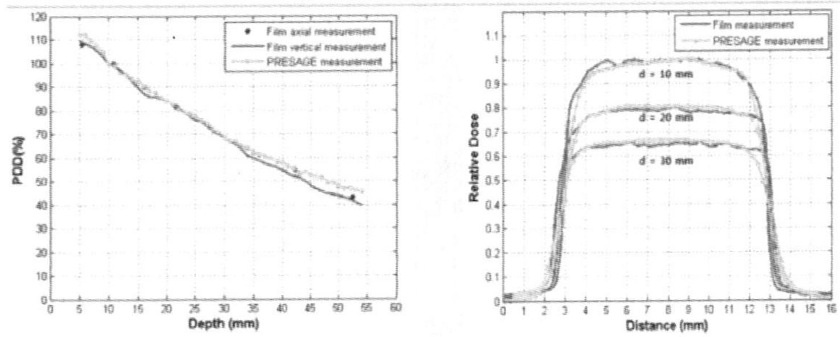


**Figure 3.** PDD and beam profile measurements for the 10 x 10 mm$^2$ field using EBT3 film and PRESAGE. Depth-dose curves were measured from 5 to 55 mm, while the beam profiles were measured at 10, 20, and 30 mm depth.

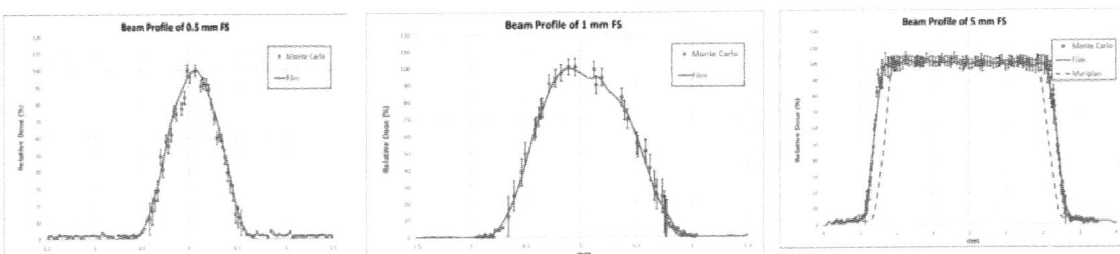


**Figure 4.** Comparison of beam profiles at 2cm depth from Muriplan, Monte Carlo and film measurement.

In Fig. 4, beam profiles of each field size from film, Monte Carlo, and Muriplan were compared. There are no Muriplan results for 0.5 and 1 mm cone size because of the resolution limitation. Error bars provide standard deviations for the Monte Carlo results.

### *3.2. Clinical small-field dosimetry*

In this study, both EBT3 films and PRESAGE are used for dose verification of clinical small-field dosimetry calculated from Eclipse Acura XB. Fig. 5 presents an experimental evaluation of the Acuros dose calculation accuracy on beam profile, PDD, and dose distribution for the 8 x 8 mm$^2$ MLC-delimited field, using EBT3 films and PRESAGE dosimeters. Figure 6 shows a comparison of dose distributions, for a single IMRT field, from EBT3 and PRESAGE. Gamma index pass rate is 99% with 2% DD (dose difference) and 1 mm DTA (distance-to-agreement) gamma criteria. Figure 7 shows dose verification along the central slice (a) and at the plane 0.8 cm from the isocenter (b), for a VMAT stereotactic radiosurgery. Gamma index pass rate is 98% for (a) and 97.1% for (b), with 2%/1mm gamma criteria. Figure 8 shows dose verification along the central slice (a) and at the plane 1.3 cm from the isocenter (b) for a pancreas SBRT using 9-field IMRT. Gamma index pass rate is 89% for (a) and 78.4 % for (b), with 2%/1mm gamma criteria.

IOP Conf. Series: Journal of Physics: Conf. Series **1305** (2019) 012023 doi:10.1088/1742-6596/1305/1/012023

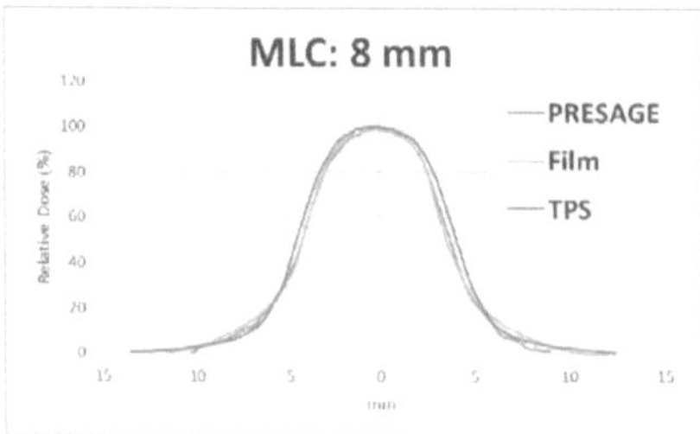


(a)

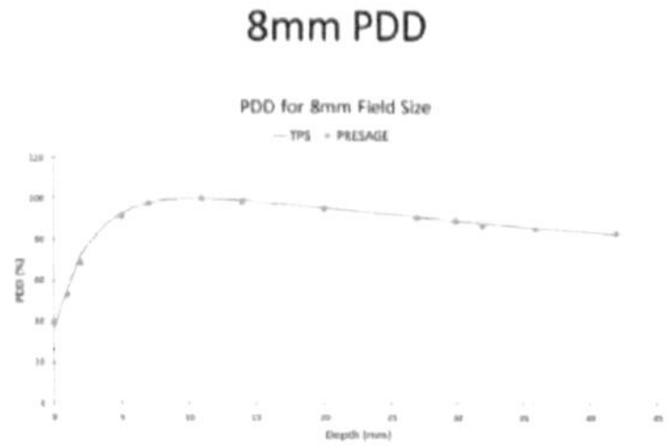


(b)

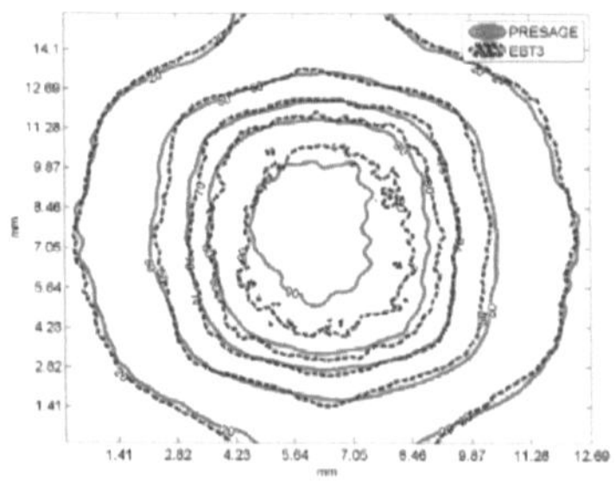


(c)

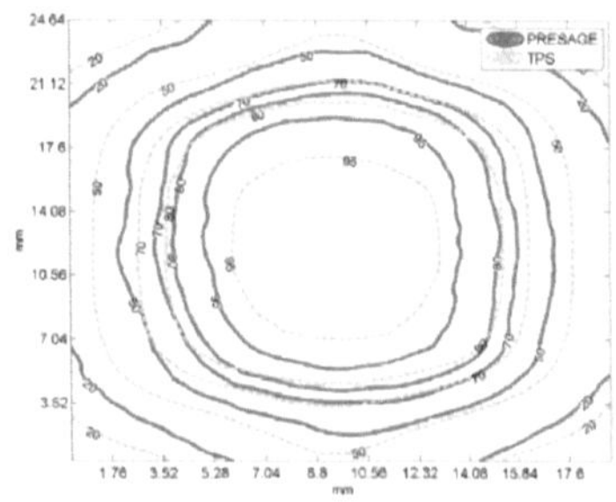


(d)

**Figure 5.** Dosimetric comparison for beam profile (a), PDD (b), and 2-D dose distribution (c, d).

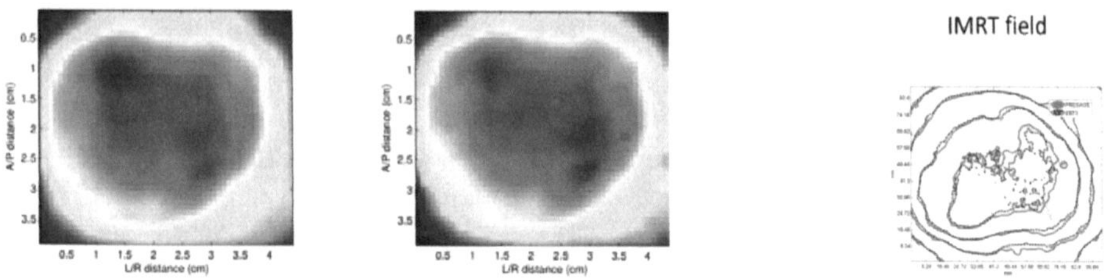


**Figure 6.** Comparison of dose distribution between EBT3 film and PRESAGE measurements for a single IMRT field. 99% gamma index pass rate, with 2% and 1 mm gamma criteria

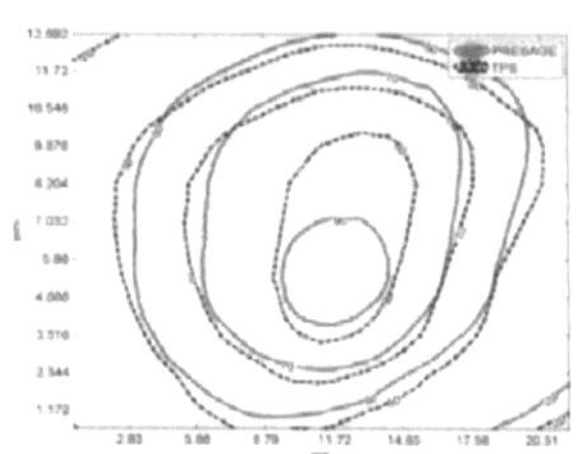

(a)

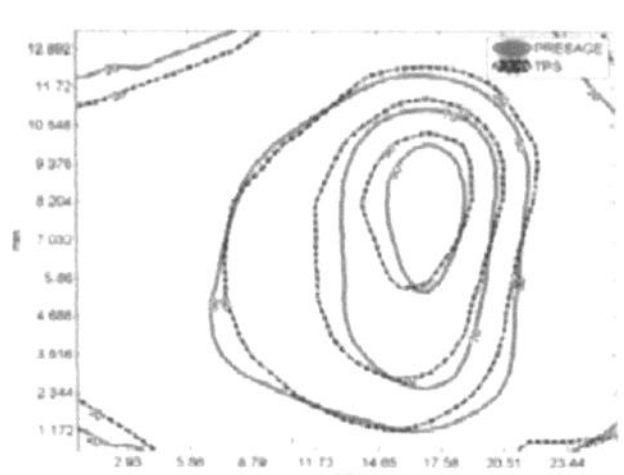

(b)

**Figure 7.** Dose verification along the central slice (a) and 0.8 cm from the isocenter (b) for a VMAT stereotactic radiosurgery. Gamma index pass rate is 98% for (a) and 97.1% for (b), with 2% and 1mm gamma criteria.

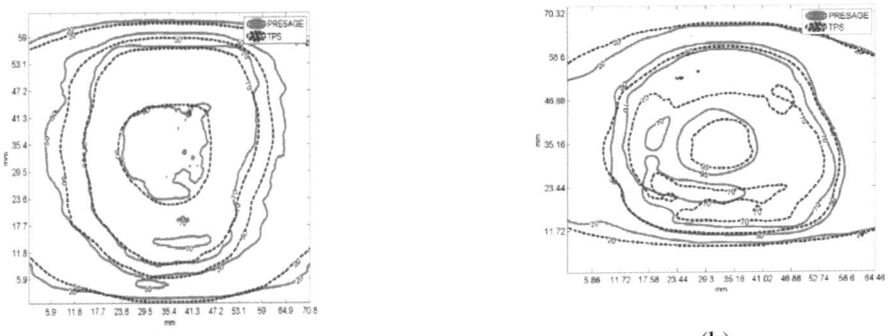


(a) (b)

**Figure 8.** Dose verification along the central slice (a) and 1.3 cm from the isocenter (b), for a pancreas SBRT using 9-field IMRT. Gamma index pass rate is 89% for (a) and 78.4 % for (b), with 2% and 1mm gamma criteria.

## 4. Discussion

### *4.1. Preclinical small field dosimetry*

Our beam data measurements indicated a maximum difference between the measured and simulated PDD of each field size as 3.3 %, 3.2 %, 0.6%, 0.3% and 1.8% for field sizes of 0.5 mm, 1 mm, 3 mm, 5 mm, and 10 mm, respectively. The maximum difference of PDD between Muriplan and simulation was 10.3%, 4.3%, 0.6%, 0.5% and 1.4% at 6 cm, respectively, and at at 2 cm depth, the difference is 3.89%, 5.29%, 0.47%, 0.88% and 0.33%. It should be noted that there was a good agreement of the PDD curves between film measurement and Monte Carlo calculation for all field sizes. Good agreement was also observed between film measurement and Muriplan for field sizes larger than 3 mm. However, Muriplan underestimated the dose for field sizes of 0.5 mm and 1 mm. This could result in overdosing deeper structures during animal irradiations. Furthermore, PDD values increased with field size due to increased scattering. This indicates that percent depth dose deceases more dramatically for small field sizes. In the comparison of PDD curves using 2 cm and 6 cm phantom thickness, no significant dosimetric difference (<±3%) was shown for 0.5, 1, 3, 5, and 10 mm FS, which indicates that the dosimetric difference due to backscattering is negligible when irradiating rodents of 2-6 cm with small field sizes.

Our results in Fig.2 and 3 indicate good agreement overall for PDD curves from PRESAGE and films at depths less than 3 cm for both 10 x 10 mm$^2$ and 3 x 3 mm$^2$ field sizes. Accurate dosimetry within this depth is sufficient for small animal treatments; however larger depths would become more of a consideration for the treatment of larger rodents such as rats. At 5 cm depth, PRESAGE measurements show a 3% higher depth dose than film vertical measurement, but agrees well with film axial measurement, for both field sizes. The slightly higher PDD values from PRESAGE at low dose region may be due to lower signal-to-noise ratio from the reconstructed images. Additionally, film axial measurements are more prone to alignment error, particularly for small fields, but this drawback is offset by the fact that they are less labor-intensive to set up. Compared with film measurements, the PRESAGE dosimeter has a limitation in acquiring data near the surface as its surface is not perfectly flat. Specifically, dosimetry data at the first 2 mm of depth are more susceptible to measurement error due to these surface irregularities. Given this limitation, however, PRESAGE measurements have the marked advantage of providing 3D dosimetry information and provide an evaluation tool that is less labor intensive to set up properly and measure. If detailed depth dose information is required in the region from the surface to 2 mm depth, it would be useful to use film measurements at these depths to supplement PDD information obtained from PRESAGE measurements.

Another discrepancy we observed in this study is related to penumbra. The penumbra values obtained from PRESAGE are larger than those from EBT3 films by about 0.1 mm for 3x3 mm$^2$ and 0.5 mm for 10x10 mm$^2$ at 2 cm depth. The larger penumbra from PRESAGE measurement is likely due to larger spatial resolution inherent to the OCT scanner and image noise. Here EBT3 film measurement can offer an advantage as its high image resolution can provide more accurate penumbra measurements in the case of small field sizes. However, PRESAGE measurements are more efficient than film as it provides more information in one 3D dosimeter and a single optical-CT scan. For FWHM measurements, PRESAGE provides comparable results as EBT3 film, with a small discrepancy of 0.01 mm for 3 x 3 $mm^2$ and 0.07 mm for 10 x 10 $mm^2$ field size at 2 cm depth. We observed in this study that PRESAGE measured beam profiles exhibit larger uncertainties in flatness and symmetry when compared to film measurements. A possible reason for this is the inherent tradeoff between signal-to-noise ratio and resolution when reconstructing PRESAGE images. More image noise is introduced when the scanning resolution is set higher. These two factors must be balanced to determine an optimized reconstruction protocol. Improvements to scanner setup parameters such as increasing the number of projections and acquired data points for each projection. Additionally, changes can be made to the image reconstruction protocol to further mitigate our resolution limitations. These aspects will be explored in future studies focused on PREGAGE evaluation of small fields.

As for beam profiles, there was good agreement between EBT3 film measurement and Monte Carlo calculations. The largest penumbra difference was 0.05 mm when field size was 0.5 mm. However, the field sizes from Muriplan were consistently smaller than film measurement and simulation. These results indicates that calculations performed by the Muriplan treatment planning system underestimate the field size compared with our measurements and simulation results.

### *4.2. Clinical small field dosimetry*

The use of small fields in radiotherapy has increased significantly, especially in SRS/SBRT and larger IMRT fields that are composed of small beamlets. As such, an experimental evaluation of the dose calculation accuracy for small fields is essential. For small field dosimetry EBT3 film has the advantages of being nearly energy independent, close to water, high resolution, and good for 2-D dosimetry. The lateral beam profiles, for 8 x 8 $mm^2$ MLC-delimited field size, shows an excellent agreement in the full width of half maximum (FWHM) and penumbra region between EBT3 and PRESAGE, and an overestimated dose in the penumbra from the Acura XB calculation algorithm. This difference in the penumbra region, between measured results and calculation, may be due to inaccurate small-field modelling of the planning system, but also possibly due to geometric uncertainty of MLC positioning. Figure 5 shows good agreement in beam profile, PDD, and dose distribution, for 8 x 8 $mm^2$ MLC field size, between EBT3 and PRESAGE. For PDD curve, the maximum difference between PRESAGE and TPS is 3.7% in the build-up region. There is a good agreement in 2-D dose distribution between EBT3 and PRESAGE, shown in Figure 5 (c), with 98.9% gamma index pass rate. The disagreement of 2-D dose distribution between PRESAGE and TPS, shown in Figure 5(d), can partly be attributed to the poor modelling of MLC leakage. 2-D planar isodose distribution comparison using radiochromic film is currently recommended for dose verification of treatment planning calculation algorithm as well as end-to-end advanced treatment techniques, especially for small fields. Figure 6 demonstrates a very good agreement of planar dose distributions between PRESAGE and EBT3 for a single IMRT field, with a 99% gamma index pass rate for 2%/1mm gamma criteria.

For the treatment planning of complex radiotherapy techniques such as the use of small-field dynamic delivery, additional tests may be required to verify the accuracy of dose calculation for the high dose gradients in a small irradiated volume. For instance, should the dose distribution be verified in one plane, in several planes or the total 3D dose distribution? Figure 7 shows the dose distribution comparison between PRESAGE and Eclipse at two different planes for a VMAT stereotactic radiosurgery in the cervical spine. The PTV has an equivalent spherical dimension of 1 cm and a volume of 0.5 cc. Therefore, the pre-treatment VMAT QA cannot be checked with ArcCheck. The main disadvantage of the 2-D detector arrays, for small field dosimetry, is that they do not offer good spatial resolution as the detector spacing is too large (7 mm). For small-field IMRT/VMAT plans radiochromic film provides a good 2-D dosimetric verification as it has probably the highest intrinsic resolution of detectors which are relatively water equivalent and no energy dependence for MV beams, and has been recommended for small field dosimetry [26]. In this case, EBT3 film was used for patient specific pre-treatment QA, and the gamma index pass rate was 97.4% with 2% DD (dose difference) and 1 mm DTA (distance to agreement) gamma criteria. PRESAGE measurement resulted in a 98% pass rate at the plane along the isocenter and 97.1% pass rate at the plane 8mm from the isocenter, providing additional information on dose verification. Presented in Figure 8, it is a pancreas SBRT using 9-field IMRT technique. The PTV has an equivalent spherical dimension of 4.6 cm and a volume of 50.4 cc. The IMRT treatment was highly modulated with a lot of small beamlets, and the IMRT QA, performed with a MapCheck, did not pass the gamma index test. Gamma index pass rate was 90.2% when EBT3 was used for IMRT QA in this case, with 2mm/1% criteria. PRESAGE measurement resulted in 89% gamma index pass rate at the plane along the isocenter and 78.4% pass rate at 1.3 cm from the isocenter. The dose verification at different planes provided useful information in this patient specific IMRT QA. The similarity between EBT3 film and PRESAGE based on the gamma index analysis indicates that PRESAGE is a suitable 3-D dosimeter for special IMRT QA when additional information is required for clinical decision, such as the small-field IMRT/VMAT presented here.

## 5. Conclusion

This study was driven by the increasing need for dosimetric characterization of image-guided small animal irradiators, as well as for accurate and precise dose verification of clinical small fields which are used for precision radiotherapy techniques.

The results in this study provide a comparison between PRESAGE and EBT3 radiochromic film evaluation of dosimetric characteristics of preclinical small fields. The principal advantage of the PRESAGE dosimeter over radiochromic film is that dose information is three dimensional in nature, allowing simultaneous evaluation of dose within the entire volume. This work also provides valuable commissioning and beam data evaluation techniques to future groups that will be using 3D dosimetry as a tool in this regard.

3-D radiochromic plastic dosimeter, PRESAGE, can be used to acquire small-field megavoltage beam dosimetric parameters precisely and accurately. Both EBT3 film and PRESAGE dosimeter can be used for dose verification of small-field IMRT/VMAT in modern radiotherapy techniques. However, our results indicate that PRESAGE is a suitable 3-D dosimeter for special IMRT QA when additional information is required for clinical decision,

## 6. References

[1] Hill R *et al* 2014 *Phys. Med. Biol.* **59** R183
[2] Na Y *et al* 2018 *Med. Phys.* **45** 3330-9
[3] Wang Y F *et al* 2018 *Phys. Med. Biol.* **63** 145001
[4] Tryggestad E *et al* 2009 *Phys. Med. Biol.* **54** 5341
[5] Newton J *et al* 2011 *Med. Phys.* **38** 6754-62
[6] IAEA 2000 Technical Reports Series No. 398
[7] Francescon P *et al* 1998 *Med. Phys.* **25** 503
[8] Das I J *et al* 2000 *J. Radiosurg.* **3** 177-185
[9] ICRU 2014 Report 91
[10] Das I J *et al* AAPM Task Group Report 155
[11] Wuu C S and Xu Y 2006 *Med. Phys.* **33** 1412
[12] Oldham M *et al* 2006 *J. Phys.: Conf. Ser.* **56** 58-71
[13] Baldock C 2009 *J. Phys.: Conf. Ser.* **164** 012002
[14] Mather M L *et al* 2003 *Phys. Med. Biol.* **48** N269-75
[15] Baldock C *et al.* 2010 *Phys. Med.Biol.* **55** R1-63
[16] Pantelis E *et al* 2008 *Med. Phys.* **35** 2312-20.
[17] Pappas E *et al* 2005 *Med. Phys.* **32** 1513-20
[18] Pappas E *et al* 2008 *Med. Phys.* **35** 4640-8
[19] Babic S *et al* 2009 *Phys. Med. Biol.* **54** 2463-81

IOP Conf. Series: Journal of Physics: Conf. Series **1305** (2019) 012023 doi:10.1088/1742-6596/1305/1/012023

[20] Vial P *et al* 2008 *Med. Phys.* **35** 4362-74
[21] Crescenti R A *et al* 2007 *Med. Phys.* **34** 1286-97
[22] Devic S 2011 *Physica Medica* **27** 122-134
[23] Dufreneix S *et al* 2014 *Eur. J. Med. Phys.* **30** e136
[24] Gonzalez-Lopez A *et al* 2015 J *Med. Phys.* **40** 61-7
[25] Morales J E *et al* 2016 *Med. Phys.* **43** 4687-92
[26] Wilcox E E and Daskalov G M 2007 *Med. Phys.* **34** 1967-74

# SRS and SBRT and advanced dosimetry: Duke experiences

**Mark Oldham**
Radiation Oncology, Duke University Medical Center, Durham, NC USA

E-mail: mark.oldham@duke.edu

**Abstract**. Stereotactic radiosurgery (SRS) and stereotactic body radiation therapy (SBRT) are among the most demanding of radiation therapy techniques in terms of requirements for high accuracy and conformality of treatment delivery. Compared to conventional treatments, they require specialized high-precision equipment (including high specification medical accelerators and alignment frames), extensive imaging (including multi-modal and motion management methods), specialized Quality Assurance protocols, and increased person-power effort. In addition, the high dose-per-fraction renders SRS/SBRT a high-stakes treatment setting, where small errors may cause significant treatment morbidity in normal healthy tissue. All of these issues combine to render SRS/SBRT one of the most opportunistically rich and clinically important areas for advanced dosimetry research. Some of the applications of advanced 3D dosimetry to SRS and SBRT and small field treatment verification are explored. Applications include base of skull IMRT, single-isocenter multi-lesion radiosurgery, pre-clinical precision treatments, interpreting the clinical significance of 3D QA data, remote dosimetry, dosimetry in magnetic fields, trigeminal neuralgia, rodent-morphic dosimetry, motion, and small field commissioning.

Stereotactic radiosurgery (SRS) and stereotactic body radiation therapy (SBRT) are among the most demanding of radiation therapy techniques in terms of requirements for high accuracy and conformality of treatment delivery. Compared to conventional treatments, these techniques require specialized high-precision equipment (including high specification medical accelerators and alignment frames), extensive imaging (including multi-modal and motion management methods), specialized Quality Assurance protocols, and increased person-power effort. In addition, the high dose-per-fraction renders SRS/SBRT a high-stakes treatment setting, where small errors may cause significant treatment morbidity in normal healthy tissue. All of these issues combine to render SRS/SBRT one of the most opportunistically rich and clinically important areas for advanced dosimetry research. In this talk we will explore some of the applications of advanced 3D dosimetry to SRS and SBRT and small field treatment verification. Applications include base of skull IMRT [1], single-isocenter multi-lesion radiosurgery [2], pre-clinical precision treatments [3-8], interpreting the clinical significance of 3D QA data [9, 10], remote dosimetry [11, 12], dosimetry in magnetic fields [13], trigeminal neuralgia [14], rodent-morphic dosimetry [15], motion [16, 17], and small field commissioning [7, 18]. Further information can be found in recent comprehensive reviews [19, 20].

## References

[1] Thomas A *et al* 2013 *J. Phys.: Conf. Ser.* **444** 012050
[2] Thomas A *et al* 2013 *Med. Phys.* **40** 121725
[3] Yoon S W *et al* 2017 *J. Phys.: Conf. Ser.* **847** 012004

[4] Kodra J *et al* 2017 *J. Phys.: Conf. Ser.* **847** 012054
[5] Cramer C K *et al* 2015 *PLoS One* **10** e0143208
[6] Rankine L J *et al* 2013 *Phys. Med. Biol.* **58** 7791-801
[7] Newton J *et al* 2011 *Med. Phys.* **38** 6754-62
[8] Yoon S W *et al* 2017 *Med. Phys.* **44** 6008-17
[9] Oldham M *et al* 2012 *Int. J. Radiat. Oncol. Biol. Phys.* **84** 540-6
[10] Norris H A *et al* 2013 *J. Phys.: Conf. Ser.* **444** 012022
[11] Mein S *et al* 2017 *Med. Phys.* **44** 6018-28
[12] Juang T *et al* 2014 Med. Phys. **41** 071706
[13] Rankine L J *et al* 2017 *Int. J. Radiat. Oncol. Biol. Phys.* **97** 1095-104
[14] Wang Z *et al* 2010 *J. Phys.: Conf. Ser.* **250** 012058
[15] Bache S T *et al* 2015 *Med. Phys.* **42** 846-55
[16] Thomas A *et al* 2013 *J. Phys.: Conf. Ser.* **444** 012049
[17] Brady S L *et al* 2010 *Phys. Med. Biol.* **55** 2187-201
[18] Clift C *et al* 2010 *Phys. Med. Biol.* **55** 1279-93
[19] *Clinical 3D Dosimetry in Modern Radiation Therapy*. 2017: CRCPress, Taylor and Francis Group.
[20] Oldham M 2014 *Methods and Techniques for Comprehensive 3D Dosimetry.* Advances in Medical Physics, Devon Godfrey et al., eds. Madison, WI: Medical Physics Publishing

IOP Conf. Series: Journal of Physics: Conf. Series **1305** (2019) 012025 doi:10.1088/1742-6596/1305/1/012025

# Verification of stereotactic cranial radiotherapy treatments with MR-based gel dosimeters: practical aspects

**Filipa Costa[1], Evanthia Kousi[2], Anne Gasnier[3], Emma Wells[4], Caroline Lamb[4], Maria A Schmidt[2] and Rollo Moore[4]**

[1]Joint Department of Physics, The Institute of Cancer Research and The Royal Marsden NHS Foundation Trust, London SM2 5NG, UK

[2]CR-UK and EPSRC Cancer Imaging Centre, Royal Marsden NHS Foundation Trust, London SM2 5NG, UK

[3]Radiotherapy Physics department, Georges Pompidou European Hospital, Paris, France

[4]Radiotherapy Physics, Royal Marsden NHS Foundation Trust, London, SW3 6JJ, UK

E-mail: filipa.costa@icr.ac.uk

**Abstract:** Image Guided Radiotherapy (IGRT) systems allow delivering high doses of radiation to a tumour with great precision and accuracy. In this work, we use patient-specific head MR-based gel dosimeters as an end-to-end test to commission IGRT delivery systems, for stereotactic cranial radiotherapy, and investigate possible sources of errors associated with practical aspects of the measurements. A CT scan of a patient is used to 3D print the shape of the cranial bones to create head phantoms with MR-based gel inside. Each phantom was used as a patient, following the radiotherapy workflow which includes: taking a CT scan of the phantom, calculating the dose distribution using a treatment planning system (TPS) and delivering the radiation calculated by the TPS (to target three different lesions: Big, Medium and Small). Each phantom was then scanned on an MR scanner to obtain a T2 map (linear with dose), which was then rigidly registered with the CT scan of the phantom based on the phantom structures. Good agreement was obtained between the normalized R2 (=1/T2) values and TPS simulations. The relationship between R2 values and dose was investigated based on 3D regions including each lesion. Good linearity was found, but different R2-dose values relationship was obtained depending on the selected regions. The impact of applying a registration based on each lesion (and not the phantom structures) was also tested. Results show that the registration has an impact on the relationship between R2 values and dose, and although absolute dose measurements are still not possible with these MR-based gel dosimeters, they provide very detailed geometrical 3D information to validate IGRT radiotherapy treatments with high level of accuracy.

## 1. Introduction

Magnetic resonance (MR)-based gel dosimetry is of great interest to Stereotactic Ablative Radiotherapy (SABR) and Stereotactic Radiosurgery (SRS) [1-4]. CyberKnife and BrainLab delivery systems use image guided radiotherapy (IGRT) to deliver high doses of radiation to the tumour in very few fractions. IGRT requires complex interactions between different systems which need to be verified to ensure that the correct dose distribution is being delivered. A great level of geometrical and dosimetric accuracy is required for these treatments to be effective in improving tumour control and reducing toxicity [5]. This work was part of a commissioning program for image guided stereotactic cranial radiotherapy of

IOP Conf. Series: Journal of Physics: Conf. Series **1305** (2019) 012025 doi:10.1088/1742-6596/1305/1/012025

multiple metastases. Patient-specific MR-based 3D gel phantoms were used as an end-to-end test to validate the accuracy of IGRT systems. An identical phantom with inserts for ionization chambers was used for absolute dose measurements. Procedures associated with MR-based gel dosimetry and possible sources of errors associated with practical aspects of the measurements were investigated.

## 2. Material and Methods

### *2.1. IGRT treatment planning systems*

Two different treatment planning systems (TPS) software were used to calculate a plan to treat multiple metastases in the brain using different IGRT delivery systems. BrainLab Elements (v1.5) and Accuray Multiplan (v5.2.1) were used to calculate the doses delivered. Relevant differences between the two platforms include that Elements automatically performs the majority of user subtasks in optimizing the beamlet placement and weighting, while Multiplan has a comparable automated approach once the collimator apertures are defined by user choices. A CT scan of a patient with three approximately spheroid-shaped metastatic lesions (termed “Big” (5.4 $cm^3$), “Medium” (0.57 $cm^3$) and “Small” (0.15 $cm^3$)), separated by at least 50mm, was used to create a treatment plan using the two TPS.

### *2.2 3D printed patient-specific phantoms and plan irradiation*

Three identical patient-specific hollow head phantoms (with sub-millimetre spatial accuracy) were 3D-printed based on the selected patient planning CT scan (RTsafe, Athens, Greece). RTSafe uses 3D printing techniques to replicate the shape of the cranial bones with a material of bone-like density, with a 2 mm thick skin-mimicking layer to seal and to provide rigidity. Two phantoms (gel phantoms) were filled with VIPAR radiosensitive polymer MR-based gel [6], that polymerizes with radiation and the changes in spin-spin (T2) relaxation times can be measured with a MR scanner. One phantom (ionization chamber phantom) contains three ionisation chamber inserts to perform absolute dose measurements by placing a chamber in the middle of each planning target volume (PTV). A semiflex ionization chamber (type 31010, PTW, Germany), with sensitive volume of 0.125$cm^3$, was placed in the middle of the Big lesion, while two pinpoint ionization chambers with sensitive volumes of 0.015$cm^3$ (type 31006 & 31014, PTW, Germany) were placed in the middle of the Medium lesion and Small lesion. This phantom is reusable and needs to be filled with water before being irradiated.

The treatment plans created on the patient CT were calculated so that the delivered doses would always be below 12 Gy (for doses > 12 Gy T2 values no longer vary linearly with dose [3]). The plans were re-calculated on the CT of the phantom (gel phantom or ionization chamber phantom depending on the availability). Each gel phantom and ionization chamber phantom pair were irradiated as per the patient plan, using stereoscopic planar x-ray on a Cyberknife or ExacTrac on a Varian LE Clinac.

### *2.3 MRI data acquisition and dose readout*

Each gel phantom was scanned in the standardized radiotherapy treatment position for head and neck using a thermoplastic mask as per the treatment. This was the same set-up used to irradiate the gel phantoms and the ionization chamber phantom for each treatment. The MR scanner was performed at 1.5T with a Siemens Aera, 18-channel surface coil (Siemens, Erlangen, Germany) 24 hours after irradiation. A 2D fast spin echo sequence with multiple echo times was employed (TR/TE=2000ms / 36, 436, 835, 1230ms, voxel size = 1.4 x 1.4 x 2 $mm^3$, bandwidth = 780 Hz / px) and a T2 relaxation time was calculated on a pixel-by-pixel basis. A 3D map of gel R2 values can be obtained, by applying the inverse of the T2 relaxation times (R2 = 1 / T2 map).

### *2.4 Dosimetric analysis*

The T2 map and the CT scan of the gel phantom (or the ionization chamber phantom) together with the calculated treatment plan and TPS dose distribution was sent to the manufacturers. RTsafe provides a data analysis service which reports the agreement between the measured doses and the TPS simulated doses, after registering the T2 map with the CT scan provided.

An independent analysis was performed in this work using the same T2 map and the CT scan of the gel phantom that were provided to the manufacturers. Matlab® was used to analyse the data and to compare the gel R2 values with TPS dose values. Normalization can be performed interactively based on a region of interest (ROI) within one of the lesions and a background area selected on a low dose region within the phantom. A square region within the Big lesion was used to normalize both gel R2 and TPS dose values, and 1D gamma criteria of 5%, 2mm was applied to assess the agreement. A comparison was also performed based on normalized local differences.

The relationship between gel R2 values and TPS doses was investigated by applying four different registrations for each of the two phantom gels used. The T2 map was rigidly registered with the phantom CT on 3D slicer based on the phantom structures of the gel and using the RTsafe registration as a starting point. A local rigid registration between the T2 map and TPS dose map was also performed to obtain three registered T2 map images, one for each lesion (registration by lesion).

### *2.5 MRI analysis*

In addition to the MR scan recommended by the gel manufacturer (Post1: within 48h of irradiation), we also performed scans prior to irradiation (Pre scans) and at later time points (Post2: one week after irradiation and Post3: one month after irradiation). With this additional data, the uniformity of the gel pre-irradiation was investigated, as R2 values drifts can in principle be associated with gravity or imperfections of the preparation. The stability of the R2 values over time was also investigated.

## 3. Results and discussion

### *3.1 Dosimetric analysis*

For the phantom gels irradiated, good agreement was reported by RTsafe according to their analysis. Similarly, our independent analysis showed a good agreement overall. Normalized R2 values from two different gel phantoms irradiated and planned with Accuray Multiplan (delivered with a Cyberknife with stereoscopic planar x-ray) and Brainlab Elements (delivered with a Varian LE Clinac with ExacTrac) are shown in Figure 1a (phantom Gel 1) and b (phantom Gel 2), respectively. Good agreement is obtained by taking a randomly selected profile (±1mm) in the central region of each lesion, as shown in Figure 1. 1D Gamma of 5%, 2mm is also shown for quantification of the agreement. Lower normalized R2 values were observed for the Medium and the Small lesions for phantom Gel 1 in comparison with the simulated profiles (Figure 1a). In Figure 1b, a small shift of about 1mm is visible in the Medium lesion but normalized values agree well with simulations in all lesions for profiles taken in different directions. Sagittal and coronal slices are not shown but the same results were observed.

Subtracting the normalized TPS dose to the measured dose gives important visual information. In Figure 2, one can see the extension of the shift observed in Figure 1b by looking at both axial and coronal slices. A small shift on the small lesion was also observed. Differences found here can reach 20%, but will not necessarily have a clinical impact if the defined gamma criteria would only fail in a small percentage of points, or if a good agreement can still be found within the defined distance to agreement. However, local differences help to visualize and identify regions where a discrepancy might exist, and draw the attention of the users to regions which must be carefully assessed.

In Figure 1b, depending on the slice and directions where the profile is taken the shift in the anterior-posterior direction can reach 2mm. These differences might change with the applied rigid registration. However, assessing the quality of the registration might be difficult as different modalities are being registered (CT and MRI) [7].

IOP Conf. Series: Journal of Physics: Conf. Series **1305** (2019) 012025 doi:10.1088/1742-6596/1305/1/012025

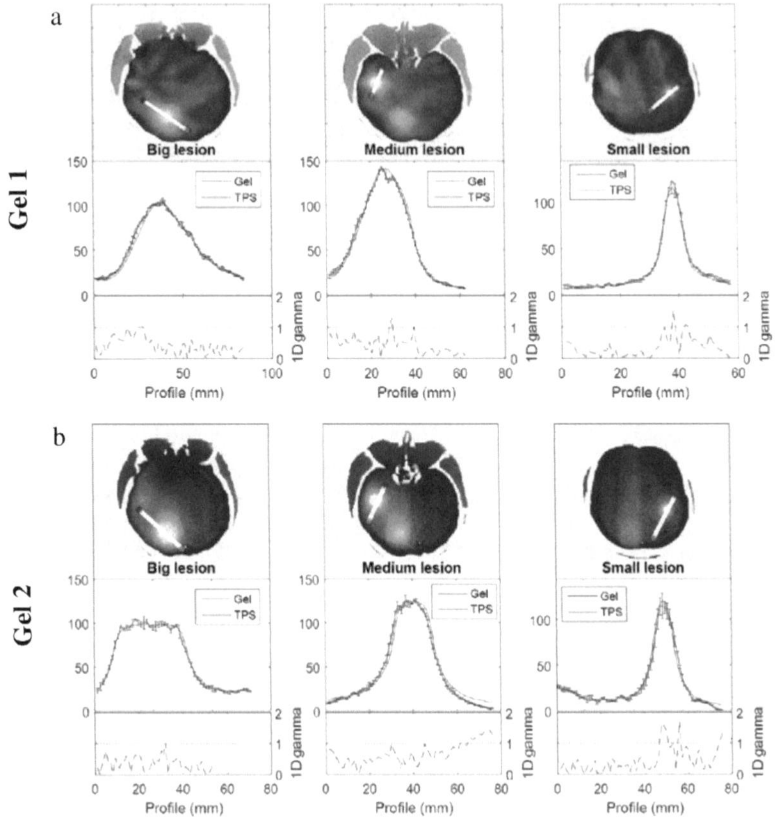


**Figure 1.** (a) Graphical display of Gel 1 phantom normalized R2 map values showing a middle axial slice crossing the Big, Medium and Small lesions. Respective normalized 1D gel phantom profiles taken in each lesion are shown in comparison with the 1D TPS. Registration based on the phantom head structures was applied here and both R2 map and TPS dose distribution were normalized to the same ROI within the Big lesion. Red to blue directions represent left to right direction of the plotted profiles. One standard deviation is shown. The 1D gamma (5%, 2 mm) is also shown. (b) The same is shown for Gel 2.

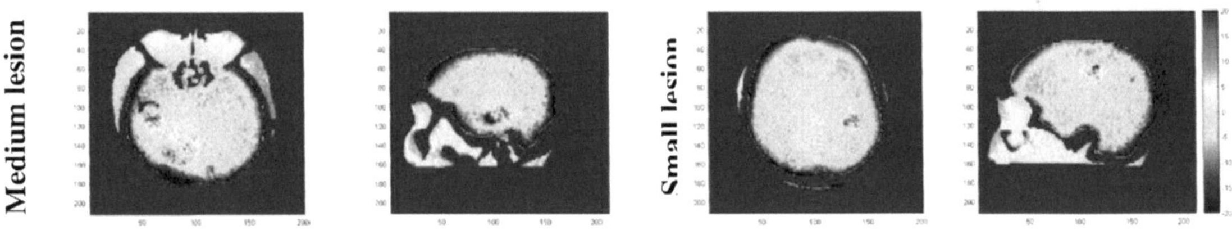


**Figure 2.** Difference between the Gel 2 phantom R2 values and TPS dose values after both being normalized to the Big lesion. Axial and sagittal slices are shown in regions were the central Medium lesion and Small lesion are visible.

In Figure 3a and c TPS dose values were plotted against the R2 values by selecting an independent 3D region around each lesion (Big, Medium and Small), for Gel 1 and Gel 2, respectively. When 3D slicer rigid registration is applied based on the phantom structures, some data spread is visible. In Figure 3b and d, the TPS dose values were also plotted against the R2 values, using the same 3D regions, but for each region a different registration was applied (registration by lesion). The spread of the data is reduced (Figure 3b and d) when compared to the spread obtained with the phantom structures based registration (Figure 3a and c) and a clear linear relationship between R2 values and dose is visible. This suggests that the registration has an effect on the relationship between dose and R2 values.

A calibration curve was fitted for each of the four plots in Figure 3. Distinct calibrations curves were obtained from the three regions using the registration by the phantom structures (Figure 3a and c). The average difference between the overall (identified as "All" in Figure 3) calibration curve gradient, which includes all data from the three regions, and the individual calibrations is 5.6% for Gel 1 and 1.3% for Gel 2. By applying a registration by lesion, the spread of the values is reduced as explained, but the difference between the calibrations curves did not improve much (4.7% difference for Gel 1 and 1.2% for Gel 2).

For Gel 1, the difference in the calibration curve could be due to a non-uniform response of the gel to radiation or MR-scanner based non-uniformities. This difference might explain why in Figure 1a, the normalized results for Gel 1 were lower for the Medium and the Small lesions when compared with the TPS simulations. For Gel 2, the spread is quite visible for the Medium lesion, but it is likely due to the shift found. An improvement is clear after registration by lesion. Despite the spread, good uniformity was found within the entire Gel 2 phantom, giving confidence for the use of only one normalization value to analyse the entire phantom.

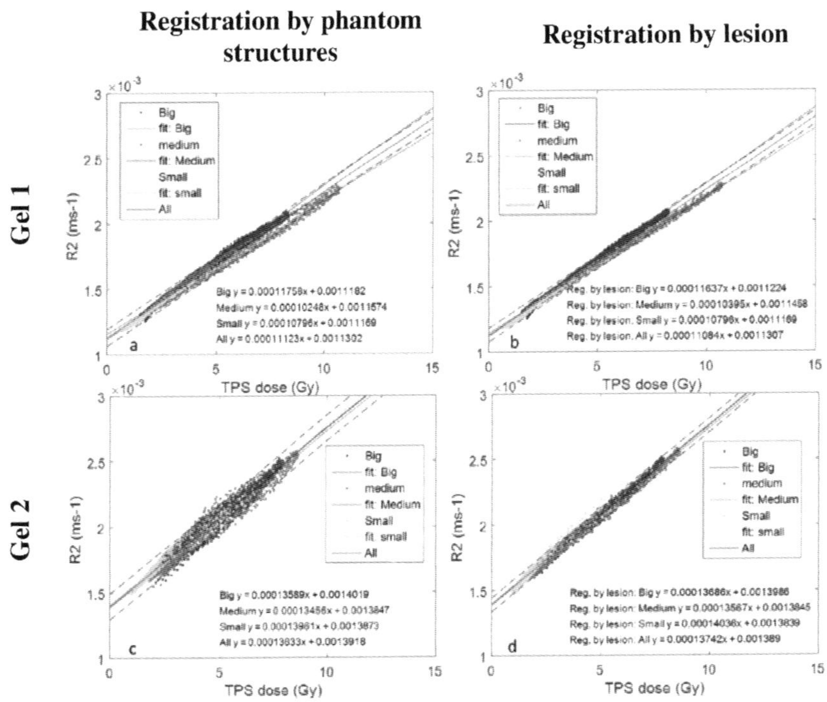


**Figure 3.** TPS dose values against R2 values taken from a 3D region around each lesion based on a registration by phantom structures ((a) for Gel 1, (c) for Gel 2) and based on an independent registration for each lesion ((b) for Gel 1, (d) for Gel 2). The upper and the lower 95% confidence bound are also shown (dashed lines) based on the data from all regions (All).

For an additional assessment of the radiotherapy treatments, absolute dose measurements were obtained for each lesion with the ionization chamber phantom. Differences between measured dose and the mean dose within the PTV calculated with each TPS are shown in Table 1.

| **Diff. (%) Meas - TPS** | Multiplan (n=2) | Elements (n=2) | Mean ± sd (n=4) |
|---|---|---|---|
| Small | -5.8 | -4.6 | -5.2±2.3 |
| Medium | -1.0 | -1.1 | -1.0±1.0 |
| Big | 0.9 | -0.9 | 0.0±1.4 |

**Table 1.** The difference between ionization chamber measurements and TPS mean dose in the PTV is shown for each lesion. TPS dose was calculated from two different TPS software, Accuray Multiplan and Brainlab Elements.

Small differences were found between measured and simulated doses, with the exception of the Small lesion, which could be explained by the ionization chamber volume averaging. Uncertainties on the estimate at the center of the PTV could be due to positional error and charge to dose conversion.

IOP Conf. Series: Journal of Physics: Conf. Series **1305** (2019) 012025 doi:10.1088/1742-6596/1305/1/012025

*3.2 MRI analysis*

Pre irradiation measurements showed a good uniformity of the gel phantom with an average R2 value prior to irradiation of 0.00100 ± 0.00002 $ms^{-1}$. No drifts were observed and all variations in R2 were attributed to noise. Comparison between the phantom MR scanned R2 values in different times are shown in Figure 4. On areas receiving high doses, Post2 and Post3 R2 exhibited an increase (13% and 19%, respectively for all lesions) when compared to Post1 R2 (a day after irradiation). This suggests gradual changes occur to the phantom gel post-irradiation. In addition, one month after irradiation (Post3 time point), it was possible to detect gel deformation, as the lesions appeared to be mismatched with the initial dataset (Post1). Further investigations with localised co-registration of each lesion suggest that shifts of up to 2mm may have occurred. This is in agreement with the manufacturers recommendations of scanning the phantom within 48h in order to avoid changes.

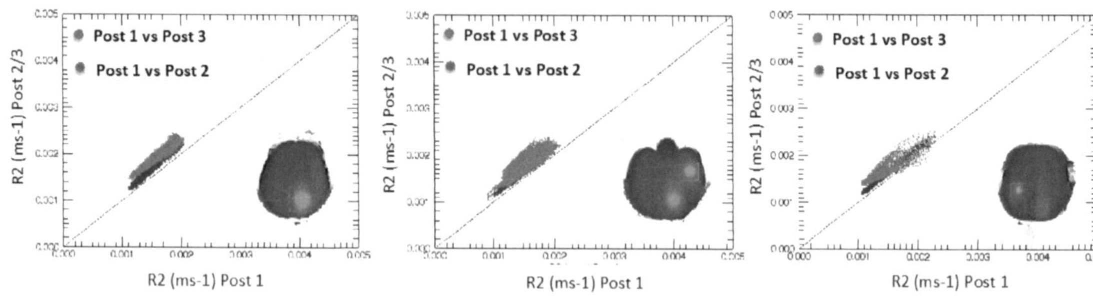


**Figure 4.** Post1 (within 48h after irradiation) R2 against Post2 (one week after irradiation) and Post3 (one month after irradiation) R2 values for all three lesions. Resulting pattern suggests misregistration at the Post3 time point, confirmed by further processing (not shown).

## 4. Conclusion

In this work we irradiate commercial patient-specific MR-based gel phantoms with high doses of radiation using complex IGRT treatment plan delivery systems. An overall good agreement was found between the TPS doses and the gel R2 values after normalization. An approximately linear relationship between dose and R2 values was found, but the gradient of the linear relationship was dependent on the selected region within the phantom. This behaviour is well-suited to relative localized dosimetric measurements and gives valuable geometrical information to assess complex radiotherapy plans. Additional independent irradiations using an ionization chamber phantom were essential to have confidence on the absolute dose values delivered. The spatial precision and dosimetric accuracy were both found adequate for stereotactic radiotherapy. Discrepancies recorded illustrate the achievable practical limits in our clinic. Further refinement is necessary before absolute dose values can be derived from the gel phantoms alone.

## 5. Acknowledgments

The Radiotherapy work was co-funded as a collaborative project by SEPnet Royal Marsden Hospital and RTsafe. The MRI work was partially funded by the National Institute for Health Research (NIHR) Biomedical Research Centre at The Royal Marsden NHS Foundation Trust and The Institute of Cancer Research, and by CR-UK (grant number C33589/A19727). The views expressed are those of the author(s) and not necessarily those of the NHS, the NIHR or the Department of Health. We acknowledge RTsafe for all the support given throughout this work.

## 6. References

[1] Liu H *et al* 2016 *J. Phys.: Conf. Ser.* **17** 142-56
[2] Pappas E *et al* 2001 *Phys. Med. Biol.* **46** 783-97
[3] Pappas E *et al* 1999 *Phys. Med. Biol.* **44** 2677
[4] Baldock C 2009 *J. Phys.: Conf. Ser.* **164** 012002

[5] National Radiotherapy Implementation Group Report; Stereotactic Body Radiotherapy; Guidelines for Commissioners, Providers and Clinicians in England 2011
[6] Pappas E *et al* 2008 *Med. Phys.* **35** 4640-8
[7] Ulin K *et al* 2010 *Int. J. Radiat. Oncol. Biol. Phys.* **77** 1584-9

IOP Conf. Series: Journal of Physics: Conf. Series **1305** (2019) 012026 doi:10.1088/1742-6596/1305/1/012026

# Assessing CBCT-based patient positioning accuracy on the Gamma Knife Icon$^{TM}$ via Presage® 3D absolute dosimetry

**Andy Y. Xu[1], Yi-Fang Wang[1], John Admovics[2], and Cheng-Shie Wuu[1]**
[1]Department of Radiation Oncology, Columbia University, New York, NY, USA
[2]Department of Chemistry, Rider University, Lawrenceville, NJ 08648

E-mail: yx2010@cumc.columbia.edu

**Abstract.** Cone beam computed tomography (CBCT) imaging has been implemented on the Leksell Gamma Knife Icon$^{TM}$ for repeated patient positioning in mask-based Gamma Knife radiosurgery. The purpose of this study was to evaluate the accuracy of the CBCT-based patient positioning on the Gamma Knife Icon$^{TM}$. Two Presage phantoms of 15 cm diameter and 10 cm height were irradiated with identical shot placements on an Acoustic Neuroma target with the same prescription dose following standard mask-based treatment workflow according to two different fraction schedules: a single fraction of 7.5 Gy and 5 fractions with 1.5 Gy fraction dose. On the top and the bottom portions of each phantom, 8 single 16 mm collimator shots were delivered with maximum doses from 2 Gy to 20 Gy for dose sensitivity calibration. The irradiated Presage phantoms were scanned and analyzed using an OCTOPOUS optical CT scanner. Both the absolute dose distributions and the relative dose distributions for the Acoustic target on each phantom were compared with those from the treatment planning system. The relative dose distribution from the single fraction irradiation agrees better with the planning system than the 5 fraction irradiation, indicating noticeable change in the dose distribution caused by the phantom positioning/repositioning process. No difference between the absolute dose distributions from the two phantoms could be identified because of the large uncertainty in the experiment data.

## 1. Introduction

The Leksell Gamma Knife Icon $^{TM}$ is the newest radiation therapy treatment machine manufactured by Elekta AB (Elekta AB, Stockholm, Sweden) for stereotactic radiosurgery on intra-cranial diseases [1]. The Icon$^{TM}$ Gamma Knife provides the capacity to perform single fraction and fractionated Gamma Knife radiosurgery using a thermal plastic mask-based head immobilization and the CBCT imaging for repeated patient positioning. The CBCT imaging technique introduced by Jaffrey et al [2] has become a standard medical imaging modality for patient positioning in external beam radiation therapy using linear accelerators. The on-board CBCT imaging module provides a treatment machine-based measurement tool that allows for a 3D visualization of the patient positioning on the treatment table. The CBCT imaging technique has been applied to many patient immobilization methods including the mask fixation that is now implemented on the Gamma Knife Icon $^{TM}$ [3].

Several dose sensitivity calibration methods have been studied for optical CT-based 3D radiation dosimetry [4]. In the early investigations of gel dosimetry, the dose responses of the dosimeters were usually obtained from measurements across a series of small size dosimeters irradiated to graded doses [5-7]. The resultant dose response curve could then be used to calibrate the dose response of large gel

IOP Conf. Series: Journal of Physics: Conf. Series **1305** (2019) 012026 doi:10.1088/1742-6596/1305/1/012026

phantoms. Several groups have observed dose deviations between calibration vials and larger phantoms originating from the same batch of gel with different potential causes [8-9]. In particular, De Deene *et al* shown that the temperature history during and after the gel fabrication process has a significant influence on the dose response and it may be difficult to keep the temperature history after fabrication similar for both the gel dosimeter phantom and the calibration vials because of the differences in phantom size [9]. Attempts were also made to implement calibration approaches utilizing multiple small photon fields irradiated to grades dose or electron depth dose curve from a calibration phantom of the same size as the dose verification phantom [10-12]. These methods were intended to minimize the temperature effect resulting from the difference in phantom size but introduced additional uncertainty in the dose calibration process such as scattered dose in the multiple field approach [11] or energy dependence of the gel dose response. A commonly used method for 3D radiation dose comparison is to do a relative dose comparison which normalizes a measured 3D dose distribution to two points of a planned 3D dose distribution [13].

In this study, we will perform a preliminary evaluation of the CBCT-based patient positioning accuracy on the Gamma Knife Icon$^{TM}$ using a dose sensitivity calibration method that does not require any calibration phantom.

## 2. Materials and Methods

The Presage dosimeters (Heuris Phanma LLC, Skillman, NJ) were formulated with tetrabromomethane, polyurethane and poured into a silicone mold to cure under 50 psi for 5 days. Two cylindrical Presage phantoms at approximately 1.8 kg with 15 cm diameter and 10 cm height from the same batch were used for this study. The Presage phantoms were imaged using a GE LightSpeed CT simulator (General Electric Co, Boston, MA) with 1 mm slice thickness. The CT images were transferred to the Gammaplan treatment planning system and defined as preplanning preference for mask-based treatment. The shot placement in a Gamma Knife treatment plan for an Acoustic Neuroma patient was used for the phantom study. The stereotactic coordinates and the weighting factors for 9 treatment shots were reproduced on both Presage phantoms. The prescription doses for the two phantoms were 7.5 Gy to the 50% isodose line in one fraction and in 5 fractions. On the top and the bottom portions of each phantom, 8 single 16 mm collimator shots were planned with maximum doses from 2 Gy to 20 Gy for dose sensitivity calibration purpose. The overall contribution from these calibration shots to the Acoustic Neuroma target in the central region of the phantoms is less than 3%. A head cushion and a mask were made for one phantom and used during the irradiation of both phantoms. For the single fraction irradiation, a reference CBCT was acquired and registered with the planning CT images for the definition of the stereotactic coordinate system. A setup CBCT image set was acquired subsequently for couch alignment during the dose delivery. For the 5 fraction delivery, the mask and the phantom were taken out of the adaptor after the first fraction and the process repeated from setting up the phantom on the adaptor.

The irradiated Presage phantoms were scanned using an OCTOPOUS optical CT scanner with same scanning parameters including 1 mm slice thickness and 1 mm resolution. On the center slice of the reconstructed images for each 16 mm shot, 5 readings were taken from the central region of the dose distribution and an averaged image intensity value was calculated. A one to one correspondence between the averaged image intensity and the delivered maximum dose was then established to obtain a calibration curve. The calibration curve was then used for the calculation of the absolute dose distribution within the phantom. For comparative purpose, relative dose comparison was also made between the two phantoms and the planning system.

IOP Conf. Series: Journal of Physics: Conf. Series **1305** (2019) 012026 doi:10.1088/1742-6596/1305/1/012026

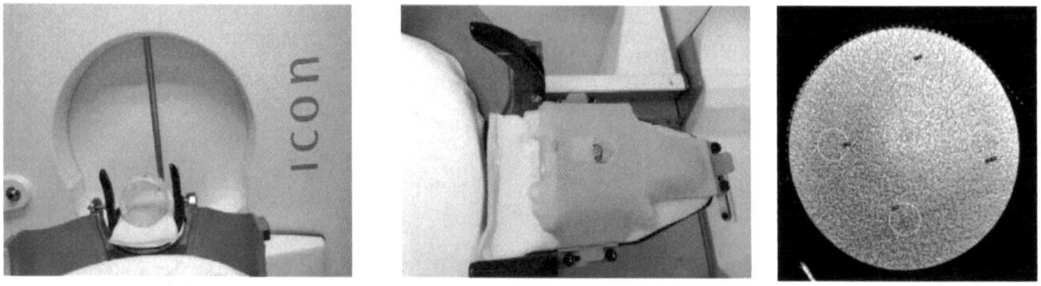


(a) (b) (c)

**Figure 1.** Irradiation of the Presage dosimeters on the Gamma Knife Icon $^{TM}$: a) phantom on the adaptor; b) mask on the phantom; c) a CBCT image for the top portion of the phantom showing the positions of 4 single 16 mm shots.

## 3. Results and Discussion

Figure 2 shows the dose response curves for the two Presage dosimeters. Several single 16 mm shots were too close to the top surface and it was difficult to find a region of uniform dose to take 5 readings from these shots because of the unevenness of the phantom surfaces. Therefore, 3 data points for the calibration curve on each phantom were rejected. Both calibration curves show good dose response linearity. When fitted to straight lines, the slopes of the two calibrations curves are different by 4.8%. The background image intensity is significantly different even though the two phantoms were from the same batch and the scanning parameters were the same.

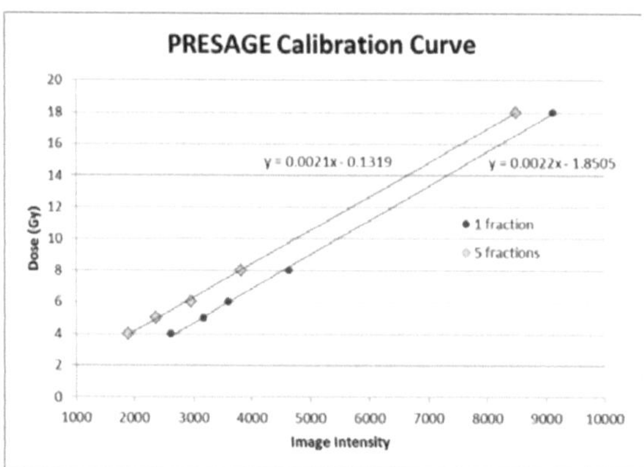


**Figure 2.** Calibration curves for the 2 phantoms.

Figure 3 compares the absolute dose distributions from the Presage phantoms to those from the planning system for the central slice in axial plane. Both the single fraction and the 5 fraction isodose lines are in reasonable agreement with the planning system. The single fraction absolute dose distribution (blue lines in Figure 3a) appears less noisy than the 5 fraction absolute distribution (green lines in Figure 3b) but the shape of the two set of isodose lines, especially the 7.5 Gy prescription isodose lines are similar.

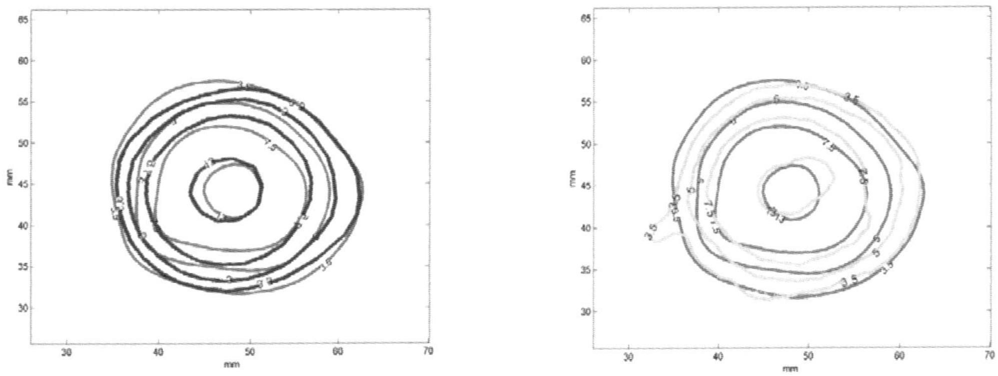

(a) (b)

**Figure 3.** Comparison of the absolute isodose lines (13 Gy, 7.5 Gy, 5 Gy, 3.5 Gy) from the Gammplan planning system (red lines) and the Presage phantoms for the central transverse slice; one fraction irradiation: blue lines; 5 fraction irradiation: green lines.

Figure 4 compares the relative dose distributions from the Presage phantoms to those from the planning system for the central slice in axial plane. The isodose lines from the one fraction experiment clearly agree better with the planning system than the 5 fraction experiment.

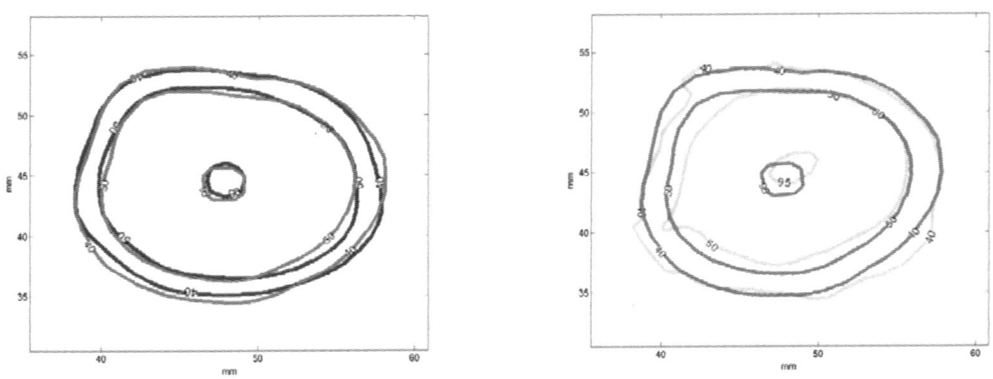

(a) (b)

**Figure 4.** Comparison of the relative isodose lines (95%, 50%, 40%) from the Gammplan planning system (red lines) and the Presage phantoms for the central transverse slice; one fraction irradiation: blue lines; 5 fraction irradiation: green lines.

Two comments can be made about the results from this preliminary study. First, there seem to be some non-negligible differences between the isodose distributions from the single fraction irradiation and the 5 fraction irradiation. The differences may be mainly related to the cumulative effect from multiple times of CBCT-based phantom repositioning. The change in isodose distribution, especially the 50% prescription isodose line may have significant clinical impact for certain treatments. Therefore, more studies are needed to quantify this effect for optimal clinical outcome. Second, the change in the isodose distribution was not detected by the absolute dose comparison because the uncertainty in the resultant absolute dose was larger than the dose distribution change from the phantom repositioning. One possible cause of the large uncertainty is that the doses delivered to the Presage dosimeters were not high enough. The reconstructed images for the two Presage phantoms were quite noisy indicating a suboptimal signal to noise ratio. The non-uniform distribution of the chemical components inside a

Presage phantom and/or the differences in the temperature history at different regions of the phantom may also contribute to the large uncertainty in the absolute dose distribution.

## 4. Conclusion
We have performed a preliminary evaluation of the CBCT-based patient positioning accuracy on the Gamma Knife Icon$^{TM}$ using optical CT-based Presage dosimetry. The relative dose comparison shows a clear effect in the dose distribution that can be attributed to the phantom positioning/repositioning process. No difference between the absolute dose distributions from the fractionated and single fraction treatment were observed because of the large uncertainty in the experiment data.

## 5. References
[1] Zeverino M *et al* 2017 *Med. Phys.* **44** 355-63
[2] Jaffray DA *et al* 2000 *Med. Phys.* **27** 1311-23
[3] Li W *et al* 2016 *Int. J. Radiat. Oncol. Biol. Phys.* **96** 214-20
[4] Baldock C *et al* 2010 *Phys. Med. Biol.* **55** R1-63
[5] Maryanski M J *et al* 1997 *Phys. Med. Biol.* **42** 303-11
[6] Baldock C 2009 *J. Phys.: Conf. Ser.* **164** 012002
[7] Mather M L *et al* 2003 *Phys. Med. Biol.* **48** N269-75
[8] Dumas E *et al* 2006 *J. Phys. Conf. Ser.* **56** 239-41
[9] De Deene Y *et al* 2007 *Phys. Med. Biol.* **52** 2719-28
[10] Oldham M *et al* 2003 *Med. Phys.* 2003 **30** 623-34
[11] Xu Y *et al* 2010 *Med. Phys.* **37** 861-8
[12] Olding T *et al* 2011 *J. Med. Phys.* **36** 3-14
[13] Xu Y *et al* 2004 *Med. Phys.* **31** 3024-33

IOP Conf. Series: Journal of Physics: Conf. Series **1305** (2019) 012027 doi:10.1088/1742-6596/1305/1/012027

# Initial commissioning measurements of respiratory gated liver VMAT stereotactic ablative body radiotherapy

**KM Alexander[1], A Kerr[1,2] and T Olding[1,2]**
[1]Department of Physics, Queen's University, Kingston, Ontario, Canada, K7L3N6
[2]Cancer Centre of South Eastern Ontario at Kingston Health Sciences Centre, 76 Stuart Street, Kingston, ON, Canada K7L 2V7

E-Mail: kevin.alexander@kingstonhsc.ca

**Abstract**. A dosimetric evaluation of a respiratory gated VMAT SABR technique was performed at two different beam energies. Dynamic ion chamber, EBT3 film and Fricke-xylenol orange-gelatin (FXG) gel measurements were acquired using a motion phantom with custom inserts for each dosimeter. Ion chamber and gel dosimeter measurements show good agreement between the measured and calculated plan dose within the planning target volume (PTV). Lower agreement than expected was observed between calculated plan dose and measured film dose, particularly for the 10 MV flattening filter free beam plan, a result that warrants further investigation.

## 1. Introduction

Patients with liver metastases may be treated with ablative radiation therapies as an alternative to surgical resection. There are a number of different options available for motion management within the broad categories of motion-encompassing, respiratory gated, breath-hold, forced shallow breathing, and real-time tumour tracking methods [1]. Our centre employs a motion-encompassing volumetric modulated arc therapy (VMAT) technique in lung stereotactic ablative body radiotherapy (SABR), a breath hold technique for left-sided breast radiotherapy, and is now investigating the use of a respiratory-gated technique for liver SABR treatments. The intention is to use a high dose rate 10 MV flattening filter-free (10FFF) beam (on a Varian TrueBeam linear accelerator, Varian Medical Systems, Palo Alto, CA) to decrease total time of delivery, as is done at other centres, but also to evaluate the 6 MV (6X) beam for possible future use.

The purpose of this work is to perform a dosimetric validation of a gated VMAT SABR technique on a well-characterized motion phantom using different dosimeters that each uniquely contribute to the strength of the validation process. These dosimeters include a 0D ion chamber, 2D EBT3 Gafchromic film, and 3D Fricke-xylenol orange-gelatin (FXG) gel dosimeter, following a similar test methodology as before [2]. Different approaches have been employed for dose verification of respiratory-gated Truebeam VMAT deliveries e.g. [3,4] and the study of respiratory gated deliveries with gel dosimetry is not new e.g. [5,6]. The results reported in this work are intended to add to this previous experience, using high resolution dose data from both film and gel dosimetry in combination with well-established point ion chamber measurements to evaluate this complex delivery process in our specific clinical environment prior to technique rollout, and establish an acceptable clinical protocol for use. This report also is part of our ongoing efforts to show the application of 3D gel dosimetry to selected clinical problems and highlight the practical challenges associated with its use.

## 2. Materials and Methods

### *2.1. Delivery Validation*

The dosimetry tools used for delivery validation in this investigation were: a) point single ion chamber (PR-05P 0.07 cc, Capintec, Ramsey, NJ), b) 2D Gafchromic EBT3 film (ISP, Wayne, NJ), and c) 3D FXG gel dosimeter with optical CT readout. Dynamic 4D computed tomography (4DCT) scans were acquired of a Quasar motion phantom (Modus Medical Devices Inc, London, ON, Figure 1a) with custom inserts for each of the dosimeters. A GE Lightspeed scanner (GE Healthcare, Maple Grove, MN) was used for CT data acquisition, along with a Varian real-time position management (RPM) motion management system (Varian Medical Systems, Palo Alto, CA) ) for phase binning of the CT images. A similar cedar insert was used for the EBT3 film (Figure 1b). The scans were then repeated with a custom gel insert (Fig. 1c). A 20 mm peak-to-peak amplitude 5 breath-per-minute (BPM) sinusoidal breathing cycle was used for all 4DCT scans. This amplitude was chosen as a representative clinical tumour motion. While a 5BPM sinusoidal waveform is not a true representation of a patient' normal breathing pattern, it is adequate to test the gating system for use in the clinic.

For VMAT SABR planning, two coplanar clockwise/counterclockwise 6X partial arcs were employed, ranging from just outside the approximated edge of the contralateral lung anteriorly to 180 degrees posteriorly (figure 1c upper right). The collimator angle on the arcs was set to ±30 degrees. A RapidArcTM VMAT plan was optimized and calculated in EclipseTM v.13.6 (Varian Medical Systems, Palo Alto, CA) on the 2.5 mm slice thickness 4DCT-derived average (of phases 60 to 40) scan of the Quasar phantom with each respective insert. The plan had a target objective placed on the overall monitor units (MU) allowed in the optimization stage of treatment planning, to limit multi-leaf collimator (MLC) motion and force the MLC aperture open throughout the arc. A 10FFF plan was also optimized and calculated on the same anatomy. A similar process was followed to produce optimized 6X and 10FFF plans on the ion chamber and gel 4DCT-derived average (again, phases 60 to 40) phantom datasets. Film and ion chamber plans were normalized to a prescription dose of 2 Gy (100% prescription dose covers 95% of PTV target volume) unless otherwise specified, while gel plans were normalized to a prescription dose of 1.8 Gy for optimal optical CT FXG gel dose readout. A screen capture of the 6X ion chamber plan is shown in figure 1c.

All treatment plans were delivered on a Varian TrueBeam linear accelerator (Varian Medical Systems, Palo Alto, CA), with cone beam CT (CBCT) setup verification. Plans were delivered within the 60 to 40 phase thresholds used to generate the average planning CT scan, or with a slightly larger phase threshold up to 70 to 30. EBT3 film measurements were acquired using an Epson 10000XL flat-bed scanner (Epson Canada, Markham, ON). A calibration set of 6MV 10x10 cm2 films were delivered under machine reference conditions to doses spanning the range of 0-4.5 Gy. Gel dosimeters were calibrated using a single field electron beam under machine reference conditions [7], and imaged using a VistaTM optical CT scanner (Modus Medical Devices Inc., London, ON). Measured film and gel dose were compared against calculated Eclipse treatment planning system (TPS) dose using 3D Slicer (www.slicer.org) [8].

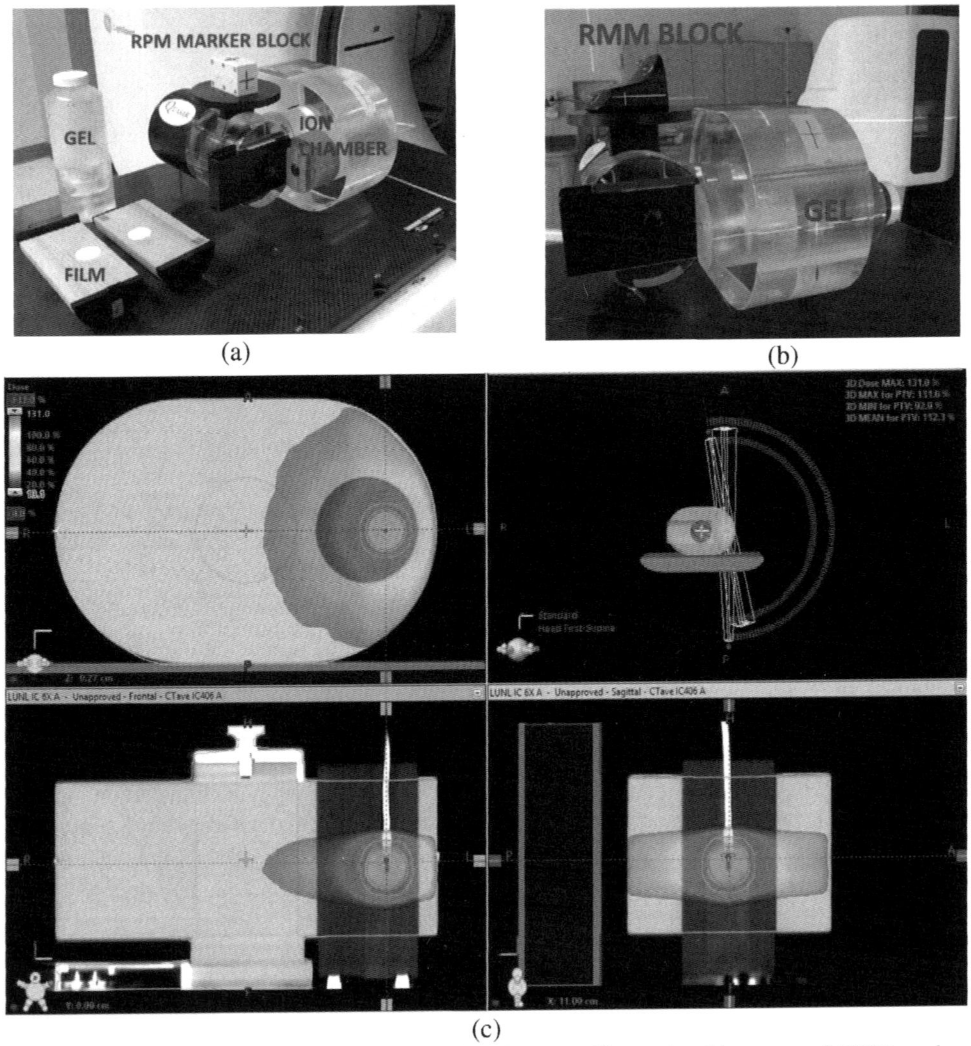


**Figure 1.** (a) Quasar motion phantom with ion chamber, film and gel inserts, and RPM marker block used at CT, (b) Quasar gel delivery setup with respiratory motion management (RMM) block at the treatment unit, and (c) Eclipse screen capture of the 6X ion chamber plan used in this study.

## 3. Results

### *3.1. Dosimeter-Plan Dose Comparison*

Results for the dosimeters used in this work are summarized in table 1. All ion chamber measurements were in good agreement with TPS calculated dose. Film and gel gamma comparison results are also reported in table 1, with selected profiles and 2D gamma comparison slices shown in figures 2 and 3.

**Table 1.** Summary: measured vs. Eclipse doses.

| Dosimeter | Plan | Prescription (Gy) | Ratio of measured to calculated doses (%) | % Gamma ≤1 (3%/3mm) |
|---|---|---|---|---|
| Ion Chamber | 6X (60-40 gated) | 2 | 99.1 ± 0.5 | - |
| Ion Chamber | 6X (70-30 gated) | 2 | 98.8 ± 0.5 | - |
| Ion Chamber | 10FFF (60-40 gated) | 2 | 99.9 ± 0.5 | - |
| Ion Chamber | 10FFF (60-40 gated) | 12 | 101.3 ± 0.5 | - |
| EBT3 Film | 6X (60-40 gated) | 2 | - | 90.1 |
| EBT3 Film | 6X (65-35 gated) | 2 | - | 85.0 |
| EBT3 Film | 10FFF (60-40 gated) | 2 | - | 64.6 |
| FXG Gel | 6X (60-40 gated) | 1.8 | - | 96.6 |
| FXG Gel | 6X (65-35 gated) | 1.8 | - | 95.0 |
| FXG Gel | 10FFF (60-40 gated) | 1.8 | - | 97.9 |

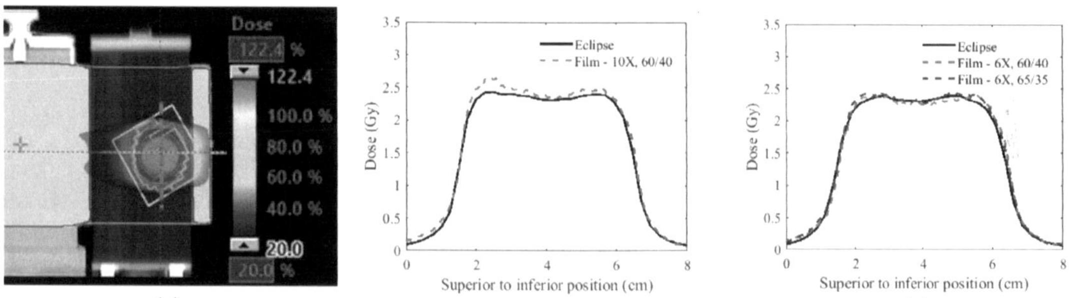


(a) (b) (c)

**Figure 2.** (a) Eclipse screen capture of the 10FFF film plan with red dashed line along axis of motion showing location of profile measurements for the (b) 10FFF and (c) 6X plan deliveries.

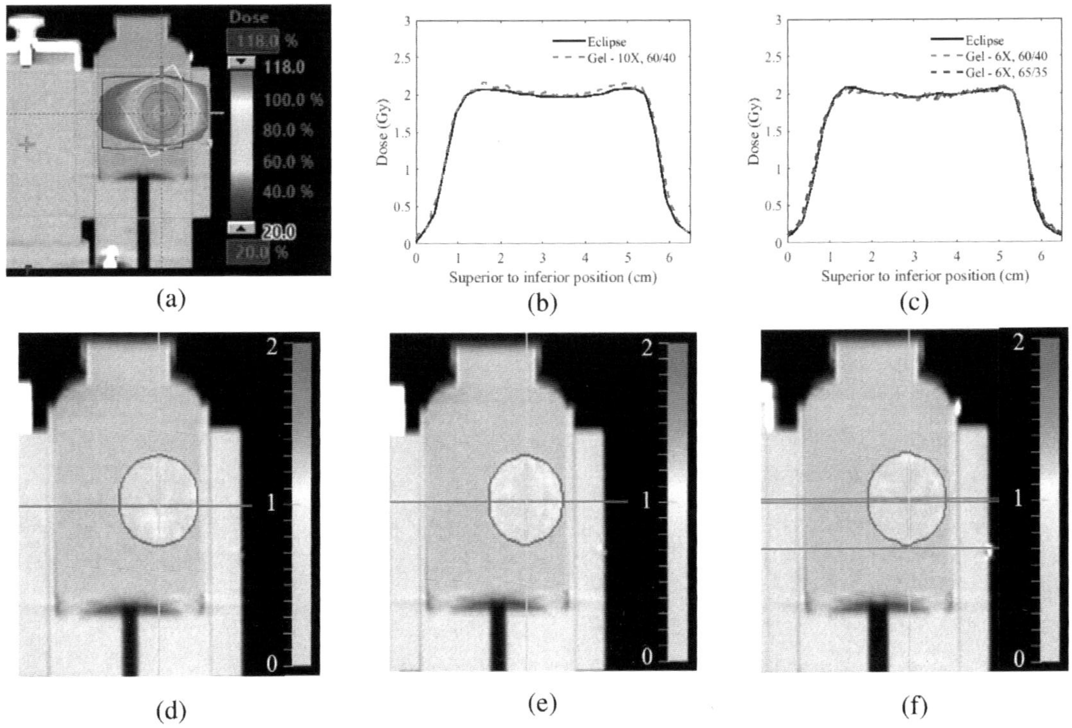


**Figure 3.** (a) Eclipse screen capture of the 10FFF gel plan with red dashed line along the axis of motion showing location of profile measurements for the (b) 10FFF and (c) 6X plan deliveries. (d)-(f) Coronal slices of the 3%, 3 mm gamma comparison results for the 10X 60 to 40, 6X 60 to 40, and 6X 65 to 35 phase threshold gel deliveries, respectively.

## 4. Discussion & Conclusions

With reference to the results summarized in table 1, increasing the phase limits on the delivery beyond the average scan-planned 60 to 40 phase window to a slightly larger 70 to 30 phase window had only a slight effect in lowering the recorded ion chamber reading. This was expected, as this corresponded to approximately ± 5 mm range of motion whereby the ion chamber was still located within the mostly open MLC-defined field aperture. The gamma comparison results from the EBT3 film analysis were lower than expected, particularly for the 10FFF beam (see table 1). Gel dosimeter measured dose showed good gamma agreement with Eclipse-calculated dose for both delivery energies. The gel dosimeter phantom anatomy does not have significant inhomogeneities near to or within the target volume, so it is not surprising that better agreement is observed from the gel gamma comparison. In the clinic, the inhomogeneous film anatomy would be a better representation of the situation where a target volume in the liver is near the diaphragm, with lung tissue in proximity. This underscores the importance of using the most realistic phantom anatomy combined with multiple dosimeters to assess new techniques prior to clinical implementation. A closer look at the 10FFF profile measurements for both film and gel (see figures 2b and 3b) show larger "horns" than calculated by the planning system on both anatomies near the field edge in the direction of motion (i.e. with and without inhomogeneities). This issue could be related to the 10FFF beam model in Eclipse, which has not yet been released for clinical use. Further evaluation is warranted, with repeat measurements and comparison to beam commissioning water tank data. Overall, the initial measurements are promising, but more work is required to fully establish a clinical protocol for respiratory gated liver VMAT SABR at our centre, including a better understanding of the source of delivery disagreement in the inhomogeneous film insert CT anatomy.

## 5. Acknowledgements
This work was funded by the Canadian CIHR funding agency, project MOP-115101.

## 6. References
[1] Keall P *et al* 2006 *Med. Phys.* **33** 3874
[2] Olding T and Alexander KM 2017 *J. Phys.: Conf. Ser.* **847** 012032
[3] Qian J *et al* 2011 *Phys. Med. Biol.* **56** 4827
[4] King R B *et al* 2016 *Phys. Med. Biol.* **61** 5529
[5] Ceberg S *et al* 2008 *Phys. Med. Biol.* **53** N387
[6] Jong W L *et al* 2016 *J. Phys.: Conf. Ser.* **694** 012015
[7] Olding T, Darko J and Schreiner LJ *J. Phys.: Conf. Ser.* **250** 012028
[8] Alexander K M *et al* 2017 *J. Phys.: Conf. Ser.* **847** 012061

IOP Conf. Series: Journal of Physics: Conf. Series **1305** (2019) 012028 doi:10.1088/1742-6596/1305/1/012028

# Chemical dosimeters

**John A Adamovics and Robert J Coakley**
Departmaent of Chemistry, Biochemistry and Physics, Rider University
2083 Lawrenceville Road, Lawrenceville, NJ 08648-3099, USA

E-mail : jadamovics@rider.edu

**Abstract.** Measuring radiation exposure has long been a priority in radiation therapy. Over the last 115 years numerous chemical-based dosimeters have been developed to measure radiation exposure. These dosimeters contain either metal complexes or leuco dyes. Among the factors which influence dose sensitivity and color stability in leuco dye based dosimeters are chemical structure and the solvent content in the formulation.

## 1. Introduction
Since the early 1900s color changes have been utilized as a means of quantitatively measuring radiation. Chemical-based radiochromic dosimeters utilize either metal complexes or leuco dyes.

## 2.Metal Complexes
One of the first popular dosimeters was developed by Sabouraud and Noire in1904 [1]. This dosimeter emloyed a small disk (pastille) of barium platinocyanide, which after irradiation changes from a green to a dark yellow-orange. These dosimeters were widely utilized until the 1930s. More recently, a ferrous hexacyanide variant (also known as Turnbull dye or Prussian Blue) was evaluated as a 3D gel dosimeter with a maximum wavelength absorbance of 690 nm (Figure 1). [2].

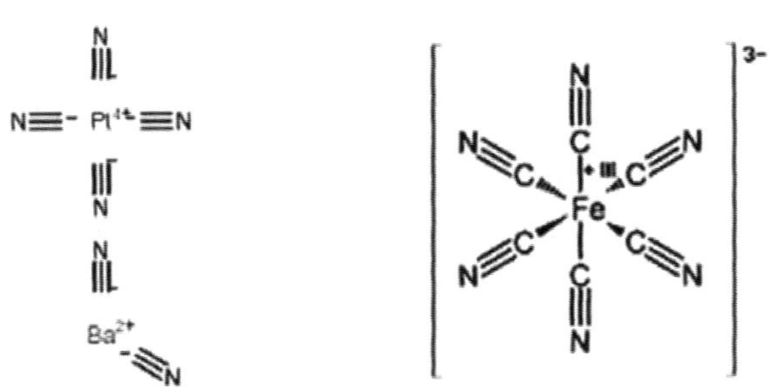

**Figure 1.** Structure of barium platinocyanide (left) and ferrous hexacyanide (right).

By 1927 Fricke and Morse developed a radiochromic dosimeter incorporating a ferrous sulfate solution (Fricke solution). Upon irradiation, ferrous ($Fe^{2+}$) ions are oxidized to ferric ions ($Fe^{3+}$) [3]. During irradiation water is decomposed to reactive **HO·** and **H·** radicals which further react with oxygen to produce the hydroperoxy radical. This oxidizes the ferrous ions (Scheme 1). Other metals such as ceric sulfate have also been used in an aqueous dosimeter [4].

In order to stabilize the geometric dose information in the Fricke dosimeter, aqueous based gel matrices containing the chelator xylenol orange (XO) are used [5]. When analyzed spectrophotometrically, a non-irradiated ferrous/agarose/ XO gel shows visible-light absorption at 440 nm; after exposure to ionizing radiation, the maximum absorption shifts to 585 nm.

$$H\cdot + O_2 \rightarrow \mathbf{HO_2\cdot}$$
$$\mathbf{HO_2\cdot} + Fe^{2+} \rightarrow HO_2^{-} + Fe^{3+}$$
$$HO_2^{-} + H^{+} \rightarrow H_2O_2$$
$$HO\cdot + Fe^{2+} \rightarrow HO^{-} + Fe^{3+}$$
$$H_2O_2 + Fe^{2+} \rightarrow HO^{-} + Fe^{3+} + HO\cdot$$

**Scheme 1.** Ionizing radiation reactions in the Fricke dosimeter.

Seventeen different chelators have been studied as potential alternatives to XO with methyl thymol blue being a potential alternative [6] (Fig 2). Since 1927, over 5,500 dosimeter references to ferrous sulfate dosimetry have been reported [7].

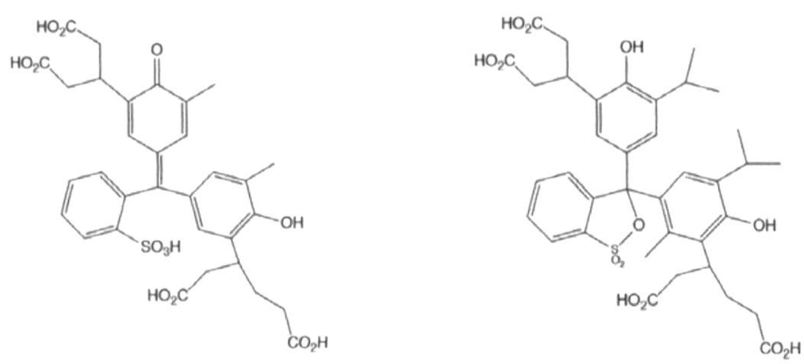

**Figure 2.** Molecular structure of the chelators XO (left) and methyl thymol blue (right).

## 3. Leuco Dyes

Leuco color formers undergo a switch between two chemical forms, one of which is colorless. This transformation occurs by the input of energy either from heat, light, ionizing radiation or change in pH. This redox process is reversible [8]. There are at least 7 leuco dye structural families each with distinctive structures [8].

### *3.1 Tetrazolium*

One group is tetrazolium salt derivatives, which have been increasingly employed in both 3D dosimeters [9-12] and 2D films [13] over the past two years.

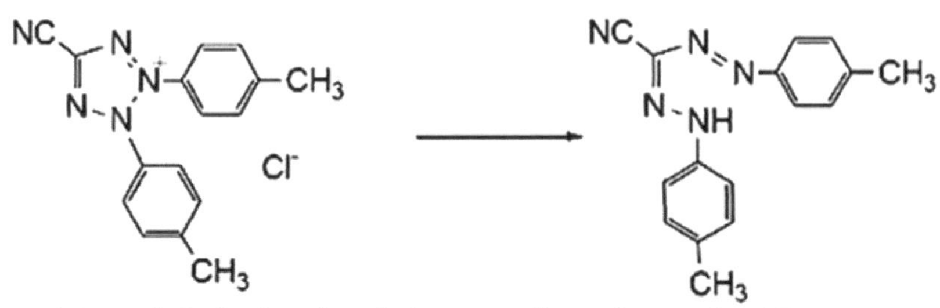

**Scheme 2.** Reduction of a colorless tetrazolium salt to a colored Formosan.

### *3.2. Triarylmethanes (TAM)*

TAM leuco dyes, formulated primarily in organic solvents, have been thoroughly studied [14]. In the early 1950s the US Army issued glass vials filled with bromocresol purple (a pH indicator) dissolved in chloroform to determine possible radiation exposure from a nuclear explosion [15]. TAMs have also been incorporated into polyvinyl chloride and other polymer matrices for high dose 2D film for non-clinical dosimeter applications [16] and more recently in polyurethane sheets for 2D clinical applications [17].

For the last 14 years, we have utilized TAMs as color formers (1-2%) in a clear polyurethane matrix containing 0.5% radical initiator, with the remainder organic solvents that are used to solubilize the TAM and initiator [18]. We focused primarily on leuco malachite green (LMG, Scheme 3, and #1, Table 1) derivatives, a N,N,N',N'-tetramethyl triarylmethane (DTM) which we synthesized [18] , and secondarily on leuco crystal violet (LCV, #16, Table 1). In order for the TAM based dosimeter to be sufficiently reactive to a clinical radiation dose, a radical initiator is required. The most effective class of initiators are halocarbons [23, 24]. The dose sensitivity was found to be consistent with the bond energy of the carbon-halogen bond. The observed sensitivity was in the order $R_3C$-I > $R_3C$-Br > $R_3C$-Cl [24].

The first step upon irradiation is the homolytic cleavage of the carbon-halogen bond, creating a halo radical which subsequently abstracts the methine proton, providing a planar carbocation which absorbs at 633 nm.

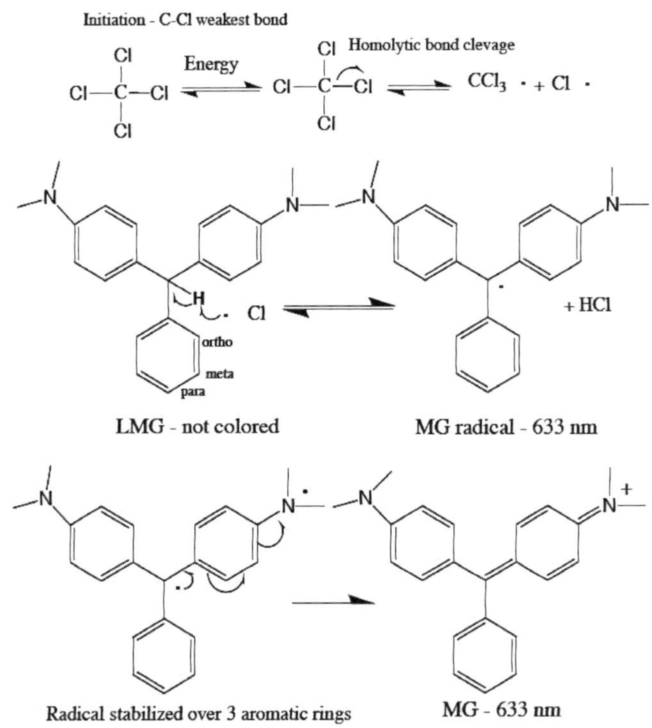


**Scheme 3.** Mechanism of TAM oxidation by ionizing radiation.

IOP Conf. Series: Journal of Physics: Conf. Series **1305** (2019) 012028 doi:10.1088/1742-6596/1305/1/012028

**Table 1.** Synthesized DTBs and their LMG (**1**) relative radiation dose sensitivity (RRDS)

| DTB | RRDS | DTB | RRDS | DTB | RRDS | DTB | RRDS |
|---|---|---|---|---|---|---|---|
| 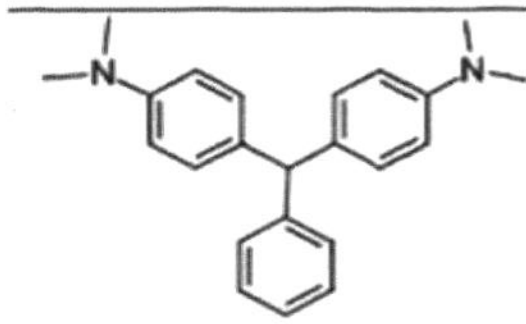 **1** | 100 | 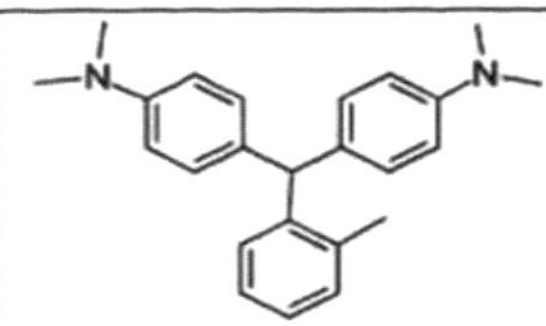 **5** | 350 | 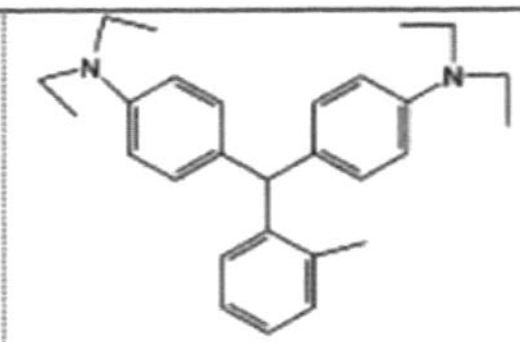 9 | 430 | 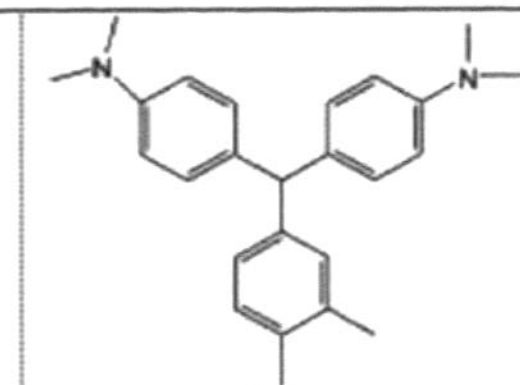 13 | 110 |
| 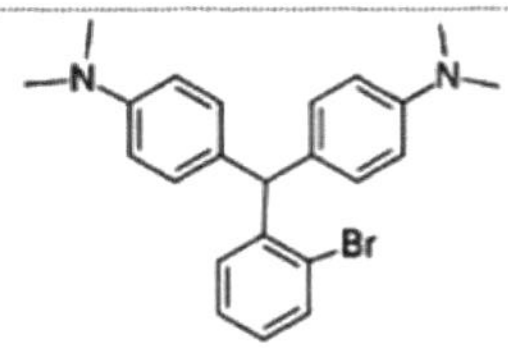 **2** | 450 | 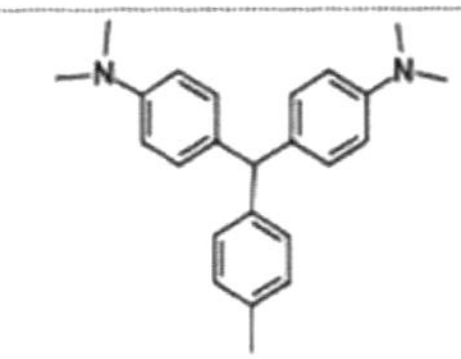 **6** | 275 | 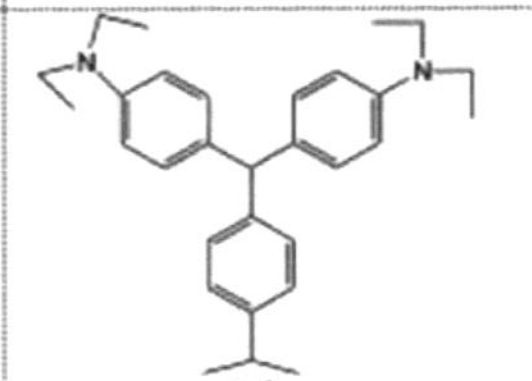 10 | 280 | 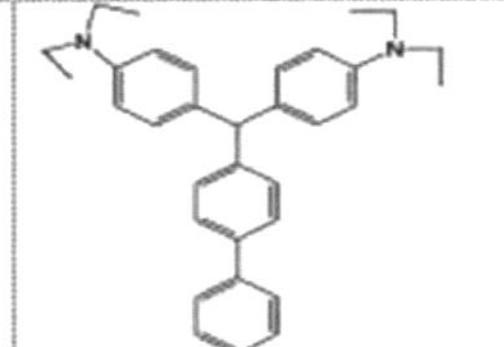 **14** | 90 |
| 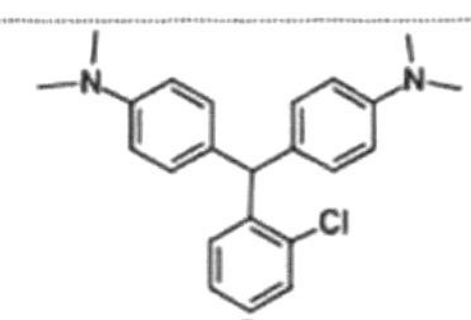 **3** | 340 | 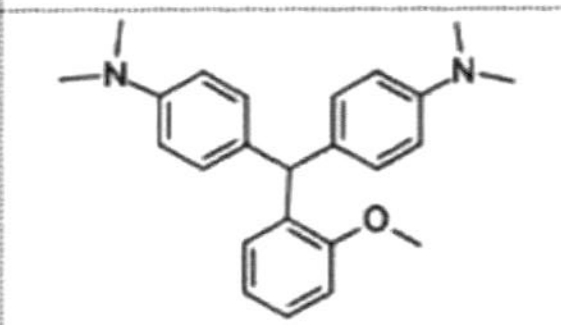 **7** | 200 | 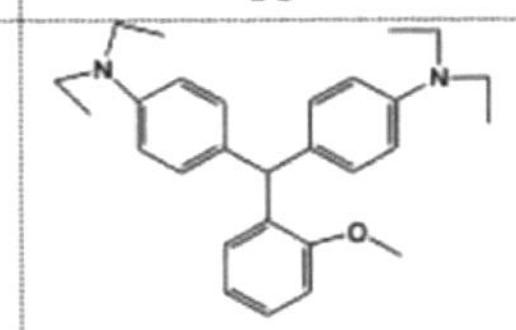 11 | 325 | 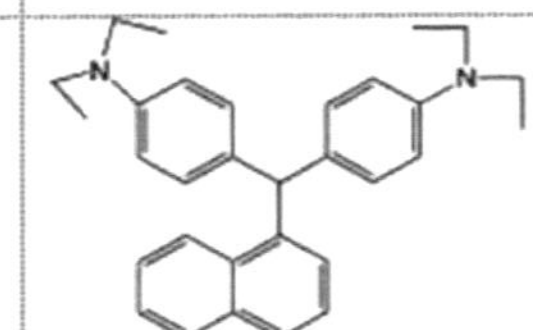 15 | 280 |
| 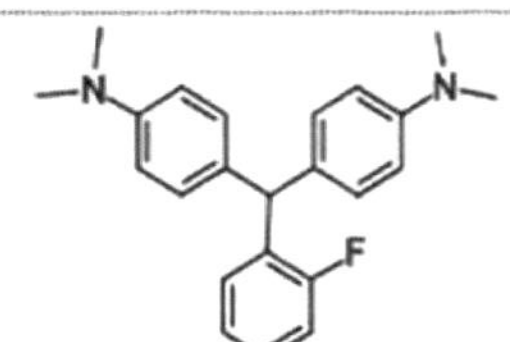 **4** | 60 | 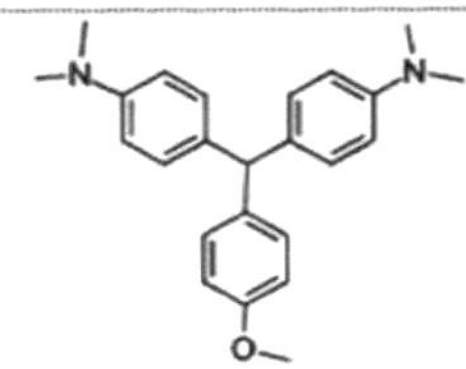 **8** | 320 | 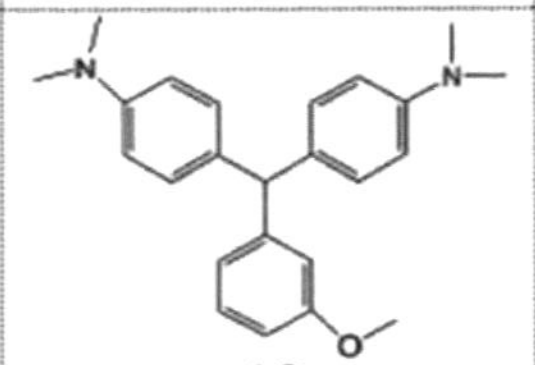 12 | 40 | 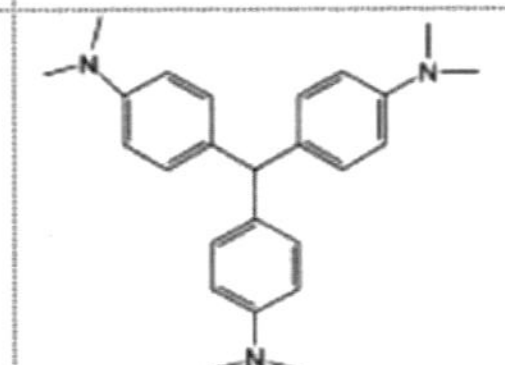 16 | 600 |

The relative optical response of sixteen TAMs is tabulated in Table 1. The TAM molecular structure has significant impact on the dose sensitivity. After irradiation, the methine radical (Scheme 3) is delocalized over the entire triarylmethane [25]. Radical stability is largely due to steric protection [25] of the central carbon, which is consistent with what is observed for the radiation dose sensitivity of the TAMs. The sensitivity varied from 3-4 times greater than LMG for the most hindered bromide derivative **2** to 0.6 less dose sensitivity than LMG for ortho-fluoro derivative **4** (Table 1). This is also consistent for the ortho-methyl derivative **5** being more dose sensitive than the para-methyl derivative **6**. There are electronic contributions of the para-methyl **6** in stabilizing the radical relative to **1** which has no para substituent. For the ortho- and para-methoxy derivatives, **7** and **8,** the interpretation of the steric and electronic contributions is not as straight forward since **8** is more dose sensitive than **7** and almost that of **5** while in **12** the meta substituent is the least dose sensitive of the DTMs tested**.** In general the N,N-diethyl substituents are more dose sensitive than the N,N-dimethyl DTMs (**5** vs **9**) which may be rationalized by additional greater hinderance of the central carbon or increase in the nitrogen basicity. Overall, the most dose sensitive was **16** which has an additional N,N-dimethyl substituent but is not sterically hindered.

Another critical dosimeter characteristic is post irradiation stability. In general, the more hindered TAMs are more stable post irradiation but that is also influenced by the amount of solvent in the formulation. The hindered TAM, 2,4 dimethyl-LMG (Structure 13, Table 1) shows a stable radiochromic response over 500 hours with 4% solvent in the formulation, while the color is totally bleached after 320 hours when formulated with 15% solvent. The mechanism is not clear.

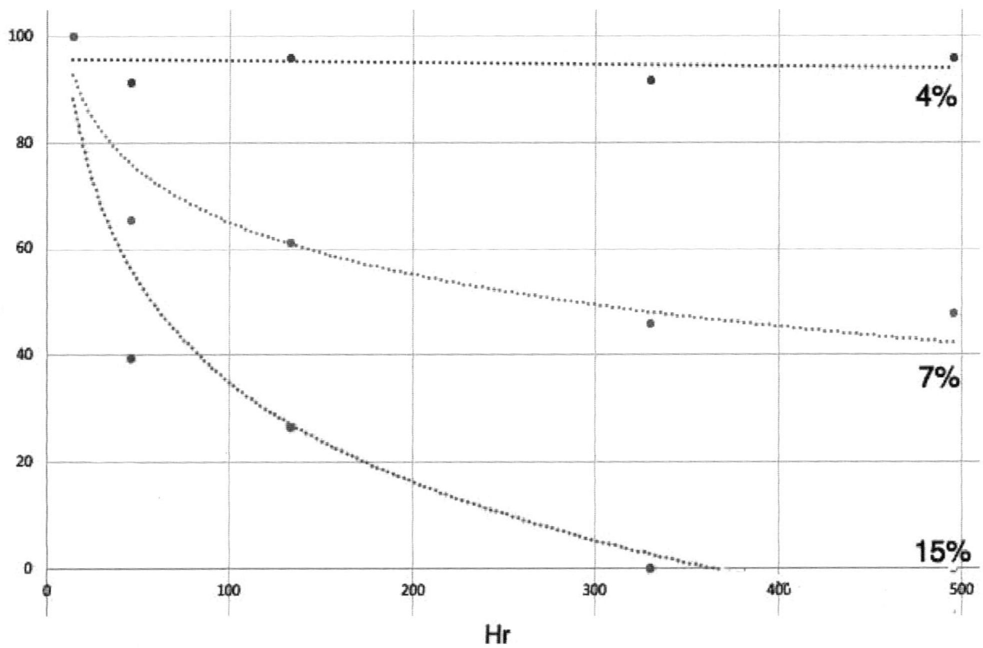


**Figure 3.** Post irradiation stability of 2,4 dimethyl LMG at 4%, 7% and 15% solvent content.

There has been a long-term interest in TAMs containing cyanide where the methine H is replaced by cyanide which makes the bond at the central atom more labile [26, 27] but to date only 2D dosimeter films have been developed. This is also true of the most common 2D Gafchromic film which is based on the polymerization of diacetylenes.

## 4.Conclusions

This review is intended to provide the reader with a comprehensive set of references from the two major categories of chemical-based dosimeters - metal complexes and leuco dyes and an update on what is understood of the mechanism of reaction.

## 5.Acknowledgements

JA is grateful to the numerous students who have worked on this project and to Dr. Richard Wood for his review of this manuscript. The research reported was supported by the National Institutes of Health under award number 1R41CA217421-01 A1.

## 6.References

[1] Frame P W 2004 *Health Phy.*. **87** 111
[2] Kozubikova P *et al* 2015 *J. Phys.: Conf. Ser.* **573** 012049
[3] Fricke H and Morse, S 1927 *Am. J. Roentgenol. Radium Ther. Nucl. Med.* **18** 430
[4] Matthews R 1982 *Inter. J. Appl. Rad. Isotop.* **33** 1159
[5] Schreiner L J 2004 *J. Phys.: Conf. Ser.* **3** 9-21
[6] SilvaAlves A V *et al* 2018 *Rad. Phy. Chem.* **150** 151
[7] *Google Scholar* June 2018 search - ferrous sulfate dosimetry
[8] Muthyala R ed. 1997 *Chemistry and applications of leuco dyes.* (Plenum Press: NY),
[9] Kwiatos K *et al* 2018 *Phys. Med. Biol.* **63** 095012
[10] Penev K I *et al* 2017 *J. Phys.: Conf. Ser.* **847** 012048
[11] Grossman M S and Courter E J 2017 *Can. J. Phys.* **95** 725-30
[12] Ascencion Y *et al* 2017 *J. Phys.: Conf. Ser.* **847** 012049
[13] Abdel-Fattah A A *et al* 2017 *Rad. Measurements* **100** 18
[14] Chalkley L1929 *Chem. Rev.* **6** 217
[15] Taplin G V *et al* 1951 *Nucleonics* **9** 73-7
[16] Kattan M 2011 *Appl. Radiation Isotopes* **69** 377
[17] Jordan K 2016 *Med. Phys.* **43** 3589
[18] Adamovics J 2017 *Beilstein J. Org. Chem.* **13** 1325
[19] Babic S 2009 *Phys. Med. Biol.* **54** 2463
[20] Babic S 2009 *Phys. Med. Biol.* **54** 6791
[21] Brindha S 2004 *Phys. Med. Biol.* **49** N353
[22] Jordan K 2009 *J. Phys.: Conf. Ser.***250** 012043
[23] Adamovics J 2006 *J. Phys.: Conf. Ser.* **56** 172
[24] Alqathami M 2012 *Radiat. Phys. Chem.* **81** 867
[25] Hicks RG 2007 *Organic Biomolecular Chemistry* **5** 1321
[26] Chalkley L 1961 *J. Org. Chem.* **26** 408
[27] Humpherys K C 1977 *Rad. Physics Chem.* **9** 737

# 3-D dosimetry readout techniques

**Simon J Doran**
[1]CRUK Cancer Imaging Centre, Division of Radiotherapy and Imaging, Institute of Cancer Research, London, SW7 3RP, UK

E-Mail: Simon.Doran@icr.ac.uk

**Abstract.** This invited technical review will discuss the numerous options available for making 3-D measurements of radiation dose, including both the physical principles underlying them and the potential sources of error involved.

One of the joys of being involved from the very beginning in the DOSGEL and IC3Ddose conference series has been the opportunity to follow the evolution of the different imaging modalities. Back in 1999, Magnetic Resonance Imaging (MRI) was the method of choice, thanks to Gore and Kang's seminal publication [1] and the work of the early "gel pioneers" [2-4]. Its popularity waned somewhat as it became clear that the level of technical understanding and careful experimentation required to obtain good results were non-trivial. In the 1990s and early 2000s access to MRI equipment for radiotherapy was also problematic, but now, with the advent of true MRI-guided radiotherapy and the possibility of real-time, MR-based 3-D dosimetry [5], it is perhaps set for a renaissance.

At the time of that first DOSGEL meeting in Kentucky, optical computed tomography (CT), although first described in 1996 [6], still felt very much like the "new kid on the block". Subsequently, the relatively low bar to entry into the field, in terms of scanner cost, led to significant innovation from a number of research groups and commercial companies, making this a dynamic and exciting area to work in. By the second DOSGEL meeting in 2001, x-ray CT as a readout technique had also already become an established field of work [7, 8], and the ubiquity and relative ease of access to hospital CT scan were encouraging signs.

Many physical properties are capable of being used to record radiation dose. So, along with these three stalwarts of the gel dosimetry field, other techniques have emerged — sometimes briefly, sometimes in a more sustained fashion — and each has its own unique advantages and problems. I will provide a brief introduction to each of the techniques, but the archives of the DOSGEL and IC3Ddose conferences, as well as previous review articles (e.g., [9]) provide a wealth of information for the interested reader to explore further.

An important initial distinction to make is the difference between 3-D methods that involve *integrating* dosimeters coupled with a two-step measurement (in which a sample is first exposed to radiation that causes a change in its physico-chemical properties, and then subsequently imaged); and *real-time imaging*, where the dosimeter gives off an instantaneous signal, captured by the imaging device, that corresponds to the radiation dose absorbed. Historically, the DOSGEL conferences focused on integrating dosimeters and these methods will be examined first. With the change in conference title to IC3Ddose, came the acknowledgement that 3-D dosimetry involved other types of method and these will be presented subsequently.

**Acknowledgments**
SJD acknowledges CRUK and EPSRC support to the Cancer Imaging Centre at ICR and RMH in association with MRC and the Department of Health C1060/A10334, C1060/A16464, and NHS funding to the NIHR Biomedical Research Centre, Clinical Research Facility in Imaging and Cancer Research Network. This report is independent research funded by the National Institute for Health Research. The views expressed in this publication are those of the author and not necessarily those of the NHS, the National Institute for Health Research or the Department of Health.

**References**
[1] Gore J and Kang Y 1984 *Phys. Med. Biol.* **29** 1189
[2] Appleby A *et al* 1987 *Med. Phys.* **14** 382
[3] Olsson L E *et al* 1989 *Phys. Med. Biol.* **34** 43
[4] Schulz R *et al* 1990 *Phys. Med. Biol.* **35** 1611
[5] Lee H J *et al* 2018 *Phys. Med. Biol.* **63** 045021
[6] Gore J *et al* 1996 *Phys. Med. Biol.* **41** 2695
[7] Hilts M *et al* 2000 *Phys.Med. Biol.* **45** 2559
[8] Trapp J *et al* 2001 *Phys. Med. Biol.* **46** 2939
[9] Baldock C *et al* 2010 *Phys. Med. Biol.* **55** R1

IOP Conf. Series: Journal of Physics: Conf. Series **1305** (2019) 012030 doi:10.1088/1742-6596/1305/1/012030

# Optimization for stability of the deformable FlexyDos3D radiation dosimeter and curing effects

**M J Wheatley[1,5], A S Balatinac[1,5], J T Booth[2,3] and Y De Deene[1,4,5]**
[1]School of Engineering, Macquarie University, North Ryde, Sydney, Australia
[2]Northern Sydney Cancer Centre, Royal North Shore Hospital St Leonards, Sydney, Australia
[3]Institute of Medical Physics, Faculty of Science, University of Sydney, Sydney, Australia
[4]Department of Radiation Oncology and Experimental Cancer Research, Ghent University, Belgium
[5]The authors M J Wheatley, A S Balatinac, and Y De Deene contributed equally to the work presented in this paper.

E-mail: Morgan.Wheatley@hdr.mq.edu.au

**Abstract**. Previous formulations of the FlexyDos3D dosimeter have shown significant changes in the dose-response over time. In this study, various formulations of the dosimeter were created and tested to see if this stability could be improved. A dosimeter that was stable over a three-day period was found. Rapid manufacture of this dosimeter for patient-specific validation of radiotherapy treatments is desirable. The use of 3D printing manufacturing techniques for thermosetting polymers require high temperatures to cure the polymer within a reasonable time. The effect of different curing temperatures and times were investigated for the FlexyDos3D radiation dosimeter for its effect on stability. No significant difference in the dose-response was found for dosimeters cured for different curing times beyond an hour. A significant dose-response offset was found between dosimeters cured at different temperatures, but the dose-response sensitivity was the same.

## 1. Introduction

Different 3D dosimeters currently exist, from Fricke gels, to polymer gels, and radiochromic dosimeters [1-3]. Whilst many of these dosimeters have been shown to provide accurate determination of the 3D dose-distribution within the dosimeter, the hardware requirements and difficulty to manufacture and image the dose-distribution of these dosimeters is a potential barrier to their wider implementation in clinical settings.

The FlexyDos3D dosimeter can be manufactured at standard atmospheric pressure and temperature, allowing it to be produced in any standard laboratory setting. Manufacturing the dosimeter at room temperature requires two days to cure and so is currently infeasible for rapid production. Heating the dosimeter at 65 °C has previously been shown to allow it to be cured in two hours [4]. However, these temperatures still only allow for castings of the dosimeter to be produced. Higher temperatures are required for 3D printing of the dosimeters, which would allow for complex structures to be created, such as the lungs and heart. The effect of higher temperatures on the dosimeter had not previously been tested.

Previous formulations of the FlexyDos3D dosimeter had poor stability characteristics, with the dose-response of the dosimeter changing significantly over a period of three days. In this study, the effect of different concentrations of the silicone curing agent and base mixture on the stability of the dosimeter were tested. Once a stable formulation was found, the effect of different curing temperatures and times were tested for their effects on the stability of the dosimeter.

## 2. Methods and materials

### *2.1. Dosimeter fabrication*

The FlexyDos3D dosimeter consists of an optically-transparent silicone elastomer (Sylgard® 184, Merck), chloroform, and a leucodye, leucomalachite green (LMG). The silicone elastomer initially consists of a separate base and curing agent (CA), which cures when mixed together. The silicone elastomer serves as a matrix for the chloroform and LMG, and gives the dosimeter deformable properties. The chloroform aids in dissolving LMG in the elastomer and provides radiation-induced radicals that initiate the reaction with LMG. The LMG changes colour from crystal clear to green upon reaction with the initiator radicals, forming malachite green (MG). The amount of MG formed from the LMG is related to the cumulative radiation dose which causes an absorbance change that can then be read out optically.

### *2.2. Irradiation*

In fundamental studies, samples were irradiated with UVC light. For the final dosimetry tests, samples were irradiated with high-energy X-rays on a linear accelerator (linac).

*2.2.1. UVC irradiation.* The dosimeter had been previously found to be sensitive to UVC light [5]. This allowed for testing of the dosimeter in the lab using a UV light source for irradiation. The cuvette samples were irradiated with a 9 W UVC germicidal lamp (TUV PL-S 9 W, Philips). The light intensity at 100 mm from the end of the lamp was measured with a UVC light reader (UVC-254, Lutron Electronic Enterprise Co., LTD) to be 5.22 W / m$^2$. From the manufacturer's specifications, the wavelength of the UV light source was predominantly 255 nm.

The cuvettes were placed 100 mm away from the UV lamp which was suspended in a polystyrene box. The box's internal surface was covered with aluminium foil to homogenize the UV flux. The cuvettes were enclosed on all sides but the one facing the UV lamp.

*2.2.2. Linac irradiation.* The cuvettes were irradiated with high-energy photon beams using a clinical Varian 21EX linac (Varian Medical Systems, Palo Alto, USA) with a beam quality of 6 MeV. All samples were irradiated together within a 10 cm × 10 cm field size at an SSD of 100 cm and were irradiated with doses varying from 0 Gy to 10 Gy in steps of 2.5 Gy and from 10 Gy to 40 Gy in steps of 5 Gy. All the cuvettes were placed in a solid water phantom with an SSD of 100 cm and the centre of the cuvettes positioned at 1.4 cm depth (Dmax).

### *2.3. Dosimeter stability study*

Table 1 provides an overview of the composition of the different samples that were used in the stability study. Equal concentrations of chloroform and LMG were used for all samples whilst the curing agent of the Sylgard® 184 was varied from 1 % to 15 % [w/w] of the overall mass of the mixture. The amount of Sylgard base was adjusted so that the total amount of elastomer was kept the same at 97 % [w/w].

The mixtures were thoroughly stirred before being poured into PMMA cuvettes. The cuvette samples were then placed in a vacuum desiccator and a cycle of vacuum pumping and releasing to atmospheric pressure was used until all visible air bubbles were removed from the samples. The cuvette samples were cured at 60 °C for 1 hour and then allowed to cool at room temperature for an hour.

The cuvettes were exposed to UV light for different exposure times ranging from 0 s to 160 s. Spectroscopic measurements were taken after each UV exposure and then every 24 hours for the next three days. The samples were kept in the dark and at room temperature when not being irradiated or measured.

**Table 1.** Concentrations of Sylgard base and curing agent for the stability study. The optimal composition of FlexyDos3D dosimeter is indicated in bold font and is used in the following experiments.

| Sample Name | Sylgard Curing Agent [% w/w] | Sylgard Base [% w/w] | CHCl3 [% w/w] | LMG [% w/w] |
|---|---|---|---|---|
| FlexyDos3D 1 % CA | 1 | 96 | 3 | 0.03 |
| FlexyDos3D 2 % CA | 2 | 95 | 3 | 0.03 |
| FlexyDos3D 3 % CA | 3 | 94 | 3 | 0.03 |
| **FlexyDos3D 4 % CA** | **4** | **93** | **3** | **0.03** |
| FlexyDos3D 4.5 % CA | 4.5 | 92.5 | 3 | 0.03 |
| FlexyDos3D 5 % CA | 5 | 92 | 3 | 0.03 |
| FlexyDos3D 7.5 % CA | 7.5 | 89.5 | 3 | 0.03 |
| FlexyDos3D 10 % CA | 10 | 87 | 3 | 0.03 |
| FlexyDos3D 12.5 % CA | 12.5 | 84.5 | 3 | 0.03 |
| FlexyDos3D 15 % CA | 15 | 82 | 3 | 0.03 |

### *2.4. Dosimeter curing temperature and time dependence study*

Two samples of the stable composition dosimeter were fabricated and poured into quartz cuvettes, with one sample cured for 60 minutes at 120 °C and the other for 60 minutes at 60 °C. Quartz cuvettes were used as quartz glass has a low absorbance of UV light, allowing UV light to penetrate the cuvette and irradiate the dosimeter, and quartz can withstand higher temperatures than PMMA. After cooling for an hour, the cuvettes were exposed to the UVC light for increasing exposure times and absorbance spectra were acquired with the spectrophotometer (Ocean Optics, USB4000).

A broader range of curing times was investigated on 8 glass cuvette samples that were filled with the optimal composition of FlexyDos3D 4 % CA and that were irradiated with high-energy X-rays on a linac. The samples were all cured at different temperatures and times: two curing temperatures of 60 °C and 120 °C were used, and curing times of 30 minutes, 60 minutes, 120 minutes, and 180 minutes were used for each temperature.

All cuvettes were cooled for 2 hours before irradiation using the linac and spectral scans were performed between each irradiation.

### *2.5. Optical spectroscopy and absorbance measurements*

LMG is transparent but when irradiated it converts to MG which shows an increase in absorbance at red wavelengths of light. The wavelengths measured for absorbance were 630 nm and 480 nm, with 630 nm undergoing a maximal change in absorbance when LMG is converted to MG, whilst at 480 nm the absorbance is the same for both LMG and MG. The total dose delivered to the dosimeter could then be calibrated against the difference in these two wavelengths:

$$\Delta OD = OD_{630\text{nm}} - OD_{480\text{nm}} \quad (1)$$

Optical spectra were acquired with a high-resolution spectrophotometer (USB4000, Ocean Optics) equipped with a cuvette holder. All optical absorbance measurements were calibrated against a cuvette filled with de-ionised water.

## 3. Results

### *3.1. Dosimeter stability*

The dose response of 10 % and 4 % curing agent FlexyDos3D samples are shown in figure 1a and 1b, respectively. From figure 1a, the dose sensitivity decreases significantly after the initial irradiation of the dosimeter. However, figure 1b shows negligible change in the sensitivity of the dose response to radiation. Only a small offset in the absorbance between the first measurement and the measurements on the following days can be seen in figure 1b which may be attributed to an incomplete curing of the sample as from day 1 onward no further measurable change is observed. Another sample with the same 4 % curing agent composition was made and was cured for 2 hours at 60 °C and is shown in figure 1c. It is shown that the offset has significantly decreased, likely the result of further curing within the sample.

Figure 1d shows that the change in sensitivities of the dosimeter do not change significantly after three days for curing agent concentrations less than 4.5 %. Greater changes in sensitivity are seen with increasing curing agent concentrations.

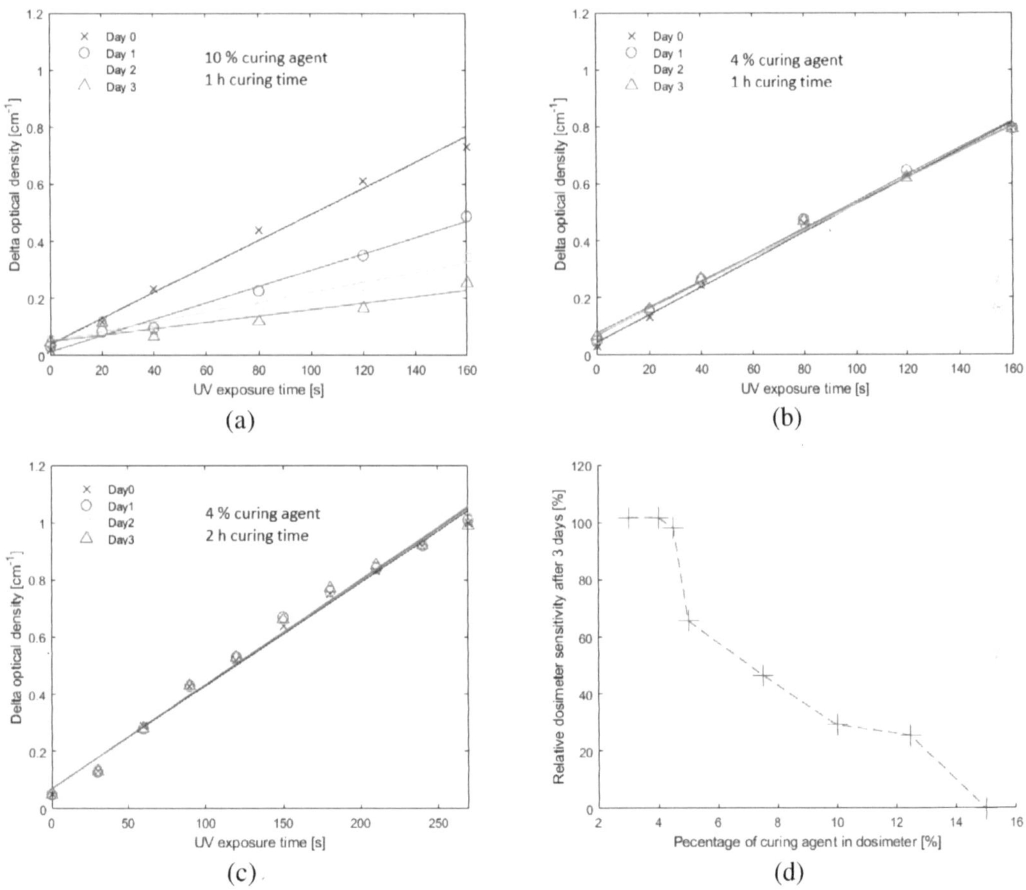


**Figure 1.** Stability of the dose response for (a) FlexyDos3D 10 % mixture and (b) FlexyDos3D 4 % mixture cured for 1 hour at 60 °C, and (c) FlexyDos3D 4 % mixture cured for 2 hours at 60 °C. Note that (c) was investigated for a longer exposure time and with more sample points. (d) Sensitivity of dosimeters with different curing agent concentrations after 3 days relative to the first measurement.

### *3.2. Curing temperature dependence for dose-response*

The effect of curing temperatures and curing times upon the FlexyDos3D dose-response is shown in figure 2a and 2b and for UV exposure and high energy X-ray radiation, respectively. A linear dose-response is observed for the X-ray radiation, whilst a non-linear response is observed in the UV irradiated samples. This is attributed to a greater attenuation of the UV light in the dosimeter as compared to the X-rays and a likely saturation of the LMG reactions in the layers closest to the UV source.

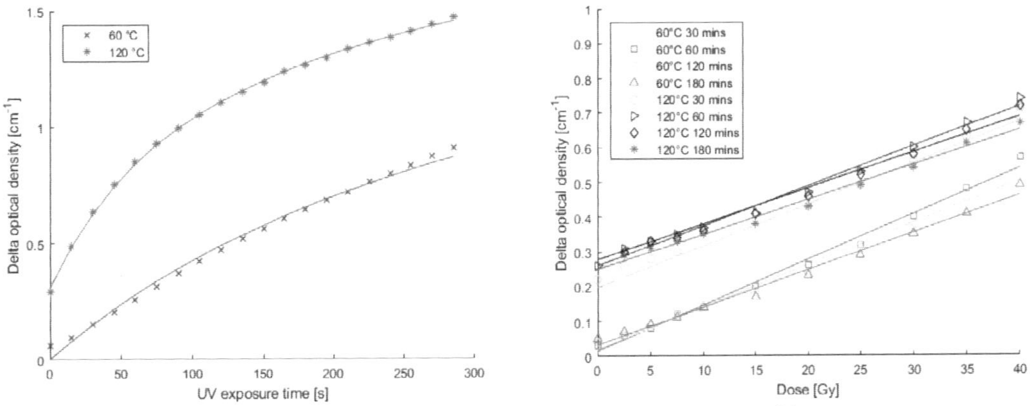


(a) (b)

**Figure 2.** Optical density difference between 630 nm and 480 nm versus UV exposure time for two FlexyDos3D 4% C samples cured at different temperatures for 60 minutes (a). Optical density difference between 630 nm and 480 nm versus radiation dose for different curing temperatures and times (b). The use of the difference in optical densities at the two wavelengths can compensate for some of the artefacts in the cuvettes caused by the contraction of the dosimeter after cooling.

The difference in the unirradiated absorbance is significant between the 120 °C and 60 °C cases. However, the curing time appears to have little effect on the dose-response of the 120 °C samples. In figure 2b, all the 120 °C samples have a similar dose-response. For the 60 °C samples, the dose-response does show a decrease in sensitivity with increasing curing times. The sensitivity of the 60 °C samples approach the same rate of change in absorbance as the 120 °C samples as the curing time is increased whilst still maintaining a large optical density offset.

These results demonstrate that incomplete curing of the dosimeters during irradiation is likely the cause of some dose-response effects and that the curing temperature does affect the initial absorbance of the dosimeter. However, the dose-response is unaffected by curing times once the dosimeter has cured for a sufficiently long time, which is more than 30 minutes at 120 °C. The incomplete curing effects have also been shown in [4] for 65 °C temperatures with curing times between 30 minutes and 5 hours, agreeing with our results.

## 4. Discussion

The FlexyDos3D dosimeter displayed a stable dose-response for compositions with curing agent concentrations between 3 % and 4.5 % over a period of 3 days after manufacture. Curing agent concentrations at 1 % and 2 % resulted in silicone matrices that were not fully cured which would be unsuitable as a solid 3D dosimeter. For concentrations above 4.5 % the dosimeter became unstable with increasing instability for increasing concentrations of the curing agent. Previously it was suggested that the elastic modulus of the dosimeter would be able to be tuned to the organ of interest, by varying the ratio of the curing agent and base concentrations of the silicone mixture to simulate different tissues within the body. It was believed that this would have a negligible effect upon the dosimetric properties [4, 6]. However, the current results show that silicone matrix concentrations affect the stability of the dosimeter and so tuning of the elastic properties of the dosimeter may be restricted to the range in which the dosimeter displays a stable response.

Building upon the results of [4], further investigation of the curing temperature found that higher temperatures had a significant effect on the zero-dose absorbance of the dosimeter but the dose sensitivity was found to be independent of curing temperature. It was found that the curing time had no effect on the dose-response for dosimeters for curing times larger than an hour at 120 °C. This demonstrates that additive manufacturing techniques, such as 3D printing, could be potentially used to manufacture FlexyDos3D dosimeters. The current study was targeted towards the effect of possible heterogeneities in temperature history during 3D printing which could lead to inconsistencies in the dose response. Our studies have shown that these factors are not significant for the curing times and temperatures investigated.

## 5. Conclusion

A stable composition of a deformable 3D dosimeter, FlexyDos3D, was found that consisted of 0.03 % LMG, 3 % chloroform, and Sylgard® 184 with 4 % curing agent and 93 % base. Investigations into the effect of higher curing temperatures upon the dosimetric properties of this stable dosimeter composition revealed that by increasing the curing temperature, the curing time could be significantly reduced without significantly distorting the dose-response sensitivity.

## 6. Acknowledgements

The authors would like to thank V Calliet and E Hewson for their help with the irradiation of the dosimeter.

## 7. References

[1] Appleby A and Leghrouz A 1991 *Med. Phys.* **18** 309-12
[2] Baldock C *et al* 2010 *Phys. Med. Biol.* **55** R1-R63
[3] Adamovics J and Maryanski M J 2004 *J. Phys.: Conf. Ser.* **3** 172-5
[4] De Deene Y *et al* 2015 *Phys. Med. Biol.* **60** 1543-63
[5] De Deene Y et al 2015 *J. Phys.: Conf. Ser.* **573** 012025
[6] Kaplan L P *et al* 2017 *Phys. Med. Biol.* **62** 5612-22

# Influence of the components of a radiochromic PVA – Iodide gel dosimeter on the optical dose response

**Shin-ichiro Hayashi[1)], Kaoru Ono[2)], Keisuke Fujino[2] and Sachie Fujimoto[2)]**
[1]Department of Clinical Radiology, Faculty of Health Sciences, Hiroshima International University, Higashi-Hiroshima, Hiroshima 739-2695, Japan
[2]Hiroshima Heiwa Clinic, Hiroshima, 730-0856, Japan

E-mail: rin@hs.hirokoku-u.ac.jp

**Abstract**. A novel radiochromic gel dosimeter based on a poly(vinyl alcohol)-iodide complex (PVA-I) has been developed with the aim of using it in optical computed tomography (CT). The PVA-I gel dosimeter exhibits excellent characteristics such as high sensitivity, dose rate independence, a wide dose range, and especially reusability. The standard PVA-I gel dosimeter is composed of poly(vinyl alcohol) (average degree of polymerization 1000 and degree of saponification 88 mol%), iodide (potassium iodide), reducing sugar (fructose), gelling agent (gellan gum), and water. In this study, the influence on dose response is investigated upon substitution of the components by PVAs with different degrees of polymerization (500, 1500, 3500) and saponification (80 mol%, 98 mol%), different iodide salts (LiI, NaI, $NH_4I$), and different reducing sugars (glucose, maltose, lactose). The results show that these substitutions have little effect on the dose rate dependence, while the iodide salt affects the sensitivity.

## 1. Introduction

In the past decade, various radiochromic gel dosimeters employing radiation sensitive dyes have been studied. Among them, the most typical dye gel dosimeter is a micelle gel dosimeter using Leuco Malachite Green [1] or Leuco Crystal Violet [2] as the dye. Others using a genipin gel [3], a tetrazolium salt gel [4], and more have also been reported.

Recently, a radiochromic gel indicator based on a poly(vinyl alcohol)-iodide complex (PVA-I) was reported by Sunagawa et al. [5, 6]. The indicator changes from colorless to red by radiation with high sensitivity but is also decolored by heating in the presence of a reducing sugar, thus being reusable. We have applied this system to a novel three-dimensional (3D) radiochromic gel dosimeter [7]. The PVA-I gel dosimeter investigated in our previous work is composed of partially saponified PVA (degree of saponification (DS) of 86-90 mol%, average degree of polymerization (DP) 1000), iodide (potassium iodide, KI), reducing sugar (fructose), gelling agent (gellan gum, GG), and distilled water. The reactions in this system are represented by the following scheme.

| | | | | | |
|---|---|---|---|---|---|
| 1) | $H_2O$ | $\rightarrow$ | R· | | (Radiolysis) |
| 2) | $3I^-$ (+ R· ) | $\rightarrow$ | $I_2 + I^-$ $\rightarrow$ | $I_3^-$ | (Oxidation) |
| 3) | PVA + $I_3^-$ | $\rightarrow$ | PVA-I | | (Complexation, colored) |
| 4) | PVA-I | $\rightarrow$ | PVA + $I_3^-$ | | (Dissociation, decolored) |
| 5) | $I_3^-$ (+ RS) | $\rightarrow$ | $3I^-$ | | (Reduction) |

IOP Conf. Series: Journal of Physics: Conf. Series **1305** (2019) 012031 doi:10.1088/1742-6596/1305/1/012031

An iodide ion is oxidized by the radical (R· ) generated by the radiolysis of water, creating an iodide molecule ($I_2$). Although the iodide molecule is insoluble in water, the triiodide ion, which is soluble, is immediately formed upon binding with another iodide ion. The triiodide ion forms a complex with the maldistributed group of a residual acetyl group in PVA, resulting in red colorization. This PVA-I complex dissociates to give PVA and a triiodide ion upon heating. Subsequently, the triiodide ion is reduced to mono-iodide ions in the presence of a reducing sugar (RS), resulting in decolorization.

Besides its reusability, the PVA-I gel dosimeter also has some excellent characteristics such as high sensitivity, dose rate independence, a wide dose range, and more [7]. In the present study, the influence on the dose response was investigated upon substitution of the various components (except for the gelling agent) with a different iodide salt (LiI, NaI, $NH_4I$), a different reducing sugar (glucose, maltose, lactose), and PVAs with different DP (500, 1500, 3500) or DS (80 mol%, 98 mol%).

**Table 1.** The components of the PVA-I gel dosimeter prepared in the present work. Each component of the standard gel dosimeter (marked in bold) was substituted one at a time. M is molarity (mol/1000g$H_2O$).

| **Base solution** | | | | | | | |
|---|---|---|---|---|---|---|---|
| PVA | Degree of polymerization | | 500 | **1000** | 1500 | 3500 | 1.0 wt% |
| | Degree of saponification (mol%) | | 80 | **88** | 98 | | |
| Gellan gum | | | | | | | 0.4 wt% |
| Water | | | | | | | 98.6 wt% |
| **Additives** | | | | | | | |
| Iodide salt | | LiI | NaI | **KI** | $NH_4I$ | | 100 mM |
| Reducing sugar | | | glucose | **fructose** | maltose | lactose | 100 mM |

## 2. Materials and methods

### 2.1. Gel preparation

The compositions of the PVA-I gel dosimeters prepared in the present work are shown in Table 1. All reagents were purchased from Wako Pure Chemical Industries Ltd (Japan). Solutions of each type of 10 wt% PVA and a solution of 0.8 wt% GG were prepared in advance and mixed with distilled water at room temperature to prepare base solutions containing 1 wt% PVA and 0.4 wt% GG. After heating the base solution to 60℃ while stirring, an iodide salt and a reducing sugar were added. The resulting solution was subdivided into PMMA cuvettes (4.5 mL, 1 cm path length) and cooled to room temperature. Before irradiation, the samples were initialized by annealing them at 45℃ for 12 hours in an incubator.

### 2.2. Irradiation

Sample irradiation was performed using a 6 MV photon beam of a medical linear accelerator (Varian/BrainLAB Novalis Tx, USA). Samples were irradiated at a depth of 5 cm in a tough water phantom at room temperature. The axis of the cuvette was perpendicular to the beam. An irradiation field size of $15 \times 15$ cm$^2$ was used with a 100 cm source-to-axis distance (SAD). The beam was delivered to the cuvettes up to 20 Gy, with the dose rate fixed at 600 cGy/min. Different dose rates (100, 200, and 400 cGy/min) at 10 Gy were also utilized to investigate the dose rate dependence.

### 2.3. Measurement

Absorption spectra of the samples were measured one day after being irradiated at different doses. The measurements were performed at room temperature using an UV-Vis spectrometer (UV-1600PC, SHIMADZU, Japan) over the wavelength region of 350-800 nm. Absorbance (Abs) was calibrated using a reference cuvette filled with distilled water. The change in the absorbance ($\Delta$Abs) was obtained by subtracting the absorbance value of the non-irradiated sample from that of the irradiated sample:

IOP Conf. Series: Journal of Physics: Conf. Series **1305** (2019) 012031 doi:10.1088/1742-6596/1305/1/012031

$$\Delta \text{Abs} = \text{Abs(i)} - \text{Abs(n)} \tag{1}$$

where Abs(i) and Abs(n) are the absorbances of the irradiated and non-irradiated samples, respectively.

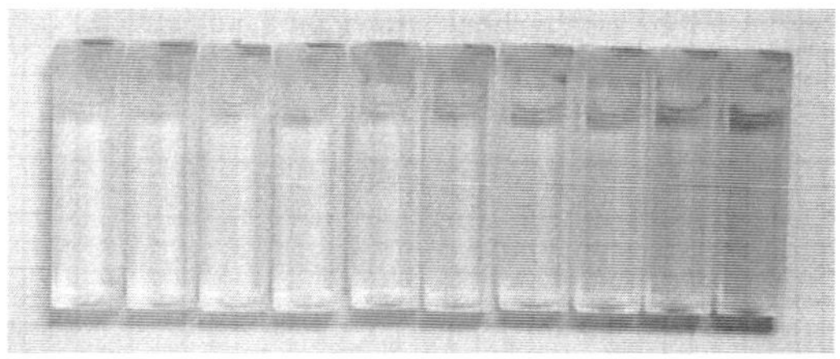

**Figure 1.** PVA- (Na)I gel samples irradiated up to 20 Gy.

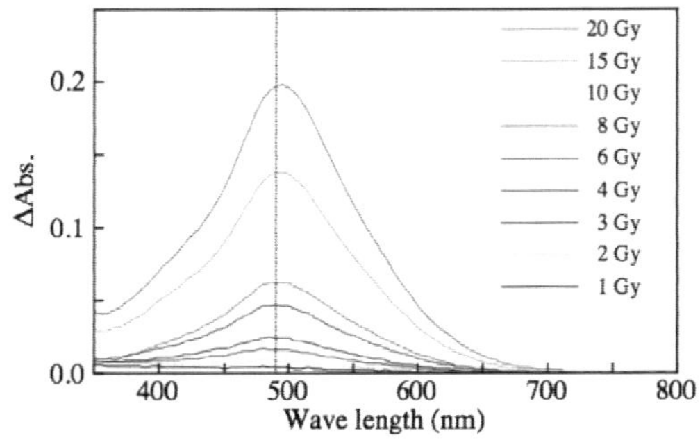


**Figure 2.** Absorption spectra of PVA- (Na)I gel samples irradiated up to 20 Gy.

## 3. Results and Discussion

Figure 1 shows the PVA-I gel dosimeters as they were irradiated up to 20 Gy. As the absorbed dose increased, they gradually changed from colorless to red. Figure 2 shows the absorption spectra of the gels. Single peaks were almost exclusively observed, centered around 490 nm.

### *3.1. Dose-absorbance responses*

#### *3.1.1. Degrees of polymerization (DP) and saponification (DS) of PVA*

Figure 3 (left) shows the dose-absorbance responses at 490 nm of the PVA-I gel dosimeters fabricated with PVAs with different degrees of polymerization and saponification. The results show that the sensitivity of the PVA-I gel dosimeters was almost independent of the DP, although the sensitivity of the dosimeter incorporating PVA with a DP of 3500 was somewhat low. The sensitivity of the dosimeter using PVA with a DS of 80 mol% was slightly higher than it was for the devices employing the other PVAs. In the PVA-I gel dosimeter using the PVA with a DS of 98 mol%, the peak was not observed. These results are consistent with those of the experiment using partially saponified PVA and an iodine-potassium iodide ($KI/I_2$) solution [8]. In addition, the PVA-I gels with DS of 80 mol% or DP of 3500 exhibited higher viscosities than the other gels. Therefore, it was easier to prepare a gel dosimeter using PVA with a DP of 500 - 1000 and a DS of 88 mol%.

#### *3.1.2. Reducing sugar*

Figure 4 (left) shows the dose-absorbance responses at 490 nm of the PVA-I gel dosimeters incorporating different reducing sugars. The sensitivity of the dosimeter containing fructose was somewhat higher than it was for the others. This may be due to the ketose-aldose equilibrium of fructose, which has a ketone group, although the cause has not been confirmed.

#### *3.1.3. Iodide source*

Figure 5 (left) shows the dose-absorbance responses at 490 nm of the PVA-I gel dosimeters using different iodide salts. The sensitivity of the dosimeter containing KI was the highest. The induction region in the lower dose range was observed for the dosimeter containing LiI. The compositions of the dosimeters being compared in this section are almost the same, differing only in the counter cations ($K^+$, $Na^+$, $Li^+$, $NH_4^+$). These cations essentially act as physical cross-linkers for the gelation of gellan gum. However, it is interesting that the influence of the salt on the dose sensitivity was larger than the influence of the other components. The ionic radius, hydration force, or pH may all affect the structure of the maldistributed group of the residual acetyl group in PVA, which forms the complex with iodide.

### *3.2. Dose rate dependence*

Figures 3 (right), 4 (right), and 5 (right) show the dose rate dependencies of the absorbance at 10 Gy in the PVA-I gel dosimeters containing each corresponding component. It is seen that the dose rate independence of the standard PVA-I gel dosimeter was not affected by substitution of the different components.

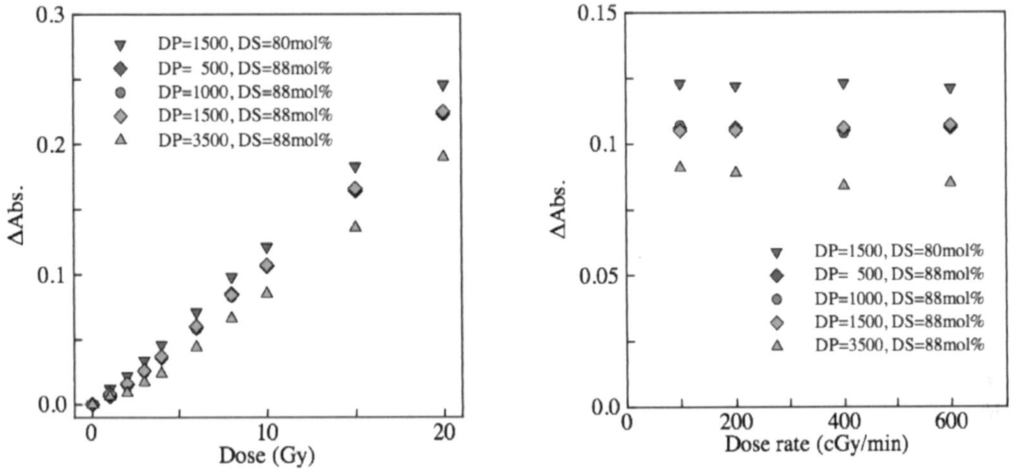


**Figure 3.** Dose-ΔAbs (at 490 nm) responses (left) and the dose rate dependence of the absorbance at 10 Gy (right) in the PVA-I gel dosimeters containing PVAs with different DPs and DSs. KI and fructose are used as a source of iodide ions and a reducing sugar, respectively. The data points of DP=500, 1000, 1500, and DS=88mol% almost overlap each other.

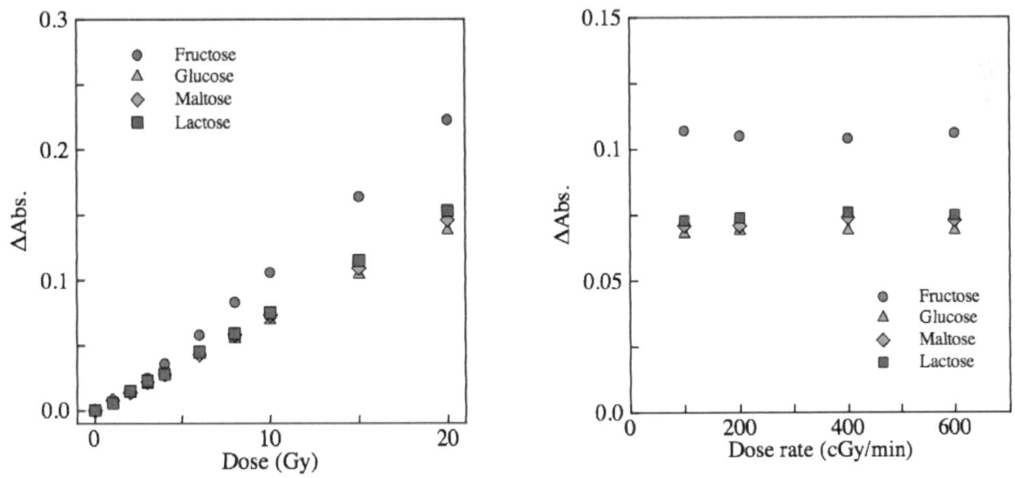


**Figure 4.** Dose-ΔAbs (at 490 nm) responses (left) and the dose rate dependence of the absorbance at 10 Gy (right) in the PVA-I gel dosimeters containing a different reducing sugar. PVA (DP=1000, DS=88 mol%) and KI are used as a PVA and a source of iodide ions, respectively. The data points of glucose, maltose, lactose almost overlap each other.

IOP Conf. Series: Journal of Physics: Conf. Series **1305** (2019) 012031 doi:10.1088/1742-6596/1305/1/012031

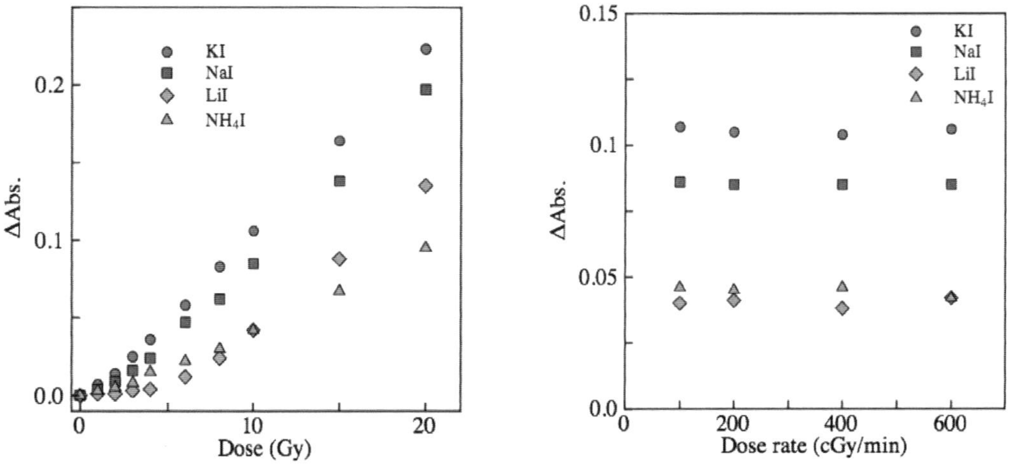


**Figure 5.** Dose-ΔAbs (at 490 nm) responses (left) and the dose rate dependence of the absorbance at 10 Gy (right) in the PVA-I gel dosimeters containing a different iodide salt. PVA (DP=1000, DS=88 mol%) and fructose are used as a PVA and a reducing sugar, respectively.

## 4. Conclusion

In this study, the influence of the chemical components on the dose response behavior of a novel radiochromic gel dosimeter based on a poly(vinyl alcohol)-iodide complex was investigated. Substitution of the components did not inhibit the favorable characteristics of this dosimeter, except for decreasing the sensitivity of the gel dosimeter containing LiI. Based on these results, the standard PVA-I gel dosimeter is expected to be a more useful dosimetric tool in radiotherapy. However, further optimization of the composition and evaluation of its fundamental characteristics, such as thermal and spatial stabilities, are needed. Efforts to address these topics are currently underway.

## 5. Acknowledgment

We would like to thank Professor Takeyoshi Sunagawa and his students of Fukui University of Technology for their suggestions. This work is supported by JSPS KAKENHI (Grant Number 17K09072).

## 6. References

[1] Jordan K *et al* 2009 *Phys. Med. Biol.* **54** 6773-89
[2] Nasr A T *et al* 2015 *Phys. Med. Biol.* **60** 4685-704
[3] Davies J B et al 2013 *Radiat. Phys. Chem.* **83** 19-27
[4] Penev A I *et al* 2017 *J. Phys.: Conf. Ser.* **847** 012048
[5] Sunagawa T *et al* 2017 *Memoirs of Fukui University of Technology* (in Japanese) **47** 105-10
[6] Aoki Y *et al* 2017 *Proceedings of the 25th International Conference on Nuclear Engineering (ICONE25-67618)* 1-5
[7] Hayashi S *et al* 2017 *Jp. J. Med. Phys.* **37** Sup.3 171. *Proceedings of the 114th Scientific Meeting of Japan Society of Medical Physics (8th Japan-Korea Joint Meeting on Medical Physics).*
[8] Hayashi S *et al* 1963 *KOBUNSHI KAGAKU* (in Japanese) **20** 303-311

IOP Conf. Series: Journal of Physics: Conf. Series **1305** (2019) 012032 doi:10.1088/1742-6596/1305/1/012032

# Preliminary investigation of a reusable radiochromic sheet for radiation dosimetry

**Cielle Collins[1], Jacob Kodra[1], Suk Whan Yoon[1], Robert Coakley[2], John Adamovics[2] and Mark Oldham[1]**
[1]Duke University, Durham, North Carolina, USA
[2]Rider University, Lawrenceville, New Jersey, USA

E-mail: cielle.collins@duke.edu and mark.oldham@duke.edu

**Abstract**. This work investigates a novel reusable radiochromic sheet developed as an economic film substitute or as a radiochromic bolus. The radiation-induced optical density (OD) change of the sheet is read-out with a commercial flat-bed optical scanner. An optimized readout procedure was developed including an optimal scan-time window. Fundamental radiochromic properties of the sheets were characterized including temporal decay of OD, dose sensitivity and consistency through repeated irradiations, and temperature sensitivity. The radiation induced OD change in the sheets was found to decay to baseline within ~24 hours, after which the sheet could be reused. The sensitivity of subsequent re-irradiations was found to be consistent within ±5% (coefficient of variation of 4.5%) for at least 6 irradiations and potentially many more. Importantly, the sheets were not observed to carry any detectable memory of previous irradiations within measurement uncertainty. In conclusion, the Presage sheets show promise as an economic multi-purpose alternative for film applications and potentially as a radiochromic bolus. Further work is required to test the sheet in diverse clinical applications, and to develop a softer material for bolus applications.

## 1. Introduction
Presage is a well-established dosimetric material composed of a transparent polyurethane matrix, containing a leuco dye with a radiochromic response when exposed to ionizing radiation [1-5]. The radiation induced optical density (OD) change can be imaged and used to measure radiation dose distributions. Permanent radiochromic response has typically limited Presage dosimeters to one-time use [1]. In this work, we characterize a new reusable formulation of Presage and evaluate its potential for use primarily as an economic radiochromic film alternative and secondarily as a radiochromic bolus. In this formulation, the Presage is cast in sheets for 2D evaluation and readout with a flatbed scanner. Extension to reusable 3D dosimetry [6] may be possible, but has not been explored here. Bolus has been used for many years, with recent improvements including the capability for 3D printed bolus [7-9]. This work extends these efforts by developing a bolus that is radiochromic, reusable, and can aid in treatment verification. Reusable Presage is suitable for manufacturing into sheets of variable thickness (1-10mm) and variable stiffness for bolus applications.

## 2. Methods
The reusable Presage sheets are manufactured by Heuris, INC. The sheets use 2% cumin-LMG-DEA (see figure 1) with varied plasticizer (25% by weight), depending on applications. The sheets consist of a uniform active layer with no coatings or substrates, which differentiates them from traditional films.

The active leuco dye molecules are uniformly distributed throughout the sheet leading to a nominal uniform response. The consistency of thickness of sheets is confirmed using an electronic caliper. The current investigations were performed with a standard reusable Presage formulation and sheet thickness of 5 mm.

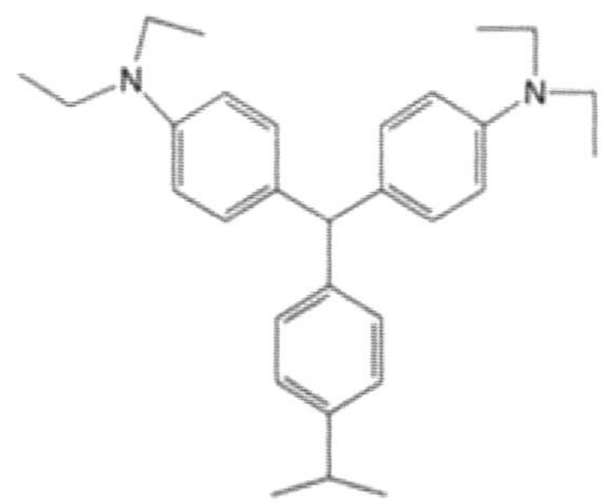


**Figure 1.** Structural formula of leuco dye cumin-LMG-DEA.

All Presage sheets were scanned without enhancements using an Epson Expression 10000XL flatbed scanner and subsequently analyzed to yield change in OD measurements, which are proportional to delivered dose. To avoid air bubbles, each sheet was pressed into a 1 cm diameter drop of mineral optical fluid, applied to the scanner bed. The scanner was set to 48-bit color, resolution 50 dpi, document size 5x5 in. Each scanned image had a 0.5 mm resolution. As in EBT film analysis, the red channel of the RBG image from the scanner was selected for image analysis [10]. A pre-scan was taken before each irradiation and a readout scan was taken soon after. Sheets were then removed from the scanner and stored in an opaque dark bag at room temperature until needed for the next irradiation. The OD change was calculated by taking the log of the pre-scan divided by the post-scan. As the change in OD radiochromic response of reusable Presage is linear with dose, these images are dose maps.

The temporal decay of the radiation induced change in OD in the sheets was studied by irradiating four 0.5 cm thick 6 $cm^2$ sheets to 6, 4, 2, and 1 Gy. They were irradiated at dmax (depth of maximum dose) in a solid water phantom stack with 6 cm solid water beneath the sheets and 1.5 cm above. Before the irradiation the sheets were pre-scanned and then after the irradiation they were scanned every 10 minutes for an hour and every hour for 9 hours, with a final readout taken at 26 hours post-irradiation.

To assess the Presage reusability, two experiments were conducted. The first experiment was a repeat pyramidal experiment where each of a series of four concentric square irradiations (at dmax, solid water phantom stack) were delivered on consecutive days to the same 0.5 cm thick, 12 $cm^2$ sheet with the field size increasing every day. This created pyramidal steps of cumulative dose, with time for the signal to fade-out between each irradiation. Each field was delivered centered on the sheet with a dose of 4.5 Gy. The Presage was pre-scanned before irradiation and scanned for readout 8 minutes after each irradiation. For analysis, a 5 mm median filter was applied to the readout scans to minimize the effect of sheet imperfections. After each irradiation in the series, a line profile through the center of the sheet, crossing each concentric square field, was analyzed, quantified, and plotted against each line profile from the previous layered irradiations.

The second experiment was a repeat irradiation series with the same field overlaid each day, for a greater number of irradiations. A 0.5 cm thick, 6 $cm^2$ Presage sheet was irradiated at dmax in a solid water phantom stack with a 4x4 cm 4.5 Gy field once a day. The sheet was pre-scanned once before each irradiation and post-scanned 10 minutes after each irradiation. The change in OD at 10 minutes was quantified and compared each day.

## 3. Results and Discussion

Figure 2 presents the decay of the reusable Presage radiation induced color change with time. A final readout scan was taken 26 hours after irradiation. This scan demonstrated cleared signal for all doses, with less than 2% of signal remaining. Temporal decay curves were created for 6 Gy, 4 Gy, 2 Gy, and 1 Gy irradiations and the control sheet. The decay constants for the curves were 6.5 hours$^{-1}$ for 6 Gy, 5.7 hours$^{-1}$ for the 4 Gy data, 5.6 hours$^{-1}$ for the 1 Gy and 2 Gy data, and a constant change in OD (within scanner error range) along the zero ΔOD line for the control sheet.

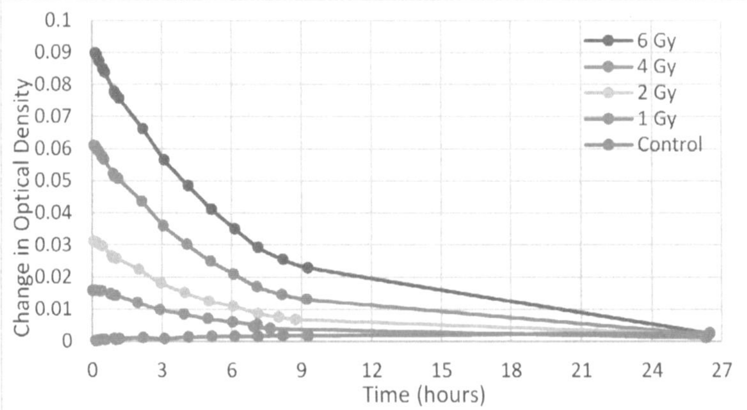


**Figure 2.** Temporal decay of reusable Presage signal in a 26 hour period. Five different sheets were irradiated to four doses: 1 Gy (green), 2 Gy (yellow), 4 Gy (orange), 6 Gy (red), and an un-irradiated control (blue). The sheets were scanned together periodically for 26 hours.

This data was used to find a clinical scanning time window. Fifteen minutes post irradiation was determined to be a clinically feasible scan time. Using the 6 Gy data, which has the steepest decay, a suggested readout window of 1-24 minutes post irradiation for doses less than or equal to 6 Gy was developed. This was based on the elapsed time before and after 15 minutes when the change in OD had diverged by less than ±3% from the change in OD at 15 minutes.

The pyramid style test was used to determine whether the reusable Presage sheets carried any memory effects from prior irradiations. In figure 3, the illustration explains the irradiation pattern used in the pyramid test. Figure 4 presents the results from this experiment. In this figure, each line profile is a smooth straight line across the irradiated section and each line reaches up to about the same change in OD value, with a 3.5% percent difference between the centers of the two most different line profiles and a 1.52% coefficient of variation. In an analysis of the scanner error and known machine error from calibration standards, an overall expected error range of ±5 % was determined. The error found in this analysis falls well within this ±5% expected error range. The overall finding that each line profile in figure 4 falls within 3.5% of every other profile with no evidence of reduction in signal in the regions of the profile where previous irradiations occurred indicates that the sheets do not carry a memory of previous irradiations.

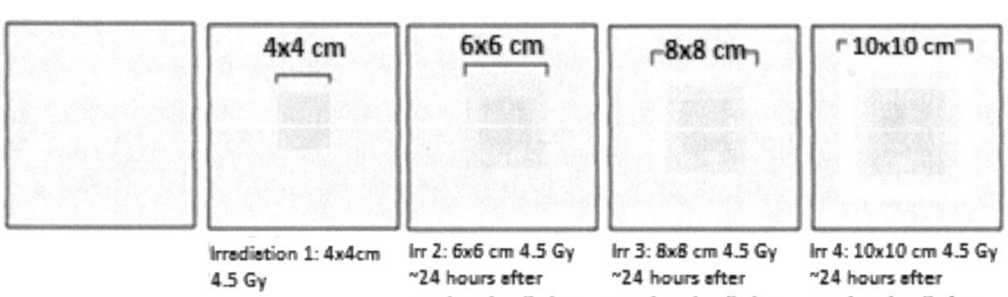


**Figure 3.** Graphic of the pyramid test experimental setup where a series of four concentric 4.5 Gy irradiations was delivered consecutively to the same sheet, with the field size increasing every day, to create pyramidal steps of cumulative dose. Signal was allowed 24+ hours to clear after each irradiation. The yellow bar in each diagram of figure 2 marks the line profiles collected 10 minutes after irradiation.

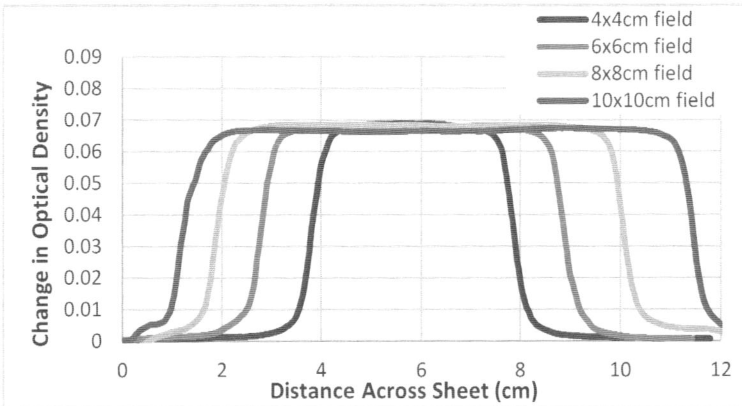


**Figure 4.** Pyramid style reusability test. The plot presents line profiles from the center axis of the reusable Presage sheet after each irradiation. Each irradiation delivered 4.5 Gy to the sheet with the first filling a 4x4 cm field (blue), second a 6x6 cm field (green), third an 8x8 cm field (yellow) and fourth a 10x10 cm field (red).

To investigate the extended reusability of the Presage sheets, one sheet was irradiated to the same dose with the same field on six different days. This process was also repeated 3 times on a different sheet with a 2 Gy dose instead of 4.5 Gy. For the six 4.5 Gy irradiations, there was a standard deviation of $7.4 \times 10^{-4}$ and a coefficient of variation of 4.5% (figure 5). The 2 Gy repeated irradiation dataset had a standard deviation of $2.9 \times 10^{-4}$ and a coefficient of variation of 1.95% (figure 5). These percentages were considered in reference to the same ±5 % expected error range from scanner error and known machine error from calibration standards.

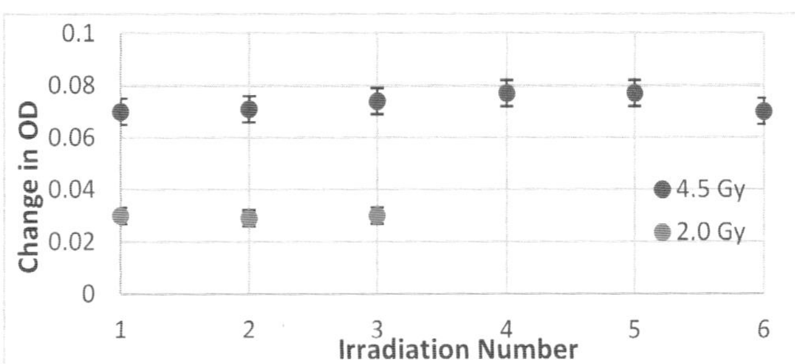


**Figure 5.** Repeatability of radiation induced color change upon repeated use. The same reusable Presage sheets were irradiated with a 4 $cm^2$ field to 4.5 Gy on six different days (green) and a different sheet to 2 Gy on three different days (orange). Sheets were scanned at 10 minutes post-irradiation and all change in OD values fell within the established machine and scanner error range (±5 %).

## 4. Conclusion

Reusable Presage shows promise as an economic multi-purpose alternative for film applications and as a radiochromic bolus. The results of this work correspond to sheets of 5 mm thickness. This may be an optimal thickness for many applications considering that the sensitivity and stiffness decreases with thinner sheets. To date studies have been limited to 6 re-irradiations, and have shown that repeat dose sensitivity is consistent within ±5%. Some uniform background increase in OD was observed which may limit the number of repeat irradiations, however further modifications to the formulation may reduce this limitation. Future work will investigate reusable Presage applications in QA practices and as a radiochromic bolus.

## 5. References

[1] Guo P *et al* 2006 *Med. Phys.* **33** 1338-45
[2] Palmer A L *et al* 2013 *Med. Phys.* **40** 061707

[3] Guo P *et al* 2005 *Int. J. Radiat. Oncol. Biol. Phys.* **63** S206
[4] Adamovics J *et al* 2003 *Med. Phys.* **30** 1349
[5] Sakhalkar H S *et al* 2009 *Med. Phys.* **36** 3371-7
[6] Baldock C et al 2010 *Phys. Med. Biol.* **55** R1-63
[7] Pugh R *et al* 2017 *J. Radiother. Prac.***16** 319-25
[8] Yang K *et al* 2017 *Radioth. Oncol.* **123** S636
[9] Fujimoto K *et al* 2017 *Eur. J. Med. Phys.* **38** 1-9
[10] Fiandra C *et al.* 2006. *Med. Phys.* **33** 4314-9

IOP Conf. Series: Journal of Physics: Conf. Series **1305** (2019) 012033 doi:10.1088/1742-6596/1305/1/012033

# Benzothiazole-containing tetrazolium salts as radiochromic indicators in gel dosimetry

**Kalin I Penev[1,2] and Kibret Mequanint[1,3]**
[1]Department of Chemical and Biochemical Engineering, the University of Western Ontario, London, ON, N5A 5B9, Canada
[2]Modus Medical Devices Inc., 1570 N Routledge Park, London, ON, N6H 5L6, Canada
[3]Biomedical Engineering Program, the University of Western Ontario, London, Ontario, N6A 5B9, Canada

E-mail: kmequani@uwo.ca

**Abstract.** Benzothiazole-containing tetrazolium salts (Bt-TS) have been previously described but have never been tested for dosimetry applications. Considering that such salts may provide higher dose sensitivity than previously tested tetrazolium salts, we synthesized a library of Bt-TS, three of which are reported here. Dosimeters based on the salts BtPP, BtNP and BtNpP were prepared in gellan gum gels, and compared to a previously-described gel dosimeter, based on bisnitrotetrazolium chloride (BNC) in the same medium. All three Bt-TS showed a higher sensitivity at their respective wavelengths of maximum optical attenuation post-irradiation. Two of the new Bt-TS (BtNP and BtNpP) were identified for further studies in large-scale samples and other gel-forming materials.

## 1. Introduction

Tetrazolium salts (TS) are a class of colourless or weakly coloured water-soluble compounds that have been used in a variety of dosimetry applications, owing to their ability to form coloured formazans following irradiation [1]. TS-based gel dosimetry has been reported in gelatin (5%) and gellan gum (1.25%) by our group [1] and more recently in gels containing 25% Pluronic F-127 [2]. Considering that most formazans are practically insoluble in water, the main feature of TS-based gel dosimeters is the complete lack of image diffusion over the course of hours, days and weeks post-irradiation [1, 2].

The initially reported high dose rate dependence of a commercially-available TS-gellan gum gel dosimeter (ClearView), manufactured by ModusQA (London, ON, Canada) [3] can be alleviated by increasing the concentration of the TS from 0.10 to 0.25 mM or higher (unpublished data), as also evident in the Pluronic gels which use 1.0 mM TS and show no dose rate effect [2]. However, the relatively low radiation sensitivity of ClearView and other TS-based dosimeters remains an issue.

In an effort to increase the sensitivity, we decided to extend the library of TS tested in gel dosimeters. Benzothiazole (Bt) is a side group with strong colour forming properties, which has been previously added to TS [4, 5]. However, Bt-containing tetrazolium salts (Bt-TS) are presently not widely used or commercially available, so we modified a previously-published approach [6] to obtain a library of Bt-TS in high yield (manuscript in preparation). Here we report the initial dosimetric characterization of three previously untested Bt-TS salts in gellan gum gels.

## 2. Materials and methods

### 2.1. *Tetrazolium salts*

Figure 1 presents the tested tetrazolium salts, including (a) 2-(benzothiazol-2-yl)-3,5-diphenyl-2*H*-tetrazolium chloride (BtPP), (b) 2-(benzothiazol-2-yl)-3-(4-nitrophenyl)-5-phenyl-2*H*-tetrazolium chloride (BtNP), (c) 2-(benzothiazol-2-yl)-3-(1-naphthyl)-5-phenyl-2*H*-tetrazolium chloride (BtNpP), and (d) 2,3-bis(4-nitrophenyl)-5-phenyl-2*H*-tetrazolium chloride, also known as "bisnitrotetrazolium chloride" (BNC). Materials (a)-(c) were synthesized in house, using a modification of [6]; the last material (BNC), used as a control, was purchased from TCI America (Portland, OR, USA) as 2,3-bis(4-nitrophenyl)-5-phenyltetrazolium chloride hydrate (cat. no: B1047).

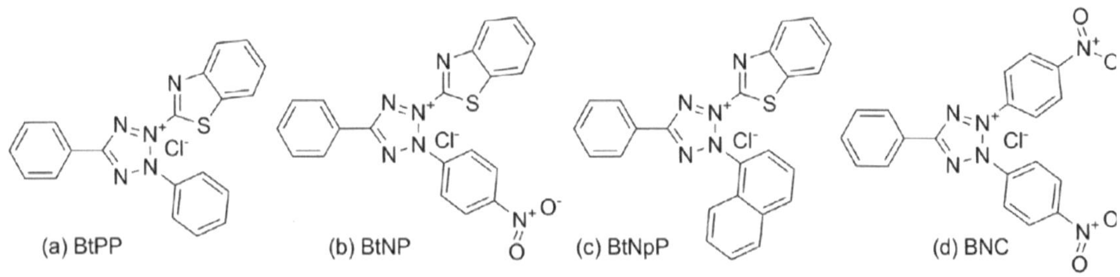


**Figure 1**. Evaluated tetrazolium salts. The dosimetric properties of BtPP, BtNP and BtNpP are reported here for the first time; BNC was used as a control.

### 2.2. *Gel preparation and irradiation*

Gellan gum (cat. no: J63423 from Alfa Aesar, Haverhill, MA, USA), propylene glycol (Caledon labs, Georgetown, ON, Canada) and each of the tetrazolium salts (TS) were dissolved in deionized water and mixed at 55 °C, to obtain the final composition containing 0.25 mM TS, 10% (v/v) propylene glycol and 1.25% (wt/v) gellan gum [1]. Samples were poured into polymethylmethacrylate cuvettes (4mL with 10 mm square cross-section) capped with polyethylene caps and refrigerated overnight to allowed for proper gel setting but otherwise used and stored at room temperature (21-23 °C) in the dark. Irradiation followed on the next day, using a Co-60 source (at approximately 90 cGy/min) at the London Regional Cancer Program in London, Canada. Optical readout (300 – 700 nm) was performed before and after irradiation on a GENESYS 10S UV-visible spectrophotometer (Thermo Scientific, Canada). Whereas the spectrophotometer measures optical absorbance (a unitless quantity), the results here are presented in units of optical attenuation ($cm^{-1}$) in order to facilitate comparison with other types of optical scanning.

## 3. Results and discussion

Figure 2 shows the background attenuation (0 Gy) and the attenuation after irradiation to a dose of 20 Gy for each of the four compositions. It can be seen that the BNC gels have very low optical attenuation in the visible part of the spectrum (above 380 nm), and appear colourless. The newly tested Bt-TS-based dosimeters absorb in the violet and blue part of the spectrum with a cut-off around 500 nm and appear yellow. After irradiation, additional attenuation is seen in the green part of the spectrum, causing the gels to become purple or brown to the naked eye.

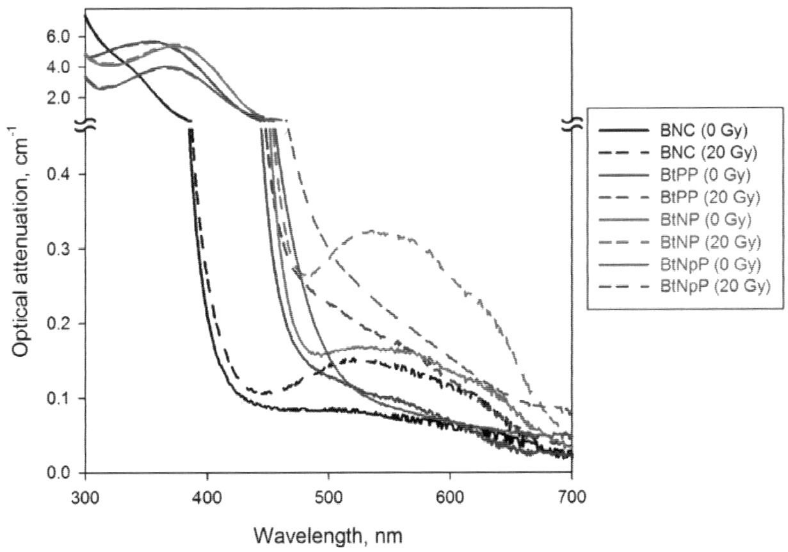


**Figure 2.** Optical attenuation one day after preparation with 0 and 20 Gy absorbed dose.

The wavelength of maximum attenuation ($\lambda_{max}$) can be determined from Figure 3 which shows the change in attenuation for doses of 5, 10 and 20 Gy. It is apparent that all of tested benziothiazole-containing tetrazolium salts (Bt-TS) show higher dose response than BNC in gellan gum. However, the attenuation maximum is blue-shifted for BtPP and BtNpP, and red-shifted only for BtNP. Comparing the Bt-TS, substitution of the simple aromatic structure (phenyl) with a condensed aromatic structure (1-napthyl) causes a moderate bathochromic shift of ~30 nm, but has little effect on the overall shape or intensity of the attenuation peak. By contrast, addition of a nitro group in position 4 of the phenyl ring causes both a marked bathochromic shift of (~60 nm) and a change of the shape and intensity of the attenuation spectrum with a secondary attenuation maximum at 575 nm. Such an effect may be beneficial from a dosimetric point of view, as at higher wavelengths both optical background and optical scatter are lower.

The magnitude of the dose response depends on the wavelength of optical readout, and is linear for all compositions within the range of approximately 450 to 650 nm. As such, choosing $\lambda_{max}$ may not always present the best dosimetric characteristics, considering factors such as background attenuation (Figure 2), lights scattering in the gel and spontaneous darkening. All gels showed some degree of spontaneous darkening at room temperature of approximate rates (at $\lambda_{max}$): $2\times10^{-3}$ $cm^{-1}$ $day^{-1}$ (BNC), $1.7\times10^{-2}$ $cm^{-1}$ $day^{-1}$ (BtPP), $4.2\times10^{-2}$ $cm^{-1}$ $day^{-1}$ (BtNP) and $1.5\times10^{-2}$ $cm^{-1}$ $day^{-1}$ (BtNpP). Thus, BtNP forms both the most sensitive and least stable of all tested dosimeters, while BtNpP may provide the best compromise between sensitivity, stability and optical properties.

## 4. Summary

All of the newly tested benzothiazole-containing tetrazolium salts (Bt-TS) showed higher dose sensitivity than the previously reported BNC in gellan gum gel dosimeters. However, that is at the expense of a higher optical background and a higher rate of spontaneous darkening. Future work includes testing Bt-TS for gel dosimetry into other gel forming materials, such as gelatin and Pluronic F-127.

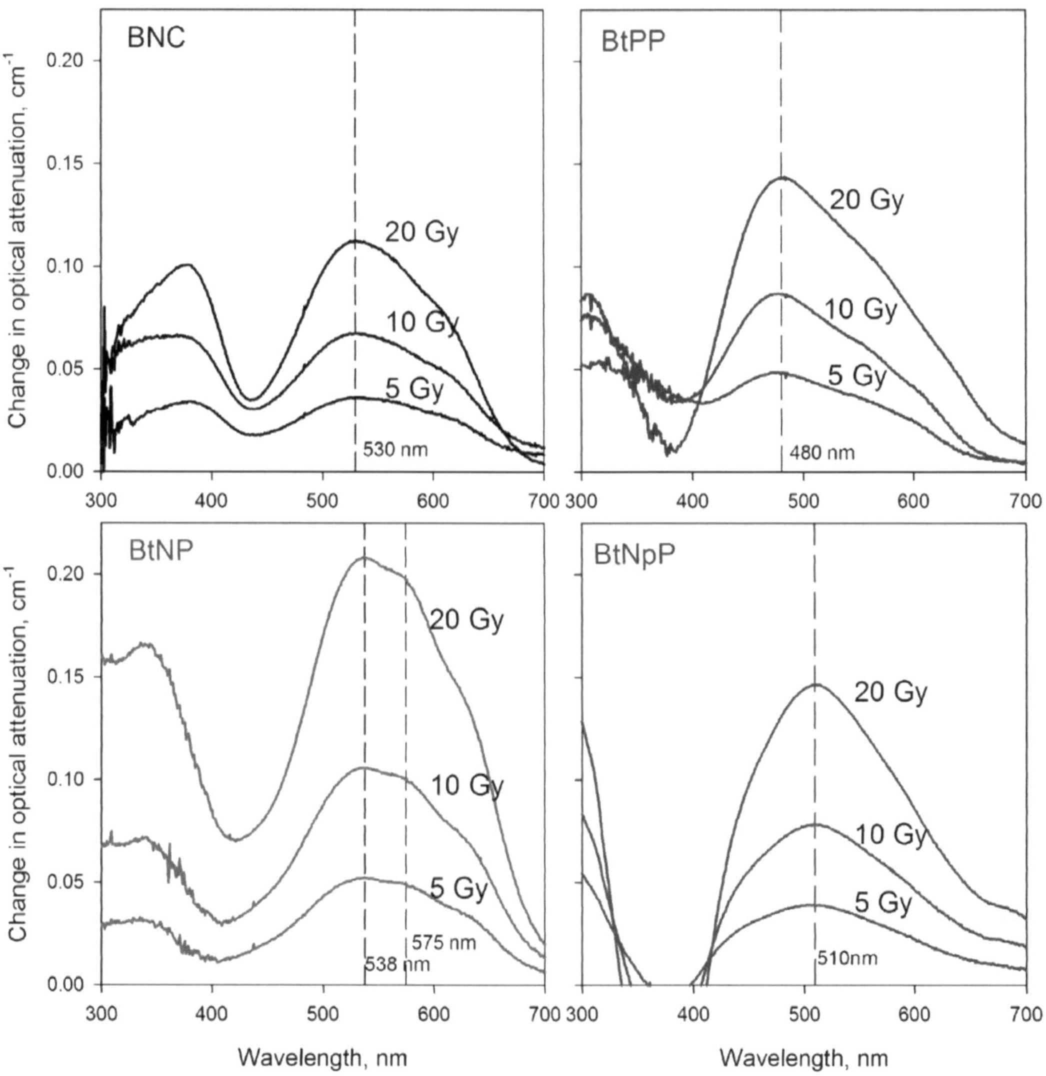


**Figure 3.** Change in optical attenuation for the four tested compositions following Co-60 irradiation to 5, 10 and 20 Gy, compared to0 Gy. Same scale; λmax presented with vertical dashed lines.

## 5. Acknowledgements
This research was funded by a Collaborative Research Health Program grant (CHRP). Disclosure: the authors have a licensing agreement with ModusQA related to gel dosimeters. The materials described in this paper may be commercialized.

## 6. References
[1] Penev K I *et al* 2017 *J. Phys.: Conf. Ser.* **847** 012048
[2] Kwiatos K *et al* 2018 *Phys. Med. Biol.* **63** 095012
[3] Ascencion Y *et al* 2017 *J. Phys.: Conf. Ser.* **847** 012049
[4] Karmarkar S S *et al* 1960 *J. Org. Chem.* **25** *575*
[5] Ishiyama M *et al* 1996 *Anal. Sci.* **12** *515*
[6] Penev K I and Mequanint K 2015 *J. Het. Chem.* **53** 1655

IOP Conf. Series: Journal of Physics: Conf. Series **1305** (2019) 012034 doi:10.1088/1742-6596/1305/1/012034

# Development of a reusable PVA-GTA-I gel dosimeter for 3D radiation dose assessments

**J Taño[1,4], S Hayashi[2], S Hirota[1], CA Gonzales[1,3,4] and H Yasuda[1]**
[1]Department of Radiation Biophysics, Graduate School of Biomedical and Health Sciences, Research Institute for Radiation Biology and Medicine, Hiroshima University, 1 Kasumi 2-3, Minami-ku, Hiroshima 734-8553, Japan
[2]Department of Clinical Radiology, Faculty of Health Sciences, Hiroshima International University, Higashi-Hiroshima, Hiroshima 739-2695, Japan
[3]Department of Radiation Oncology, Graduate School of Biomedical and Health Sciences, Hiroshima University, Kasumi 1-2-3 Minami-ku, Hiroshima 734-8551, Japan
[4]Phoenix Leader Education Program (Hiroshima Initiative) for Rennaissance from Radiation Disaster, Hiroshima University, Japan

E-mail: hyasuda@hiroshima-u.ac.jp

**Abstract**. A novel formula of crosslinked polyvinyl alcohol (PVA) with glutaraldehyde (GTA), a tri-iodide complex, and glucono-δ-lactone (GDL) acid for gel dosimetry was investigated in the present study. The objectives of this study were to evaluate the formula's dose response properties through spectrophotometry and possible reusability by reannealing. After production, the gel samples were irradiated from 1 to 70 Gy of gamma-rays from a Cs-137 source with a constant dose rate of 0.857 Gy/min. Spectrum data were obtained using a UV-Vis spectrophotometer and analyses were done for dose linearity, dose sensitivity versus GTA concentration, and absorbance profile versus time. The resulting unirradiated gel samples were colorless and transparent, while the irradiated samples turned to a reddish-brown hue with a peak absorbance response at 490 nm. Dose linearity results indicated $R^2$ values of 0.9, 0.986 and 0.8 for GTA concentrations of 7 mM, 15 mM and 30 mM, respectively. Moreover, dose sensitivity is higher for lower concentrations of GTA. Time progression results indicated that the absorbance decreases within one day after irradiation and increases subsequently. Through reannealing for 24 hours at 45°C in an oven, a colorless material with an absorbance value identical to the unirradiated samples was finally made for samples irradiated from 1 to 20 Gy, while the 70 Gy irradiated sample had a significant decrease in color and absorbance peak. This radiochromic gel dosimeter is promising and should be useful for 3D radiation dose assessments. Further investigation of the formula and preparation techniques are suggested for future experiments.

## 1. Introduction

In the recent years, various advancements in gel dosimetry have been made in line with the continuous progress of radiotherapy technologies and complication of the treatment plans that need precise dose distributions in the bodies of patients [1-8].

There have been notable developments of Fricke-gel dosimeters containing polyvinyl alcohol (PVA) crosslinked with glutaraldehyde (GTA) as base matrix. The crosslinking of PVA and GTA allows the construction of transparent gels at room temperature and the reduction of diffusion in the gel matrix due to the ease of reactions between PVA hydroxyl groups and GTA. The diffusion coefficient of the gel is defined by the concentration of the GTA [4, 9].

Some gel dosimeters have also been investigated. Kozicki *et al* [5] reported that radiochromic gels based from a Pluronic co-polymer matrix and leuco malachite green was effective for measuring 3D distribution of doses from both radiotherapy beams. Zhang *et al* [10] developed a gel dosimetry system for radiation that was composed of 2-hydroxyl acrylate (HEA) monomer and PVA-GTA matrix combined with a photoinitiator; this material showed good stability, dose response and linearity.

And recently, Sunagawa *et al* [11] reported a radiochromic gel indicator with a PVA base matrix doped with potassium iodide. This gel indicator has high sensitivity and changes from colorless to red after exposure to radiation. Based on their result, Hayashi *et al.* have applied this system to a novel three-dimensional (3D) radiochromic gel dosimeter using gellan gum as a gelling agent [12]. However, the improvement in terms of temporal and spatial stabilities and transparency has still required.

Motivated by the aforementioned studies, in the present study, we examine the feasibility of a novel gel dosimeter composed of polyvinyl alcohol (PVA) crosslinked with glutaraldehyde (GTA) as base matrix and combined with an iodide such as potassium iodide (KI), and additives such as fructose and a mild acid such as glucono-δ-lactone (GDL).

## 2. Materials and methods

### 2.1. Gel production

All PVA-GTA-based gels were prepared using ultrapure water and analytical grade chemicals. After the water was heated to 80 °C, the PVA (1000 of polymerization degree, partially hydrolyzed, 86~90 mol% saponification, Wako Pure Chemical Co., Japan) was added and completely dissolved for approximately 30 minutes using a magnetic stirrer at 80 °C. The mixture was then allowed to cool down to 25°C, after which, the KI (Wako Pure Chemical Co.) and fructose (Wako Pure Chemical Co., Japan) were added until fully dissolved. Then, the GDL (Wako Pure Chemical Co., Japan) and GTA (25% aqueous, Nacalai Tesque, Inc., Japan) were put into the solution for final incorporation. The solution was stirred constantly throughout the mixing process. In order to test the dependence of sensitivity with respect to GTA concentration, three sets of samples were prepared with various GTA molarities. The summary of the reagent concentrations is shown in Table 1. The final mixture was poured into 10 x 10 x 45 mm PMMA cuvettes with a parafilm cover for contamination from air and evaporation to air be prevented. All cuvette samples were placed inside an oven with the temperature at 45°C for 12 hours to allow gelation.

**Table 1.** Reagent concentrations of the gel dosimeter. All samples were initially prepared with 10 wt.% partially saponified PVA dissolved in ultrapure $H_2O$.

| Solution | potassium iodide (M) | fructose (M) | GDL (mM) | GTA (mM) |
|---|---|---|---|---|
| A | 0.1 | 0.1 | 25 | **7** |
| B | 0.1 | 0.1 | 25 | **15** |
| C | 0.1 | 0.1 | 25 | **30** |

### 2.2. Cs-137 irradiation

A Gammacell40 Exactor Low Dose Rate Research Irradiator (Best Theratronics Ltd., Canada) was used to irradiate the samples. This equipment is located in the Radiation Isotope Facility at the Kasumi campus of the Hiroshima University; it has dual Cs-137 radiation sources with total activity of 178 TBq. This equipment has been regularly checked for its dose rate and has dose uniformity of ±7% in the whole area of a sample container with 260 mm in diameter and 100 mm in height. As of June 2018, the dose rate was 0.857 Gy/min. The samples were irradiated with 1,2,4,6,8,10,15,20 and 70 Gy (for water) and one sample was kept unirradiated as control.

### *2.3. UV-Vis Spectrophotometry*

The absorbance data of the samples were obtained through a NanoDrop™ 2000c UV-Vis Spectrophotometer (Thermo Fisher Scientific Inc., USA). This device is located in the same campus as the Cs-137 irradiation equipment. All samples were examined 0.125 days (4 hours) after irradiation. Spectrum data were analysed for dose response linearity, dose sensitivity versus GTA concentration, absorbance profile versus time and post-annealing absorbance value.

## 3. Results and Discussion

The resulting gel dosimeters were transparent if unirradiated and turns to a reddish-brown hue after exposing to gamma-rays, with UV-Vis absorbance peak at 490 nm. Figure 1 shows the increasing gradient corresponding to the doses from 0 Gy (left) to 70 Gy (right).

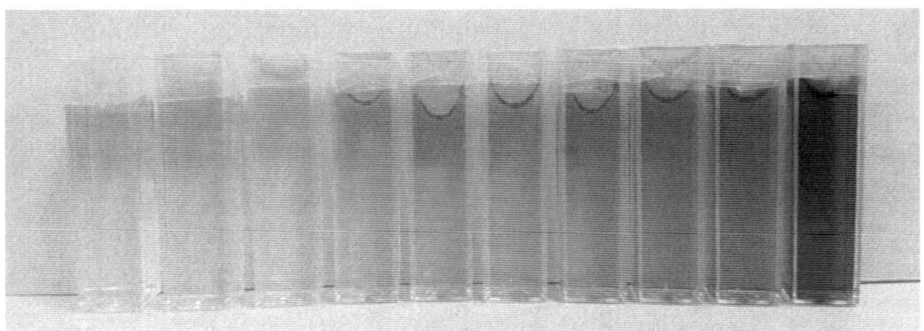

**Figure 1.** Photograph of the PVA-GTA-I samples (GTA=7mM). The dose levels from left to right were: 0 (Control),1, 2, 4, 6, 8, 10, 15, 20 and 70 Gy.

### *3.1. Dose response*

The dose response plots and linear regression curves of the different samples at 490 nm are shown in Figure 2. These plotted points were generated as the means of the measured data for each sample set during the first measurement or 4-hours post-irradiation. It was seen that the solution having lower GTA concentration (Solution A) exhibited a good linearity in the dose response up to 70 Gy ($R^2$ = 0.999). While, both Samples B and C showed lower $R^2$ values; it is attributable to the absorbance values of the samples exposed to the two lowest doses (i.e. 1 and 2 Gy) in both Solutions B and C, wherein the measured data are approximately the same with those of the control sample's value.

### *3.2. Sensitivity and Minimum Detectable Dose*

The data obtained above were evaluated using the method developed by Marini, *et al* [4]. The sensitivity of the gel dosimeter was obtained from the slope of the linear regression lines. As shown Figure 2, the sensitivity of the gel dosimeter significantly increased as the GTA concentration decreased; the sensitivities of Samples A, B and C in a 10 mm optical path were $2.5\text{x}10^{-2} \pm 0.001$ (1 s.d.), $1.7\text{x}10^{-2} \pm 0.001$ (1 s.d.) and $1.5\text{x}10^{-2} \pm 0.001$ (1 s.d.) $Gy^{-1}$, respectively. Table 2 shows a comparative summary of the sensitivity changes with previously reported gel dosimeters. According to the results on our samples, it is hoped that the sensitivity of the PVA-GTA-I gel dosimeter will be improved through further optimization.

**Table 2.** Sensitivity comparison between various gel dosimeters.

| Matrix | Reference | Sensitivity ($Gy^{-1}$)[a] | Uncertainty |
|---|---|---|---|
| PVA-GTA-I (A) | *b* | 0.025 | ± 0.001 |
| PVA-GTA-I (B) | *b* | 0.017 | ± 0.001 |
| PVA-GTA-I (C) | *b* | 0.015 | ± 0.001 |
| PVA-GTA-Fricke | [4, 9] | 0.073 | ± 0.001 (1 s.d.) |
| 20% XO-PVA | [13] | 0.014 | Not reported |
| 20% PVA hydrogel | [3] | 0.046 | Not reported |

[a]Sensitivity through 10 mm optical path; [b]Authors from this study.

IOP Conf. Series: Journal of Physics: Conf. Series **1305** (2019) 012034 doi:10.1088/1742-6596/1305/1/012034

The minimum detectable dose (MDD) was obtained through the following equation:

$$MDD = \frac{3\sigma}{s} \tag{1}$$

where $\sigma$ represents the standard deviation of the absorbance values of the unirradiated (control) samples and $s$ is the linear regression slope of each sample in the dose response curve. Using the equation above, the computed values of MDD were derived as $5.5\times10^{-2} \pm 0.001$, $7.8\times10^{-2} \pm 0.001$ and $9.3\times10^{-2} \pm 0.001$ Gy for samples A, B and C, respectively.

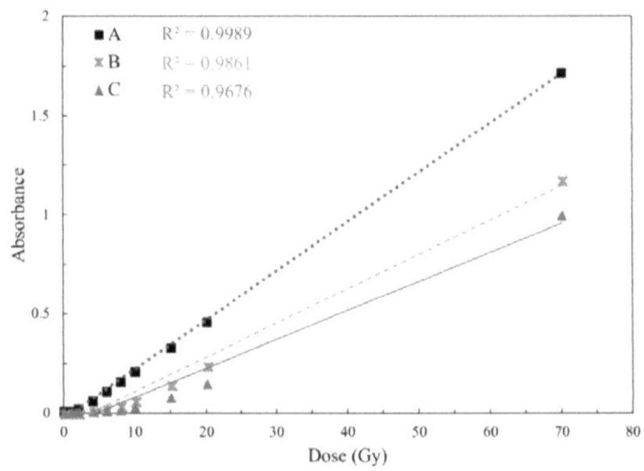

**Figure 2.** Dose response of the PVA-GTA-I gel with varying GTA concentrations.

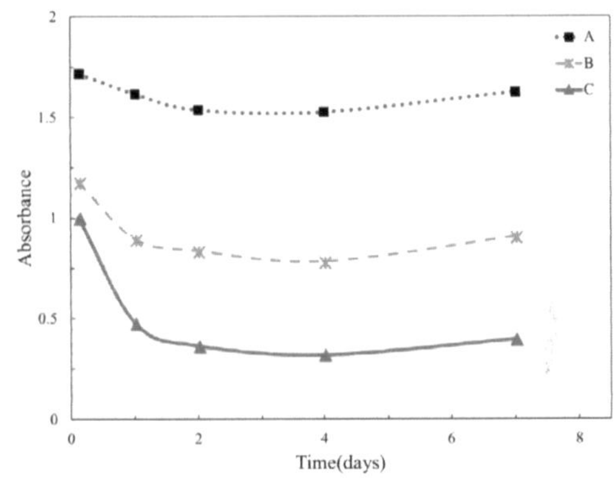

**Figure 3.** Absorbance profiles versus time of the 70 Gy-irradiated samples with different concentrations.

### *3.3. Time progression*

Absorbance profiles measured after 0.125, 1, 2, 4, and 7 days post-irradiation are shown in Figure 3. It was seen that the absorbance decreased within one day after irradiation and then gradually increased afterwards. The decrease during the initial period may be caused by the reduction of the polyiodide ion to mono-iodide ion due to the excess GTA or fructose. On the other hand, the increase of absorbance is attributable to the oxidation of the iodide ion or reaction with the light. These hypotheses should be investigated in future experiments.

### *3.4. Reannealing*

The feasibility on the reuse of the gel was initially examined by using the cuvette samples that were irradiated using Gammacell40 (Figure 1) and then reannealed for 24 hours at 45°C in the oven. As result, the samples irradiated from 1 to 20 Gy became colourless, while the 70 Gy irradiated sample converted to a lighter hue, as shown in Figure 4. This colour change was confirmed with the absorbance spectra also, as shown in Figure 5. The absorbance values of the samples irradiated from 1 to 20 Gy were comparable to that of the control sample. On the other hand, the 70 Gy sample demonstrated a lower absorbance peak at 490 nm compared to its previous value in Figure 2. These results suggest the possibility of reusing the gel dosimeter through annealing. Further examination on the reproducibility, sensitivity and stability of the gel dosimeter is recommended for future studies.

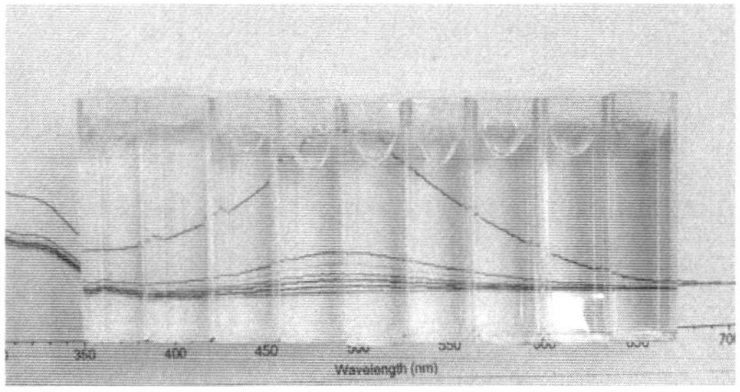


**Figure 4.** Photograph of irradiated PVA-GTA-I samples (GTA=7mM) after re-annealing for 24 hours. The dose levels from left to right were: 1, 2, 4, 6, 8, 10, 15, 20 and 70 Gy.

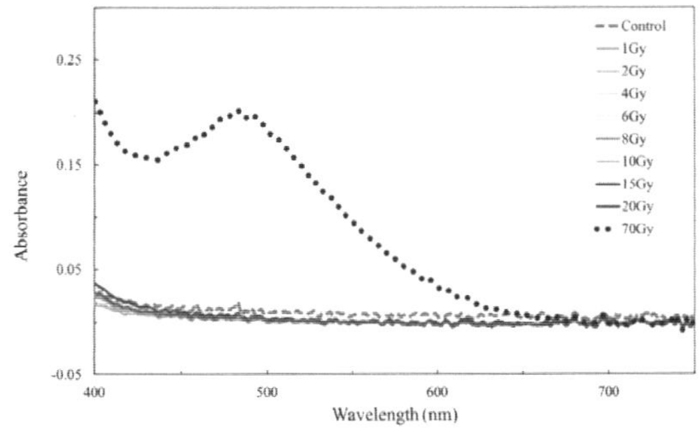


**Figure 5.** Spectrum data of the samples after 24 hours of re-annealing.

## 4. Conclusions

A novel gel dosimeter based on a formula of crosslinked PVA-GTA mixed with KI and GDL was produced. Its characteristics after exposure to Cs-137 gamma-rays were examined with the spectrophotometry techniques. The utilization of the of PVA-GTA as a base matrix resulted to highly transparent gels. Exposure to gamma-rays converts the colour of the gel dosimeter to a reddish-brown hue which allows dose relationship evaluation through spectrophotometry or other optical techniques. The results of the investigation on linearity in dose response and sensitivity indicated that the formula with the lowest concentration of GTA performed the best with $R^2$=0.999, sensitivity = 0.025 ± 0.001 $Gy^{-1}$ and MDD=0.545 ± 0.001 Gy. Time progression results show that the trend of the absorbance curve in time starts to fade in the first day post-irradiation and then increases subsequently. The formulas with the higher concentrations of GTA were less sensitive and had higher absorbance fading compared to the 7mM GTA samples. Annealing could make the gel transparent with absorbance values approximately similar to that of an unirradiated sample. Finally, the data acquired from this study suggest a promising future of the radiochromic gel dosimeter for radiation dose assessments. Further investigation of the formula's properties and preparation techniques are recommended for future experiments such as: increasing the annealing temperature and utilizing other cross-linkers other than aldehydes.

## 5. Acknowledgements

The authors would like to thank Prof. Francesco d'Errico, PhD for his insights and guidance in the utilization of crosslinked PVA and GTA in gel dosimetry. This work is supported by JSPS KAKENHI (Grant Number 17K09072).

## 6. References

[1] Hayashi S *et al* 2012 *Radiat. Phys. Chem.* **81** 884-8
[2] Fernandes J P *et al* 2009 *J. Phys.: Conf. Ser.* **164** 53-8
[3] Chu K C *et al* 2000 *Phys. Med. Biol.* **45** 955-69
[4] Marini A *et al* 2017 *Radiat. Meas.* **106** 618-21
[5] Kozicki M *et al* 2018 *J. Photochem. Photobiol. A Chem.* **351** 197-207
[6] Oldham M et al 2017 in *Clinical 3D Dosimetry in Modern Radiation Therapy* (Ed. Mijnheer B) (ISBN: 9781315118826)
[7] Mather M L *et al* 2003 *Phys. Med. Biol.* **48** N269-75
[8] Baldock C 2009 *J. Phys.: Conf. Ser.* **164** 012002
[9] D'Errico F *et al* 2017 *Radiat. Meas.* **106** 612-7
[10] Zhang W *et al* 2016 *J. Photochem. Photobiol. B Biol.* **163** 100-4
[11] Sunagawa T *et al* 2017 *Mem Fukui Univ Technol (in Japanese)* **47** 105-10
[12] Hayashi S *et al* 2017 *Proc 114th Sci Meet Japan Soc Med Phys (8th Japan-Korea Jt Meet Med*

*Physics)* **37** 171
[13] Smith S T *et al* 2015 *J. Phys.: Conf. Ser.* **573** 6-10

# Three-dimensional radiochromic and polymer gel dosimeters with Pluronic F-127 matrix – a review of current research

**M Kozicki[1,2], M Jaszczak[3], K Kwiatos[1], P Maras[4], S Kadlubowski[5], R Wach[5] and M Dudek[3]**

[1]Department of Man-Made Fibres, Lodz University of Technology, Lodz, Poland
[2]GeVero Co., Lodz, Poland (http://polygevero.com)
[3]Institute of Materials Science and Engineering, Lodz University of Technology, Lodz, Poland
[4]Medical Physics Department, Copernicus Hospital, Lodz, Poland
[5]Institute of Applied Radiation Chemisty, Lodz University of Technology, Lodz, Poland

E-mail: marek.kozicki@p.lodz.pl, gevero@polygevero.com (MK)

**Abstract**. A co-polymer of poly(ethylene oxide)-*block*-poly(propylene oxide)-*block*-poly(ethylene oxide) (Pluronic F-127, PEO-PPO-PEO) was proposed as a physical gel matrix, substituting gelatine for three-dimensional polymer gel dosimeters and as a matrix for the preparation of new radiochromic gel dosimeters. Two polymer gel dosimeters and seven radiochromic gel dosimeters were obtained with this new matrix. In this review we summarise the main advantages of using Pluronic F-127 for manufacturing radiation dosimeters and the best performing new Pluronic dosimeters.

## 1. Introduction

Pluronic F-127 was used for two purposes in our previous research. The first was to propose the substitution of a natural gelatine polymer with a synthetic co-polymer matrix in polymer gel dosimeters. The second purpose was to propose new radiochromic gel dosimeters, with respect to the existing ones [1-4], on the basis of this co-polymer and tetrazolium salts. The goals were to achieve dosimeters of improved thermal and optical properties with respect to gelatine-based dosimeters, and in case of radiochromic dosimeters, to obtain systems recording dose distributions with no diffusion of irradiated parts over time after irradiation. In most similar system referred to in the above publications, radiochromic dosimeters are prone to losing dose distribution information over time.

A few gelling agents have been used for manufacturing three-dimensional (3D) polymer gel and radiochromic dosimeters [5]. In 1993, Maryanski *et al* 6] proposed the first polymer gel dosimeter with agarose that fixed a dosimetric solution by the formation of a physical gel. Soon afterwards, it was discovered that gelatine outperformed agarose by better transparency of the gel matrix, easier preparation of the gel in lower temperatures than in the case of agarose and the formation of a more even gel [7]. Since that time, most researchers have used gelatine as a matrix for polymer and radiochromic gel dosimeters. Nevertheless, the gelatine matrix has one disadvantage, namely, a low gel–sol transition temperature (~30 °C) that limits the temperature range in which non-irradiated and irradiated samples can be stored and transported. Also, it is clear from our studies that an aqueous solution of this natural polymer of gelatine degrades over time after preparation, which is inconvenient if gelatine-based dosimeters are designed for a long period of use [8, 9].

Several studies have been performed on modifications of gelatine matrix to ensure better thermal stability. For instance, a modified MAGIC [10], MAGIC–f, contained formaldehyde that shifted the melting point to 69°C [11, 12]. However, a disadvantage of formaldehyde is its toxicity [13]. In another approach, Abtahi *et al* [14] mixed gelatine with agarose in MAGIC–A, which led to the melting point of such a matrix being 60°C but decreased the dose sensitivity with respect to MAGIC. Poly(vinyl alcohol) was also used as a matrix in a PAGAT dosimeter [15], and the new gel dosimeter was called PVABAT [16]. The melting point of PVABAT increased to 80 °C. The authors reported that the linear dose range for this gel was 30–45 Gy, while for PAGAT it was 0–7 Gy.

Over the last two years, research has been performed on both substituting gelatine matrix with the physical matrix made of a synthetic co-polymer of poly(ethylene oxide)-*block*-poly(propylene oxide)-*block*-poly(ethylene oxide) (Pluronic F-127) in polymer gel dosimeters and on the development of the new radiochromic gel dosimeters with Pluronic F-127 matrix [9, 17-19]. The results of these studies are summarised in this communication.

## 2. Features of Pluronic F-127

The reasons behind the choice of the Pluronic F-127 copolymer for 3D dosimetry are as follows: (i) the exceptional transparency of Pluronic physical gels, which can be compared to that of clear water, (ii) the phase behaviour of Pluronic, which exists in sol and gel form, and is related to its concentration and presence of co-solutes [20] – this affects the manufacturing of gel dosimeters, (iii) the stability of gels over different temperature ranges, which is also related to the phase behaviour and concentration of Pluronic, (iv) Pluronic F-127 is a synthetic, non-toxic co-polymer that is used in biomaterials research [21] and is US Food and Drug Administration (FDA) approved [22], and (v) Pluronic does not degrade over time, unlike gelatine aqueous solutions [8, 9].

The exceptional transparency of Pluronic gel was assessed with UV-Vis spectrophotometric measurements. In brief, Pluronic gels (23 and 25%) exhibited much lower absorbance from gelatine solutions (5, 7.5 and 8%) over the wavelength range of 300-700 nm [9, 17]. The absorbance of Pluronic gels was six-to-ten times lower than gelatine gels at 400 nm and 1.5-times lower at 730 nm. Gels consisting of 23% and 25% Pluronic showed similar absorbance. In addition, gelatine in solution is susceptible to degradation. The aging of gelatine was a subject of studies by other researchers [23-25], who reported gelatine to be a very sensitive compound to environmental changes. Our findings of gelatine aging are reported elsewhere [8, 9].

The phase diagram for Pluronic F-127 [20] revealed that the higher the concentration of this co-polymer, the broader the temperature range over which its solution remains as a physical gel. For instance, a 25% gel is stable in the temperature range of ~20-85 °C [20]. The gel-sol transition occurs below and above these temperatures. This phase diagram is affected by other substances present in Pluronic gel. The specificity of the sol-gel-sol transition of Pluronic gels makes it attractive in terms of keeping a gel dosimeter with a Pluronic matrix stable at temperatures far above room temperature in contrast to gelatine gel dosimeters.

## 3. Polymer gel dosimeters with Pluronic F-127

The following polymer gel dosimeters were examined for substitution of gelatine with a Pluronic F-127 matrix: methacrylic acid containing MAGAT gels [26], acrylamide containing PAGAT gels [15, 26], *N*-isopropylacrylamide containing NIPAM gels [27] and *N*-vinylpyrrolidone containing VIP [28] and VIC [29].

Substitution of gelatine with Pluronic was impossible in MAGAT and VIC polymer gel dosimeters due to the incompatibility between methacrylic acid (MAGAT) and Pluronic, or a too high concentration of dosimeter components (VIC). Alternatively, PAGAT (6 and 9% T), NIPAM (different % T) and modified VIP allowed proper Pluronic-based polymer gel dosimeters to be obtained. In contrast to Pluronic-containing PAGAT (6% T) and modified VIP gels, NIPAM containing a Pluronic matrix did not respond to irradiation. A comparison of the new Pluronic-based polymer gel dosimeter properties are shown in Table 1. Their main feature is that the dose sensitivities of both new gels were ~33% higher than their corresponding compositions containing a gelatine matrix. The Pluronic matrix ensures improved transparency and stability in contrast to the gelatine-based gel dosimeters. The new gel dosimeters' thermal stabilities assessed by $T_{gel}$ values (Table 1) are 4.4–56.7 °C (VIP3- Pluronic) and 10.1–56.7 °C [9]. Note that the thermal transitions start and end at lower and higher temperatures, respectively, than $T_{gel}$. Differential scanning calorimetry (DSC) results showed two gelatine-based dosimeters, a VIP gel dosimeter [28] and VIC [29] to be stable at 0–26 °C and 0–29.5 °C, respectively (the upper values correspond to initiation of the gel-sol transition) [8, 9]. Our recent studies showed also that substituting oxygen scavengers in VIC dosimeters (copper sulfate and ascorbic acid) with tetrakis(hydroxymethyl)phosphonium chloride (THPC) causes chemical crosslinking of gelatine, which impacts on the thermal stability of this gel (VIC-T). DSC studies showed its stability at about 0–80 °C [8].

**Table 1**. Comparison of VIP3-Pluronic with PAGAT2-Pluronic: 0.47 T NMR derived calibration characteristics and DSC measured thermal properties (VIP3-Pluronic: 25% Pluronic, 4% *N,N'*-methylenebisacrylamide, 4% *N*-vinylpyrrolidone, 10 mM THPC, 4% tert-butanol; PAGAT-2-Pluronic: 23% Pluronic, 3% acrylamide, 3% *N,N'*-methylenebisacrylamide, 10 mM THPC).

| | Threshold dose [Gy] | Linear dose range [Gy] | Dynamic dose range [Gy] | $R_2 = a \times D + R_{2,0}$ (D is dose [Gy]) | | |
|---|---|---|---|---|---|---|
| | | | | a | $R_{2,0}$ | $R^2$ |
| VIP3-Pluronic | 0.5 | 0.5–30 | 0.5–>50 | 0.1010 | 1.249 | 0.999 |
| PAGAT2-Pluronic | 0.5 | 0–20 | 0–>50 | 0.0913 | 1.052 | 0.988 |
| DSC | $T_{gel, onset}$ [°C] | $T_{gel}$ [°C] | $T_{gel, end}$ [°C] | $T_{gel, onset2}$ [°C] | $T_{gel2}$ [°C] | $T_{gel, end2}$ [°C] |
| VIP3-Pluronic | 1.6 | 4.4 | 13.8 | 55.2* | 56.7* | 58.8* |
| PAGAT2-Pluronic | 5.7 | 10.1 | 21.4 | 55.2* | 56.7* | 58.8* |

*Data for 25% Pluronic F-127. No such results for second transition was observed for VIP3-Pluronic and PAGAT2-Pluronic [9]. The upper temperature of thermal stability for VIP3-Pluronic and PAGAT2-Pluronic is assumed to be around $T_{gel2}$ given in Table 1 for 25% Pluronic F-127.

## 4. Radiochromic gel dosimeters with Pluronic F-127

New radiochromic gel dosimeters were manufactured with the Pluronic F-127 matrix and the following tetrazolium salts: 2,3,5-triphenyltetrazolium chloride (TTC), tetrazolium violet, blue tetrazolium chloride, nitro blue tetrazolium chloride (NBT), tetranitro blue tetrazolium chloride, thiazolyl blue tetrazolium bromide and a representative of leuco dyes - leuco malachite green (LMG) [17-19]. The main feature of the tetrazolium salt-Pluronic radiochromic gels is the conversion of the salts into water insoluble formazans of different colours inside the Pluronic matrix (Figure 1). This is in contrast to LMG-Pluronic, where a water soluble dye is formed. However, the Pluronic matrix prevents the formazan molecules from diffusion, aggregation and sedimentation. This, in turn, relates to an extraordinary maintaining of a 3D dose distribution integral over a long time after irradiation with the diffusion coefficient equal to 0 $mm^2/h$ [17, 19].

The first tetrazolium salt-Pluronic radiochromic dosimeter reported was that with TTC [17]. The following experiments revealed that different tetrazolium salts can be used for manufacturing such dosimeters. However, some of them were unstable over time after preparation or revealed low dose sensitivity. The one of the highest dose sensitivity was NBT-Pluronic (Figure 1), which was composed of 0.0818% NBT (1 mM), 25% Pluronic F-127 and $0.136 \times 10^{-2}$% sodium formate. This gel was equivalent to water.

The most important characteristic of the NBT-Pluronic gel was: (i) the average dose sensitivity of $0.0047 \pm 0.1 \times 10^{-4}$ (Gy cm)$^{-1}$, (ii) insensitivity to changes in the dose rate for photons of different energies, (iii) ~12% difference in the dose sensitivity between the gel irradiated with 12 MeV electrons and photons, (iv) linear and dynamic dose ranges between <1 and ⩾150 Gy and (v) the dose threshold below 1 Gy [18]. DSC results showed that this NBT-Pluronic gel is stable in the temperature range of ~9.5–57 °C.

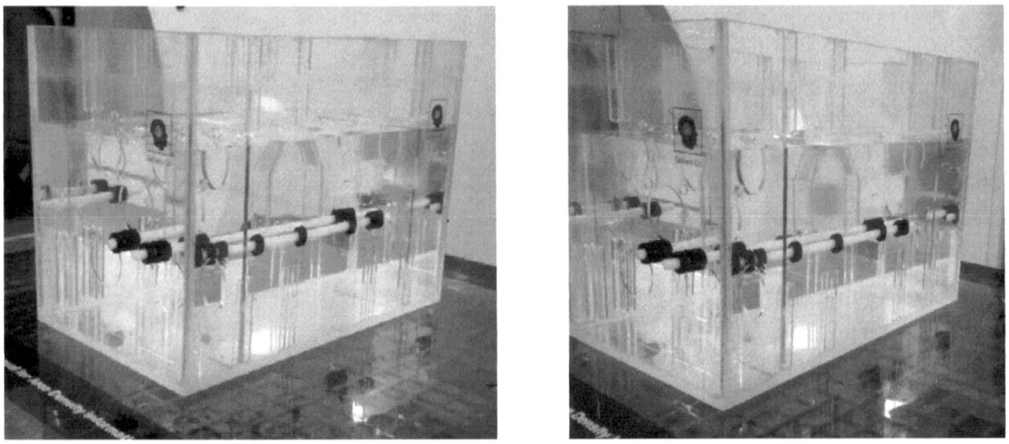

(a) (b)

**Figure 1.** A view of NBT-Pluronic gel dosimeter in a water phantom before (a) and after (b) irradiation to 60 Gy (field size 2x2 cm$^2$). A region of NBT conversion after irradiation can be seen (pointed with a red arrow) as a rectangular shape of red-violet colour coming from the formation of water insoluble NBT formazan.

## 5. Conclusions

Pluronic F-127 has been shown to be a very good physical matrix for manufacturing some 3D polymer gel dosimeters and 3D radiochromic dosimeters based on tetrazolium salts. The best compositions for polymer gel dosimeters are VIP3-Pluronic and PAGAT2-Pluronic, whereas for radiochromic gel dosimeters – NBT-Pluronic.

Pluronic matrix ensures improved optical properties and thermal stability for both types of dosimeters and improved dose sensitivity of polymer gel dosimeters with respect to their corresponding gelatine-based compositions. By using a Pluronic synthetic matrix, the longevity of dosimeters is enhanced with respect to gelatine gel dosimeters, where gelatine degrades in solution over time after preparation.

## 6. References

[1] Babic S *et al* 2009 *Phys. Med. Biol.* **54** 6791
[2] Jordan K and Avvakumov N 2009 *Phys. Med. Biol.* **54** 6773
[3] Nasr A T *et al* 2015 *J. Phys.: Conf. Ser.* **573** 012037
[4] Schreiner L J 2015 *J. Phys.: Conf. Ser.* **573** 012003
[5] Baldock C *et al* 2010 *Phys. Med. Biol.* **55** R1-63
[6] Maryanski M J *et al* 1993 *Magn. Reson. Imaging* **11** 253
[7] Maryanski M J *et al* 1994 *Phys. Med. Biol.* **39** 1437
[8] Jaszczak M *et al* 2018 *Phys. Med. Biol.* (submitted)
[9] Jaszczak M *et al* 2018 *Phys Med Biol.* **63** 175010
[10] Fong P M *et al* 2001 *Phys. Med. Biol.* **46** 3105
[11] Fernandes J P *et al* 2009 *J. Phys.: Conf. Ser.* **164** 012004
[12] Pavoni J F and Baffa O 2012 *Rad. Meas.* **47** 1074
[13] Songur A *et al* 2010 In *Reviews of Environmental Contamination and Toxicology* (Ed. Whitacre D

M) Springer, New York
[14] Abtahi S M *et al* 2016 *J. Radioanal. Nucl. Chem.* **307** 855
[15] Venning A J *et al* 2005 *Phys. Med. Biol.* **50** 3875
[16] Abtahi S M and Abandansari H S 2017 *Australas. Phys. Eng. Sci. Med.* **40** 651
[17] Kozicki M *et al* 2017 *Phys. Med. Biol.* **62** 5668
[18] Kozicki M *et al* 2018 *J. Photochem. Photobio. A* **351** 197
[19] Kwiatos K *et al* 2018 *Phys. Med. Biol.* **63** 095012 (10pp)
[20] Alexandridis P and Hatton T A 1995 *Colloids Surf. A* **96** 1
[21] Diniz I M *et al* 2015 *J. Mater. Sci., Mater. Med.* **26** 153
[22] Almeida H *et al* 2012 *J. Pharm. Pharmaceut. Sci.* **15** 592
[23] Duconseille A *et al* 2018 *Food Hydrocolloid.* **77** 777
[24] Duconseille A *et al* 2017 *Food Hydrocolloid.* **66** 246
[25] Duconseille A *et al* 2017 *Food Hydrocolloid.* **66** 378
[26] Brindha S *et al* 2004 *Phys. Med. Biol.* **49** N353
[27] Chain J N M *et al* 2011 *Phys. Med. Biol.* **56** 2091
[28] Kozicki M *et al* 2007 *Macromol. Symp.* **254** 345
[29] Kozicki M *et al* 2017 *Phys. Med. Biol.* **62** 986

IOP Conf. Series: Journal of Physics: Conf. Series **1305** (2019) 012036 doi:10.1088/1742-6596/1305/1/012036

# Initial performance evaluation of a 3D gel dosimeter based on modified tetrazolium compounds

**Rubin Hazarika[1], Kalin I Penev[2,3], Kibret Mequanint[3,4] and Kevin Jordan[1,5]**
[1]London Regional Cancer Program, London Health Sciences Centre, London, ON, N6A 4L6, Canada
[2]Modus Medical Devices Inc., 1570 N Routledge Park, London, ON, N6H 5L6, Canada
[3]Department of Chemical and Biochemical Engineering, The University of Western Ontario, London, ON, N5A 5B9, Canada
[4]Biomedical Engineering Program, The University of Western Ontario, London, ON, N5A 5B9, Canada
[5]Dept of Medical Biophysics, The University of Western Ontario, London, ON, N6A 5B9, Canada

E-mail: kevin.jordan@lhsc.on.ca

**Abstract**. Tetrazolium-based 3D gellan gum gel dosimeters have been reported previously and tested at several sites. They show no image diffusion and excellent stability; however, the dose sensitivity of the current commercial product, ClearView™, is relatively low for routine clinical applications. Here we report initial testing of a second generation product with increased sensitivity and retained non-diffusive properties. This formulation has a 25% increase in sensitivity relative to ClearView™. However, the sample has a higher initial attenuation and the background attenuation increases faster than ClearView™. The composition was tested in a 1-liter sealed container using a 2x2 cm, 6MV, x-ray beam, and show agreement between Monte Carlo simulation and measurements.

## 1. Introduction

Non-diffusing radiochromic hydrogels based on the reduction of colourless and water-soluble bisnitrotetrazolium chloride (BNC) to highly-coloured, water-insoluble formazan dye in a gellan gum matrix has been previously investigated as a potential robust hydrogel that can be shipped globally [1]. This material has been evaluated by several research groups and early formulations were found to have dose rate effects that limited quantitative 3D radiation dosimetry [2, 3]. Current formulations have higher concentrations of BNC (from 0.1 to 0.25 mM) and dose rate limitations appear to have been eliminated (in-house results, manuscript in preparation). While the current formulation is promising, higher dose sensitivity is preferred for many applications. Several new tetrazolium salts have been synthesized to achieve a more sensitive dosimeter (manuscript in preparation).

Two new tetrazolium compounds with increased dose sensitivity, based on 1 cm cuvette measurements, were selected for evaluation in one litre volumes for 3D dosimeter performance.

## 2. Methods

Gel samples were prepared to fill 10 cm diameter custom cylindrical PETE vessels. Optical cone beam computed tomography scans were performed with a Vista16 scanner provided by ModusQA with a 530 nm LED light source.

### *2.1. Gel preparation*

Hydrogels formulations were similar to previously reported ClearView™ gels with the substitution of the dye precursor 0.25 mM of either 2-(benzothiazole-2-yl)-3-nitrophenyl-5-phenyl-2*H*-tetrazolium chloride (BtNP) or 2-(benzothiazole-2-yl)-3-(1-naphthyl)-5-phenyl-2*H*-tetrazolium chloride (BtNpP), synthesized in house. The gels contain 1.25% gellan gum and 10.2% propylene glycol by mass and have a mass density of ~1.02 g $cm^{-3}$ and optical cone-beam computed tomography (CBCT) scans were performed in aqueous solutions containing 12% propylene glycol. Sample vessels were custom cylinders, constructed in house from 0.25 mm thick PETE film. The vessel fill port was sealed with aluminized tape to make an oxygen-impermeable vessel. Samples were stored at 21 °C in the dark.

#### *2.1.1. Optical CT scanning*

Samples were scanned with a Vista16 optical CBCT scanner, modified by placing a 1.5 cm diameter aperture at the diffuse source to reduce stray light. A 12% propylene glycol - 88% water solution was placed in the scanner aquarium to maximize FOV for optical CT scanning. After placing the sample in the scanner a 15-minute interval was used to let the aquarium liquid to relax to an optically uniform condition. Pre-irradiation and post-irradiation scans consisted of a 1000 projections acquired over 360 degree rotations and reconstructed at 0.05 $cm^{-3}$ voxel size with OSC-TV iterative algorithm within the VistaScan software package.

#### *2.1.2.* Irradiation

A Varian 21iX series, medical, linear accelerator was used to irradiate samples. The AP beam was incident on the base of the upside-down gel. Beam conditions: 6 MV, dose rate 400 MU/minute, dose 2642 MU (~ 20 Gy), SSD = 90 cm, jaws set to 2x2 cm.

## 3. Results

The first gel, containing BtNP was formed yellow but turned brown within a day of preparation. It contained nonuniformities in refractive index that resulted in a low transmission, scattering material that produced noisy reconstructions. The second sample contained BtNpP and resulted in a clear, yellow gel that is the focus of this report. Immediately following irradiation, a purple image of the dose distribution was formed. The absorption spectrum maximum is at approximately 510 nm, and is included in the results presented in another paper at this conference, Penev et al. At the wavelength of interest (530 nm), dose sensitivity increase of 25% was recorded compared to current ClearView™ formulations from ModusQA. Figure 1 shows the normalized mean central depth dose for beam irradiations 2 and 4 days after gel preparation and corresponding Monte Carlo calculation. The attenuation coefficient plots versus depth were the same, indicating no change in dose sensitivity from top to base of gel over the 2 day period at 21 °C.

Figure 2 compares jaw-defined x profile at depth of maximum dose in gel versus Monte Carlo calculation at same depth in water. Both curves represent the average of 10 profiles from -0.25 to 0.25 cm in y direction. Each point can also be considered as a voxel size (x,y,z) = (0.05x0.5x0.05 cm). Each curve was normalized at beam centre and a constant background was subtracted from the gel reconstruction. Note the small under-response of gel at -2.5 cm is likely an artifact due to proximity of vessel wall.

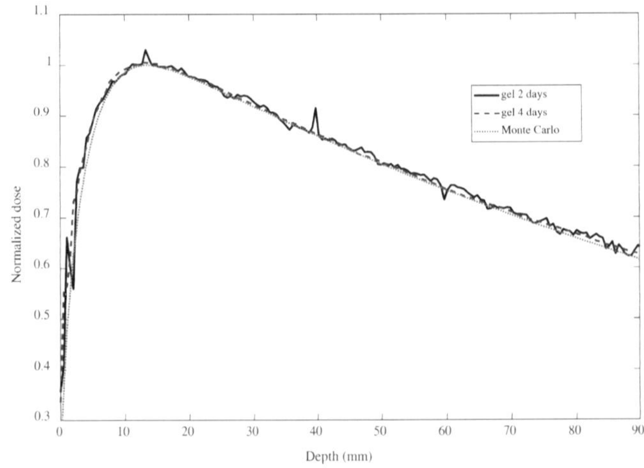


**Figure 1.** Plot of normalized dose for BtNpP gel, 6 MV, 2x2 cm jaw defined, SSD= 90 cm, dose ~20 Gy. (solid line) irradiated 2 days post preparation, (dashed line) irradiated 4 days post preparation, (dotted line) Monte Carlo calculation.

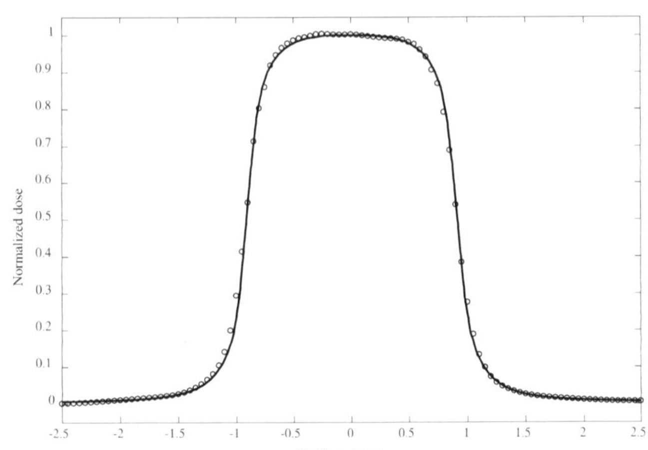


**Figure 2.** Plot of x direction profiles at depth of 1.4 cm for 6 MV, SSD=90 cm, field size jaws=2x2 cm (at 100 cm). (circles) gel reconstruction, (solid line) Monte Carlo calculation, voxels size same.

## 4. Discussion

Agreement between Monte Carlo depth dose and gel reconstructions demonstrate uniform dose sensitivity for the bottom (beam entrance) to the top of the sample. Agreement between day 4 and day 2 indicate stability of the gel when stored in dark at 21 °C. Longer times were not evaluated due to observation of mould growth sites on day 4. It should be noted that the biological contamination issue can be addressed in large-scale manufacturing more easily than on the bench top by using clean manufacturing practices, and mould growth is not expected to occur in future samples. In contrast to the colourless ClearView™ gel based on BNC tetrazolium, the current formulation is yellow and the initial attenuation increases at a rate of 0.013 $cm^{-1}$ per day vs ~$10^{-3}$ $cm^{-1}$ per day for ClearView™ gel. This initial background is likely related to the increased dose sensitivity. The gel and Monte Carlo beam profiles are nearly the same providing a more stringent test of dose linearity, dose rate and energy independence of the gel formulation compared to central axis depth dose comparisons.

## 5. Summary

The BtNpP gel provides a 25% increased dose sensitivity compared to ClearView™ formulation and has and increasing background of 0.013 $cm^{-1}$ per day at 21 °C. Future work will include fine-tuning the compromise between sensitivity and background attenuation.

## 6. Acknowledgements

R Hazarika was funded by Collaborative Research Health Program grant (CHRP #356860). Disclosure, K Jordan has a licensing agreement with Modus QA related to optical CT. K Penev and K Mequanint have a licensing agreement with Modus QA related to gel materials. The Vista16 scanner was provided by Modus QA in support of above CHRP grant.

## 7. References

[1] Penev K I *et al* 2017 *J. Phys.: Conf. Ser.* **847** 012048
[2] Ascencion Y *et al* 2017 *J. Phys.: Conf. Ser.* **847** 012049
[3] Huet C *et al* 2017 *J. Phys.: Conf. Ser.* **847** 012015

# Making and assessing 3D dosimeters

**Kevin Jordan[1] and Yves De Deene[2]**
[1]Department of Medical Biophysics, Western University London, ON, Canada
[2]School of Engineering, Macquarie University, North Ryde, Sydney, Australia

E-mail: Kevin.Jordan@lhsc.on.ca, Yves.DeDeene@mq.edu.au

**Abstract**. Several 3D dosimeters are commercially available. However, there are many circumstances that require a customized 3D dosimeter. Examples include feasibility tests of non-standard treatment modalities, inhomogeneous tissue configurations, unique shapes and sizes and teaching. In this session, general approaches for preparing radiochromic dosimeters, Fricke and polymer gel dosimeters, micelle gel and silicone dosimeters were presented. Advise will be given to developers of new 3D dosimeters. For optical readout, light absorption and scatter can limit the practical size of dosimeters. Specifically, increasing from 5 to 15 cm diameter dosimeters is optically challenging. Strategies to maximize initial optical transmission were presented. For MRI readout, the dose resolution is determined by both the dosimeter sensitivity and the pulse sequence parameters and the accuracy is determined by the sensitivity of the dosimeter to temperature and dose rate, next to imaging performance. For X-ray CT imaging, the dose resolution is determined by the sensitivity of the dosimeter which largely depends on the polymer density that can be achieved. The importance of characterizing the dosimeter in terms of dose sensitivity and stability, spatial integrity, dose rate and fractionation dependence, oxygen and ambient light sensitivity, temperature sensitivity and thermal history were emphasized. The dosimeter requirements also dictate the types of vessels and scanners appropriate for readout. For example, the preferred dosimeter formulation may include a compound that is incompatible with the preferred vessel.

## 1. Introduction

This report summarizes a gel dosimeter preparation presentation from the IC3DDose conference in Kunshan, China. Strategies for obtaining useful 3D dose data, were emphasized with reminders to read the literature, especially the previous proceedings. The audience was also reminded to assess commercially available products, before investing in gel preparation. Examples that may not be suited to commercial products include feasibility tests of non-standard treatment modalities [1-7], inhomogeneous tissue configurations [8-12], unique shapes and sizes [13-17] and teaching [18]. Likely, commercial suppliers are willing to place their materials in custom vessels supplied by the users, if they are chemically compatible. General approaches for preparing radiochromic dosimeters [19-21], Fricke [22, 23] and polymer gel [24-27] dosimeters, micelle gel [28, 29] and silicone dosimeters [30-32] were presented. Two applications were used to demonstrate some issues to consider. The first dosimetry example included: optical CT readout of a 15 cm diameter, cylindrical, non-diffusing, radiochromic hydrogel containing small fields with 20 Gy maximum dose. And the second example was: normoxic polymerization gels with MRI or x-ray CT readout. Small field depth doses and profiles are challenging dose distributions but simple to compare against commissioning data and many of the presentations at the conference used small fields to assess their 3D dosimeter performance.

The application, dosimeter material, vessel and readout technology are interdependent, see figure 1. For example, oxygen impermeable, plastic vessels are required for most water-equivalent, dosimetry measurements. These vessels should also be transparent, thin and of uniform thickness for optical CT readout. In other cases, the gel material may contain alcohols or chlorinated solvents which restrict compatible plastics. While there are many published studies on 3D gel dosimeters, users were reminded they must critically assess their samples, whether purchased or made in laboratory, for accurate measurements of 3D dose distributions.

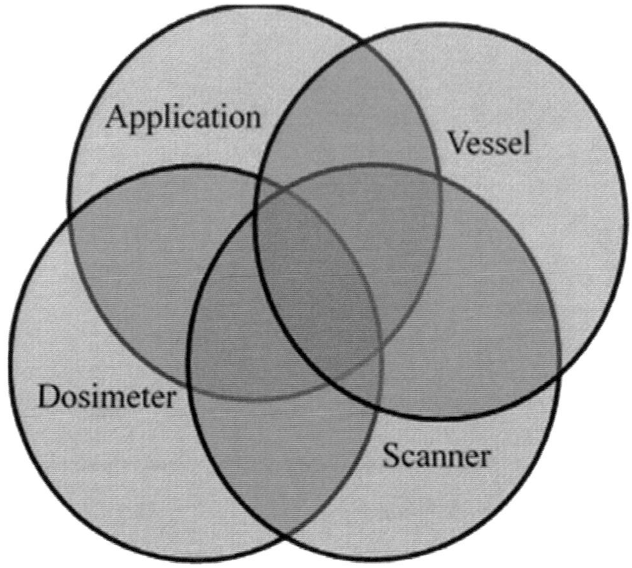


**Figure 1.** Diagram showing interdependence of factors to consider for 3D dosimetry measurements.

## 2. Presentation

Many of the key points were made by both authors. These include: minimize thermal and chemical gradients, prepare gels at lowest practical temperatures, minimize exposure to light at photosensitive wavelengths.

### *2.1 Vessels*

Laboratory glassware is generally used for gel preparations because of its chemical inertness and low oxygen permeability. Dosimeter vessels are generally plastic with polyethylene terephthalate (PET) the most common material due to low oxygen transport, low scatter and good optical quality. Barex has been used for anoxic polymerization gels because of its exceptionally low oxygen permeability (figure 2).

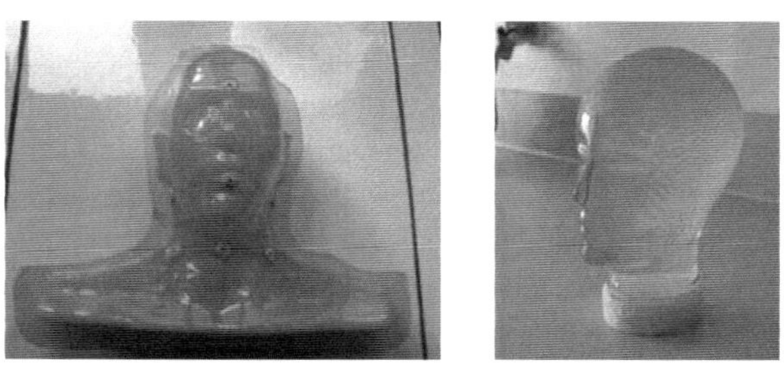

**Figure 2.** Antropomorphic shaped Barex phamtom (a) and glass phantom (b) filled with polymer gel.

Effectiveness of methods to seal vessels must be verified. This is generally performed by measuring spatial uniformity of sample's dose sensitivity. A change in sensitivity near the seal generally indicates an oxygen leak or oxygen in the gas phase above the gel. As storage times increase it is likely the oxygen concentration will decrease due to its reactivity with the dosimeter chemicals, producing organic peroxides. When comparing litre sized or larger samples with smaller samples such as 1 cm pathlength polymethylmethacrylate (PMMA) disposable cuvettes, users should be aware that oxygen permeability of PMMA is relatively large leading to well oxygenated cuvette samples and larger volumes with likely lower oxygen concentrations.

*2.2 Laboratory equipment*

The first polymer gel dosimeters were fabricated in a specialized laboratory that required perfusion of the sol with nitrogen and a glove-box filled with nitrogen to avoid any oxygen infiltrating in the phantom container upon filling (figure 3).

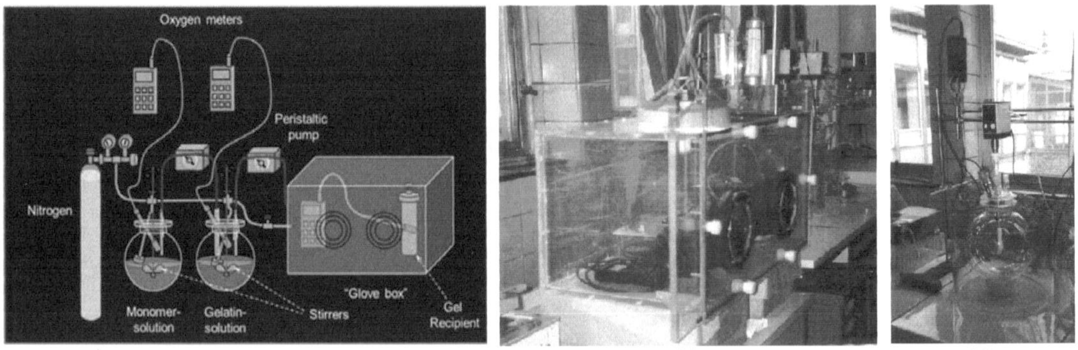


**(a)** **(b)** **(c)**

**Figure 3.** Anoxic polymer gel dosimeters are constructed in an oxygen tight environment, requiring closed vessels, oxygen meters, peristaltic pumps and a glove box (a). Photographs of a laboratory set-up showing a glove box (b) and closed reaction vessels (c).

Oxygen radicals that are formed during radiation can also be captured by use of anti-oxidants which makes the fabrication of polymer gel dosimeters less complicated. However, caution is still required as an excess of oxygen can still inhibit the polymerization reaction [50, 52].

Access to compressed air and vacuum ports is very useful. Compressed air is effective for blow drying vessels to avoid spots and films remaining after evaporation. Blow drying also avoids scratching the plastic surfaces that occurs from mechanical wiping. Vacuum filtration (figure 4) is an effective method to remove debris in both the gelatin solution prior to adding radiation sensitive chemicals and the multiple use, refractive index optimization liquids required for optical CT scanning. Plastic containers are useful for storage and handling of liquids and disposal of gels since several litre volumes are required in the larger scanners. Certain gel formulations can easily be disposed of by chopping the gels into small pieces and leaving to dehydrate and then placing in landfill, for example ferrous xylenol orange or leucocrystal violet gelatin gels. An even better practice is to use redox chemistry or electrochemistry to mineralize the dyes prior to disposal.

**Figure 4**. Photograph of vacuum filter and flask, filtering refractive index optimization solution of propylene glycol and water.

Temperature control is generally performed with water baths. This approach provides uniform temperatures during preparation, a key factor in obtaining reproducible batches of gel.

**Figure 5**. Photograph of water bath with immersion heater, not shown is the magnetic stirrer below the stainless-steel tank.

Figure 5 shows beaker of dissolved gelatin in water and an irradiated blue gel being melted to empty and reuse the vessel.

Generally, gel dosimeters are kept in a fridge around 4 °C to minimize chemical reactions and biological growths to maximize storage time. Shown in figure 6, are three samples of a radiochromic gel prepared with the same concentrations of starting materials but by slightly different steps. Note the differences in scatter and initial colour for the same chemical composition as a caution that preparation details can make a difference. Note also, this particular gel formulation has most uniform response for same day gel samples. For longer times this spatially changing dose response is likely a result of the gelation kinetics of this formulation.

**Figure 6.** Photograph inside storage refrigerator of three 10 cm diameter gels, same formula but different batches, demonstrating marked changes in initial colour and scatter.

For optical CT readout, the initial background scatter can limit the size of the dosimeter that can be scanned quantitatively. In general, optical CT laser scanners that employ a narrow laser beam have the greatest rejection of stray scattered light and broad beam scanner geometries the least rejection. Colourless dosimeters with transmission >10% relative to water along the diameter should provide quantitative 3D attenuation coefficients with most scanner geometries. For optical readout, light absorption and scatter can limit the practical size of dosimeters [33-35]. For example, a material that is acceptable for a 10 cm diameter sample may be of little value for 15 cm diameter samples. A simple test of scatter is to observe a back-illuminated, transparent ruler, see figure 7.

**Figure 7.** Photograph of back-illuminated transparent plastic ruler viewed through a 10 cm diameter radiochromic hydrogel in a water tank.

In this example, a 10 cm diameter, previously irradiated, gelatin gel is photographed in a tank of water. This image represents a low scatter sample. Note, that near the inside edge of the vessel wall some of the ruler gradations are missing. This is due to the refractive index mismatch between the dosimeter in this custom PET vessel and the surrounding water. A small, fiducial ink dot on the vessel wall can serve a similar purpose when assessing projection images acquired with the system. See dot, beside seam 5 cm above vessel bottom end. Note that most polymer gels exhibit a dose dependent change in refractive index which results in reconstruction artefacts for optical CT scanning.

For MRI readout, the dose resolution [36] is determined by both the dosimeter sensitivity and the pulse sequence parameters [37] and the accuracy is determined by the sensitivity of the dosimeter to temperature [38, 39] and dose rate [40], next to imaging performance.

For X-ray CT imaging, the dose resolution is predominantly determined by the sensitivity of the dosimeter which largely depends on the polymer density that can be achieved [41-43].

### *2.3 Gel preparation practical tips*

Two step by step procedures for a radiochromic and a normoxic polymer gel with gelatin were presented and the cautionary notes were similar. For example, start by sprinkling gelatin onto cold water, to wet before heating to dissolve. After dissolving suspension in 40 to 50 °C water bath with stirring, cool to 30 to 32 °C before adding radiation sensitive chemicals. Remember that oxygen solubility is temperature dependent and that exposure to atmosphere can equilibrate oxygen concentrations before sealing dosimeter vessels. Thermal gradients during gelation, scanning and irradiation can all have measurable effects on 3D response. As the dosimeters increase in size, several hours may be needed to uniformly change the temperature from storage to scanning temperature. For example, a 2.5 litre gel cylinder may require 6 hours to equilibrate from 4 to 20 °C. In order to achieve reproducible initial performance of gelatin dosimeters, the gelatin-water solution can be evaluated for residual oxidizing species by adding ferrous xylenol orange solution to aliquots prior to addition of radiation sensitive chemicals. Specifically, the gelatin solution can be kept warm, for example at 50 °C until oxidizing species have thermally decomposed. This is shown in figure 6 where the solution is sampled until the gel remains yellow orange with the addition of the test FX solution.

### *2.4 Gel calibration strategies*

The importance of characterizing the dosimeter in terms of dose sensitivity and stability [44, 45], spatial integrity [23, 45-47], dose rate [23, 40, 48] and fractionation dependence [49], oxygen and ambient light sensitivity [48, 50], temperature sensitivity [38, 48] and thermal history [51] were emphasized. Sample specific calibrations are generally required. For example, without additional calibration data sets, an individual dosimeter should be irradiated to a uniform dose to first establish spatial dose sensitivity. Assuming, one sample behaves like another may result is spurious results even when samples are from same batch and have the same thermal history. While each gel formulation has its own features, current experience indicates gelatin gels with lower concentrations of gelatin and same or next day usage provide the most spatially uniform dose response. Other formulations designed for long storage times may have the opposite trend, such that waiting several weeks before irradiation provides more spatially reproducible results. Much effort has been invested in designing calibration strategies, but the user must still verify performance of their specific samples and not assume similarity to published results.

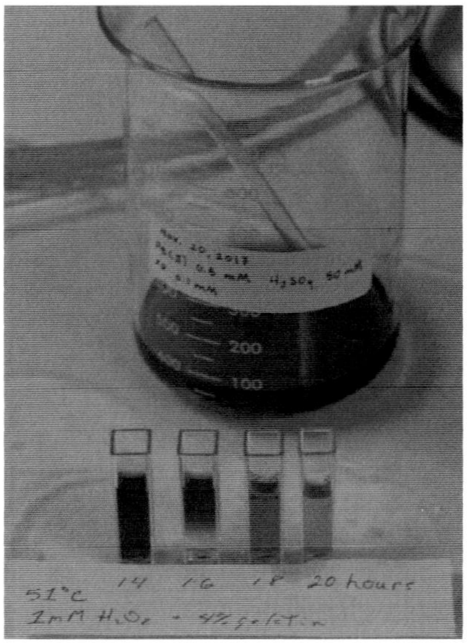


**Figure 8.** Photograph of colour changes as FX solution added to cuvettes of gelatin-water solution to measure residual hydrogen peroxide.

## 3. Summary

While gel preparation of several formulations is relatively simple and inexpensive to perform, achieving useful and reproducible results requires methodical development of preparation and calibration procedures. Essentially good lab practice is required to obtain reproducible results.

**Disclosure:** Authour, K Jordan has a licensing agreement with ModusQA concerning optical CT and radiation dosimetry.

## 4. References

[1] Farajollahi *et al* 2000 *Phys. Med. Biol* **45**, N9-N14
[2] Amin M N *et al* 2003 *Br. J. Radiol.* **76** 824-31
[3] Wuu C-S *et al* 2003 *Med. Phys.* **30** 132-7
[4] Hill B *et al* 2005 *Med. Phys.* **32** 1589-97
[5] Gear J I *et al* 2006 *Phys. Med. Biol.* **51** 3503-16
[6] Deman P *et al* 2011 *Phys. Med. Biol.* **56** 4465-80
[7] Tremblay N M *et al* 2011 *Med. Phys.* **38** 2754-61
[8] Gum F *et al* 2002 *Phys. Med. Biol.* **47** N67-N77
[9] Vergote K *et al* 2003 *Radiother. Oncol.* **67** 119-28
[10] Watanabe Y *et al* 2004 *Med. Phys.* **31** 975-84
[11] De Deene Y *et al* 2006 *Med. Phys.* **33** 2586-97
[12] De Deene Y *et al* 2013 *J. Phys.: Conf. Ser.* **444** 012026
[13] De Deene Y 1998 *Radiother. Oncol.* **48** 283-91
[14] Vergote K *et al* 2004 *Phys. Med. Biol.* **49** 287-305
[15] Trapp J V *et al* 2004 *Phys. Med. Biol.* **49** 1625-35
[16] Soliman Y S *et al* 2017 *Appl. Rad. Isotopes* **120** 126-32

[17] Iqbal K *et al* 2018 *Journal of Radiotherapy in Practice* **17** 96-103
[18] De Deene Y 2019 *Eur. J. Phys.* **40** 024001
[19] Adamovics J *et al* 2006 *Radiat. Prot. Dos.* **120** 107-12
[20] Oldham M 2018 In *Clinical 3D Dosimetry in Modern Radiation Therapy* Mijnheer B (ed) Boca Raton: CRC Press
[21] Jordan K 2010 *J. Phys.: Conf. Ser.* **250** 012043
[22] Olsson L E *et al* 1989 *Phys. Med. Biol.* **34** 43-52
[23] Schreiner L J *et al* 2004 *J. Phys.: Conf. Ser.* **3** 9-21
[24] Baldock C *et al* 1998 *Phys. Med. Biol.* **43** 695-702
[25] Baldock C *et al* 2010 *Phys. Med. Biol.* **55** R1-R63
[26] De Deene Y 2018, In *Clinical 3D Dosimetry in Modern Radiation Therapy* Mijnheer B (ed) Boca Raton: CRC Press
[27] Mather M L *et al* 2003 *Phys. Med. Biol.* **48** N269-75
[28] Jordan K *et al* 2009 *Phys. Med. Biol.* **54** 141-53
[29] Vandecasteele J *et al* 2011 *Phys. Med. Biol.* **56** 627-51
[30] De Deene Y *et al* 2015 *Phys. Med. Biol.* **60** 1543-63
[31] Hoye E M *et al* 2015 *Phys. Med. Biol.* **60** 5557-70
[32] Wheatley M J *et al* 2018 *Phys. Med. Biol.* **63** 215028
[33] Oldham M and Kim L 2004 *Med. Phys.* **31** 1093-104
[34] Bosi S G *et al* 2009 *Phys. Med. Biol.* **54** 275-283
[35] Olding T *et al* 2010 *Phys. Med. Biol.* **55** 2819-40
[36] Baldock C *et al* 2001 *Phys. Med. Biol.* **46** 449-60
[37] De Deene Y and Baldock C 2002 *Phys. Med. Biol.* **47** 3117-41
[38] De Deene Y and De Wagter C 2001 *Phys. Med. Biol.* **46** 2697-711
[39] Vandecasteele J and De Deene Y 2013 *Phys. Med. Biol.* **58** 63-85
[40] De Deene Y *et al* 2006 *Phys. Med. Biol.* **51** 653-73
[41] Hilts M *et al* 2000 *Phys. Med. Biol.* **45** 2559-71
[42] Jirasek A *et al* 2006 *Phys. Med. Biol.* **51** 2599-617
[43] Jirasek A *et al* 2010 *Phys. Med. Biol.* **55** 5269-81
[44] De Deene Y *et al* 2000 *Phys. Med. Biol.* **45** 859-78
[45] De Deene Y *et al* 2002 *Phys. Med. Biol.* **47** 2459-70
[46] Baldock C *et al* 2001 *Australas. Phys. Eng. Sci. Med.* **24** 19-31
[47] Vergote K *et al* 2004 *Phys. Med. Biol.* **49** 4507-522
[48] De Deene Y and Vandecasteele J 2013 *J. Phys.: Conf. Ser.* **444** 012015
[49] Karlsson A *et al* 2007 *Phys. Med. Biol.* **52** 4697-706
[50] De Deene Y *et al* 2002 *Phys. Med. Biol.* **47** 3441-463
[51] De Deene Y *et al* 2007 *Phys. Med. Biol.* **52** 2719-28
[52] Sedaghat M *et al* 2010 *J. Phys.: Conf. Ser.* **250** 012017

IOP Conf. Series: Journal of Physics: Conf. Series **1305** (2019) 012038 doi:10.1088/1742-6596/1305/1/012038

# 3D dosimetry for proton therapy

**S Beddar**
Department of Radiation Physics, The University of Texas MD Anderson Cancer Center, Houston, TX 77030, USA

E-mail: abeddar@mdanderson.org

**Abstract** We have been developing novel 3-dimensional (3D) detector systems using organic plastic and liquid scintillators to measure and image the dose distribution from proton therapy beams in near-real time. Proof-of-concept and initial feasibility studies using a single charge-coupled device camera have already been conducted. Our recent studies focused on the characterization of scanning proton beams used for patient treatments using a 3D liquid scintillator-based detector system with a set of scientific-complementary metal-oxide-semiconductor (sCMOS) cameras. The basic concept consists of using a large volume of a solid or liquid scintillator to measure or image the dose distributions from proton beams in 3D. We recently developed a large liquid scintillator-based detector system consisting of a 20 × 20 × 20-cm transparent acrylic tank filled with a water-equivalent, commercially available liquid scintillator that generates scintillation light when irradiated with protons. To track rapid spatial and dose variations in spot-scanned proton beams, we used 3 high-speed sCMOS cameras to image the scintillation light signals from 3 orthogonal projections in cine mode. Furthermore, we developed a new image acquisition approach that synchronized camera imaging times with dynamic pencil-beam deliveries to efficiently capture the dose and therefore enable accurate dosimetric calculations. This system was fully developed and characterized at the Proton Therapy Center at The University of Texas MD Anderson Cancer Center. We show that such systems can provide fast and accurate measurements of the range, lateral profile, and lateral position of scanning proton beams with excellent spatial resolution (0.21 mm). We also demonstrate that such detectors can rapidly measure proton beam characteristics and intensities at multiple energies, which makes them an ideal tool for scanned proton-beam systems, beam quality assurance studies, and verification of patient treatment delivery.

## 1. Introduction

While photons continue to be used for radiation therapy, the use of proton therapy has grown over the last few years. The steep dose gradient at the distal edge of the proton Bragg peak can be used to create treatment plans that accurately target tumors while minimizing dose exposure of healthy tissues. Spot-scanned or pencil-beam proton therapy is particularly attractive for delivering complex treatment plans because of its ability to precisely contour the tumor in 3 dimensions (3D) while avoiding critical organs and reducing the harmful neutron background [1]. Significant technological advances in proton beam delivery systems, however, have outpaced the development of clinical dosimeters, which play a vital role in safety and quality control for proton therapy. Ionization chambers are the prevailing gold-standard dosimeters. They are routinely used for quality assurance measurements of machines and treatment plans. However, ionization chambers provide single-point measurements that are time consuming, suffer from volume averaging effects, and are therefore poorly suited for mapping complex treatment plans. These problems can be partially overcome by using 2D configurations of ionization chambers, which have been used for mapping large-field dose variations and symmetry [2]. As alternatives to ionization chambers, Fricke and polymer gels have been studied for 3D chemical dosimetry [3], and commercially available 3D radiochromic gels have also been used for commissioning studies of small-field systems [4]. These dose-integrating dosimeters require long preparation times and are not reusable; therefore, they are not ideal for dynamic imaging applications, such as those of scanned proton-beam systems.

Organic scintillators, because of their water equivalency and energy independence, have been shown to be a good fit for precision dosimetry applications [5]. Early work with volumetric organic scintillators for 3D dosimetry highlighted several promising features, including high spatial resolution, good temporal resolution, and flexibility for applications involving photons, protons, and other heavy-ion particles [6,7]. These works focused on the development of a large-volume liquid scintillator (LS) coupled to optical cameras for 2D dosimetry of photons [8] and proton beams [6,7,9]. Organic scintillators, however, exhibit reduced light output at the Bragg peak due to their high ionization density. This phenomenon is called ionization quenching and is quantified as linear energy transfer for charged particles. We previously developed an ionization quenching correction for organic scintillators using the Birks equation [10]. In this work, we describe the design and development of a 3D detection system based upon an LS detector. We describe the detector design and evaluate its performance for proton-beam therapy.

## 2. Materials and methods

### *2.1. LS detector design*

The 3D detector consisted of a LS-filled tank (20 × 20 × 20 cm) constructed of acrylic. The transparent tank surfaces facing the cameras allowed the scintillation signal to be captured by cameras, while the opaque surfaces minimized the influence of ambient light on measurements. Optiphase HiSafe3 (PerkinElmer, Waltham, MA) was the choice of LS (0.986 g/cm$^3$) and doubled as a phantom and an active element for measuring the radiation field without perturbation. The reference coordinate system was oriented as shown in figure 1.a. As illustrated in the figure, camera *X* was positioned such that it pointed directly into the front surface of the tank perpendicular to the proton beam direction (the *z*-axis). Two mirrors located on the top and one side of the tank redirected the scintillation light to cameras Y and Z, respectively, for generating additional beam perspectives (Figures 1.a and 1.b). Cross-hair markers imprinted on the tank surfaces allowed geometric calibrations to correct for optical artifacts such as light refraction, lens perspective, lens distortion, and rotational/translational shifts between the camera and the tank [9]. The tank and cameras were securely mounted on an optical breadboard to make the system portable and robust.

### *2.2. Optical system design*

Three Zyla 5.5 scientific-complementary metal-oxide-semiconductor cameras (Andor Technology Ltd., Belfast, UK) were used for imaging. The camera sensors were equipped with 2560 × 2160 pixels, with an individual pixel size of 6.5 μm. A 16-bit analog-to-digital converter digitized the light signal. The cameras were capable of sustaining a full-frame transfer rate of 30 frames per second (fps) using a USB3.0 data transfer interface. Smaller regions of interest, such as the 1100 × 1100-pixel region of interest selected for this study, further increased the imaging rate (96 fps), resulting in better time resolution (10.33 ms) for the 3D detector. The cameras were fitted with 20.5-mm fixed-focal-length objective lenses (Schneider Optics, Van Nuys, CA). The lens selection was based on calculations of depth of field [7] and field of view. With a 750-mm working distance and f/8 aperture setting, a 20-cm focal depth situated around the center of the tank was obtained.

All the experiments were conducted using spot-scanned proton beams at the Proton Therapy Center at The University of Texas MD Anderson Cancer Center at Houston.

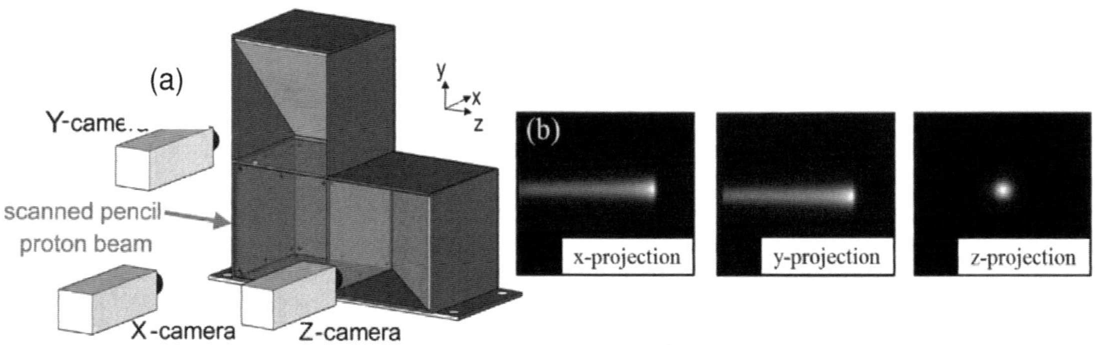


**Figure 1.** (a) Schematic of the 3D detector consisting of a liquid scintillator-filled tank, mirrors, and 3 sCMOS cameras. Two mirrors placed at 45° angle to the top and right tank surfaces redirected the scintillation light to cameras Y and Z, respectively. (b) Integrated light projections of a spot-scanned proton beam captured by the detector.

*2.3. Techniques to synchronize camera acquisitions with proton beam deliveries*

To capture the complete dose delivered to the detector, camera imaging cycles must be coordinated with dynamic proton-beam delivery intervals. However, the beam delivery system at our institution does not produce a compatible signal that can be used to directly trigger cameras. We therefore developed an Arduino microcontroller-based camera-triggering technique that used a combination of signals obtained from the beam delivery system to acquire images only during the duration of the beam delivery. These synchronization techniques offered the flexibility to either capture all the pencil-beam spots with identical energies in a single camera frame (the "spill-based" synchronization technique) or to capture individual beam-spots (the "spot-based" synchronization technique).

The spill-based camera synchronization technique used synchrotron-generated radiofrequency extraction START and STOP signals to trigger and terminate camera cycles (Figure 2) [11]. While the synchrotron produced proton energies ranging from 72.5 MeV to 221.8 MeV, each energy was delivered in a different "spill" of protons, and each spill could contain more than a single spot. This imaging approach combined all spots with identical energies into a single spill, which could last up to 4.4 s, in a camera frame. A 2.1-s inter-spill duration allowed the camera sufficient time to read and digitize the collected signal and be ready for the next spill. This synchronized imaging approach captured all the delivered spots and used significantly fewer camera frames (# frames = # spills).

The spot-based camera synchronization technique used an additional signal in the form of dose monitor (DM) pulses to allow imaging of individual proton beam spots (Figure 2). The DM pulses provided a count of the number of spots delivered by the treatment plan. A Schmitt trigger was used to eliminate the noise on the DM pulses before the microcontroller unit counted them. The microcontroller used an interrupt mode to monitor the DM pulse deliveries every 1 ms and thereby check the status of the spot delivery. The DM pulse readout by the microcontroller indicated active spot delivery, and the absence of DM pulses indicated completion of the spot delivery. At the end of spot delivery, the microcontroller sent a trigger signal to the camera to terminate the current frame and initiate a new frame. Using this method, the number of spots delivered by the synchrotron could be reliably counted and imaged.

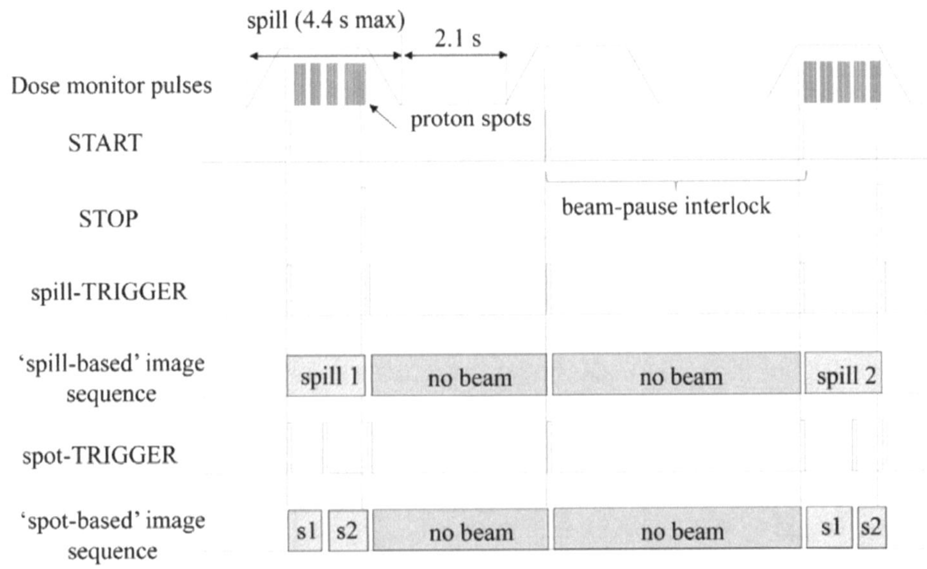


**Figure 2.** Timing diagram showing 2 techniques developed in this work for synchronizing camera acquisitions to beam deliveries. The proton beam was delivered in short bursts of doses (referred to as spots). The synchrotron-generated dose monitor (DM), START, and STOP signals were used for monitoring the status of the beam delivery. The spill-based imaging technique used the START and STOP signals to combine all the spots delivered during a beam spill into a single camera frame. The spot-based imaging technique used DM pulses to separate an individual proton spot or a user-defined combination of proton spots into a single camera frame. Errors introduced by the beam delivery system, indicated as a beam-pause interlock, could be accounted for with the developed techniques.

## 3. Results/Discussion

We conducted several studies using spill-based camera synchronization technique to analyze and quantify the performance of the 3D LS-based detector. The results are described in detail by Darne *et al.* (2017) [11]. Here, we briefly summarize the highlights of this detector. The detector had submillimeter imaging resolution (0.21 mm) for all 3 cameras and was found to be useful for generating complete light distribution profiles for the proton beams, which is especially useful around the high-gradient Bragg peak region. The synchrotron delivered pencil beams within a typical duration of 1 to 10 ms. A temporal resolution of 10.33 ms for the cameras was therefore critical for real-time imaging of dynamically delivered proton spots. Amongst other features, the detector demonstrated good stability (1%) over a period of 3 weeks, a linear signal-to-noise ratio ranging from 20 dB at 5 mGy to 85 dB at 10.6 Gy, and an excellent measurement linearity response (99.6%) over a 3-orders-of-magnitude change in the scintillation light intensity. The utility of the detector to support near-real-time measurements of proton ranges was also evaluated over all 94 synchrotron-generated beam energies. The average range measurement accuracy across all the beam energies was found to be $0.07 \pm 0.03$ mm [11]. The range measurement uncertainty for doses below 1 cGy was found to be $\pm$ 0.36 mm, indicating good precision for low dose measurements. The detector was able to rapidly calculate the beam ranges for all 94 energies in a clinically feasible time frame (< 10 min). The detector was also able to determine the size of each pencil spot from the full-width-half-maximum of the lateral beam profiles. Finally, the detector used proton beam perspectives collected from all 3 cameras to facilitate 3D location of the beam position within the detector with an average accuracy of 1 mm.

For an actual treatment plan, each energy layer can comprise a large number of individual spots delivered at different locations. If all these proton spots were combined into a single frame, the image analysis would become overly complicated. To overcome this problem, we developed the spot-based synchronization technique. The ability to isolate and analyze individual beam profiles within a treatment plan in 3D would help to detect errors. Furthermore, we found that the spot-based synchronization technique was able to capture the complete dose delivered by the beams, which could be lost with non-triggered image acquisitions because of unsynchronized imaging and the presence of camera deadtime. In addition, we found signal-to-noise ratio dependence such that low-intensity unsynchronized imaging pixels could lose as much as 10% ± 0.4% mean signal intensity compared to triggered acquisitions. Thus, synchronization prevented signal loss and improved the ability of the detector to measure dose in 3D. It also simplified image analysis and reduced the size of the generated imaging data sets.

We previously conducted a simulation study to demonstrate a 3D reconstruction of scintillation light emission using limited camera projections [12]. Our future work will involve imaging a patient treatment plan using a spot-based triggering technique with the developed maximum *a posteriori* algorithm for 3D reconstruction of the scintillation light. This light will be converted into precise dose by using an improved energy density by secondary electrons model for the ionization quenching correction [13]. In the future, we will apply this model to all 94 beam energies for accurate 3D dosimetry.

## 4. Conclusion

This study showed that an LS-based detector can be used for fast (10.33 ms), high-resolution (0.21 mm), accurate (< 0.2 mm) range prediction and precise mapping of dose distributions in real time. The camera-triggering schemes provided additional flexibility in collecting the pencil beams to simplify the image analysis. The proposed system, therefore, has the potential to enable efficient, high-resolution 3D quality assurance studies of clinical proton beams.

## 5. Acknowledgements

We would like to thank Chinmay Darne, Fahed Alsanea, Daniel Robertson, and Narayan Sahoo for their contributions to this project. We would also like to acknowledge Amy Ninetto from the Department of Scientific Publications at UT MD Anderson Cancer Center. The research reported in this publication was supported by the National Cancer Institute of the National Institutes of Health under Award Number R01CA182450.

## 6. References

[1] Xu XG *et al* 2008 *Phys. Med. Biol.* **53** R193
[2] Poppe B *et al* 2006 *Med. Phys.* **33** 1005
[3] Lepage M and Jordan K 2010 *J. Phys.: Conf. Ser.* **250** 012055
[4] Kevin J 2010 *J. Phys.: Conf. Ser.* **250** 012043
[5] Beddar AS *et al* 1992 *Phys. Med. Biol.* **37** 1883
[6] Beddar S *et al* 2009 *Med. Phys.* **36** 1736
[7] Archambault L *et al* 2012 *Med. Phys.* **39** 1239
[8] Ponisch F *et al* 2009 *Med. Phys.* **36** 1478
[9] Hui CK *et al* 2015 *Biomed. Phys. Eng. Express* **1** 025204
[10] Robertson D *et al* 2014 *Phys. Med. Biol.* **59** 23
[11] Darne CD *et al* 2017 *Phys. Med. Biol.* **62** 5652
[12] Hui C *et al* 2014 *Phys. Med. Biol.* **59** 4477
[13] Alsanea F and Beddar S 2017 *J. Phys.: Conf. Ser.* **847** 012022

IOP Conf. Series: Journal of Physics: Conf. Series **1305** (2019) 012039 doi:10.1088/1742-6596/1305/1/012039

# Comparison of low dose proton and photon irradiation induced polymerization processes in advanced nMAG gels using Raman spectroscopy

**N Šeperienė and D Adlienė**
Physics Department, Kaunas University of Technology, Studentu St.50, Kaunas, LT51368 Lithuania

E-mail: Diana.adliene@ktu.lt

**Abstract**. A rapid development of the new radiation based cancer treatment technologies that are more individualized and patient oriented request precise 3D treatment dose verification methods in high dose gradient fields. Polymeric dose gels are excellent 3D dosemeters that allow analysis of spatial dose distribution in the irradiated target and also in the regions out of the target since the investigation of low dose induced secondary effects (secondary cancer, inflammations) stands on the priority list. Since the sensitivity of dose gel to different types of irradiation depends on radiation induced polymerization processes, in this work we present Raman analysis results related to advanced nMAG dose gels polymerization upon its irradiation with high energy photons and protons to doses $\leq 5$ Gy.

## 1. Introduction

A growing up variety of different radiation based cancer treatment methods accelerates the development of the new dosimetry methods, new detectors for dose assessment and materials for these dosimeters. Well-known methods and dose registration devices (dosimeters) are also applicable if their characteristics meet the new requirements. It is known [1-3] that being nearly tissue equivalent polymer gel dosemeters can provide precise and high-resolution verification of complex 3D dose distributions in clinical radiotherapy, if adjusted to the type of radiation. A number of formulations of the radiation sensitive polymer gels that might be used as 3D dosimeters have been proposed indicating high potential for dose gel applications for 3D dosimetry in photon radiotherapy and are still being explored [4]. However it is evident that in special cases particle therapies (protons, light ions, neutrons) are more effective for cancer treatment. Proton therapy is becoming increasingly useful as an external beam radiotherapy modality because of its characteristic dose distribution. Taking into account that neutral photons interact with matter in different way as the protons do, at least in the sense of different LET (linear energy transfer effect), a question arises whether well-known dose gels [5] are sensitive enough to provide information on proton induced 3D dose distribution in these materials. Direct response to proton beam LET in novel polymer gel dosimeters has been already reported [6] however the problem of dose gel sensitivity to proton irradiation remained. In this frame investigation of proton induced polymerization processes in dose gels is of great importance. The usual dose response analysis in irradiated gel dosimeters is performed using certain modalities including MRI, X-ray computed tomography, optical CT, ultrasound. Raman spectroscopy might be also used for the evaluation of dose response in polymeric gels, but this technique is more dedicated and has high potential for the analysis of radiation induced changes in polymerized gel structures [2, 3]. The polymerization sensitivity of the irradiated gel can be calculated from the intensity of the characteristic Raman bonds. This parameter provides irradiation dose related information on the efficiency of monomer consumption and polymer formation within the gel. Polymerization sensitivity for photon irradiated acrylamide or methacrylic acid based gels have been reported in [7, 8], however in the case of proton irradiation the most research is focused on the investigation of dose - LET dependencies in irradiated polymer gels and evaluation of the other irradiation parameters (RBE) [6, 9, 10].

Despite the availability of a certain amount of the data related to proton irradiated polymer gels, low dose (up to 10 Gy) induced changes in the formed polymeric structures are not investigated in details yet. Also the assessment of the dosimeter's suitability for low dose evaluation requires more targeted research of proton induced polymerization processes in gels, since fractionation and proton doses per fraction are different as compared to standard 2 Gy dose per fraction in photon radiotherapy.

The aim of this work was to compare photon and proton beam induced polymerization process in Methacrylic acid based advanced normoxic dose gels (nMAG) using Raman spectroscopy and assess their feasibility to serve as a 3D dosimeters also for the evaluation of low dose distributions in the "out of the target" regions irradiated during treatment of cancer patients. Raman spectroscopy has been chosen for the analysis since it is non-invasive method, is not affected by the presence of water in evaluated samples and therefore is advantageous when analyzing soluble and/or biochemical substances.

## 2. Materials and methods

### *2.1. Dose gel fabrication*

Using UV-VIS spectrometry nMAG gels containing of 8% w/w of Methacrylic acid (MMA, 99%, Sigma-Aldrich); 2 mM of (hydroxymetyl) phosphonium chloride (THPC, Sigma-Aldrich); 6% w/w of gelatin (300 Bloom Type A, Sigma-Aldrich) and 86% of pure water were identified as being most sensitive to photon irradiation. Due to this, gels of the same composition were prepared for the investigation of photon and proton irradiation induced polymerization processes following usual gel fabrication procedure. Prepared gels were poured into PMMA standard disposable cuvettes (BRAND®, 45mm x 12.5mm x 12.5mm), left to settle for a 5 min and then tightly closed and left for 24 hours in dark and cool place.

### *2.2. Irradiation of gels*

*2.2.1. Proton irradiation of gels.* Experimental samples of nMAG dose gels were irradiated in proton accelerator IBA PROTEOS C230 to 230 MeV at OncoRay Clinic in Dresden/Germany. Proton irradiation geometry was realized keeping gantry and treatment table orientations at $0^{\circ}$ position. Farmer chamber PTW30013 with build-up cap was fixed in a central position of the irradiation field, which was formed applying 15.0 cm snout. This ionization chamber was used for the independent registration of proton irradiation dose which was assessed using electrometer UNIDOS E (PTW). Dose correction factor $k_0$=1.022 was calculated for this chamber taking into account atmospheric conditions: p=1002.1 hPa and T=23.1°C. In each irradiation experiment two couples of cuvettes belonging to different batches with the same type of gel were positioned on the treatment table parallel to the ionization chamber on both sides of it. Arranged cuvettes were surrounded by 1.50 cm thick PMMA blocks and covered by the additional 10.3 cm thick PMMA block (corresponds to the thickness of 11.19 cm in water) in order to secure 10.0 cm width of modulated SOBP with a total proton beam penetration depth of 17.0 cm. Gel samples were irradiated to doses from the interval 0.5 - 5.0 Gy.

*2.2.2. Photon irradiation of gels.* The irradiation of gel samples with photons was performed in medical linear accelerator Clinac DMX (Varian) at Oncology hospital of Lithuanian University of Health Sciences. Following irradiation parameters were used: max dose depth was 1.5 cm, source to surface distance was set to 100 cm, field size 10×10 $cm^2$, dose rate - 3 Gy/min. 1 monitor unit (MU) of the Linac corresponded to 0.01 Gy on average, however using calibrated ionization chamber the value of MU in terms of Gy was adjusted for each treatment procedure separately [11].

*2.3. Evaluation of gels by Raman spectroscopy*

In contrast to small - molecule compounds, in polymer molecules the atoms are linked together to form longer chains. The presence of such long chains causes additional vibrational modes of the whole chain that might be assessed using FT- Raman spectroscopy. It is well-known [12], that Raman spectroscopy is more sensitive to highly polarizable C-C and C=C groups, transformation of which during irradiation of gels play the most important role in gel polymerization process. FT- Raman scattering spectrometer (RENISHAW in Via Raman microscope) equipped with DPSS laser ($\lambda$=532 nm) was used for gel evaluation. Stokes line measurements were performed within the range of 250 cm$^{-1}$ and 3600 cm$^{-1}$, resolution of measurements was better than 1 cm$^{-1}$. Silicon wafer was used to calibrate the Raman spectrometer for both Raman wavenumber and spectral intensity.

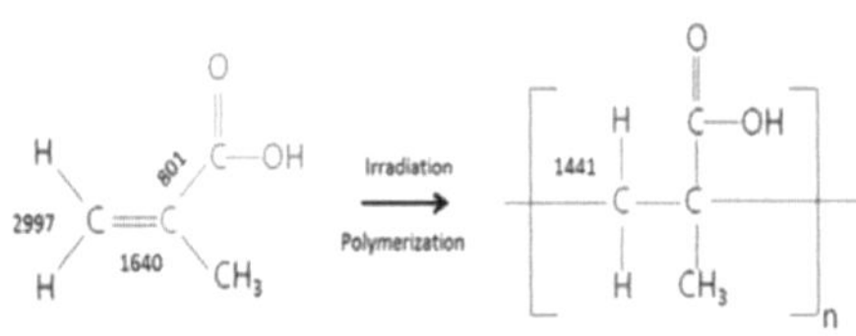


**Figure 1.** Basic polymerization process in MAG dose gels (adapted from [6]).

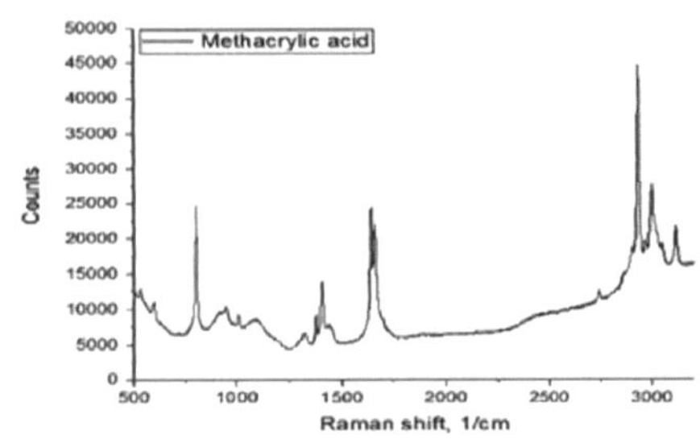


**Figure 2.** Raman spectrum of pure MAA (used for preparation of dose gels) with corresponding peaks: 801cm$^{-1}$- ν (C-COOH), 1403 cm$^{-1}$- δ ($CH_2$), 1441 cm$^{-1}$- ν $(CO)_s$, 1640 cm$^{-1}$- ν (C=C), 2937 cm$^{-1}$ and 2997 cm$^{-1}$- ν $(CH_2)_s$.

Irradiated nMAG gels were analysed 2 days after the irradiation. This time was needed for transportation of gel samples to host institution (OncoRay Clinic, Dresden, Germany) for proton irradiation and back, and was sufficient enough to complete photon or proton radiation induced polymerization process within the gel samples. Since nMAG gel does not contain any cross linker, two monomers of methacrylic acid (MAA) are interacting to start polymerization and precipitation of polymethacrylic acid (PMAA) upon irradiation (figure 1) thus forming small insoluble polymeric structures embedded in gelatine matrix. In order to follow polymerization process, characteristic frequencies (Raman peaks) of gel constituents before irradiation were evaluated analysing Raman spectra (figure2) or using information provided by other researchers [13, 14].

## 3. Results and discussion

It is known that carbon covalent bonds are tending to break into carbon single bonds upon irradiation (figure 1). Due to this the number of carbon single bonds increases, while the number of covalent carbon bonds decreases with the increasing radiation dose, indicating ongoing polymerization process. Analysis of dose dependent Raman peak intensity changes allows for the estimation of dose gel sensitivity parameters. It should be noted, that in order to enhance the sensitivity of experimental dose gels relative high amount of MAA (8%) was added to gel and the amount of gelatin was reduced to 6%. It was done taking into account that irradiated to higher doses gelatin starts interacting with methacrylic acid and forms polymer derivatives complicating the evaluation of dose gels. Also the polymerization level of dose gel is reduced due to additional consumption of methacrylic acid monomers in reactions with gelatin [15].

Raman spectra of the advanced nMAG dose gels after their irradiation to photon and proton beams are provided in figure 3 and figure 4 respectively.

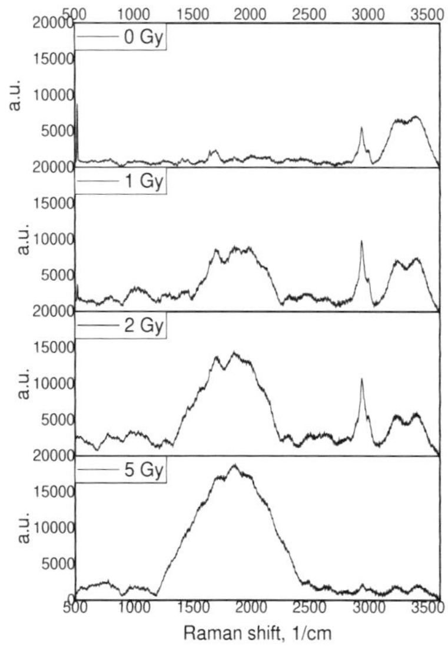


**Figure 3**. Raman spectra of photon irradiated nMAG gels

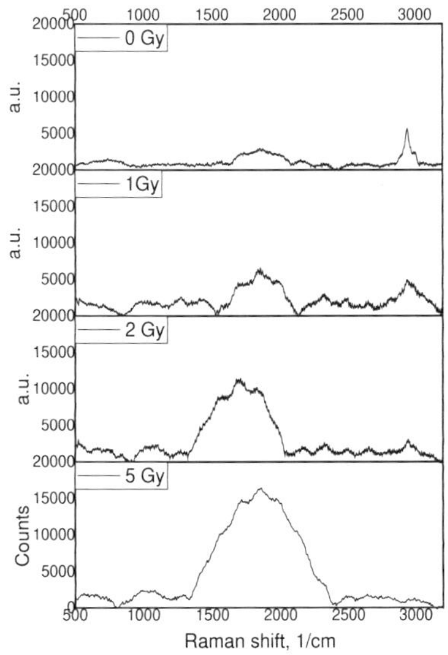


**Figure 4.** Raman spectra of proton irradiated nMAG gels

It was found that the most important Raman peaks of irradiated gels were slightly shifted towards higher Raman wavelength values and their exact positions were slightly altering with the dose. The overall shape and Raman spectrum development tendencies after gel irradiation were similar for both: photon and proton irradiated gels. A broad band seen in spectra between 1200 cm$^{-1}$ and 2500cm$^{-1}$ was attributed to gelatin and imparted structures; another one seen between 3100 cm$^{-1}$ and 3500cm$^{-1}$ was related to O-H vibrational modes in water. Growing Raman intensity with increasing irradiation dose indicated growing level of polymerization in irradiated gel. The broadening of the "gelatin" band indicated rearrangements in the gel structure and growing amount of formed polymerized derivatives embedded in the gelatin matrix. The increase of "gelatin" band intensity and its broadening was less expressed for gels irradiated to different proton doses, thus indicating lower sensitivity of gels to proton irradiation. It was explained by the fact that the decreased sensitivity of polymer gels may be attributed to a larger density of created polymer radical chains close to the proton tracks that favors a faster termination of the polymerization reactions. Occurrence of a sharp high intensity peak observed in initial gels around 2940 cm$^{-1}$ was related to relative high amount (8%) of MAA, which was added to gel in order to increase its overall sensitivity. The difference in Raman spectra (at 0 Gy) for reference samples thought for proton and photon irradiation was caused by the fact that all (incl. reference) to proton irradiation facility transported samples underwent additional X-ray exposure during inspections at the airport. Raman spectrum of the transported reference samples was later on used for the evaluation of radiation induced changes in proton irradiated gel samples.

### *3.1. Evaluation of photon irradiated nMAG gels using Raman spectroscopy*

Relatively well pronounced Raman peak at 809 cm$^{-1}$ observed in all photon irradiated nMAG gel samples (figure 3) records symmetric stretching mode of C-COOH representing a bond between carboxylic carbon and double bond carbon. The bond connecting two carbons is formed by the presence of a sp$^2$ hybrid orbital. Due to the energy supply during photon irradiation it changes to a sp$^3$ hybrid orbital and becomes weaker. This transformation records monomer consumption which is present when radiation induced gel polymerization process starts. In general monomer consumption lead to the reduction of the vibrational mode intensity with the increasing dose, however the overall peak intensity does not drop to zero, especially at low irradiation doses, since a certain amount of monomer is still left

IOP Conf. Series: Journal of Physics: Conf. Series **1305** (2019) 012039 doi:10.1088/1742-6596/1305/1/012039

in the gel after irradiation. Another band used for verification of monomer transformation to polymer was caused by C=C stretching vibrational mode and was observed at 1687 $cm^{-1}$ [7, 16]. A C=C double bond changes to a single C-C bond when polymerization proceeds and is followed by the gradual decrease of the vibrational mode intensity. The identification of this C=C Raman peak might be complicated since the broad band (1200 $cm^{-1}$ - 2500 $cm^{-1}$) corresponding to gelatin and its constituents in gels can overlap it. This problem was not acute analyzing photon irradiated gels since well-established Raman peak at 1687 $cm^{-1}$ was found in all analyzed nMAG gels irradiated to different doses. It was decreasing smoothly with the increasing radiation dose, indicating effective consumption of monomer. Spectral information regarding photon induced polymerization behavior of methacrylic acid was also obtained analyzing intensive band appearing at 2937 $cm^{-1}$, which was assigned to symmetric vinyl stretching mode $\nu(CH_2)_s$. This vibrational mode is the only one which does not overlap with other bands [16] thus providing "true" information about polymerization process. The peak at 1450 $cm^{-1}$ (figure 3) most probably may be attributed to the $CH_2$ vibrational bending mode of polymethacrylic acid (PMAA). As expected the intensity of this band was increasing with the increasing photon irradiation dose, since it records formation of polymeric structure.

### *3.2. Evaluation of proton irradiated nMAG gels using Raman spectroscopy.*

A week peak at 771 $cm^{-1}$ corresponding to symmetric stretching mode of C-COOH was observed in proton irradiated gels (figure 4). Since the location of this peak was also altering with a dose, it was impossible to provide evidence based statement regarding monomer consumption in proton irradiated nMAG gels using this particular band. Decreased intensity of the 1695 $cm^{-1}$ peak was attributed to the C=C stretching vibrational mode changes to a single C-C bond. Only modest changes of 1695 $cm^{-1}$ peak intensity with the increasing dose were observed due to the waste contribution of the overlapping "gelatin" band. Well-expressed peak at 2940 $cm^{-1}$, which was assigned to symmetric vinyl stretching mode $\nu(CH_2)_s$, was the only one which clearly indicated peak intensity decrease with the increasing dose thus allowing the evaluation of monomer consumption due to polymerization in proton irradiated nMAG gels. However a sharp peak was observed only for initial gels. It became broader after irradiation indicating formation of a number of polymer radical chains that favor a faster termination of the polymerization reactions. A band with modest increasing Raman peak intensity due to increase of irradiation dose was found at 1410 $cm^{-1}$. It was assigned to the $CH_2$ vibrational bending mode of PMAA, whose changes record formation of polymeric structure.

### *3.3. Tendencies of monomer consumption /polymer production in irradiated nMAG gels.*

The tendencies for monomer consumption and polymer production in photon and proton irradiated nMAG gels are provided in figure 5 and figure 6 respectively.

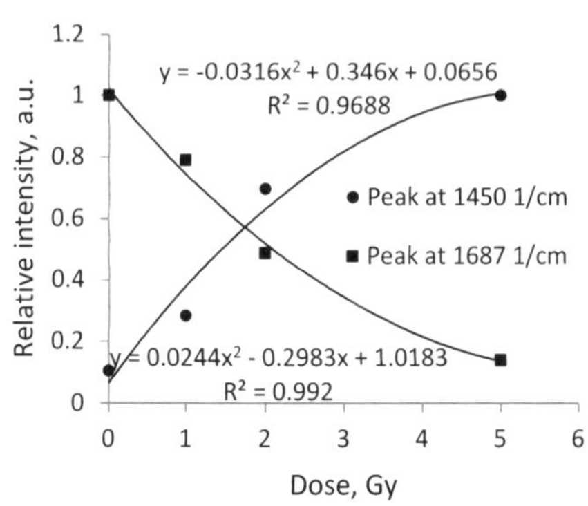


**Figure 5.** Monomer consumption (peak at 1687 $cm^{-1}$) and polymer formation (peak at 1450 $cm^{-1}$) in photon irradiated nMAG gels.

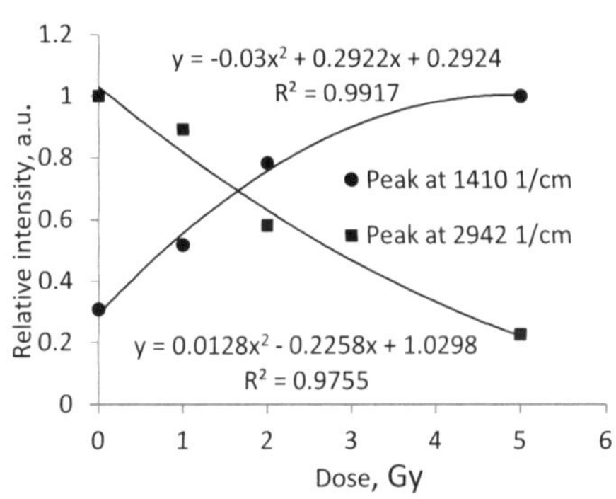


**Figure 6.** Monomer consumption (peak at 2942 $cm^{-1}$) and polymer formation (peak at 1410 $cm^{-1}$) in proton irradiated nMAG gels.

Investigation revealed that even at low irradiation doses monomer consumption and polymer formation processes in irradiated gels were not following linear tendency. It was found that monomer consumption and polymer formation rates were almost the same in photon irradiated gels and were slightly higher than those estimated for proton irradiated gels. This indicates relative higher dose sensitivity of nMAG gel to irradiation with 15 MeV photons.

Faster consumption of monomer as compared to polymer production in proton irradiated gels was observed. It was explained by the fact that polymerization takes place very locally in the vicinity of proton path. Due to the high LET, monomer consumption in proton irradiated nMAG gels is not uniform (figure 6). The relative intensity of characteristic Raman peaks drops down due to the significant changes in the bonding structure of dose gels along the high energy proton track and possible local polymerization. However not all proton irradiation produced chemical fragments are tending to form polymeric chains. Some of them may participate in a reverse process of monomer formation since a large number of produced fragments are concentrated very locally. Dose (absorbed energy) dependent competing processes: polymerization and monomer formation contribute to the overall modest polymer formation rate as compared to monomer consumption.

## 4. Conclusions

Performed investigation revealed that the advanced nMAG gels containing 8% of methacrylic acid and 6% of gelatin were sensitive enough to assess gel polymerization caused by low dose photon and proton irradiation, however proton irradiated gels were less sensitive to irradiation due to the specific high LET proton interaction processes within gels. It was found that even at low irradiation doses polymerization level was not linearly increasing with the increasing dose. Monomer consumption and polymer production rates were almost the same in photon irradiated gels. In the case of proton irradiation monomers were destroyed (breaking of carbon double bonds) relatively fast, but not all chemical fragments were tending to form polymer chains. Presence of two competing processes in proton irradiated gels: monomer formation and polymerisation, led to for relative modest polymer formation rate. The reverse process of monomer formation was possible due to the fact, that a large number of radicals were produced very locally in the nearest vicinity of proton tracks.

## 5. Acknowledgements

This work was partly supported by the research grant S-MIP-17-104 of the Lithuanian Research Council. Authors are grateful to prof. Wolfgang Enghardt (OncoRay TU Dresden) for sample irradiation with protons.

## 6. References

[1] Massillon G and Minniti R 2010 *App. Rad. Isot.* **68** 144-54
[2] Baldock C and De Deene Y 2010 *Phys. Med. Biol.* **55** R1-R63
[3] De Deene Y 2018 Chapter 5 Polymer gel dosimetry in *Clinical 3D Dosimetry in Modern radiation Therapy* 674 p. ed B Mijnheer (Boca Raton: CRC Press)
[4] Baldock C 2009 *J. Phys.: Conf. Ser.* **164** 012002
[5] Lopatiuk-Tirpak O et al. 2012 *Technology in Cancer Research and Treatment* **11** 441-5
[6] Jihye B and Kyuseok C 2011 *Bull. Korean Chem. Soc.* **32** 625-9
[7] Adenan G M Z *et al* 2015 *Advanced Materials Research* **1107** 103-7
[8] Gustavsson H *et al* 2004 *Phys. Med. Biol.* **49** 3847-55
[9] Jihye Bong and Dongho Shin 2014 *Nucl. Instr. Meth. A* **735** 314-18
[10] Marshall T I *et al* 2016 *Int. J Rad.Oncol. Biol. Phys.* **95** 71
[11] Adliene D, Jakstas K and Vaiciunaite N 2014 *Nucl. Instr. Meth. A* 741 88-94
[12] Snively C M and Koenig J L 1999 Polymer applications of IR and Raman spectroscopy. In:

*Vibrational, rotational and Raman spectroscopy*. CWR University, Cleveland, USA
[13] Adenanm Z *et al* 2014 *J Phys: Conf. Series* **546** 012011
[14] Jirasek A and Duzenli C 2002 *Med. Phys.* **29** 569-577
[15] Todica M *et al* 2015 *Acta Phys. Polonica* **128** 128-35
[16] Murli C and Song Y 2010 *J. Phys. Chem. B* **114** 9744-50

IOP Conf. Series: Journal of Physics: Conf. Series **1305** (2019) 012040 doi:10.1088/1742-6596/1305/1/012040

# Clear micelle gel dosimeter with nanoclay

**K Hayashi[1], M Nemoto[2], T Takanashi[3], Y Kang[4], H Togo[4], J Kotoku[4], T Kobayashi[4], M Mihashi[1], S Hayashi[5] and H Gotoh[1]**

[1]Department of Chemistry and Life Science, Graduate School of Engineering Science, Yokohama National University, 79-1 Tokiwadai, Hodogaya-ku, Yokohama, Kanagawa, 240-8501, Japan

[2]Department of Radiology, Jichi Medical University Hospital, 3311-1 Yakushiji, shimotsuke-shi, Tochigi, 329-0498, Japan

[3]3D Gel Dosimeter Research Laboratory, Cluster for Science, Technology and Innovation Hub, RIKEN, 2-1, Hirosawa, Wako, Saitama 351-0198, Japan

[4]Graduate School of Medical Care and Technology, Teikyo University, 2-11-1 Kaga, Itabashi-ku, Tokyo, 173-8605, Japan

[5]Department of Clinical Radiology, Faculty of Health Sciences, Hiroshima International University, Higashi-Hiroshima, Hiroshima, 739-2695, Japan

E-mail: gotoh-hiroaki-yw@ynu.ac.jp

**Abstract**. We built a clear micelle gel dosimeter with nanoclay. Jordan et al. reported that Laponite RD clay nanoparticles when added to radiochromic leucomalachite green micelle gels eliminate diffusion and increase the dose sensitivity by roughly ten folds. However, owing to the cloudiness of the sample, there was a problem in reading the optical computed tomography (CT). In this study, we constructed a nanoclay-added micelle gel dosimeter by changing the type of gelatin. As a result, in addition to yielding a clear gel and making the optical CT readable, diffusion and temporal stability were improved more than the gel without using nanoclay, and the dosimeter showed the same level of sensitivity and diffusion as the one based on Laponite-added micelle gel reported by Jordan et al.

## 1. Introduction

Radiochromic gel dosimeters - micelle gel dosimeter, polymer gel dosimeter, and Fricke gel dosimeter - are one of the promising tools for verification of three-dimensional (3D) dose distribution [1]. Micelle gel dosimeter has several advantages over other types of dosimeters, such as, no or low diffusion, insensitivity to oxygen, lower toxicity, and suitability for optical computed tomography (CT) read-out. Such dosimeters based on both leucocrystal violet (LCV) and leucomalachite green (LMG) have been reported [2, 3]. However, dose sensitivities of these gels are low. To resolve this problem, addition of clay to radiochromic micelle gel has been studied. It has been reported by Jordan *et al* that clay nanoparticles eliminate diffusion and increase dose sensitivity [4]. However, owing to the cloudiness of the sample, there was a problem in reading the optical CT.

The aqueous solution with nanoclay is transparent, and a high concentration solution with nanoclay directly forms a colorless gel [5]. It has been reported by Jordan *et al* that, when preparing micelle gel dosimeter by adding Laponite RD clay nanoparticles, the micelle gel produces scattering signals. The more the number of scattering signals, the more is their influence, as seen in a polymer gel dosimeter. This influence cannot be ignored and it is considered that this might interfere with the measurement of

IOP Conf. Series: Journal of Physics: Conf. Series **1305** (2019) 012040 doi:10.1088/1742-6596/1305/1/012040

an accurate 3D dose distribution, resulting in, for example, observation of cupping artifact [6]. In this study, in order to prepare a new micelle gel dosimeter by adding nanoclay and with the aim of obtaining a clear read-out, we changed the type of gelatin, and analysis of the gelatin was conducted to determine its influence on the gel dosimeter. Moreover, we investigated the sensitivity, diffusion, and temporal stability of this gel after subjecting it to X-ray irradiation.

## 2. Materials and methods

### 2.1. *Gel fabrication*
In this study, LCV (Wako Pure Chemical Inc., Japan) was used as a dye for radiochromic micelle gel. We used two types of gelatin: one was procured from Sigma-Aldrich (porcine, 300 bloom, USA), which was used in previous reports, and another was the guaranteed reagent of gelatin supplied by Nacalai Tesque Inc. (cattle bones, 150 bloom, Japan). The gel formulation comprised of 5% gelatin, 0.16% Laponite, 0.3 mM LCV, 50 mM dichloromethane, and 50 mM Triton X-100. Laponite XLG (BYK japan KK) was added to water and stirred to form a clear colorless solution in an oil bath at 50°C. The gelatin was slowly added to the Laponite solution and stirred in the same condition for 1 h. After the gelatin dissolved, the solution was cooled at 35°C. The dye was added to dichloromethane and mixed under light shielding. The dye solution was poured into Triton X-100 (Alfer Aesar, USA) and stirred at room temperature. The gelatin solution was added slowly to the surfactant solution and poured into 1-cm PMMA cuvettes or polyethylene terephthalate jars (900 mL) supplied by Modus Medical Devices. The samples were stored at 4°C for gelation.

### 2.2. *Gelatin analysis*
For investigating the composition of the two types of gelatin, element analysis (Elementar, Germany) and amino acid analysis using HPLC were carried out.

### 2.3. *Irradiation*
The samples were irradiated by 10-MV X-rays from a linear accelerator (Clinac iX Linear accelerator, Varian). For studying the sensitivity, some of the cuvettes were irradiated at 0–10 Gy. Additionally, another LCV sample was irradiated with 1000 MU (10 Gy) at the half-beam irradiation field with equal power of the upper and lower irradiation.

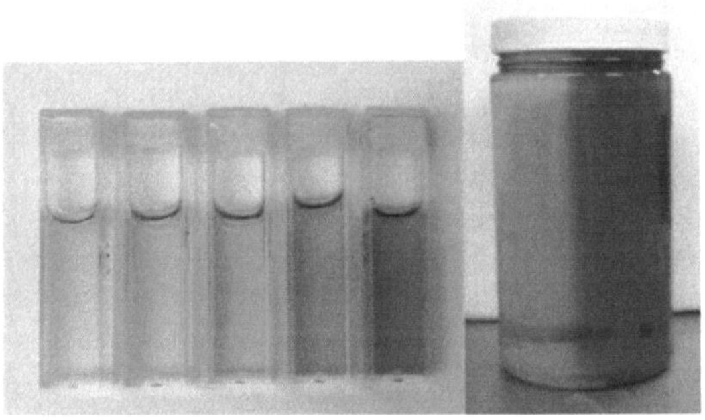

**Figure 1.** Right: 1-cm PMMA cuvettes irradiated by X-rays. Left : polyethylene terephthalate jars (900 mL) irradiated by X-rays.

### 2.4. *UV-vis absorption measurement*
For investigating the sensitivity and scattering signals, all the irradiated cuvettes were studied using UV-vis absorption spectrometer (JASCO, Japan). The reference was measured using water.

IOP Conf. Series: Journal of Physics: Conf. Series **1305** (2019) 012040 doi:10.1088/1742-6596/1305/1/012040

### 2.5. *Optical CT scanning*

Optical CT scanning was conducted with a commercial optical cone-beam scanner, model Vista 15 from Modus Medical Devices. In all the measurements, the reference and data scans were taken with 590 nm and 633 nm LED illumination, which is the maximum absorption wavelength of crystal violet (CV), and a 1024 × 768 pixel CCD camera. 342 image sets were reconstructed using a software provided by Vista15 scanner. The samples were stored at 4°C and brought to room temperature at the time of measurement.

## 3. Results and discussion

In the measurement of the scattering signals for UV-vis absorption, absorbance of the samples using the gelatin provided by Nacalai Tesque Inc. was weak at 590 nm, which is the maximum absorption wavelength of CV, and the values were similar to those without using nanoclay. On the other hand, the absorbance of the samples using the gelatin made by Sigma-Aldrich, which was used by Jordan et al., was generally stronger (figure 2).

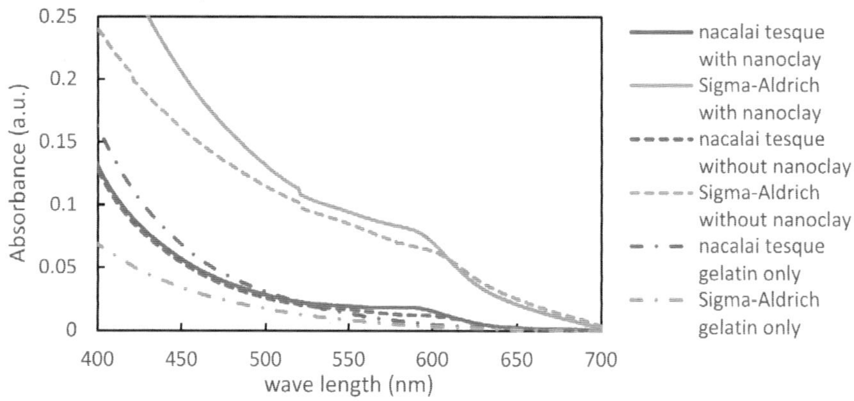


**Figure 2.** Difference in absorbance for different types of gelatin.

As the measurement wavelength became shorter, the difference in absorbance between the two types of samples increased. These results indicated that the samples with gelatin provided by Nacalai Tesque Inc. have lower scattering signals than those by Sigma-Aldrich. The absorbance of pure gelatin samples of both Sigma-Aldrich and Nacalai Tesque Inc. were weak. A comparison of the photographs of the two samples suggest that the sample with gelatin provided by Sigma-Aldrich was more cloudy (figure 3).

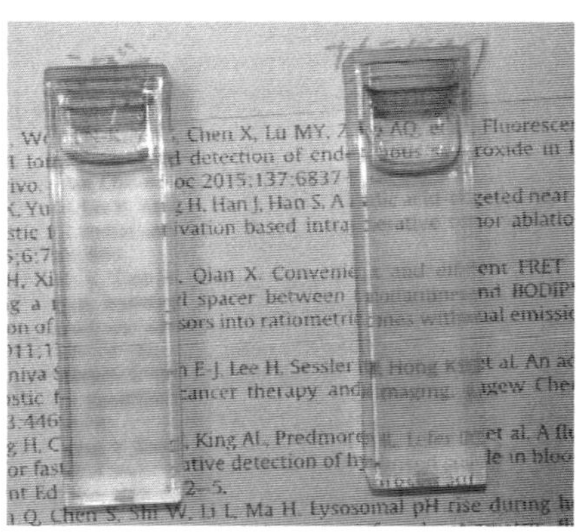

**Figure 3.** Photograph of the two samples. Left: Sigma-Aldrich with nanoclay Right: Nacalai Tesque Inc. with nanoclay.

IOP Conf. Series: Journal of Physics: Conf. Series **1305** (2019) 012040 doi:10.1088/1742-6596/1305/1/012040

These results indicate that difference in amino acid composition of the gelatins was caused by the cloudiness resulting from mixing with other substances (LCV, dichloromethane, and/or Triton X-100). The results of elemental analysis of gelatin (figure 4 left) and amino acid analysis (figure 4 right) are shown.

| | C (%) | N (%) | H (%) | Others (%) |
|---|---|---|---|---|
| Sigma Aldrich | 43.94 | 16.40 | 7.237 | 32.42 |
| Nacalai Tesque | 44.04 | 16.01 | 7.163 | 32.79 |

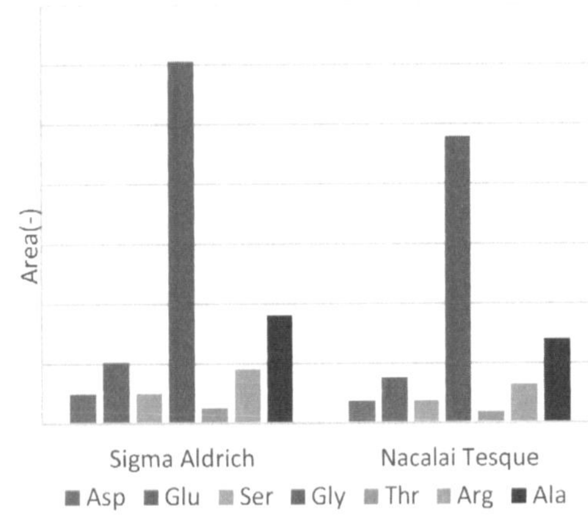


**Figure 4.** Results for two gelatins. Left: Elemental analysis and Right: Amino acid analysis.

The elemental analysis showed no significant difference between the two types of gelatins. Moreover, there is a possibility that the method of fabrication and type of nanoclay added also affect the cloudiness. In amino acid analysis as well, there were no considerable differences between the two types of gelatins in terms of the constituent amino acid type and its ratio. These observations suggest that impurities such as metals contained in gelatin and trace components derived from raw materials may affect the properties of the gel dosimeter.

Sensitivity to radiation was up to 10 Gy (5 Gy/min), as shown in figure 5. Compared with the case without nanoclay, the sensitivity was improved, and linearity was also observed.

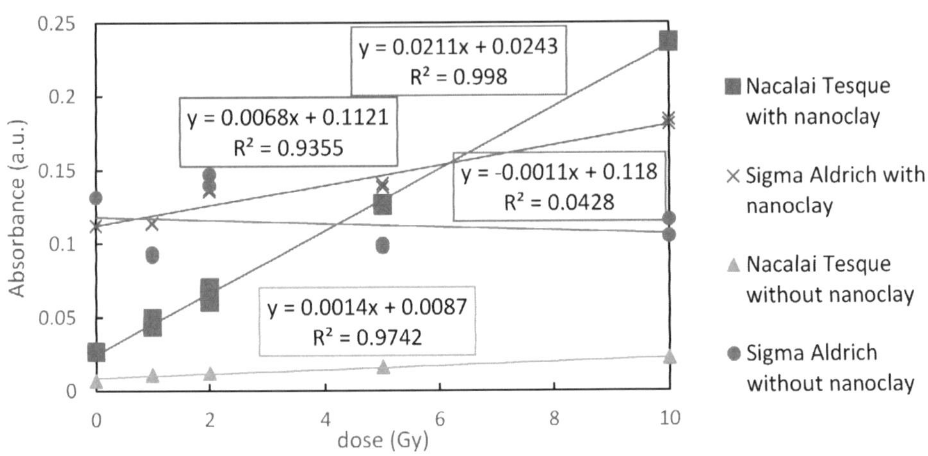


**Figure 5.** Dose response of micelle gel dosimeters.

The sensitivity to X-rays obtained in this experiment was of about the same level as that reported by Jordan et al. In experiments with different dose rates (5 Gy/min vs 1.25 Gy/min), the absorbance of the samples with low dose rate was about 10% larger than that of the samples with high dose rate.

Half of the samples was irradiated in the vertical direction with respect to the container of the LCV sample, and the measurement was carried out by optical CT. The measured Vista CT at 590 nm was about 1.0 in the irradiated region, and it was practically readable. It was observed that the measurement result at the center of the irradiation area was smaller than the calculated value of the treatment planning device (figure 6, top left). This is thought to be because of the influence of the dye's scattering, but this issue is currently under investigation. On the other hand, the measured Vista CT at 633 nm was about 0.1 in the irradiated region (figure 6, bottom left). Because the influence of noise was large, the measured values of the irradiated region were observed to fluctuate, but roughly coincided with the calculated value of the treatment planning device.

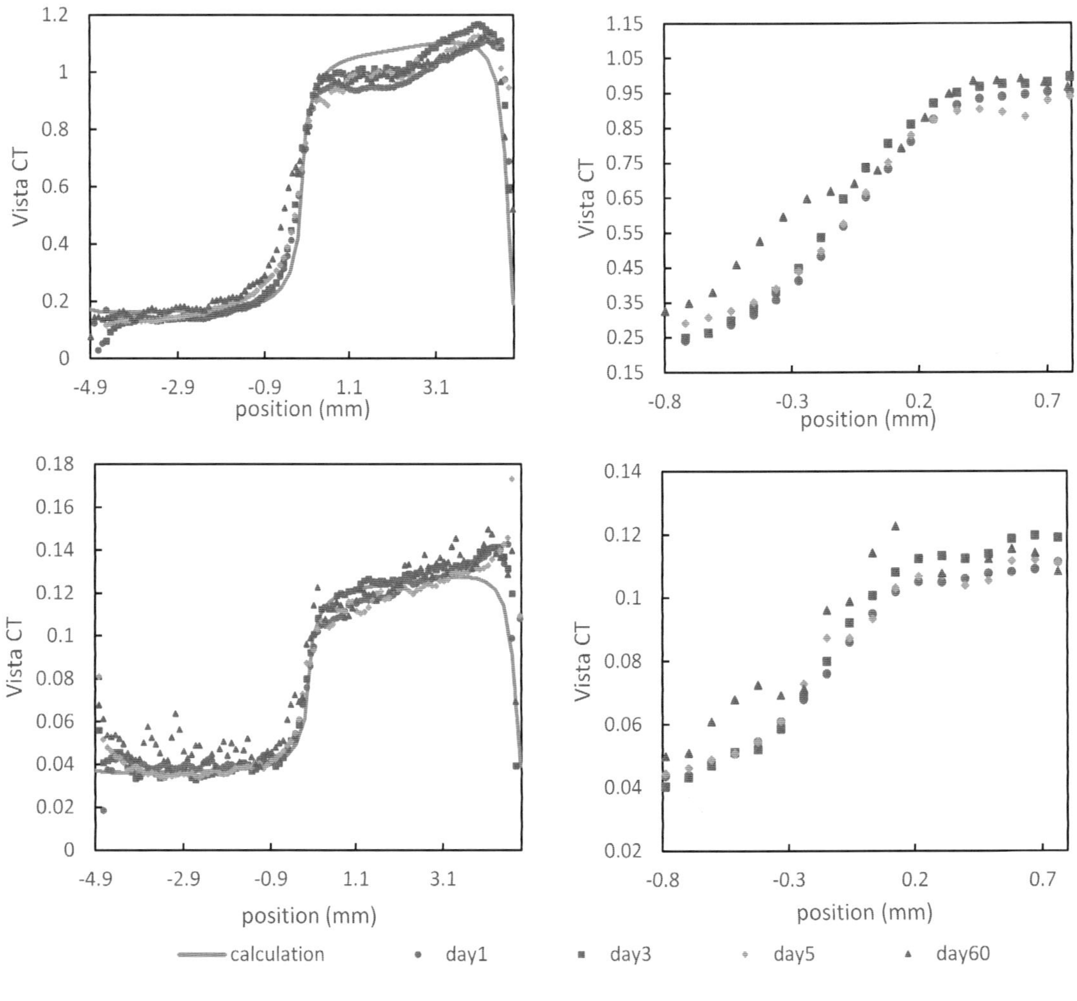


**Figure 6.** Measurement result by optical CT at 590 nm (top) and 633 nm (bottom). Left: Comparison with calculated value. Right: Enlarged view of the center. Water was used as the reference.

In order to investigate the influence of diffusion, the data of 60 days after irradiation were investigated. The calculated profile is the result of simulation with the treatment planning system (Eclipse, VARIAN Medical Systems). The results in the vicinity of the boundary between the irradiated region and the nonirradiated region at the central part of the container measured at 590 nm and 633 nm were compared (figure 6, right). The data variation during this period was smaller than that when no nanoclay was used. Hence, dye diffusion was suppressed by the nanoclay.

IOP Conf. Series: Journal of Physics: Conf. Series **1305** (2019) 012040 doi:10.1088/1742-6596/1305/1/012040

## 4. Conclusion

A clear dye gel dosimeter with nanoclay was prepared. It was found that the gel prepared in this study had high transparency and sensitivity. In addition, it was found that the influence of diffusion of the coloring region and natural coloring of the gel was nominal, and that the dose distribution evaluation was not significantly affected until about 60 days.

## 5. Acknowledgements

We thank Kawamura H. (Tsukuba International University) for useful discussion, Oe A. (Jichi Medical University Hospital), Norikazu Arai (Central radiation division, Teikyo University, Hospital), and Shinobu Kumagai (Central radiation division, Teikyo University, Hospital) for irradiation operation, and BYK-Chemie Japan and Wilbur-Ellis Japan for providing samples of nanoclays. This work was supported by Nuclear Technology and The Wakasa wan Energy Research Center.

## 6. References

[1] Baldock C *et al* 2010 *Phys. Med. Biol.* **55** R1-63
[2] Jordan K and Avvakumov N 2009 *Phys. Med. Biol.* **54** 6773
[3] Babic S *et al* 2009 *Phys. Med. Biol.* **54** 6791
[4] Jordan K and Sekimoto M 2010 *J. Phys.: Conf. Ser.* **250** 012031
[5] Maeyama T *et al* 2017 *J. Phys. Chem. B* **121** 4238
[6] Olding T *et al* 2009 *J. Phys.: Conf. Ser.* **164** 012031

IOP Conf. Series: Journal of Physics: Conf. Series **1305** (2019) 012041 doi:10.1088/1742-6596/1305/1/012041

# An Investigation of dosimetric accuracy of a novel PRESAGE radiochromic sheet and its clinical applications

**Yi-Fang Wang[1], Kevin Liu[1], John Adamovics[2] and Cheng-Shie Wuu[1]**
[1]Department of Radiation Oncology, Columbia University Medical Center, New York, NY, 10032
[2]Department of Chemistry, Biochemistry, and Physics, Rider University, Skillman, NJ, 08648

E-mail: csw6@cumc.columbia.edu

**Abstract.** A novel radiochromic PRESAGE sheet (Heuris Inc.) with variable thickness of 1-3 mm has been developed to provide 2D dosimetry. It can also be used for patient surface dosimetryor as a tool for patient-specific QA (PSQA) in 3D dosimetry. Its softness and flexibility to conform with the patient's skin make this product advantageous for future clinical applications. This study presents a comprehensive investigation into the PRESAGE sheet's dosimetric accuracy at different scanning times and its potential use for clinical applications. For the characterization of the dosimeter, temporal stability of dose rate, energy dependence, dose linearity, beam profile, and dose distribution measurements were investigated by irradiating a single sheet of PRESAGE with different doses and energies while scanning the sheet with an Epson 11000XL high-resolution scanner at different time intervals following its irradiation. Additionally, two clinical applications, including PSQA and surface dose measurement, were conducted and were compared to dosimeters currently used for clinical applications. For QA measurement, a stack of PRESAGE sheets, with EBT3 films sandwiched in-between, was used to measure the dose distribution of a pancreas SBRT treatment plan at different depths. For surface dose measurement, PRESAGE sheets, EBT3 films, and OSLDs were placed on the surface of an anthropomorphic phantom to measure the skin dose of a modulated treatment fields. When the stack of PRESAGE sheets were scanned within a period of two hours following its irradiation, the dosimeter exhibits a stable linear response to dose with negligible dose rate and energy dependence. In addition to its temporal stability, the dosimeter can provide accurate relative dose measurements comparable to those exhibited by EBT3 films. In the application of PSQA, when compared with EBT3 films using gamma test with a 2%/2 mm criteria, PRESAGE sheets have a passing rate of 99.7% measured at the isocenter and 99.1% in application of surface dose measurements. This study demosntrates the dosimetric characteristics and the potential use of a novel dosimeter, PRESAGE sheets.

## 1. Introduction

The dosimetric characteristic of PRESAGE polyurethane dosimeters, formulated with a halogenated hydrocarbon free radical initiator and leuco-dye has been investigated thoroughly [1, 2]. The primary advantages of PRESAGE over other types of dosimeters include its linear response to the absorbed dose over a wide doserange, its tissue equivalency over a wide energy range, and its capability to provide 3D dose distribution with high resolution. Additionally, the dosimeter can be fabricated to any size or shape for different purposes.

IOP Conf. Series: Journal of Physics: Conf. Series **1305** (2019) 012041 doi:10.1088/1742-6596/1305/1/012041

In 2015, a published study shows the potential use of PRESAGE thin sheets for dosimetry when analyzed using a flatbed scanner [3]. Unlike the conventional PRESAGE dosimeters, PRESAGE sheets don't require expensive equipment such as optical computed tomography (OCT) [4-6], laser fluorescence confocal microscopy (LFCM) [7] and expert users. However, the sensitivity, signal to noise ratio and temporal stability of the PRESAGE thin sheets in the study were insufficient to be used in clinical applications. Consequently, this work seeks to investigate the dosimetric accuracy of the newly developed radiochromic sheets concerning their temporal stability with regards dose linearity, energy dependence, dose rate dependence, beam profiles, and dose distribution measurements. Additionally, two clinical applications were first conducted using the PRESAGE sheets. First, a stack of PRESAGE sheets was used to perform IMRT patient-specific QA, which provides the dose distribution with high resolution at different depths. Secondly, surface dose measurements were performed by conforming the dosimeter to the contour of an anthropomorphic phantom to investigate its *in-vivo* dosimetric capability. The results of this study provide insight into the dosimetric properties of the PRESAGE sheet for its use in future applications.

## 2. Materials and Methods

### *2.1. Dosimetric Characterization*

A comprehensive characterization regarding the dosimetric accuracy of the PRESAGE at different scanning times was performed by investigating the dose rate dependence, energy dependence, dose linearity, beam profile, and dose distribution measurements of the dosimeter. As shown in Fig. 1, a single radiochromic sheet with 3.2 mm thickness was irradiated with a 6 MV photon beam using a TrueBeam (Varian Inc.) linear accelerator at six different dose points (1, 3, 5, 8, 10, and 15 Gy) within a 4 x 4 $cm^2$ field size to measure the dosimeter's linearity in its dose response. For dose rate dependence measurements, 5 Gy was delivered to the sheet using dose rates of 100, 300, and 600 MU/min using a 6 MV photon beam. Additionally, 5 Gy was also delivered to the PRESAGE sheet using a 18 MV photon beam to investigate the energy dependence of the dosimeter. Following its irradiation, the PRESAGE sheet was scanned at five different time intervals (0.5, 1, 2, 19, 24 h) using an Epson 11000XL high-resolution scanner to acquire information about the optical density (OD) development and fading of the dosimeter. To further verify the accuracy of dose distribution measurement, the beam profiles of different doses (3, 4, 5, 6 Gy) and the isodose line distributions from the PRESAGE sheets were compared with the measurements taken on the EBT3 films sandwiched in-between the PRESAGE sheets during the irradiation.

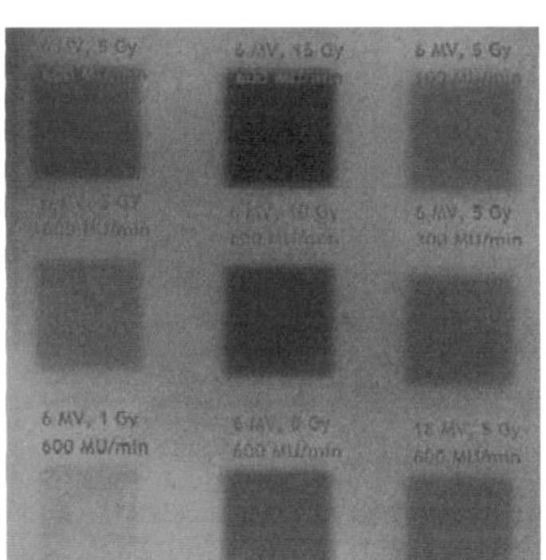


**Figure 1.** An image of an irradiated radiochromic sheet.

**Figure 2.** An image of a stack of PRESAGE sheets.

### *2.2. IMRT QA*

A stack of PRESAGE sheets with 3 mm thickness, shown in Fig. 2, was used to perform patient-specific QA of a pancreas SBRT treatment plan using modulated photon fields from a linear accelerator,

TrueBeam (Varian Inc.). EBT3 films with the same area as the PRESAGE sheets were sandwiched in-between the sheets to verify the dose distribution measurements. In addition, to compare the measured distributions with the calculated distributions from the treatment planning system, at each depth, simulated images of the dosimeter were imported into the planning system to perform dose calculation. The passing rates of gamma tests between the dose distributions measured from the EBT3 and PRESAGE sheets, as well as the EBT3 sheets and the treatment planning system, at eight different depths were calculated using DoseLab (Mobius Inc.).

### *2.3. Surface Dose Measurement*
To verify the capability of *in-vivo* surface dose measurement, as presented in Fig.3, PRESAGE sheets, EBT3 gafchromic films, and OSLDs were placed on the surface of an anthropomorhic phantom to measure the skin dose of a modulated 6 MV photon beam from a Trilogy (Varian Inc.) linear accelerator. Five OSLDs were placed with a 3 cm space separating each OSLD adjacent vertically and 4 cm space separating each adjacent OSLD horizontally to measure the relative doses at five different locations in the field. The results of relative dose measurement from OSLD and PRESAGE sheets were compared with those measured on the EBT3 films.

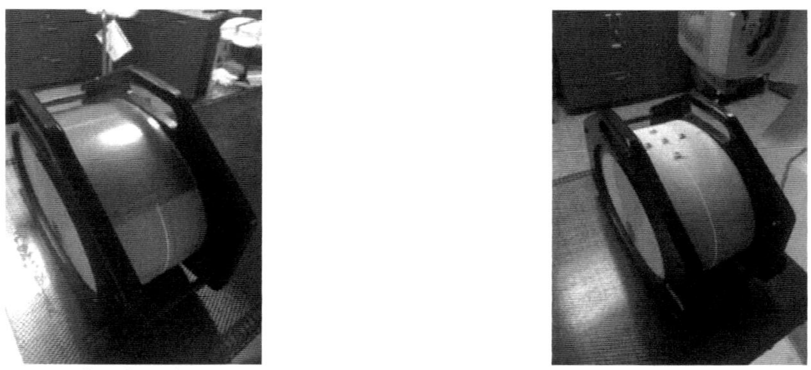

**Figure 3.** The setup of surface dose measurement. (a) PRESAGE sheets with EBT3 filmbeneath and (b) OSLDs were put on the surface of an anthropomorphic phantom.

## 3. Results

### *3.1. Dosimetric Characterization*
This study examined both the temporal stability and dosimetric accuracy of the PRESAGE dosimeter. With regards to its temporal stability, the results presented in Fig. 4 show that the PRESAGE sheet exhibits a stable linear response to dose, negligible dose rate dependence (error < 1%), and negligible energy dependence (error < 2.5%) when scanned within two hours following its irradiation. There was also an negligible difference in the measured beam profiles and dose distributions in the dosimeters when scanned within a two hour time frame following their irradiation (penumbra difference <1%). However, when scanned after a period greater than two hours, the dosimeter exhibited a nonlinear dose response, energy dependence and significant changes to its penumbra. For dosimetric accuracy, inFig. 4, beam profiles of doses greater than 4 Gy corresponded to what was measured by the EBT3 film (error <1 %) while a non-uniform color-change was observed when the dose delivered to the dosimeters was below 3 Gy, which contributes to and manifests as a ± 5% noise measured on the top portion of the beam profile. Beyond beam profile comparison,s to quantify the discrepancies between measured dose distributions from PRESAGE sheets and EBT3 films, gamma tests with a criteria of 2%/2mm were performed on the dose distributions of 4 Gy measured by the two dosimeters, which had a passing rate of 99.4%. Fig. 4 shows the isodose lines comparison of the two distributions.

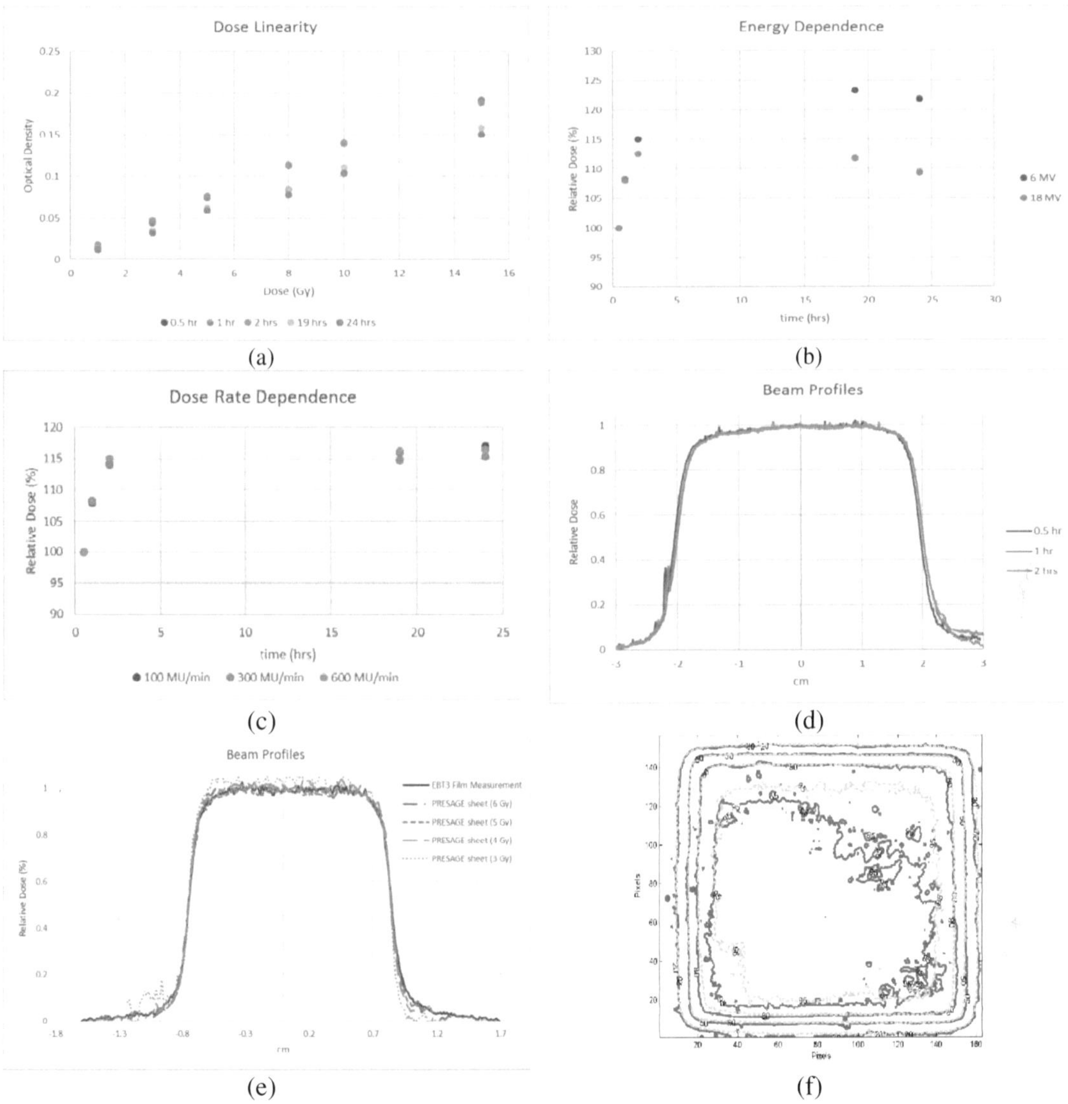


**Figure 4.** Results of dosimetric characterization. Temporal stability of (a) linear dose response (b) energy dependence (c) dose rate dependence (d) beam profile measurements was investigated. In addition, (e) dose uniformity at different delivered doses and (f) comparison of 2D distributions from EBT3 (red) and PRESAGE (green) were presented.

### *3.2. IMRT QA*

Table 1 shows the passing rates of gamma tests between the dose distributions measured from EBT3 and PRESAGE sheets, and the passing rates of gamma tests between EBT3 and the treatment planning system at eight different depths. The passing rates with a 2%/2 mm criteria between EBT3 and PRESAGE at all depths were greater than 95% with the maximum passing rate measured to be at 99.8%. On the other hand, dose distributions from the treatment planning system were not close to what from the EBT3 films due to the overmodulation of the fields. Using the 3%/3mm criteria, the lowest and the highest passing rates were measured to be 85% and 94.8%. Fig. 5 shows the difference of isodose lines at the 6 mm depth from the isocenter.

**Table 1.** Passing rates of gamma tests at multiple depths

| Passing Rates of gamma tests at multiple depths | | | | | | | |
|---|---|---|---|---|---|---|---|
| Depth of the dose plan relative to isocenter | -18 mm | -12 mm | -6 mm | Isocenter | 6 mm | 12 mm | 18 mm |
| EBT3 vs. PRESAGE sheets (2%/2mm) | 95.9% | 99% | 96.4% | 99.7% | 96.1% | 94.6% | 99.8% |
| TPS vs. EBT3 (3%/3mm) | 91.3% | 94.1% | 94.8% | 90.8% | 90% | 92.1% | 90.1% |

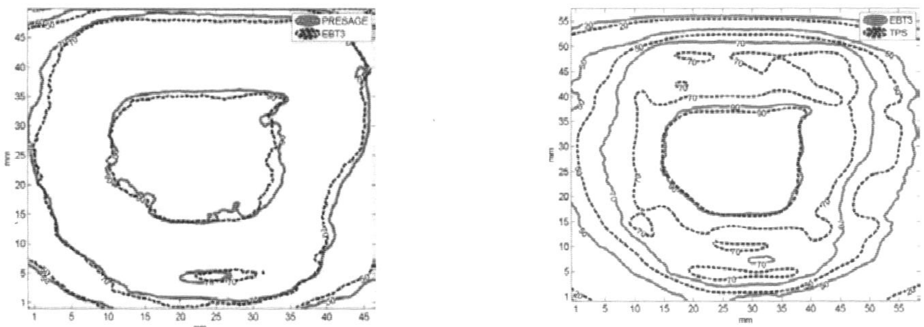


(a) (b)

**Figure 5.** Isodose lines comparison of EBT3 film, treatment planning and PRESAGE sheets at 6 mm depth from the isocenter.

### *3.3. In-vivo Skin Dose Measurement*

To evaluate the accuracy of relative dose measurement on the surface, the measured dose distribution of a modulated photon field measured from PRESAGE sheets was compared with the measured dose distribution from the EBT3 film. Gamma test with 2%/2mm criteria shows the passing rate to be at 99.1% (Fig. 6). Additionally, the relative doses at the measured points from both OSLDs and PRESAGE sheets were compared with EBT3 films. To obtain the relative dose, the measured doses at five points from the OSLDs were normalized to the OSLD placed at point 5 as shown in Fig. 7. Table 2 shows that the OSLDs exhibit a larger discrepancy than PRESAGE sheets with its maximum relative dose difference to be at 3.7% and the relative dose differences exhibited by the PRESAGE sheet to be within 1.0%.

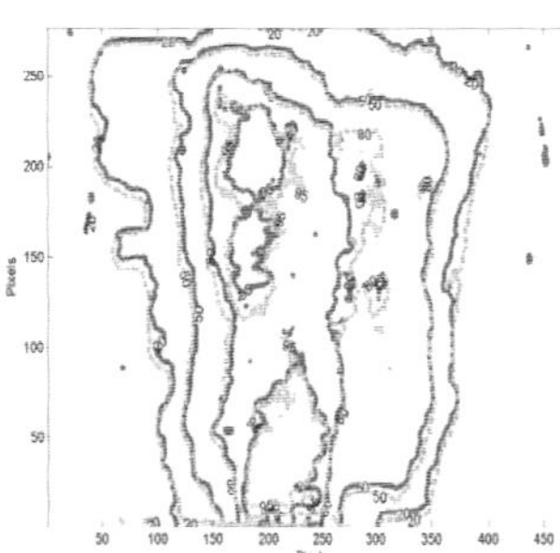

**Figure 6.** Isodose lines comparison of EBT3 film (red) and PRESAGE sheets (green).

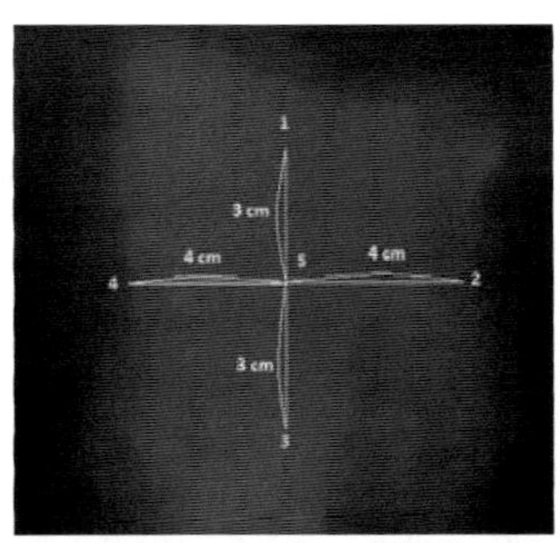


**Figure 7.** Positions of five OSLDs relative to the dose distribution.

**Table 2.** Relative dose difference of OSLDs and PRESAGE sheets at measured points compared with EBT3 film. The doses at five measured points were normalized to Point 5 to obtain relative dose.

| Relative Dose Difference Compared with EBT3 Film | | | | |
|---|---|---|---|---|
| Dosimeter | Point 1 | Point 2 | Point 3 | Point 4 |
| OSLD | 3.70% | 1.80% | 2.00% | 3.50% |
| PRESAGE Sheet | 0.61% | 0.73% | 0.85% | 0.45% |

## 4. Discussion

The results from the preceding section show that there is a negligible change to the dosimetric characteristic of the PRESAGE sheet which includes its linear dose response, dose rate independence, energy independence, and relative dose measurements when the PRESAGE sheet is scanned within a period of two hours following its irradiation. However, when scanned within a period greater than two hours, the scanned results become less accurate due to different rates of OD increase and fading at different doses. Moreover, the fading of the OD in low dose region was observed to occur at a later period than OD in the high dose region. This effect leads to the larger penumbra of the beam profiles when the sheet is scanned within a time frame greater than two hours following its irradiation. For dosimetric accuracy, PRESAGE films can provide an accurate relative dose measurement, comparable to EBT3 film, when scanned within a two hours period and delivered with sufficiently high dose.

For clinical applications, using a stack of PRESAGE sheets, patient-specific QA at different depths can be conducted using a conventional film scanner. The results of gamma tests show that the dose distributions from PRESAGE sheets correspond to what was measured using EBT3 film. In addition, the passing rates of the gamma tests between EBT3 and treatment planning system were observed to change at different depths significantly. Plans that had smoothed distributions exhibited passing rates that were higher than those with modulated fields, which indicates that the conventional IMRT QA at a certain depth could be insufficient to make a proper judgement for the small field and high gradient treatments.

In the third part, *in-vivo* surface dose measurements of PRESAGE sheets and OSLDs were compared with EBT3 films. From the results, OSLDs exhibit larger discrepancies than PRESAGE sheets due to their comparatively larger statistical and experimental uncertainty in comparison to EBT3 film and the changes in the OSLD response as a function of dose [8, 9].The greater discrepancy in the relative dose measurements between OSLDs may have been attributed to the uncertainty in the sensitivity of an individual dosimeter as well as the uncertainty in the performance of the reader used to measure the dose extracted from the OSLDs [9] The difference in sensitivity between OSLDs may be ascribed to the slight differences of the amount of dosimetry material ($Al_2O_3$:C) contained in the dosimeter [9]. This difference in sensitivity is manifested as differences in their response when exposed to the same radiation field. Screened OSLDs were reported to have a measurement uncertainty of ~ ± 2% and the uncertainty in the performance of the microstar reader (Landauer Inc.) used to measure the dose from the OSLDs in this experiment is reported to have an uncertainty between 1-2% [9]. Besides, PRESAGE sheets have a comparable resolution as EBT3. Compared to EBT3 film, PRESAGE sheets have the same dosimetric capability as EBT3 and are softer and more flexible to conform to the patient's skin, which makes it a valuable tool for *in-vivo* dosimetry.

## 5. Conclusion

This study shows the dosimetric characteristics and the potential use of a novel dosimeter. When scanned within a two hour time frame following its irradiation, the radiochromic sheets can provide accurate relative dose measurement comparable to EBT3 films. For clinical applications, the PRESAGE sheets can be used effectively in clinics to provide accurate *in-vivo* dosimetry with less statistical error and higher resolution than OSLDs. Additionally, it can be a valuable and convenient

IOP Conf. Series: Journal of Physics: Conf. Series **1305** (2019) 012041 doi:10.1088/1742-6596/1305/1/012041

tool for patient- specific QA in 3D to verify the dose distributions at multiple depths. In the future, the sheets could be produced in different thickness for other purposes, such as a replacement of bolus, and the influence of thickness difference on its dosimetric characteristic, and the depth of measurement should be investigated further.

## 6. References

[1] Adamovics J and Maryanski M J 2006 *Rad. Prot. Dosim.* **120** 107-12
[2] Gorjiara T *et al* 2011 *Med. Phys.* **38** 2265-74
[3] Dumas M and Rakowski J T 2015 *Med. Phys.* **42** 7138-43
[4] Guo P *et al* 2006 *Med. Phys.* **33** 3962-72
[5] Xu Y *et al* 2004 *Med. Phys.* **31**3024-33
[6] Oldham M *et al* 2001*Med. Phys.* **28** 1436-45
[7] Youkahana E *et al* 2016 *Biomed. Phys. Eng. Express* **2** 045009
[8] Yukihara E G and McKeever S W S 2008 *Phys. Med. Biol.* **53** R351
[9] Yahnke C J 2010 Calibrating the MicroStar. In *Light TM Landuaer, Inc.*

# Basic characteristics of an AQUAJOINT®-based VIPET polymer gel dosimeter

**Mikio Nemoto[1,2], Ayumi Oe[2], Tomokazu Kotani[2], Daniel Antonio Sahade[1,3] and Toshimasa Hamada[1,4]**

[1]3D Gel Dosimeter Research Laboratory, Cluster for Science, Technology and Innovation Hub, RIKEN, 2-1, Hirosawa, Wako, Saitama 351-0198, Japan
[2]Department of Radiology, Jichi Medical University Hospital, 3311-1, Yakushiji, Shimotsuke, Tochigi 329-0498, Japan
[3]Frontier Materials Research Department, Materials Research Laboratories, Nissan Chemical Corporation, 488-6, Suzumi-cho, Funabashi, Chiba 274-0052, Japan
[4]Advanced Materials & Planning Department, Nissan Chemical Corporation, 5-1, Nihonbashi 2-Chome, Chuo-ku, Tokyo 103-6119, Japan

E-mail: toshimasa.hamada@riken.jp

**Abstract.** The basic characteristics of a VIPET-type polymer gel dosimeter were assessed using AQUAJOINT® as a hydrogel matrix. The dosimeter exhibited a good dose response. The threshold dose was 0.5–0.6 Gy and the linear section of the response curve, in the range of 1.7–2.7 Gy, had a minimum dose response of 0.05–0.06 Gy. The dose-rate dependence was determined by the delivery of 200 MU for dose rates between 100 and 600 MU/min; the resulting value of R2 was 3.222 ± 0.036 (mean ± 1SD). A slight change in the value of R2 in the integration of the divided doses was observed; the R2 value was 3.059 ± 0.015 (mean ± 1SD) for a total irradiation of 200 MU. Gel samples of different volumes were irradiated and compared with a PAGAT polymer gel. The change in sensitivity of the AQUAJOINT®-based VIPET was smaller than that of the PAGAT.

## 1. Introduction

AQUAJOINT® is a heat-irreversible, physical hydrogel consisting of a water-soluble polymer and inorganic nanoparticles in water. Mixing the gel precursors of liquid A and B forms a self-standing hydrogel at room temperature through the electrostatic interaction of these components in water.

Normoxic *N*-vinylpyrrolidone polymer gel (VIPET) is a three-dimensional dosimetric tool with radiation-induced polymerization depending on the radiation dose [1]. AQUAJOINT®-based VIPET gel dosimeters have previously been reported, and the gels were irradiated using industrial 250 kVp X-ray equipment [2,3,4]. In this study, we present the preliminary measurement results of the basic characteristics of AQUAJOINT®-based VIPET gels irradiated by a 10 MV X-ray from a medical linear accelerator (LINAC), including the effect of container size, in comparison with a PAGAT polymer gel.

## 2. Materials and methods

### *2.1 AQUAJOINT® -based VIPET gel dosimeter fabrication*

IOP Conf. Series: Journal of Physics: Conf. Series **1305** (2019) 012042 doi:10.1088/1742-6596/1305/1/012042

We prepared a dosimeter mixture by first dissolving the monomer *N*-vinyl-2-pyrrolidone (NVP), the bifunctional monomer *N,N'*-methylene-bis-acrylamide (BIS), oxygen scavenger tetrakis-(hydroxymethyl)phosphonium chloride (THPC), and AQUAJOINT® B-5 (Nissan Chemical Corporation, water content: 86%) in deionized water. The addition of AQUAJOINT® A-5 (Nissan Chemical Corporation, water content: 81%) to the aqueous solution yielded a homogeneous system comprising a gel matrix with monomers. The resultant mixture contained 8% NVP, 4% BIS, 50 mM THPC, 11% AQUAJOINT® A-5, and 11% AQUAJOINT® B-5. The resulting gel precursor was poured into polyethylene terephthalate (PET) vials of three sizes (mouth inner diameter × body diameter × total height) — 30 mL vials (14.8 mm × 30.5 mm × 78.7 mm), 50 mL vials (14.8 mm × 34.2 mm × 89 mm), and 100 mL vials (14.8 mm × 40.2 mm × 117.8 mm) — and into 500 mL fluorine-treated polypropylene vials (45.0 mm × 77 mm × 161 mm). The gel vials were then stored at 4 °C.

### *2.2 PAGAT gel dosimeter fabrication*

PAGAT gel dosimeters [5] were prepared using 89% w/w deionized water, 3% acrylamide (AAm), 3% BIS, 5% gelatin (300 bloom), and 10 mM of THPC. Gelatin was added to stirred water in one portion at room temperature and was then stirred for 10 min, after which the solution was heated at 40 °C for 10 min. AAm and BIS were then added, in this order, to the gelatin solution which was subsequently stirred at 40 °C for 10 min. Next, THPC was added and the solution was stirred at 40 °C for 1 min. The resulting mixture was poured into the PET vials (30 mL, 100 mL, and 500 mL). The gel vials were then stored at 4 °C.

### *2.3 Sample irradiation*

A sample vial was submerged in a water phantom, used to calibrate the monitor unit of the LINAC (figure 1). After aligning the center of the vial with the isocenter and securing it with a custom-made jig, the vial was irradiated from above. The applied X-ray energy was 10 MV, and the beam delivery parameters of the LINAC (CL-iX, Varian Medical Systems, Palo Alto, CA) were as follows: the source-to-axis distance (SAD) was 100 cm, the measurement depth was 10 cm, and the irradiation field size was 10 $cm^2$.

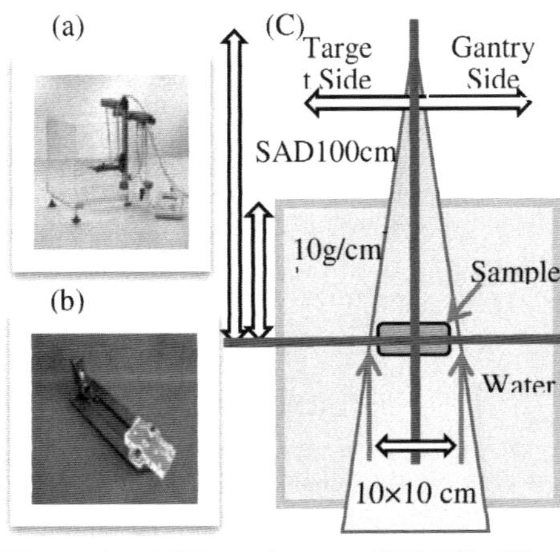

**Figure 1.** (a) Water phantom (WP-1D), (b) sample-fixing jig (custom-made), and (c) measurement

### *2.4 Validation items (basic characteristics)*

The vials of AQUAJOINT® gel (50 mL) were irradiated and the basic characteristics were verified. Validation items were as follows: dose response (0–10.2 Gy, 0–2.5 Gy, and 1.7–2.7 Gy, 600 MU/min), dose rate characteristics (total 200 MU, 100–600 MU/min), and dose reproducibility (total 200 MU divided into 1, 2, 4, 10, 20, and 40 fractions, 600 MU/min).

### *2.5 Validation items (effect of container size)*

Samples of AQUAJOINT®-based VIPET and PAGAT with different volumes (30, 100, and 500 mL) were used. All samples were left to stand at room temperature for 9 h and were then irradiated (0–8.5 Gy). Subsequently, the vials were stored in the dark at room temperature (24–25 °C) for 24 h (figure 2).

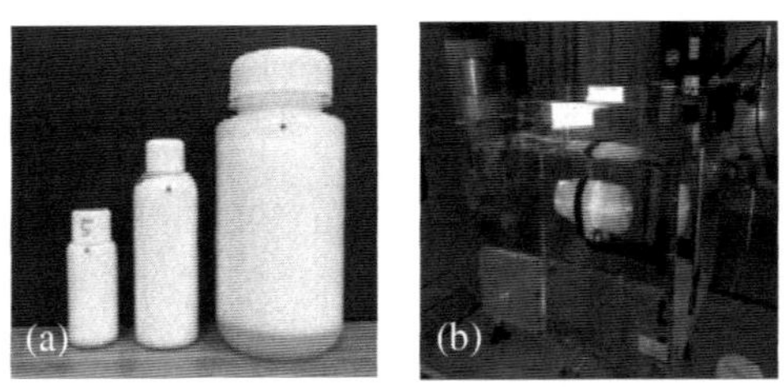

**Figure 2.** (a) Vials for verification of the effect of container size and (b) state of irradiation.

*2.6 Scan and analysis of MRI*

The magnetic resonance imaging (MRI) measurements of all samples were carried out 24 h after irradiation. The AQUAJOINT®-based VIPET and PAGAT were kept under the same conditions until they were imaged by MRI after irradiation, and the temperature returned to room temperature. Using the head coil, the isocenter plane at the time of irradiation was imaged with a T2 image using a multi-slice spin-echo sequence. The equipment and scan conditions used were as follows: 1.5-T MRI (Avant, Siemens Healthcare), TR = 4000 ms, TE1 = 30 ms, TE2 = 136 ms. The pixel size was 1.0 mm and the slice width was 3.0 mm. After converting the MRI image to the R2 image using the ImageJ software, the pixel value within the isocenter plane was read out.

## 3. Results and discussion

*3.1 Basic characteristics of AQUAJOINT®-based VIPET*

The response characteristics corresponding to the absorbed doses are shown in figure 3.

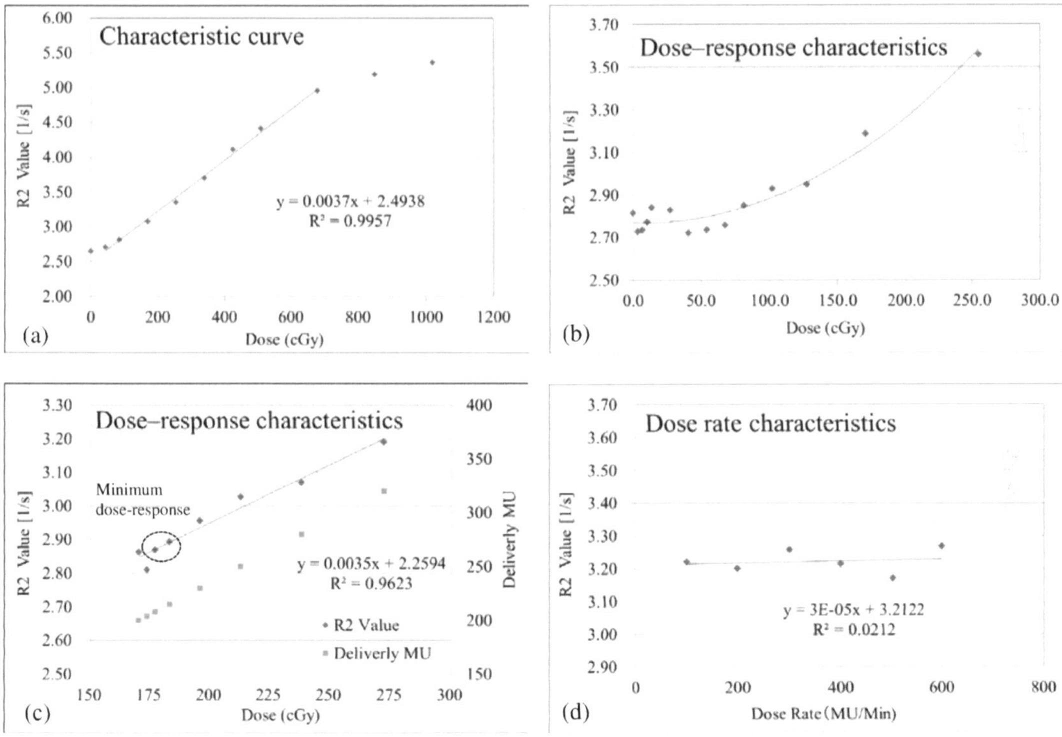


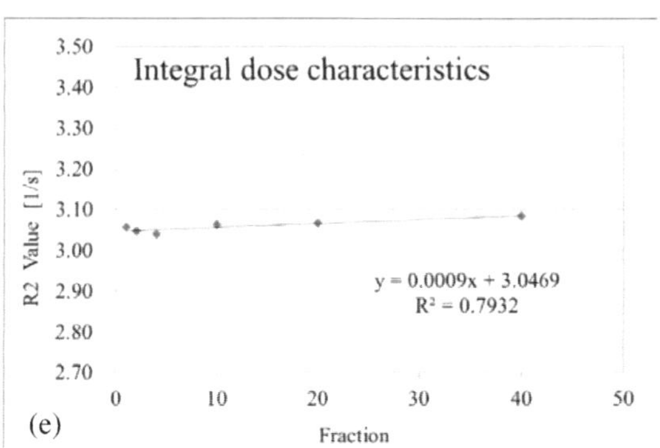


**Figure 3.** Basic properties of AQUAJOINT®-based VIPET polymer gel dosimeters.
(a) Dose–response curve.
(b) Dose–response characteristics (0–2.5 Gy)
(c) Dose–response characteristics (1.7–2.7 Gy)
The orange dots represent the irradiated MU and the blue dots represent the corresponding measured values of R2.
(d) Dose rate characteristics.
(e) Integral dose characteristics.

The dose-R2 curve can be well approximated by a linear plot between 1 and 7 Gy. The usual dose prescribed by radiotherapy is in the range of 1.4–8.0 Gy, which shows that it had a sufficient dynamic range (figure 3(a)). The R2 value was gradually increased in the low dose range, and the threshold dose was approximately in the range of 0.5–0.6 Gy (figure 3(b)). To reduce the threshold dose, we are going to optimize the chemical compositions and gel preparation procedures. When the measurement was performed in the dose range (1.7–2.7 Gy), where linearity of the response curve was ensured, the minimum dose-response was in the range of 0.05–0.06 Gy (figure 3(c)).

The dose-rate dependence was determined by delivering 200 MU for dose rates between 100 and 600 MU/min, the resulting R2 value was 3.222 ± 0.036 (mean ± 1SD). No significant dose-rate dependence was observed (figure 3(d)). The dose reproducibility was confirmed by a total irradiation of 200 MU through multiple irradiations. When the number of divisions of irradiation was increased, the R2 value was slightly overestimated compared to the value obtained with a single irradiation. The change in the R2 value was 3.059 ± 0.015 (mean ± 1 SD).

*3.2 Response characteristics of different volume samples*

In this preliminary experiment, the difference of the R2 value of non-irradiated PAGAT samples was larger than that of the AQUAJOINT®-based VIPET (figure 4). The R2 values of 500 mL were -17.2% (PAGAT) and -7.1% (AQUAJOINT®-based VIPET), compared to that of the 30 mL samples.

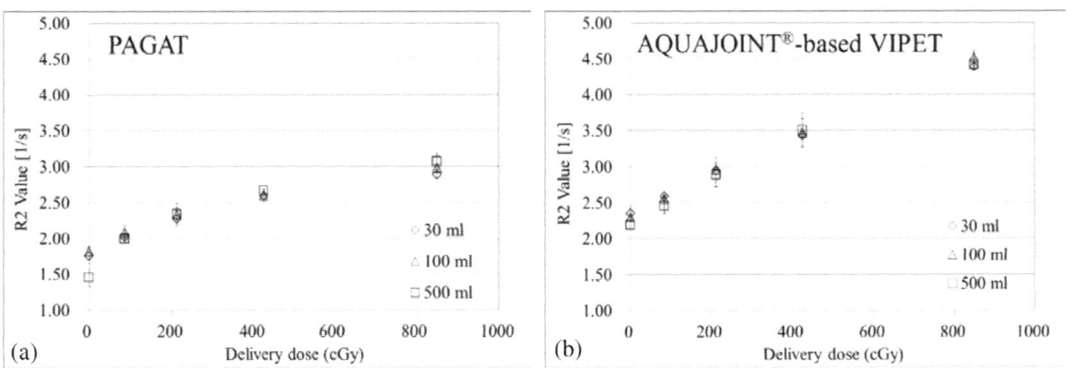


**Figure 4.** Dose-R2 response curve of different volume samples: (a) PAGAT and (b) AQUAJOINT®-based VIPET. Each value is the average of three samples.

The difference in sensitivities between samples of different volumes were reported to be caused by the effect of thermal history [6,7]. A report that verified the relationship between the temperature change inside the gel and the R2 value generated at the time of irradiation, concluded that the thermal history at the time of preparation of the gel is a cause of the sensitivity difference [8]. In contrast, as previously mentioned, the gel precursors of AQUAJOINT® liquid A and liquid B form a hydrogel at room temperature. It is considered that the thermal change during fabrication was small.

Converting the graph in figure 4 from dose vs. R2 to dose vs. ΔR2 shows that the variation of the R2 value is significant due to deference of the R2 value of un-irradiated samples. The response of the PAGAT was non-linear in the range of 0-8.5 Gy, and its sensitivity was low (figure 5(a)). Contrarily, the sensitivity of the AQUAJOINT®-based VIPET was nearly twice that of the PAGAT, and the AQUAJOINT®-based VIPET exhibited linear dose responsiveness (figure 5(b)). Clinically, phantoms in the range of 500–2000 mL can be used, and it is desirable to use smaller samples to acquire a characteristic curve. The AQUAJOINT®-based VIPET has the effect of relatively reducing the error of R2 values.

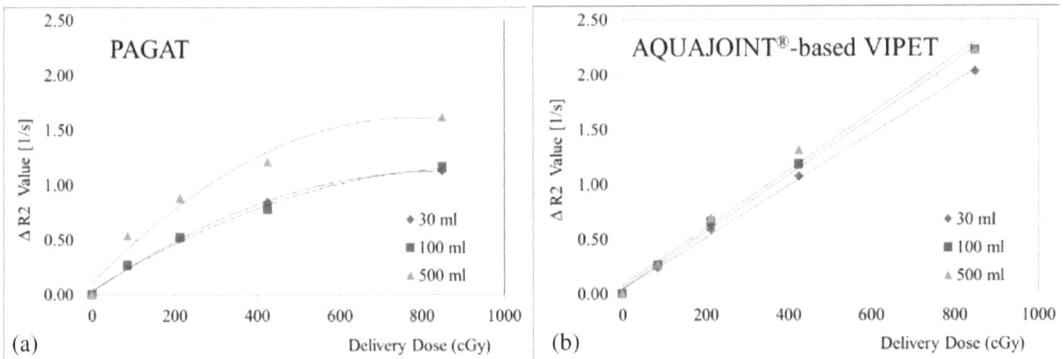


**Figure 5.** Dose-ΔR2 response curve of different volume samples: (a) PAGAT and (b) AQUAJOINT®-based VIPET. Each value is the average of three samples.

## 4. Conclusions

The fundamental characteristics of a gel dosimeter prepared using AQUAJOINT® hydrogel as a gelling agent were examined. The AQUAJOINT® -based VIPET polymer gel dosimeter exhibited good responsiveness to the irradiation dose. In addition, the dose response due to the change in gel volume was smaller than that of PAGAT.

## References

[1] Papadakis A E *et al* 2007 *Phys. Med. Biol.* **52** 5069-83
[2] Maeyama T *et al* 2015 4th 3D Gel Dosimeter Research Meeting, Teikyo University, November 2015
[3] Takanashi T *et al* 2017 *J. Phys.: Conf. Ser.* **847** 012013
[4] Maeyama T *et al* 2018 *Radiat. Phys. Chem.* **146** 121-5
[5] Venning A J *et al* 2005 *Phys. Med. Biol.* **50** 3875-88
[6] Macdougall N D *et al* 2008 *Br. J. Radiol.* **81** 46-50
[7] De Deene Y *et al* 2007 *Phys. Med. Biol.* **52** 2719-28
[8] Sedaghat M *et al* 2010 *J. Phys.: Conf. Ser.* **250** 012018

IOP Conf. Series: Journal of Physics: Conf. Series **1305** (2019) 012043 doi:10.1088/1742-6596/1305/1/012043

# Novel dual-wavelength optical-CT imaging method for gel dosimeter readout

**Yi Du[1,2,3], Xiangang Wang[2], Xincheng Xiang[2] and Yves, De Deene[3]**
[1]Key Laboratory of Carcinogenesis and Translational Research (Ministry of Education/Beijing), Department of Radiotherapy, Peking University Cancer Hospital & Institute, 52 Fucheng Road, Beijing, 100142, China
[2]Institute of Nuclear & New Energy Technology, Tsinghua University, Beijing, 100084, China
[3]School of Engineering, Faculty of Science and Engineering, Macquarie University, Sydney, NSW 2109, Australia

Email: wangxiangang@tsinghua.edu.cn

**Abstract**. Dual-wavelength scanning is a technique eliminating the need of phantom repositioning for optical-CT gel readout. To further diminish artifacts caused by phantom impurities, we hereby propose a novel dual-wavelength imaging method based on phantom impurity recognition and correction. In this method, impurities in motion trajectories during phantom rotation are recognized at the reference wavelength via motion-tracking as a binary mask, which is then applied to correct impurity-contaminated pixels at the data wavelength. Compared with the existent dual-wavelength imaging method, the proposed method is demonstrated to be capable of reducing impurity-induced artifacts and improving imaging SNR and CNR in the same process.

## 1. Introduction

With the wide application of highly conformal radiotherapy treatments featured by IMRT and VMAT, three-dimensional radiotherapy dosimetry has become a key research area where growing interest has been directed to gel dosimetry [1-4]. As a compact and cost-effective readout modality alternative to MRI [5-7] , optical-CT has been proposed to scan radiosensitive gel dosimeters. Optical density (OD) distribution quantitatively related to localized absorbed dose can be remapped by optical projections of transmitted light over 360 degrees, and then absolute dose distribution can be obtained via calibration between ΔOD and dose. In order to suppress imaging artifacts caused by phantom impurities, pre- and post-irradiation scans are often subtracted [8, 9]. However, as demonstrated in another abstract at this conference, this method is very sensitive to pre- and post-scan registration errors and prone to positive-negative pair errors. To avoid the problem of co-registration between pre- and post-irradiation scans, a dual-wavelength scanning has been proposed by De Deene in [10].

The principle behind dual-wavelength scanning is based on the property of wavelength-dependent absorption. As in [10], by scanning post-irradiation dosimeters at reference and data wavelengths (or channels) respectively, logarithmic projections can be calculated for tomographic reconstruction as:

$$Projection = -log_{10}\left(\frac{I_{data} - I_{dark}}{I_{ref} - I_{dark}}\right) \quad (1)$$

where $I_{ref}$ and $I_{data}$ are the measured transmitted light intensity at reference and data channels respectively and $I_{dark}$ is the dark current of the detector.

In this abstract, we propose a novel dual-wavelength imaging method other than Equation (1), which uses the same dual-wavelength data but reduces imaging noise and artifacts by phantom impurity recognition and correction.

## 2. Method & Materials

### 2.1. Motion-Tracking based Impurity Recognition

The key physical principle of the proposed method is that at the reference wavelength radiosensitive chemicals turn almost transparent while phantom impurities such as stain particles will attenuate the traverse light independent of the wavelength. Hence when a dosimeter phantom is scanned at the reference wavelength, the phantom rotates along the central axis with opaque impurities projected onto the detector forming *footprint-like* continuous motion trajectories. Because of measurement noise and slight light attenuation by matrix material, phantom impurities are difficult to be directly segmented from background via threshold selection. However, the impurity motion trajectories can be recognized as a binary mask via motion-tracking, which can thereby be used to correct contaminated measurements at the data wavelength.

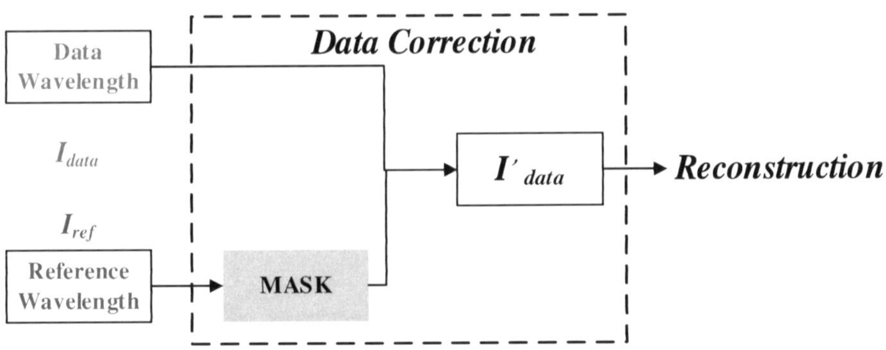


**Figure 1**. Key workflow of the dual-wavelength imaging method with impurity correction.

In this study, the binary mask of impurity motion trajectories is generated using a modified algorithm from the Matlab R2016a™ *Motion-Based Multi-Object Tracking* example. Some parameters are tuned to improve the stability and robustness of the algorithm. When the binary mask is obtained, the corresponding pixels of the data-wavelength images, which are assumed to be contaminated by phantom impurities, are thereby eliminated and replaced by four-pixel neighborhood interpolation. The logarithmic projections are then calculated with the corrected data images and blank field image.

### 2.2. Experiment Apparatus

A cylindrical silicone dosimeter [11-13] phantom was irradiated along the central axis by a square field from a 6-MV Linac with a grid-pattern attenuator placed on top of the phantom. The attenuating grid had a divergence corresponding with the beam divergence. Posterior to irradiation, the phantom was scanned on a Vista™-rebuilt dual-wavelength optical-CT scanner with diffuse light sources at 630-nm for the data projections (red) and 530-nm for the reference projections (blue) respectively [10].

## 3. Results

### *3.1. Dual-wavelength Traverse Data*

Figure 2 shows corresponding projections for the reference and data channels respectively.

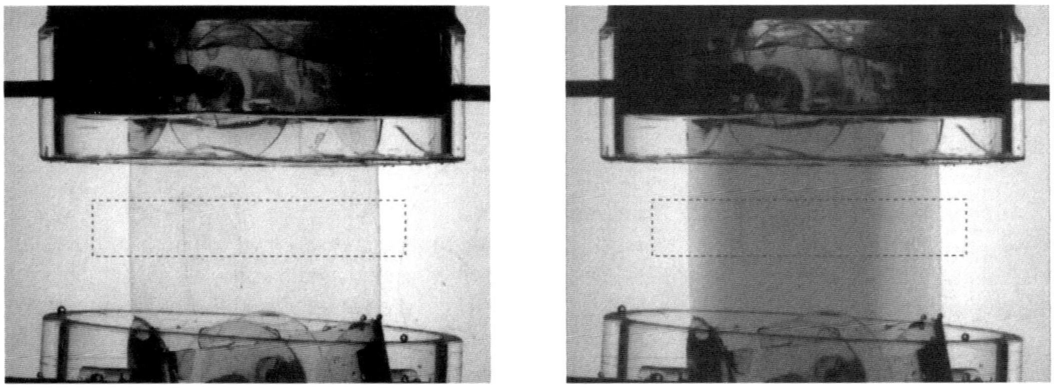

(a) (b)

**Figure 2**. Reference (a) and data (b) projection scans of the silicone dosimeter (FlexyDos3D) phantom. The red box depicts the ROI for following data processing.

### *3.2. Impurity Recognition Mask*

For illustration, Figure 3 gives one frame of the calculated impurity recognition mask together with the reference and data images at the same projection angle.

### *3.3. Reconstruction Image Evaluation*

Three different sets of OD images are reconstructed respectively by using: (1) the data channel and its blank field data, denoted as mono-wavelength at 630 nm, (2) the dual-wavelength scan data and following the method in Equation (1), and (3) the impurities recognition and correction method proposed in this study.

Figure 4 shows the reconstructed images of the three methods. A zoomed ROI (red box) is selected, where an impurity is identified. Also, a pair of ROIs (yellow boxes) inside and outside the irradiation field are selected to calculate SNR and CNR as listed in Table 1.

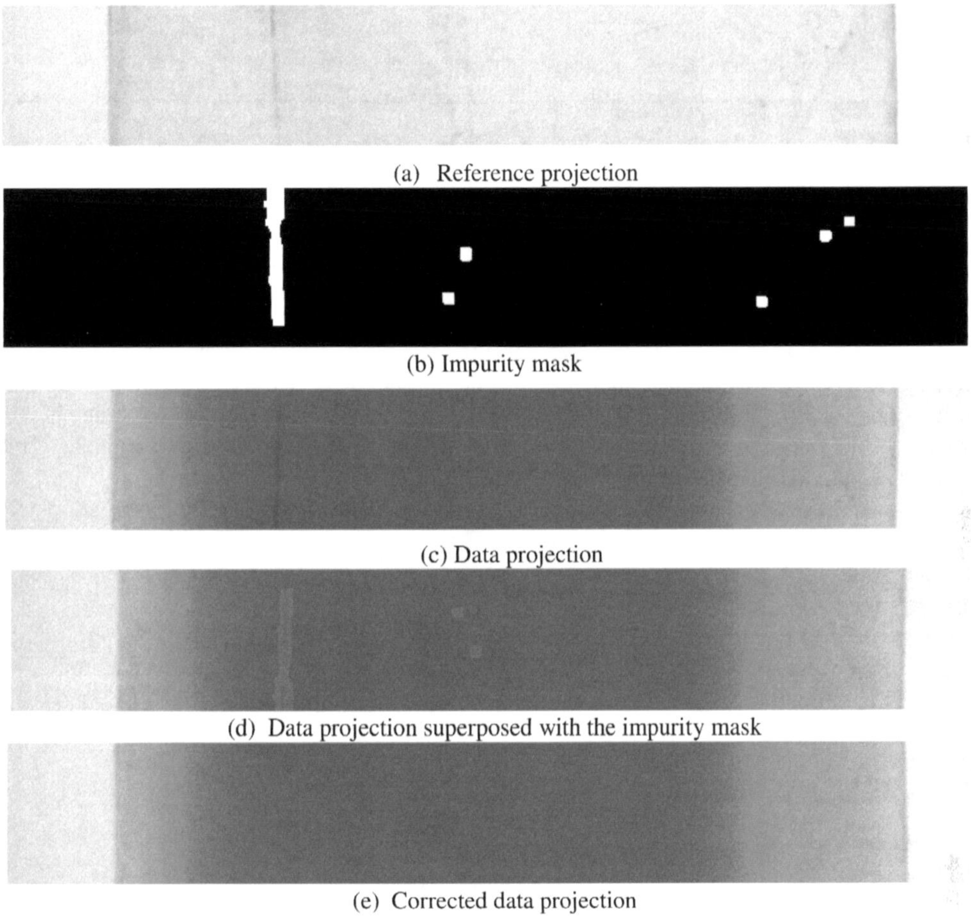


**Figure 3**. Illustration of the corresponding reference and data projections, recognized impurity mask and corrected data projection.

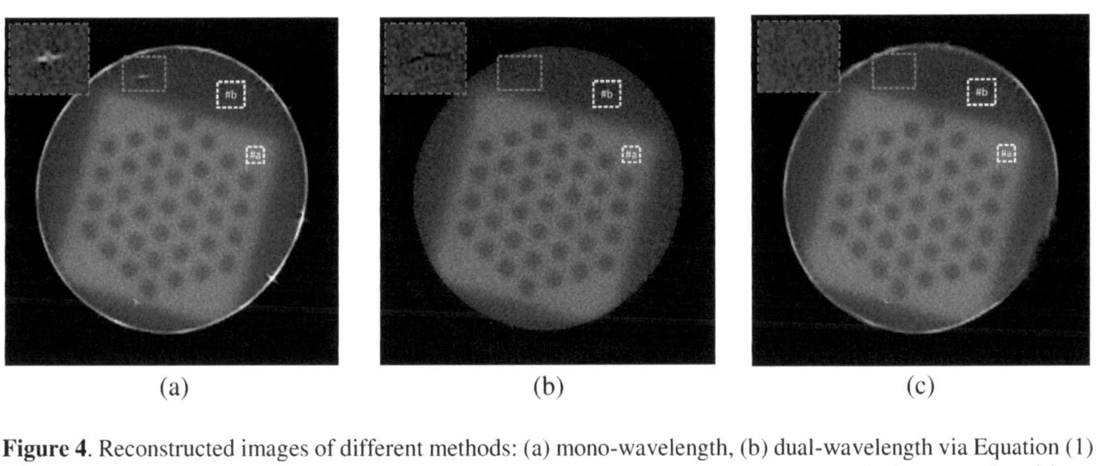


**Figure 4**. Reconstructed images of different methods: (a) mono-wavelength, (b) dual-wavelength via Equation (1) and (c) proposed method. The red box depicts the zoomed ROI with a dust stain visible in (a), replaced by a *negative scar* in (b), and clearly removed in (c). The yellow boxes ROI #a and ROI #b are selected to calculate the SNR and CNR.

**Table 1**. SNR and CNR of the images in figure 4 calculated from ROI *#a* and ROI *#b*.

| Method | SNR | | CNR |
|---|---|---|---|
| | ROI #a | ROI #b | |
| Mono-wavelength at 630 nm | 10.9659 | 5.4264 | 4.3122 |
| Dual-wavelength via Equation (1) | 9.2542 | 4.5895 | 3.8368 |
| Proposed method | 11.2453 | 5.6022 | 4.4084 |

## 4. Discussion and conclusions

Similar to the conclusion in [10], dual-wavelength scanning, which eliminates the need for pre-irradiation scan, is shown effective in reducing imaging artifacts related to phantom impurities. As originality of this study, a novel dual-wavelength imaging method via impurity recognition is proposed. Based on the same data acquisition process as in [10], this method in comparison is capable of reducing impurity-induced imaging artifacts. In the meantime, improved imaging SNR and CNR is shown, which can be attributed primarily to using a smoothed light field projection for image reconstruction and partly to reducing streaking artifacts.

## 5. Acknowledgement

Jointly Supported by Beijing Natural Science Foundation (No. 1184014, 1174016), National Science Foundation of China (No. 61571262, 11575095), China Postdoctoral Science Foundation (No. 2017M620542) and National Key Research and Development Program (No. 2016YFC0105406).

## 6. References

[1] Baldock C *et al* 2010 *Phys. Med. Biol.* **55** R1-63
[2] Schreiner L J 2017 *J. Phys.: Conf. Ser.* **847** 012001
[3] Oldham M 2017 *J. Phys.: Conf. Ser.* **847** 012006
[4] Mather M L *et al* 2003 *Phys. Med. Biol.* **48** N269-75
[5] Vandecasteele J and De Deene Y 2013 *Phys. Med. Biol.* **58** 19-42
[6] Vandecasteele J and De Deene Y 2013 *Phys. Med. Biol.* **58** 43-61
[7] Vandecasteele J and De Deene Y 2013 *Phys. Med. Biol.* **58** 63-85
[8] Doran S J 2013 *J. Phys.: Conf. Ser.* **444** 012004
[9] Wuu C-S and Xu Y 2010 *J. Phys.: Conf. Ser.* **250** 012044
[10] De Deene Y 2015 *J. Phys.: Conf. Ser.* **573** 012058
[11] Høye E M *et al* 2015 *J. Phys.: Conf. Ser.* **573** 012067
[12] De Deene Y et al 2015 *Phys. Med. Biol.* **60** 1543-63
[13] Carpentier E *et al* 2016 **847** 012052

IOP Conf. Series: Journal of Physics: Conf. Series **1305** (2019) 012044 doi:10.1088/1742-6596/1305/1/012044

# Evaporation and diffusion of chloroform with the deformable FlexyDos3D radiation dosimeter

**M J Wheatley[1,5], J T Booth[2,3] and Y De Deene[1,4,5]**
[1]School of Engineering, Macquarie University, North Ryde, Sydney, Australia
[2]Northern Sydney Cancer Centre, Royal North Shore Hospital St Leonards, Sydney, Australia
[3]Institute of Medical Physics, Faculty of Science, University of Sydney, Sydney, Australia
[4]Department of Radiation Oncology and Experimental Cancer Research, Ghent University, Belgium
[5]The authors M J Wheatley and Y De Deene contributed equally to the work presented in this paper.

E-mail: Morgan.Wheatley@hdr.mq.edu.au

**Abstract**. Chloroform in the FlexyDos3D dosimeter acts as a radical initiator which brings about the colour change of the dosimeter when the radicals react with the leucomalachite green. However, the volatility of the chloroform likely results in a rapid loss of chloroform from the dosimeter. Gravimetric analysis and NMR diffusion-ordered coherence spectroscopy were used to measure the diffusion and evaporation rates of chloroform from the dosimeter and found that both rates were both significantly large resulting in a rapid loss of chloroform. Dose maps of irradiated phantoms aged for different times found a significant difference in the dose measurements of the dosimeter, likely a result of the different chloroform concentrations within the dosimeters.

## 1. Introduction

The FlexyDos3D dosimeter consists of an optically-transparent silicone elastomer (Sylgard® 184, Merck), chloroform, and a leucodye, leucomalachite green (LMG). The chloroform is used to produce radicals when irradiated that will react with the LMG, converting the LMG to malachite green (MG). This conversion of LMG to MG will result in the dosimeter changing from crystal clear to green.

Chloroform will evaporate rapidly when it is exposed to standard atmospheric conditions in an unenclosed space. Evaporation of chloroform from the FlexyDos3D dosimeter would likely result in less radicals being produced when the dosimeter is irradiated, resulting in less colour change in the dosimeter. This would likely also lead to spatial differences in the chloroform distribution within the dosimeter.

In this current study, the evaporation and diffusion rates of chloroform within the FlexyDos3D dosimeter were measured, both by gravimetric analysis and NMR diffusion-ordered coherence spectroscopy. Phantoms were also constructed and irradiated at different times after manufacture to see the effect of the evaporation of chloroform on the dose distributions within the dosimeter.

IOP Conf. Series: Journal of Physics: Conf. Series **1305** (2019) 012044 doi:10.1088/1742-6596/1305/1/012044

## 2. Methods and Materials

*2.1 Dosimeter fabrication*

The dosimeter, known as 'FlexyDos3D', consists of an optically-transparent silicone elastomer (Sylgard® 184, Merck), chloroform, and a leucodye, LMG. The silicone elastomer initially consists of a separate base and curing agent (CA), which cures when mixed together.

*2.2 Irradiation*

In fundamental studies, samples were irradiated with UVC light. For the final dosimetry tests, samples were irradiated with high-energetic X-rays on a linear accelerator (linac).

*2.2.1 UVC irradiation.* The cuvette samples were irradiated with a 9 W UVC germicidal lamp (TUV PL-S 9 W, Philips). The light intensity at 100 mm from the end of the lamp was measured with a UVC light reader (UVC-254, Lutron Electronic Enterprise Co., LTD) to be 5.22 W / $m^2$.

The cuvettes were placed 100 mm away from the UV lamp which was suspended in a polystyrene box. The box's internal surface was covered with aluminium foil to homogenize the UV flux. The cuvettes were enclosed on all sides but the one facing the UV lamp.

*2.2.2 Linac irradiation.* The cuvettes and phantoms were irradiated with high-energetic photon beams using a clinical Varian 21EX linac (Varian Medical Systems, Palo Alto, USA) with a beam quality of 6 MeV. All cuvettes were irradiated within a 10 cm × 10 cm field size at an SSD of 100 cm with the centre of the cuvettes positioned at 1.4 cm depth (Dmax). Cuvettes were each irradiated in 2 Gy increments up to a maximum dose of 20 Gy. The phantoms were irradiated with a photon beam with a 2 cm x 2 cm field size and an SSD of 100 cm and 2000 MU corresponding with a dose of 17.2 Gy (Dmax) at a depth of 1.4 cm (Dmax).

*2.3 $^1H$ NMR diffusion study*

5 mm NMR tubes were filled with different combinations of the FlexyDos3D dosimeter, deuterated water ($D_2O$), deuterated chloroform ($CDCl_3$), and chloroform ($CHCl_3$) (table 1). $CDCl_3$ or $D_2O$ were used in the samples to lock the $^1H$ signal.

A diffusion-ordered spectroscopy (DOSY) sequence was used to measure the diffusion coefficients of the chloroform and water in the samples, and the sequence was based on a stimulated echo sequence with a gradient strength of 0.535 T / m, a gradient duration (δ) of 1 ms, and a diffusion time (Δ) of 1 s.

Samples S3 and S4 consisted of two layers of materials in the NMR tube; S3 had a layer of cured FlexyDos3D beneath a layer of deuterated water, and S4 had a layer of non-deuterated chloroform beneath a layer of deuterated water, with the deuterated water being used in both cases to lock the NMR signal. Samples S1 and S2 were used as an independent experimental control to determine the self-diffusion coefficient of pure chloroform. Sample S3 was cured in an oven for 2 hours at 60 °C.

**Table 1.** Mixtures for NMR measurements.

| NMR Sample Name | $CDCl_3$ [% w/w] | $CHCl_3$ [% w/w] | $D_2O$ [% w/w] | FlexyDos3D [% w/w] |
|---|---|---|---|---|
| S1 | 100 | 0 | 0 | 0 |
| S2 | 50 | 50 | 0 | 0 |
| S3 | 0 | N/A | 50 | 50 |
| S4 | 0 | 50 | 50 | 0 |

*2.4 Depth-dose profile over time study*

Two 100 mm tall cylindrical phantoms with 58 mm diameters were constructed; one phantom was made 14 days before irradiation and the other on the day of irradiation. Cuvettes were also filled with the same

IOP Conf. Series: Journal of Physics: Conf. Series **1305** (2019) 012044 doi:10.1088/1742-6596/1305/1/012044

batch of dosimeter and were irradiated with a set of known doses which were used to calibrate the acquired phantom images from the optical cone-beam CT scanner to three-dimensional dose maps.

*2.5 Evaporation of chloroform study*

Open-topped cylindrical glass vials were filled with different amounts of the dosimeter which allowed chloroform to evaporate from the top surface of the dosimeter. Two groups of vials were made; one group was cured for 2 hours at 60 °C in an oven and the other was left to cure at room temperature. 15 samples were made with each containing up to 15 g of the dosimeter in 1 g increments. Another vial was also filled with 10 g of chloroform to measure the evaporation rate of pure chloroform.

The mass of each cylinder was measured immediately after manufacture and at various times over the next 2 months (60 days) to determine the mass loss relative to the first measurement. Another measurement was taken after 116 days to measure the final mass of the samples.

*2.6 Measurements and imaging*

*2.6.1 Optical spectroscopy and absorbance measurements.* LMG is transparent but when irradiated it converts to MG which shows an increase in absorbance at red wavelengths of light. The absorbance was compared at 630 nm, which showed a large increase, and 480 nm, which showed a minimal change in absorbance with increasing dose. The difference in these wavelengths could then be used to determine the dose delivered to the dosimeter.

$$\Delta OD = OD_{630\mathrm{nm}} - OD_{480\mathrm{nm}} \quad (1)$$

Optical spectra were acquired with a high-resolution spectrophotometer (USB4000, Ocean Optics) equipped with a cuvette holder.

*2.6.2 Dual-wavelength optical cone-beam CT imaging.* A modified dual-wavelength cone-beam optical scanner based on the Modus Vista™ scanner (Modus Medical Devices Inc, London, Canada) was used to image the phantoms and cuvettes. Optical CT images were constructed using an in-house developed cone-beam reconstruction software using the Matlab (The MathWorks, Inc.) environment.

Red and blue LEDs provided the two sets of wavelengths which were used to image the phantoms [1].

## 3. Results and Discussion

*3.1 Diffusion NMR*

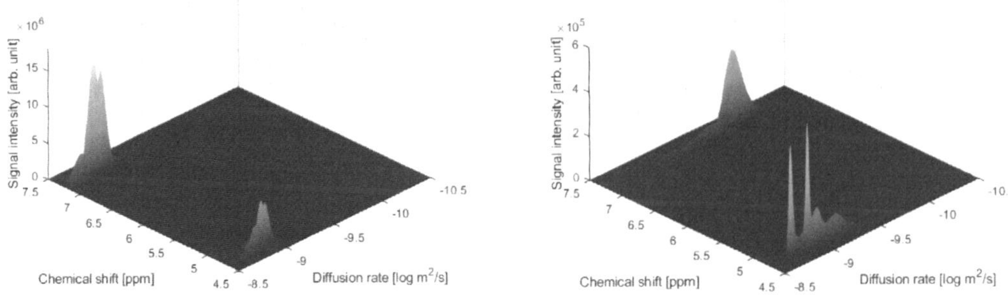


(a) (b)

**Figure 1.** Two-dimensional DOSY spectra of a sample containing 50 % chloroform and 50 % deuterated water (a) and a sample containing 50 % FlexyDos3D and 50 % deuterated water (b).

The peaks for chloroform and water can be resolved at 7.2 ppm and 4.7 ppm, respectively (figure 1). The diffusivities of the chloroform and water sample (S4) were both measured as 1.61 x $10^{-9}$ $m^2$ / s ($\sigma$ = 0.28 x $10^{-9}$ $m^2$ / s) and 1.36 x $10^{-9}$ $m^2$ / s ($\sigma$ = 0.16 x $10^{-9}$ $m^2$ / s), respectively. The corresponding diffusion coefficients in the FlexyDos3D sample (S3) were measured as 0.15 x $10^{-9}$ $m^2$ / s ($\sigma$ = 0.04 x $10^{-9}$ $m^2$ / s) and 1.48 x $10^{-9}$ $m^2$ / s ($\sigma$ = 0.52 x $10^{-9}$ $m^2$ / s), respectively (figure 1b). The measured diffusion coefficient of the water is slightly less than that found in the literature (1.69 x $10^{-9}$ $m^2$ / s at 295 K (21.85 °C) [2]).

*3.2Dose distribution of a photon beam*

A noticeable difference is seen between the dose distribution obtained in a phantom irradiated 5 hours post-manufacture and in a phantom irradiated 14 days post-manufacture (figure 2).

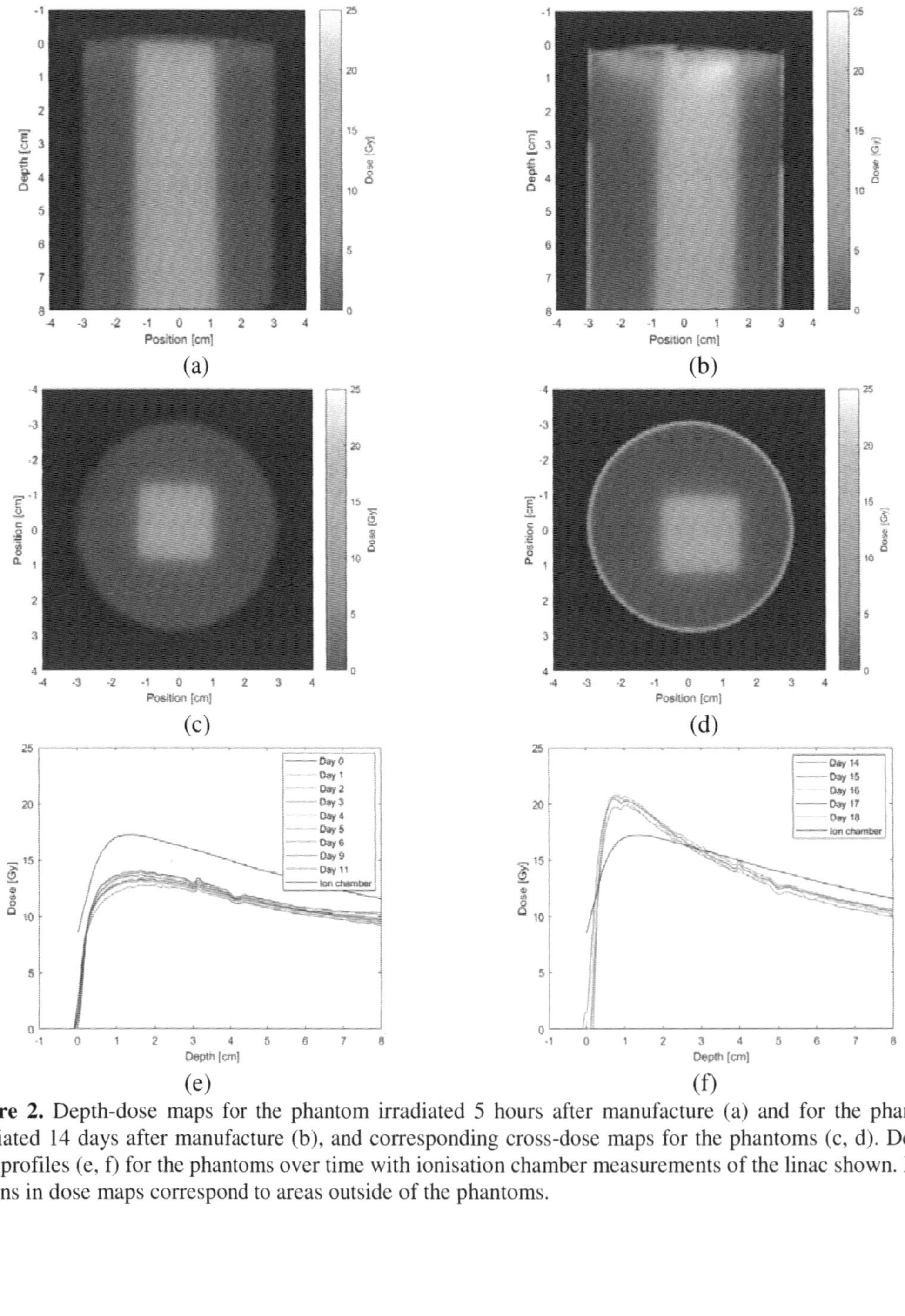


**Figure 2.** Depth-dose maps for the phantom irradiated 5 hours after manufacture (a) and for the phantom irradiated 14 days after manufacture (b), and corresponding cross-dose maps for the phantoms (c, d). Depth-dose profiles (e, f) for the phantoms over time with ionisation chamber measurements of the linac shown. Dark regions in dose maps correspond to areas outside of the phantoms.

The different responses between the two phantoms was also evident from the dose-response of both batches of calibration cuvettes (figure 3).

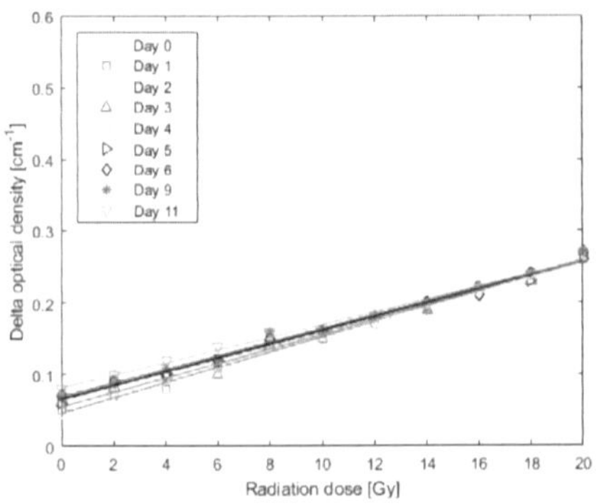


(a)

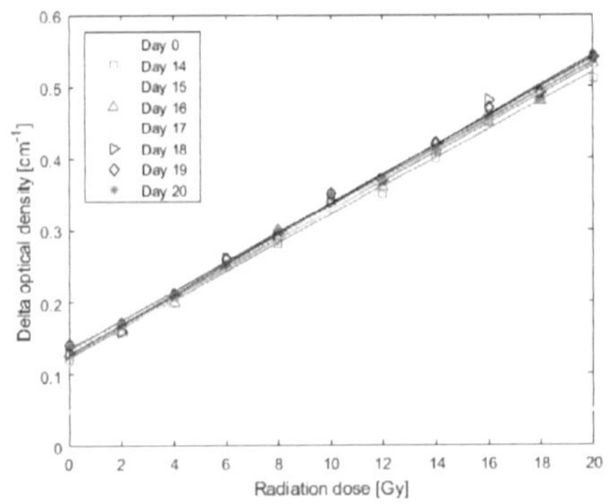


(b)

**Figure 3.** Dose-response of the cuvettes irradiated on the day of construction (a) and 14 days after construction (b) using equation 1.

The phantom that was constructed 14 days prior to irradiation still showed a significant dose-response, as shown in figure 3b. The dose-response sensitivity was twice that of the phantom constructed on the day of irradiation, shown in figure 3a. A significant dose offset was also seen in the 0 Gy dose samples between the day of manufacture and 14 days later.

The depth-dose responses of the phantoms were calibrated using calibration curves obtained from the calibration cuvettes (figure 3). Dose profiles and maps of the two phantoms are shown in figure 2. The depth-dose profile in the dosimeter constructed on the day of irradiation is significantly less than that of the expected depth-dose response (figure 2e). The phantom constructed 14 days before irradiation has a depth-dose profile that is significantly larger near the surface than that the expected depth-dose response (figure 2f). A yet unexplained artefact appears at the surface of the aged dosimeter, as shown in figure 2d.

### *3.3 Chloroform evaporation*

Measurements of the pure chloroform evaporation samples showed a constant mass loss rate over time of 0.19 g / hr. Using equation 1 and assuming the chloroform is very pure ($C_0(t)$ = 1 g / g) and that the chloroform concentration in the atmosphere is negligible at all times ($C_s(t)$ = 0 g / g), the dimension of proportionality ($\alpha$) is equal to the mass loss rate [3].

$$M_t(t) = M_\infty\left(1 - \sum_{n=1}^{\infty}\frac{2L^2\exp\left(-\frac{\beta_n^2 D t}{l^2}\right)}{\beta_n^2(\beta_n^2+L^2+L)}\right) \tag{1}$$

$$\frac{\mathrm{d}M_t(t)}{\mathrm{d}t} = \alpha(C_0(t) - C_s(t)) \tag{2}$$

$$L = \frac{l\alpha}{D} \tag{3}$$

$$\beta_n\tan(\beta_n) = L \tag{4}$$

where: $M_t(t)$ is the chloroform mass loss at time $t$ [g], $M_\infty$ is the chloroform mass loss at an infinitely large time [g], $\mathrm{d}M_t(t)$ / $\mathrm{d}t$ is the mass loss rate of chloroform [g / s], $C_0(t)$ is the concentration of chloroform within the material [g / g], $C_s(t)$ is the concentration of chloroform at the surface of the material [g / g], $\alpha$ is a constant of proportionality which relates change in mass over time of the chloroform to the concentration difference between the inside and outside of the material [g / s], $l$ is the height of the material [m], $\beta_n$ are the roots of equation 4 [arb. units], and $D$ is the diffusion rate of the chloroform [$m^2$ / s].

The samples all lost mass over time (figure 4a) and appear to approach a mass loss of 3 % which is equal to the amount of chloroform in the samples initially (figure 4b).

Apparent diffusion values were obtained by fitting the mass loss over time (figure 4a) using equations 1 – 4. The calculated diffusion values increased from 0.25 x $10^{-9}$ $m^2$ / s to 0.40 x $10^{-9}$ $m^2$ / s with higher initial masses of the dosimeter having a greater diffusion rate. The measured diffusion coefficient is dependent on the initial phantom volume indicates that the diffusion of chloroform in the elastomer is non-Fickian. No significant differences were observed between the diffusion rates of samples cured at different temperatures.

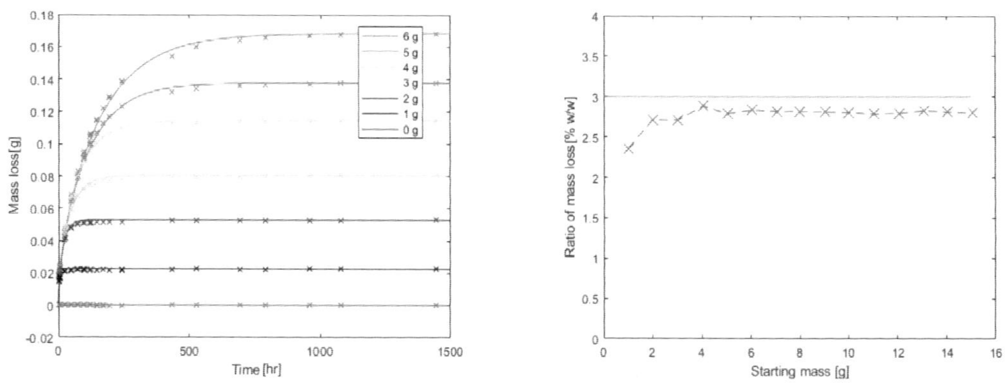


(a) (b)

**Figure 4.** Mass loss over time for the 60 °C cured samples up to 6 g (a). Fits using equations 1 – 4 are shown with the solid lines. Ratio of mass loss in the 60 °C cured sample after 116 days (b). A solid line has been added in (b) to show the 3 % mass loss value. The 120 °C results have not been shown as they are almost identical to the 60 °C results.

## 4. Conclusion

It was shown by gravimetric studies and two-dimensional NMR diffusion-ordered coherence spectroscopy that the evaporation of chloroform from the FlexyDos3D dosimeter is significant. The calibrated dose distributions of the phantoms did not match that of the expected dose distribution. The differences in the measured dose distributions in the phantoms that were irradiated at different times after manufacture were the likely result of evaporation of chloroform over time from the dosimeter.

## 5. Acknowledgements

The authors would like to thank V Calliet for his help with the irradiation of the dosimeter.

## 6. References

[1] De Deene Y et al 2015 *Phys. Med. Biol.* 60 1543-63
[2] Dahal U and Adhikari N P 2012 *J. Mol. Liqu.* 167 34-9
[3] Crank C 1975 *The mathematics of diffusion* (Oxford science publications.) Oxford

IOP Conf. Series: Journal of Physics: Conf. Series **1305** (2019) 012045 doi:10.1088/1742-6596/1305/1/012045

# Radiation induced degradation of rhodamine 6G and 7-Diethylamino-4-methylcoumarin in nano-clay gel for use in dosimeter

**T. Maeyama[1,2], A. Mochizuki[1] and T. Takanashi[2]**
[1]Department of Chemistry, School of Science, Kitasato University, Kanagawa, Japan
[2]3D Gel Dosimeter Research Laboratory, Cluster for Science, Technology and Innovation Hub, RIKEN, Saitama, Japan

E-mail: maeyama@kitasato-u.ac.jp

**Abstract**. Rhodamine 6G (R6G) and 7-diethylamino-4-methylcoumarin (7D4MC) were applied to a nanoclay gel dosimeter based on radiation-induced degradation. The radiological properties were evaluated under X-ray irradiation. The fluorescent dyes showed linear degradation with an increase in dose. In addition, the distribution of fluorescence induced by inhomogeneous irradiation was maintained for two months and a suppressed diffusion of the fluorescent dyes in the gel matrix was observed.

## 1. Introduction

The 3D gel dosimetry [1] is a useful tool for dose verification in advanced radiotherapy because a gel dosimeter can be used as phantom in the same way as in an actual treatment chain of radiotherapy [2, 3]. Several studies have contributed to the development of gel dosimeters using fluorescence spectroscopy [4-8]. Recently, we reported a nanoclay gel-based radio-fluorogenic gel (NC-RFG) dosimeter [9]. These gel dosimeters use a chemical compound, known as a radio-fluorogenic probe or a fluorescence probe, in which a fluorescent dye is produced when a non-fluorescent dye reacts with radicals formed via the radiolysis of water.

Radiation-induced degradation of fluorescent dyes is also used in dosimeters, particularly for high-dose dosimetry in the liquid state [10, 11]. When rhodamine 6G (R6G) is used as a fluorescent dye, a gelatin-based gel dosimeter shows a good dose linearity within 0-200 Gy [12].

This study reports on a nanoclay-based gel dosimeter that uses the radiation-induced degradation of a fluorescent dye. Nanoclay is easily dispersed in water without heating and forms thixotropic gel. Several kinds of gel dosimeters containing nano-sized particles have been discussed for Fricke gels [13-19] and for polymer gels [20, 21]. In the initial investigations, R6G and 7-diethylamino-4-methylcoumarin (7D4MC) [22], are explored. The fluorescence of these dyes degrades with increasing radiation dose. In addition, the distribution of fluorescence induced by inhomogeneous irradiation is maintained.

## 2. Material and methods

R6G and 7D4MC were used as the fluorescent dyes in this study. On the basis of a previously reported method [9, 16], we prepared gel samples containing 1.5-wt% nanoclay (Laponite XLG) and 1-μM fluorescent dye, which were sealed into 5-mL vials. Gel samples having an increased clay concentration

IOP Conf. Series: Journal of Physics: Conf. Series **1305** (2019) 012045 doi:10.1088/1742-6596/1305/1/012045

up to 2.5 wt% were prepared and sealed into Petri dishes (2 cm high and 9 cm in diameter) for 2D measurements. Table 1 summarizes the details of these preparation conditions. The prepared samples were irradiated using an industrial X-ray machine (250 kVp, 8 mA, with a 1-mm-thick Al filter; Rigaku Radioflex 250CG, Japan) at a dose rate of 1.64 ± 0.01 Gy/min, determined using aqueous Fricke solution dosimetry. Two or three days post irradiation, fluorescence measurements of the irradiated samples were done using a F-4500 Spectrofluorometer (Hitachi, Japan). Two months (58 days) after the inhomogeneous irradiation, 2D distribution measurements were conducted using a digital camera in a dark box under 365-nm excitation.

**Table 1.** Chemical composition of the fluorescent dye gels.

| No | Fluorescent dye | | Clay [wt%] | Dose [Gy] | Fluorescent measurement [nm] | | |
|---|---|---|---|---|---|---|---|
| | | | | | Ex | Em | Slit width Ex/Em |
| 1 | R6G | 1 μM | 1.5 | 0–50 | 500 | 562 | 2.5/5.0 |
| 2 | 7D4MC | 1 μM | 1.5 | 0–40 | 394 | 448 | 2.5/5.0 |
| 3 | 7D4MC | 0.25 μM | 2.5 | 0, 30, 60 | 365 (peak) | -- | Digital Camera |

## 3. Results and discussion

Fig. 1 shows the radiation-induced degradation in the fluorescence properties of R6G. By X-ray irradiation, degradation of a peak at 555 nm was observed in the excitation spectra (dashed lines) and a peak at 562 nm decreased in the fluorescence spectra (solid lines). A small Stokes shift (the difference between the excitation and emission maxima) compared to a R6G solution without clay was evident, and a red shift in the excitation and emission spectra was observed upon the addition of the nanoclay (the red and black dot-line in Fig. 1 (A), respectively). In contrast, good linearity in the differential fluorescence intensity with increasing dose was observed within 0–50 Gy under 500-nm excitation.

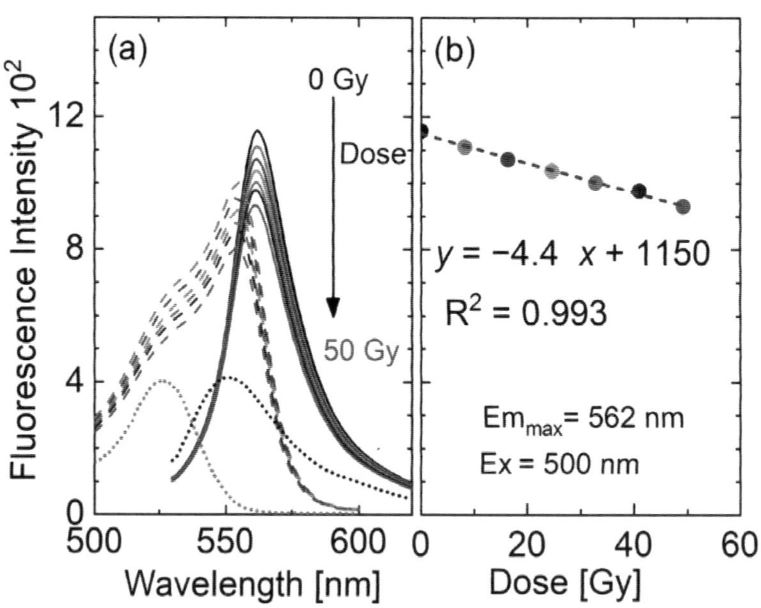


**Figure 1.** (a) Fluorescence properties of 0.1-μM R6G in 1.5-wt% clay before and after irradiation. Dashed and solid lines show the excitation and fluorescence spectra, respectively. Red and black dot-line show the excitation and fluorescence spectra of R6G in the water, respectively. (b) Dose dependence of fluorescence intensity at 562 nm using 500-nm excitation light.

In the case of 7D4MC, the fluorescence intensity also showed a linear decrease with increase in dose within 0-40 Gy (Fig. 2). In Fig. 2(a), degradation of a peak at 394 nm was observed in the excitation spectra by X-ray irradiation, whereas a peak at 448 nm decreased in the fluorescence spectra.

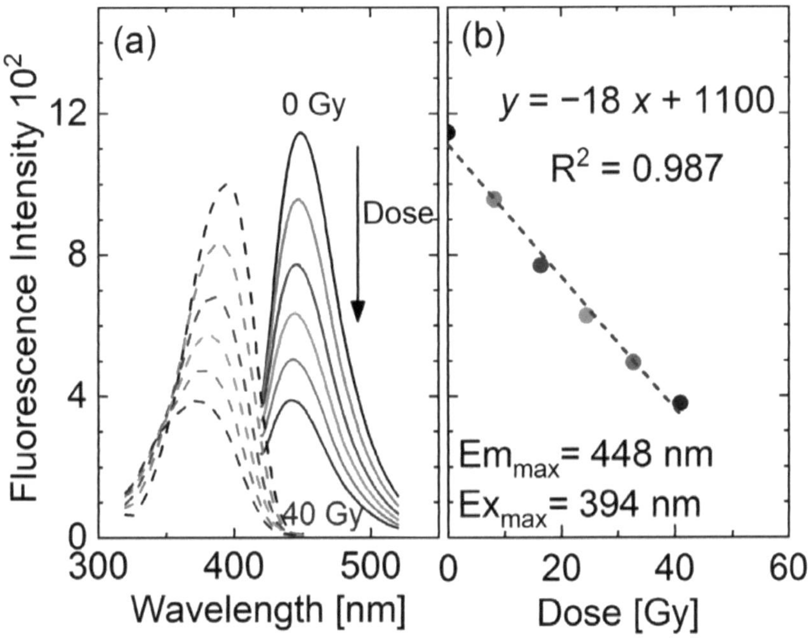


**Figure 2.** (a) Fluorescence properties of 1-μM 7D4MC in a 1.5-wt% clay before and after irradiation. Dashed and solid lines show the excitation and fluorescence spectra, respectively. (b) Dose dependence of the fluorescence intensity at 448 nm using 394-nm excitation light.

From Fig.1(b) and Fig.2(b), the radiolytic decomposition yields of the fluorescence dyes were roughly estimated from the ratio of degradation from the initial concentration (1 μM) as G(−R6G) = 0.00385±0.0001 μM/J and G(−7D4MC) = 0.016±0.001 μM/J. The radiolytic decomposition yield of 7D4MC is three times higher than that of R6G under the same experimental conditions. In addition, the large Stokes shift of 7D4MC (54 nm) is convenient to reduction of noise from excitation light in the case of evaluation of 2 or 3D fluorescence measurement.

An aqueous R6G solution shows a high radiolytic decomposition yield of 0.3 μM/J using 1 to 0.1-mM R6G and 1-M ethyl alcohol [10]. This value is 80 times higher than that of R6G using a nanoclay gel. In contrast, the sensitivity of the gelatin-based R6G gel was approximately 0.0006–0.003 μM/J, which depended on the concentration of inorganic ion radiosensitizer [12], and were of the same order as the results obtained using the nanoclay gel. Detailed discussion on the differences in yield require further experiments concerning the concentration dependence of the dyes [10], pH dependence [11], and clay effects [22].

An improvement in sensitivity is required for the use of 3D gel dosimetry in radiation therapy. For example, a nanoclay-based RFG is reported for doses within 0-5 Gy [9], which is several tens of times more sensitive than the results reported herein. Since sensitivity of radiation-induced fluorescence dye degradation is dependent on the ratio of the change in the total intensity, low sensitivities were shown than the RFG, if the measured yields by the reaction with water radiolysis radicals is the same as that in the RFG.

In Fig. 3, a nanoclay gel with 2.5-wt% clay and 0.25-μM 7D4MC was irradiated with 15 and 30 min irradiation time that are roughly equivalent to 30 and 60 Gy, respectively. This figure was obtained using a digital camera placed in a dark box using a 365-nm excitation light, two months after irradiation. We consider that the diffusion of 7D4MC is almost suppressed via adsorption of the nano-sized clay because we previously observed diffusion in a gel using anionic fluorescent dyes, such as fluorescein, uranine, and aromatic carboxylic acids (the details are not shown herein). In addition, the diffusion suppression of cationic substance in nanoclay gel was previously reported [14-16].

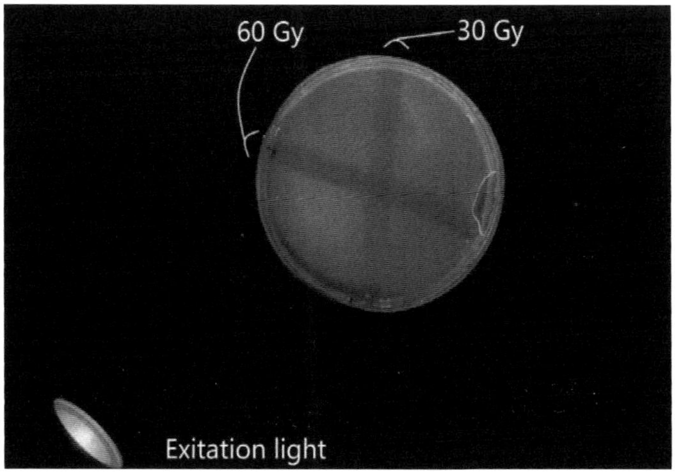


**Figure 3.** Irradiated nanoclay gel containing 7D4MC with doses 30 and 60 Gy.

## 4. Conclusion

We reported a nanoclay gel dosimeter that utilized the radiation-induced degradation of fluorescent dyes. A gel using R6G or 7D4MC showed a linear response across the radiation dose range of 0-50 Gy. The diffusion of 7D4MC was suppressed since the dye distribution was maintained for two months after irradiation. The development of a 3D evaluation method of a fluorescence gel dosimeter is ongoing.

## 5. Acknowledgments

This study was supported by MEXT of JAPAN, KAKENHI Grant Number JP17K07013. We thank S. Hase, A. Kato and N. Sato (Kitasato Univ.) for experimental support.

## 6. References

[1] Watanabe Y *et al* 2017 *World J. Rad.* **9** 112-25
[2] Baldock C 2009 *J. Phys.: Conf. Ser.* **164** 012002
[3] Baldock C et al 2010 *Phys. Med. Biol.* **55** R1-63
[4] Warman J M *et al* 2009 *Phys. Med. Biol.* **54** 3185-200
[5] Warman J M *et al* 2011 *Phys. Med. Biol.* **56** 1487-508
[6] Sandwall P A *et al* 2016 *J. Radioanal. Nucl. Chem.* **307** 2505-08
[7] Tiantian Y *et al* 2017 *Biomed. Phys. Eng. Express.* **3** 027004
[8] Sandwall P A *et al* 2018 *Bioeng.* **5** 53
[9] Maeyama T and Hase S 2018 *Radiat. Phys. Chem.* **151** 42-46
[10] Pikaev A K *et al* 1991 Uses of aqueous solutions of rhodamine dyes for measurements of high doses in radiation processing *High dose dosimetry in radiation processing* International Atomic Energy Agency (IAEA) 147-58
[11] Beshir W B *et al* 2014 *Appl. Radiat. Isot.* **89** 13-7
[12] Benevides C A *et al* 2015 *Appl. Opt.* **54** 6935-9
[13] Maeyama T *et al* 2013 1183-86. In *Proceedings of World Congress on Medical Physics and Biomedical Engineering May 26-31, 2012, Beijing, China* (Ed. Long M)
[14] Maeyama T *et al* 2014 *Radiat. Phys. Chem.* **96** 92-6

[15] Maeyama T *et al* 2016 *J. Radiat. Res.* **57** 318-24
[16] Maeyama T et al 2017 *J. Phys. Chem. B.* **121** 4238-46
[17] Babu S E S *et al* 2017 *J. Phys. Conf. Ser.* **847** 012053
[18] Gallo S *et al* 2018 *Sens. actuators. B Chem.* **272** 618-25
[19] Vedelago J *et al* 2018 *Appl. Radiat. Isot.* **141** 182-6
[20] Titus D *et al* 2016 *Appl. Microbiol. Biotechnol.* **100** 4803-16
[21] Maeyama T *et al* 2018 *Radiat. Phys. Chem.* **146** 121-5
[22] Wlodarczyk P *et al* 1996 *J. Mater. Chem.* **6** 1967-9

IOP Conf. Series: Journal of Physics: Conf. Series **1305** (2019) 012046 doi:10.1088/1742-6596/1305/1/012046

# Gel dosimetry measurement of dose enhancement bismuth-based nanoparticles in radiation therapy

**Azimeh Rajaee[1], Shi Wang[1], Lingyun Zhao[2] and Yaqiang Liu[1]**
[1]Institute of Medical Physics and Engineering, Depaqtment of Engineering Physics, Tsinghua University, Beijing 100084, China
[2]State Key Laboratory of New Ceramics and Fine Processing, School of Materials Science and Engineering, Tsinghua University, Beijing 100084, China

Email:rajaeea10@mails.tsinghua.edu.cn

**Abstract:** Recently bismuth-based nanoparticles as a promising radiosensitizer have drawn great attention in radiation therapy. To prove physical dose enhancement effect of the nanoparticles, gel dosimeters can be considered as an ideal method. This study aims to prove the applicability of bismuth ferrite nanoparticles as a magnetic localized dose enhancement agent by gel dosimetry method. Bismuth ferrite nanoparticle was synthesized by the conventional sol-gel method. Then we investigated dose enhancement property of the nanoparticles with gel dosimetry. MAGIC Polymer Gel dosimeters with nanoparticles were prepared and irradiated. According to gel dosimetry assay, for 0.5 mg/ml concentration of bismuth ferrite nanoparticles dose enhancement factor were obtained as 2 and 1.6 at 160 keV and average energy of 380 keV, respectively. Moreover, radiosensitiser effect of bismuth ferrite nanoparticles in presence of a low dose rate brachytherapy source (125-I) was investigated by Monte Carlo method. Whereas bismuth ferrite nanoparticles have magnetic property, we made a biodegradable spacer (fiducial) brachytherapy loaded with the nanoparticles for delivering nanoparticles and drug by applying an external magnetic field.

## 1. Introduction

Over the two last decades, extensive studies have been drawn on nanoparticles as radiosensitizer in radiation therapy. High atomic number nanoparticles can increase deposited dose in tumore locally due to production of photoelectric effect and agure electron. A large number of high-Z metal nanoparticles have been widely investigated as radiosensitizer such as gold, bismuth, platinum, tungsten, and gadalinium [1, 2]. In recent years, many studies have been focused on bismuth-based nanoparticles as radiosensitizer and contrast agent in radiation therapy and imaging due to high atomic number (Z=82), high photoelectric absorption, low cost, and low toxicity [3, 4]. Auger electron and photoelectrons which are produced for nanoparticles have short range, then it would be difficult to measure dose enhancement from nanoparticles by routine dosimeters. To deal with this challenge, 3D dosimeters [5, 6] can be a good choice [7].

In this study, we employed MAGIC Polymer Gel dosimeter to confirm radiosensitizer effect of bismuth ferrite nanoparticles ($BiFeO_3$). Moreover, dose amplification effect of $BiFeO_3$ nanoparticles for high concentration and bigger size at low dose rate (LDR) brachytherapy source was investigated by Monte Carlo modeling.

In spite of the large body of studies on delivering nanoparticles, delivering sufficient amount of nanoparticles concentration into the tumor is challenging. As a solution to delivering nanoparticles for internal radiation therapy (Brachytherapy), Kumar *et al* [8], and Sinha *et al* [9], suggested that the fiducial markers and spacers could be used as a carrier of nanoparticles and radiation-sensitizing drugs at brachytherapy. In radiation therapy, fiducial markers and spacers are used to determine the position of the tumor for improving special accuracy. In this study, taking the advantage of the magnetic property of bismuth ferrite nanoparticles, we tried to design a spacer (fiducial) for delivering nanoparticles and drugs. Inductive heating property of the $BiFeO_3$ nanoparticles can raise the temperature in presence of the alternative magnetic field, which increases the biodegradation rate and as a result the releasing of nanoparticles and drug from a biodegradation polymer spacer. Drug release from the biodegradable spacers was measured in presence of the magnetic field and without it.

## 2. Methods

### 2.1. *Synthesis of $BiFeO_3$ nanoparticles*

Bismuth ferrite nanoparticles ($BiFeO_3$) were prepared based on Wang *et al* work [10]. In brief, bismuth nitrate Bi $(NO_3)_3 \cdot 5H_2O$ and iron nitrate Fe $(NO_3)_3 \cdot 9H_2O$ were dissolved in ethylene glycol (EG). The mixture was stirred at 140 °C to get the gel. The gel was heated to 450 °C for 30 min. Finally, to obtain the nanoparticles, the powder was washed and dried in vacuum at 60 °C. Size of the nanoparticles was measured by Image J 1.6.0_24 G software for more than 100 particles, which were chosen randomly from SEM image. It is found that nanoparticles' size is of 28nm±5.5. Biocompatibility of the nanoparticles was confirmed at previous works [3, 11].

### 2.2. *Gel dosimetry*

To approve dose enhancement effect of $BiFeO_3$ nanoparticles as a radiosensitizer, MAGIC gel dosimeter was prepared based on the Fong *et al* [12] method with some modifications. In order to improve the sensitivity of the gel and oxygen scavenging, tetrakis (hydroxymethyl) phosphonium chloride (THPC) was added to the gel. Prepared gel was divided into two groups. One group without nanoparticles and the other with 0.5 mg/ml nanoparticles. The gels were irradiated at 160 keV energy ( RS-2000 x-ray biological irradiator 25 mA with 0.3 mm Cu filter) and the average energy of 380 keV (Iridium (Ir -192), after loading Varian GammaMed plusIX, brachytherapy source). After one week, the gels were read out by MRI scanner (Philips -3TNX, Netherlands) with a head coil (Figure 1and 2).

### 2.3. *Monte Carlo simulation*

Iodine 125 (125-I, SelectSeed 130.002) as low dose rate (LDR) brachytherapy source was chosen for evaluation dose enhancement effect of $BiFeO_3$ nanoparticles [13]. The I-125 source was modeled by general-purpose Monte Carlo N-Particle Transport (MCNP6) code. In order to validate the code, we calculated TG-43 parameters such as radial dose function g(r) and anisotropy function $F(r,\theta)$. After validation of the radiation source, 100 nm $BiFeO_3$ nanoparticles with concentrations of 7 and 15 mg were simulated in a cubic tumor with dimension of 1×1×1 $cm^3$ that was placed in 1 cm from the source. The cube was divided into cells with dimension of 2×2×2 $mm^3$ and each cell was filled with smaller cells according to the concentration of nanoparticles. Dose distribution was calculated by F6 tally (MeV/g) and simulation was performed for $10^8$ histories.

### 2.4. *$BiFeO_3$ nanoparticles-loaded brachytherapy spacer*

Biodegradable spacer (fiducial) brachytherapy loaded with nanoparticles and drug was prepared based on Kumar *et al* [7] work with some modification. To prepare spacers, 500 mg poly (lactic-co-glycolic acid) PLGA (50/50), 10% Polyethylene glycol (PEG) and 50mg of docetaxel were dissolved in 1 ml acetone. The mixture of nanoparticles and acetone was added to PLGA solution. The mixture of all materials was stirred until obtained a viscous paste. The paste was transferred to a silicone tube in diameter of 1mm by a syringe. The tube was dried at 46 °C and after that dried material in shape of the tube was cut in 5mm to get spacers. Spacers were divided in two group for measuring drug release, one with applying external magnetic field and another without it. The spacers were placed in tubes

IOP Conf. Series: Journal of Physics: Conf. Series **1305** (2019) 012046 doi:10.1088/1742-6596/1305/1/012046

with phosphate-buffered saline (PBS, pH 7.4) and the tubes were incubated at incubator-shaker (37°C, 100 rpm). Every day one group of spacers were exposed to the alternative magnetic field (AMF) for 10 min. Then each day 0.5 ml PBS of each group (with and without applying AMF) was removed and replaced with fresh phosphate-buffered saline (PBS). The removed PBS was used for measuring drug release by UV-Visible spectrophotometer.

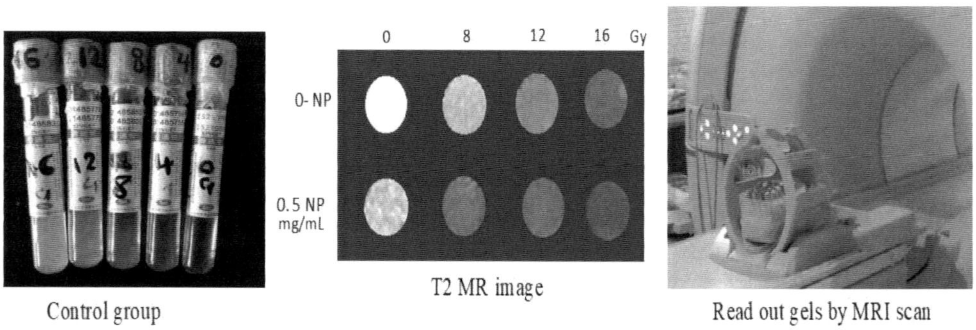


**Figure 1.** Gel dosimeter samples irradiated at 160 keV energy of photo and T2 MRI image.

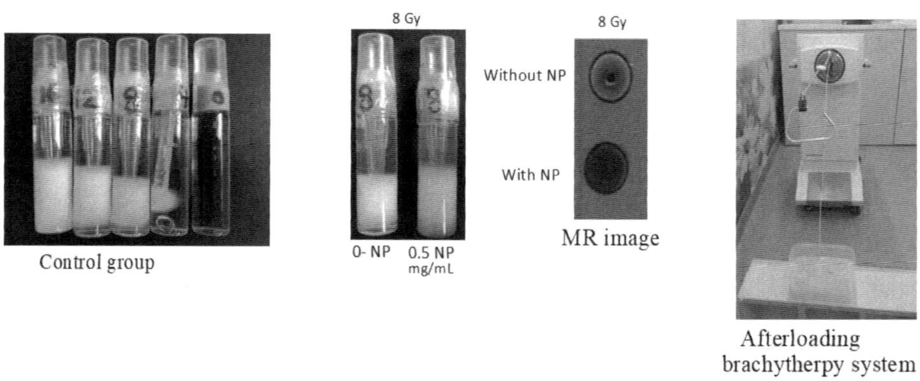


**Figure 2.** Gel samples irradiated at 380 Mean keV energy of photon iridium source by afterloading brachytherapy system.

## 3. Results and Discussion

MR imaging was performed to read polymer gel dosimeters. The spin-spin relaxation rate ($R_2$) as a function of absorbed radiation dose for gel samples with and without $BiFeO_3$ nanoparticles was shown in Figure 3 a and b. Gel data are fitted by a linear function, which indicates good correlations (R=0.99). Dose enhancement factor obtained DEF=2 at lower energy of 160 keV, while the factor is DEF=1.6 at higher average energy of 380 keV due to inverse proportion of photoelectric cross section to energy. The ability of the nanoparticles as a dose enhancement clearly is shown in Figure 1 and 2 from comparing gel tubes and MR image.

Results of Monte Carlo simulation of 125-I source for calculating the radial dose function, g(r) and the anisotropy function, F(r, θ) in this and a previous study with the same source is shown in Figure 4. The simulation results exhibited a good agreement for TG-43 parameters from the simulation and previous study [13]. Figure 5 shows enhancement in energy deposited with depth for bismuth and bismuth ferrite nanoparticles in size of 100 nm with 7 and 15 mg concentrations. Energy deposited in cells increases with concentration of nanoparticles

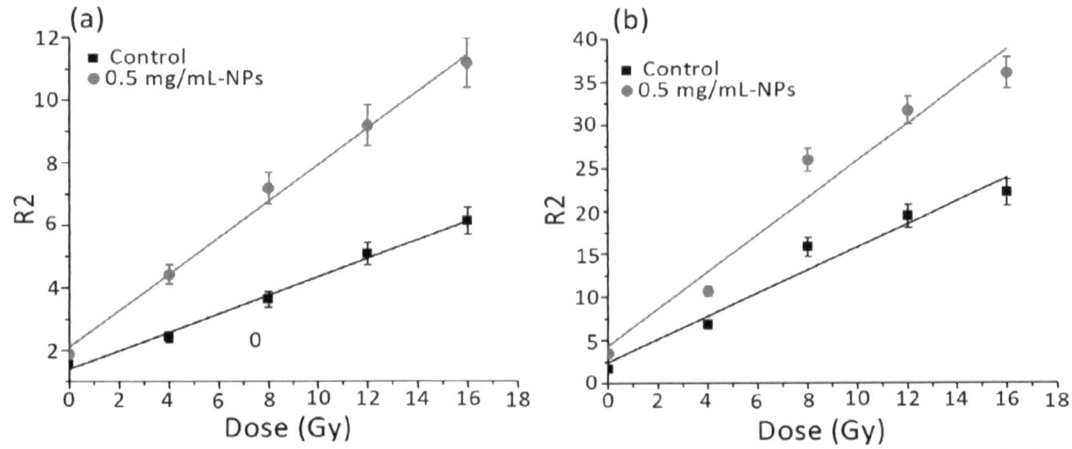


**Figure 3.** Gel dosimetry (a) dose response curves for gel pure and gel+$BiFeO_3$-Np irradiated at energy of 160 keV. (b) At average energy of 380 keV.

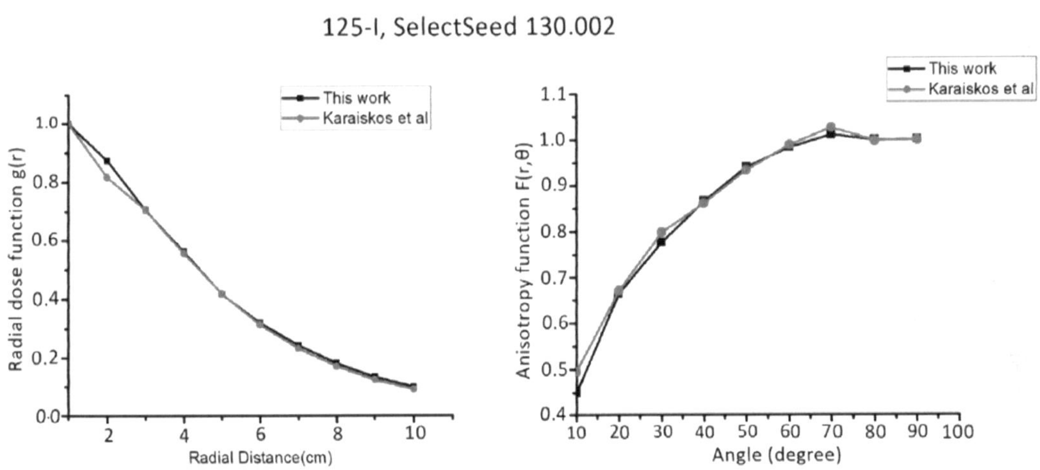


**Figure 4.** Comparison of radial dose function and anisotropy function of I-125, Selectseed 130.002 from the study and Karaiskos *et al* [13].

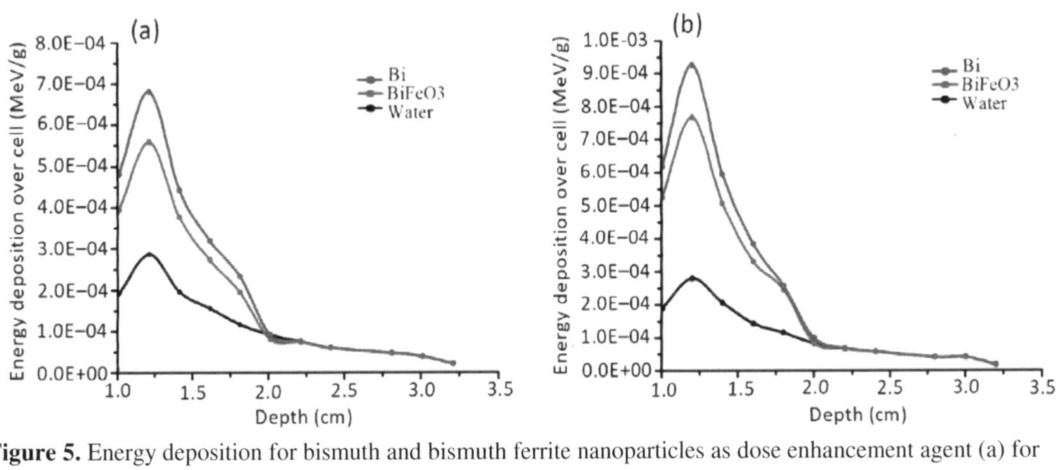


**Figure 5.** Energy deposition for bismuth and bismuth ferrite nanoparticles as dose enhancement agent (a) for 7mg and (b) for 15 mg.

Magnetic bismuth ferrite nanoparticles and docetaxel drug were loaded on PLGA-PEG spacers (Figure 6). The release profile of docetaxel drug from spacers (fiducials) with and without applying magnetic field was shown in Figure 6. After 10 days, 44% of the drug was released from spacers without applying magnetic field while it increased up to 90 % for the spacers with applying a magnetic field.

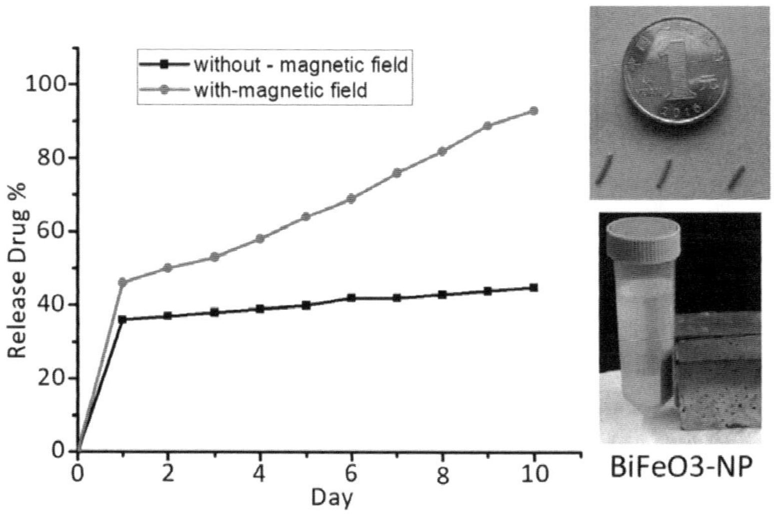


**Figure 6.** Drug release percentage from spacer (fiducial) with applying a magnetic field and without magnetic field.

## 4. Conclusion

In summary, we evaluated dose enhancement effect of bismuth ferrite nanoparticles with gel dosimetry and Monte Carlo modeling. The results proved that significant dose enhancement of the nanoparticles by polymer gel dosimeters which are one of a useful tool for the purpose. Moreover, based on inductive heating property of the $BiFeO_3$ nanoparticles in an alternative external magnetic field we designed a control delivery system for combination chemoradiation therapy.

## 5. References

[1] Kotb S *et al* 2016 *Theranostics* **6** 418-27
[2] Cooper D R *et al* 2014 *Front. Chem.* **2** 86
[3] Rajaee A *et al* 2018 *J. Biomed. Nanotechnol.* **14** 1159-68
[4] Deng J *et al* 2018 *Biomaterials* **154** 24-33
[5] Mather M L *et al* 2003 *Phys. Med. Biol.* **48** N269-75
[6] Hurley C *et al* 2006 *Nucl. Instrum. Meth. A* **565** 801-11
[7] Marques T *et al* 2010 *J. Phys.: Conf. Ser.* **250** 012084
[8] Kumar R *et al* 2015 *Int. J. Radiat. Oncol. Biol. Phys.* **91** 393-400
[9] Sinha N *et al* 2015 *Int. J. Radiat. Oncol. Biol. Phys.* **91** 385-92
[10] Wang X *et al* 2011 *Journal of Sol-Gel Science and Technology* **60** 1-5
[11] Staedler D *et al* 2015 *Nanomedicine:NBM* **11** 815-24
[12] Fong P M *et al* 2001 *Phys. Med. Biol.* **46** 3105-13
[13] Karaiskos P *et al* 2001 *Med. Phys.* **28** 1753-60

IOP Conf. Series: Journal of Physics: Conf. Series **1305** (2019) 012047 doi:10.1088/1742-6596/1305/1/012047

# Preliminary study of normoxic PAGAT polymer gel dosimeter by adding formaldehyde

**Libing Zhu[1], Yi Du[2], Xincheng Xiang[1], Song Zou[3], Xiangang Wang[1]**
[1]Institute of Nuclear and New Energy Technology, Tsinghua University, Beijing, 100084, China
[2]Key Laboratory of Carcinogenesis and Translational Research (Ministry of Education/Beijing), Department of Radiotherapy, Peking University Cancer Hospital & Institute, 52 Fucheng Road, Beijing, 100142, China
[3]Department of Chemistry, Tsinghua University, Beijing, 100084, China

E-mail: wangxiangang@tsinghua.edu.cn

**Abstract**. The performance of normoxic PAG (nPAG) gel dosimeter with respect to spatial integrity, temperature sensitivity and dose-rate dependency makes it more reliable than PAG gel dosimeter. As a 3D dose distribution measurement in dosimetry, nPAG gel dosimeter with THPC as anti-oxidant, named PAGAT was studied. 2.9% mass concentration of the formaldehyde solution was added into PAGAT to solidify the gel, named PAGAT-f. The melting point of PAGAT-f polymer gel was increased by adding formaldehyde and can be stored at room temperature after gel preparation without refrigeration. After irradiation with a 9 MV LINAC, magnetic resonance imaging (MRI) is employed as readout modality. It's found that by adding formaldehyde, $R_2$-dose response of PAGAT-f is similar to PAGAT but the dose response sensitivity of PAGAT-f is better than that of PAGAT at low doses. The results prove the feasibility and good temporal stability of PAGAT-f for dose distribution measurement.

## 1. Introduction

As a novel radiotherapy technique, intensity-modulated radiation therapy (IMRT) is able to deliver high doses of radiation to the target volume without harming the surrounding tissues. It's realized by the modulation of radiation fields with multileaf collimators that conform the radiation fields during the irradiation [1]. Due to the requirement of accurate radiotherapy, dose distribution becomes complex in IMRT and thus an effective dose measurement is necessary in quality assurance (QA).

Traditional ion chamber is one dimensional measurement and film dosimetry is two-dimensional measurement for dose distribution. Three-dimensional polymer gel dosimeters work on the principle of radiation-induced polymerization which is proportional to the absorbed dose and proportionately affects the spin-spin relaxation time of the gel. The performance of normoxic PAG (nPAG) gel dosimeter with respect to spatial integrity, temperature sensitivity and dose-rate dependency makes it more reliable than PAG and normoxic MAG gel dosimeter [2]. PAG type gel containing N, N'-methylene-bis-acrylamide, acrylamide, water and gelatin was studied by Oldham and Maryanski [3, 4], which is not normoxic. PAGAT is a promising polyacrylamide-based gel in dose distribution verification, which was proposed by Venning [5, 6]. The normoxic gel is composed of water, gelatin, N, N'-methylene-bis-acrylamide, acrylamide and THPC as anti-oxidant. De Deene [7] used ascorbic acid as anti-oxidant to compare nMAG and nPAG type gel dosimeters.

Temperature is a factor that may disturb the ability of gel-based dosimeter to precisely measure dose distributions in that at low temperatures (~25°C) the gel will melt, and three-dimensional dose information is lost [2]. PAGAT gel was stored in a 4 °C refrigerator to solidify [6, 8]. Formaldehyde was used to increase the melting point of MAGIC type gel [9-11] and MAGAT type gel [12] as both of them were not solid at room temperature [13]. 3% in volume comcentration of the formaldehyde solution was added into MAGIC gel and the melting point of it increased from 25 °C to 69 °C [9]. In this paper, we add 2.9% in mass concentration of the formaldehyde solution into PAGAT to solidify the gel, named PAGAT-f. Dose response of PAGAT-f and PAGAT was compared to show the potential of PAGAT-f in clinical application.

## 2. Materials and methods

### *2.1. Gel preparation*

4.9 % gelatin (Type A, Bloom 300, Sigma-Aldrich) was firstly mixed with 86.4 % ultra pure distilled water in a flask stirred magnetically and heated to 50 °C. When the gelatin was completely dissolved, the heater was turned off and the gel was cooled to about 38 °C. After that, 2.9% acrylamide (Energy Chemical) and 2.9% N, N'-methylene-bis-acrylamide (Energy Chemical) was added into the flask. After the dissolution of acrylamide and bis, 2.9 % formaldehyde solution and 10 mM THPC (Energy Chemical) were added. Stirring was keeping during the whole mixing process. The gel was pour into plastic bottles for dose response measurement. PAGAT-f gel was stored at room temperature (~30°C).

**Table 1.** PAGAT-f gel composition.

| Component | Mass concentration |
|---|---|
| Ultra pure deionized water | 86.4% |
| Gelatin (Type A, Bloom ~300, Sigma-Aldrich) | 4.9% |
| Acrylamide | 2.9% |
| Bis | 2.9% |
| THPC | 10 mM |
| Formaldehyde (a water solution with 37% minimum and stabilized with 10% methanol) | 2.9% |

### *2.2. Irradiation and MRI acquisition*

The gel was irradiated with a 9 MV beam LINAC. Dose range was 0, 5, 10, 15, 20, 30 Gy. Irradiation was performed with dose fraction 5 Gy/fraction and 10 Gy/fraction. The time interval between fractions was about 3 minutes. The irradiation was eight days after gel preparation. Figure 1 (a) shows the gel irradiation setup.

Magnetic resonance (MR) images were acquired five days after irradiation. MR spin echo images to evaluate $R_2$ were obtained with a 3.0 T scanner (*Philips Achieva TX*) with a head coil as shown in Figure 1 (b). The scanning room temperature was 18 °C. A 16-echo multi-spin echo (MSE) sequence was used for MR images. Detailed parameter is shown in Table 2. The gel was scanned 4 days and 12 days after irradiation in order to assess the temporal stability of it.

**Figure 1.** (a) Gel irradiation setup and (b) MRI facility with a head coil.

**Table 2.** Sequence parameter of MRI.

| Parameter | Units | Value |
|---|---|---|
| Echo spacing | ms | 22.5 |
| Maximum echo time (TE) | ms | 360 |
| Repetition time (TR) | ms | 3000 |
| Field of view (FOV) | mm | 180 |
| Matrix size (MS) | pixels | 512×512 |
| Slice thickness | mm | 4 |
| Number of excitations (NEX) | | 2 |

## 3. Results & Discussion

### *3.1. Dose response*

Figure 2 shows the gel tube after irradiation. The gel becomes white opaque from initial transparent state. $T_2$ weighted map was obtained after MR imaging as shown in Figure 3. A circle with a square of 250 $mm^2$ was chosen for mean $R_2$ calculation. $R_2$ was measured with two slices at 5 mm and 10 mm from bottom respectively. $R_2$ dose response was then fitted and compared with the $R_2$-dose curve of Zhao [14] and Pourfallah [15] as shown in Figure 4. It's observed that the dose response sensitivity of PAGAT-f is better than that of PAGAT. Fitted $R_2$-dose relationship for 5 mm and 10 mm is $R2=-0.001048d^2+0.08074d+1.658$ and $R2=-0.001224d^2+0.08683d+1.687$ respectively, where $d$ is dose. R-square for each fitting curve is shown in Figure 4.

**Figure 2.** Gel tube after irradiation (from left to right: 0, 5, 10, 15, 20, 30 Gy).

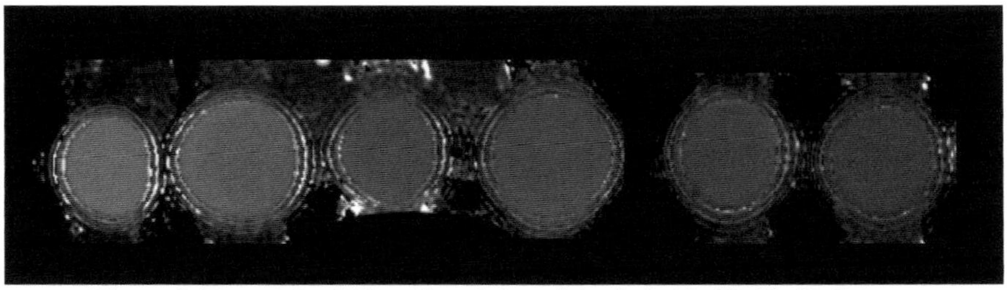

**Figure 3.** $T_2$-weighted Map (from left to right: 0, 5, 10, 15, 20, 30 Gy).

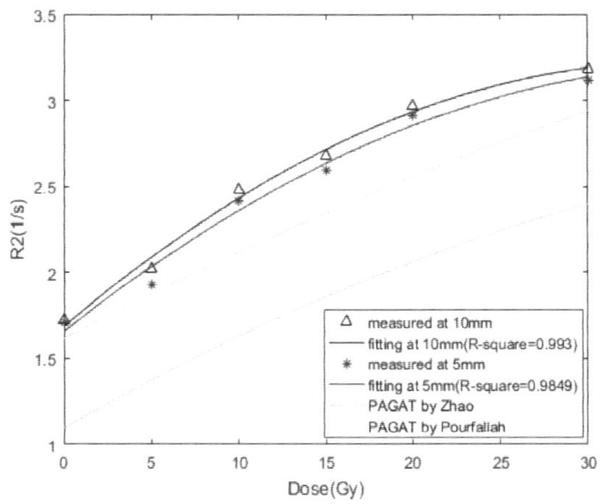


**Figure 4.** $R_2$ Dose response for PAGAT-f gel irradiated with 9 MV beams.

### 3.2. *Dose response sensitivity*

Dose response sensitivity of PAGAT-f at 5 mm and 10 mm high slice from bottom, which is represented by d$R2$/d$d$, is plotted as a function of absorbed dose as shown in Figure 5. The results are also compared with PAGAT sensitivity from Zhao [14] and Pourfallah [15]. $R_2$ sensitivity of PAGAT-f is better than that of PAGAT from 0 Gy to about 18 Gy. $R_2$ sensitivity of PAGAT-f is then less than that of PAGAT, making the saturation of $R_2$ dose response comes earlier for PAGAT-f.

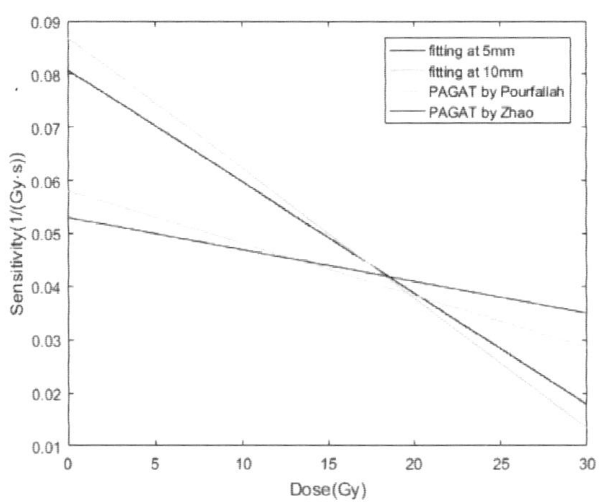


**Figure 5.** $R_2$ sensitivity versus absorbed dose.

### 3.3. *Temporal stability and deviation*

PAGAT-f gel dosimeter has great temporal stability as shown in Figure 6 (a). The $R_2$ was measured 4 days and 12 days after irradiation. The greatest deviation is only 3%. This proves good temporal stability of PAGAT-f gel dosimeter.

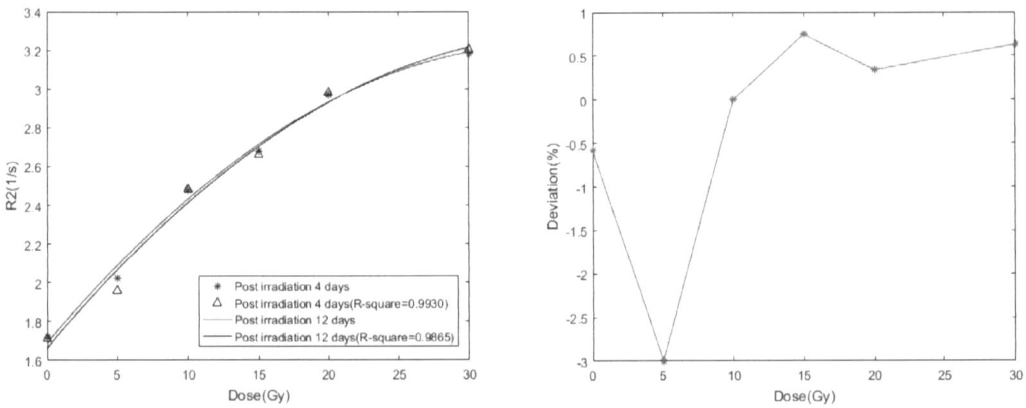


**Figure 6.** (a) Temporal stability of PAGAT-f gel dosimeter and (b) deviation between different measuring time.

## 4. Conclusion

This paper adds formaldehyde into PAGAT to solidify the gel, named PAGAT-f. By adding formaldehyde, the melting point of PAGAT-f was increased and can be stored at room temperature. It's found R2-dose response of PAGAT-f is similar to PAGAT but the dose response sensitivity of PAGAT-f is first better compared with PAGAT at low doses (<~18 Gy). R2 sensitivity of PAGAT-f is then less than that of PAGAT, making the saturation of R2 dose response comes earlier for PAGAT-f. The results preliminarily show the feasibility and good temporal stability of PAGAT-f for dose distribution measurement and clinical application. Further study will be focused on temperature dependency and dose fraction influence on PAGAT-f gel dosimeter.

## 5. Acknowledgement

This work was jointly supported by the National Key Research and Development Program (No.2016YFC0105406) and National Natural Science Foundation of China (No.61571262, No.11575095, No.61171115).

## 6. References

[1] Vial P *et al* 2008 *Med. Phys.* **35** 4362-74
[2] De Deene Y *et al* 2006 *Phys. Med. Biol.* **51** 653-73
[3] Maryanski M J *et al* 1994 *Phys. Med. Biol.* **39** 1437-55
[4] Oldham M *et al* 1998 *Phys. Med. Biol.* **43** 2709-20
[5] Venning A J *et al* 2004 *J. Phys.: Conf. Ser.* **3** 155-8
[6] Venning A J *et al* 2005 *Phys. Med. Biol.* **50** 3875-88
[7] De Deene Y *et al* 2002 *Phys. Med. Biol.* **47** 3441-63
[8] Abtahi S M *et al* 2014 *J. Radioanal. Nucl. Chem.* **300** 287-301
[9] Fernandes J P *et al* 2008 *Phys. Med. Biol.* **53** N53-8
[10] Pavoni J F *et al* 2012 *Radiat. Measure.* **47** 1074-82
[11] Marques T *et al* 2010 *J. Phys.: Conf. Ser.* **250** 012012
[12] Aziz S A *et al* 2013 *Proc. Global Engineering, Science and Technology Conference* ISBN: 978-1-922069-32-0
[13] Fong P M *et al* 2001 *Phys. Med. Biol.* **46** 3105-13
[14] Zhao Y *et al* 2011 *Atomic Energy Science and Technology* **45** 360-4
[15] Pourfallah T A *et al* 2009 *Appl. Radiat. Isotop.* **67** 186-91

IOP Conf. Series: Journal of Physics: Conf. Series **1305** (2019) 012048 doi:10.1088/1742-6596/1305/1/012048

# Iodine contrast nPAG for radiologically visible target region in deformable dosimeters

**C J Watson[1], A U Yeo[2], J R Supple[3], M Geso[4], T Kron[2] and R D Franich[1]**
1School of Science, RMIT University, Melbourne, Australia
2Physical Sciences, Peter MacCallum Cancer Centre, Melbourne, Australia
3Australian Clinical Dosimetry Service, ARPANSA, Yallambie, Australia
4School of Health & Biomedical Sciences, RMIT University, Bundoora, Australia

E-mail: callum.watson@rmit.edu.au

**Abstract**. The utility of gel dosimeters is sought to be improved upon in this study which proposes a target region of different X-ray CT contrast that is dose sensitive. The changes in the physico-chemical makeup of nPAG caused by the addition of the X-ray imaging Iodine based contrast agent Isovue are explored. The impact of this change on dose measurements is also discussed. The increase in HU as it correlates with increasing Isovue concentration is detailed, along with the dosimetric changes that occur, namely the steepness of the dose response curve and general shape of the percentage depth dose curve. It is noted that diffusion of Isovue from one gel region to another has significant dosimetric impact and the experimental method was constructed and conducted with this in mind. Further refinement and optimisation of the Isovue nPAG formulation will lead to a target region dosimeter that can be contoured on X-ray CT and used in the improvement of planning protocols, especially in cases that involve motion and deformation of target volumes.

## 1. Introduction

Polymer gel dosimeters provide a valuable measurement tool for the assessment of 3D conformal radiotherapy [1], with the dose distribution that is delivered to a dosimeter being readily visible and resolvable within the volume [2]. The shape of the deposited dose in the volume is stable for a length of time and if the dosimeter moves or changes conformation this will be evident in the dose distribution visible in the dosimeter volume. The polymer gel is a material that is able to change shape while undergoing irradiation and still measure a spatially accurate dose distribution which has seen polymer gels used for the study of motion and deformation in radiotherapy [3]. The entire dosimeter volume is often considered as an irradiated volume, or a plan aligned and delivered to the dosimeter with that intent being to assess the agreement of the total dose delivered and localized variations investigated. The current generation of polymer gel dosimeters generally lacks any internal features that are resolvable visibly or by X-ray CT. Adding such a feature may offer a pathway for increasing the utility of gel dosimeters for treatment verification [4] by clearly allowing assessment of dose difference regions with respect to target boundaries.

By introducing a target into a gel dosimeter this 'gel-in-gel' geometry can simulate a target present within an irradiated volume. The inner gel can be treated as a target volume and this feature presents opportunities for bettering the assessment of motion and deformation using gel dosimeters. The role of a target region inside a gel dosimeter would be to benchmark and improve dose calculations, image guidance, tracking strategies and setup strategies in treatments involving motion and deformation. This target volume is proposed to be comprised of nPAG containing an X-ray imaging contrast agent, namely Isovue. It is expected that due to the addition of Isovue in only a certain volume of the dosimeter a readily visible and imageable target region will be apparent.

The increase in the HU of nPAG due to adding Isovue is investigated, the intent being to provide a clearly visible and recognizable boundary in X-ray CT. From this study the working concentration of Isovue was identified. It is then logical to identify the dosimetric properties of the iodinated nPAG(InPAG), it's dose response curve, dynamic range and depth dose behavior. The insertion or manufacture of the composite dosimeter is detailed and challenges with the production of target regions in gel dosimeters were also identified.

An investigation of several imaging and dosimetric properties of InPAG were undertaken with the intent of refining the formulation for optimal utility in the form of a deformable target region dosimeter. The visibility of the target region is considered vitally important for the dosimeter's utility and the necessary concentration of Iodine contrast agent required for a well-defined target volume was observed to impact the dosimetric properties of the gel, which informed some decisions when optimizing the method laid out here.

## 2. Method

### *2.1. Iodine concentration required for X-ray contrast*

The nPAG used in this study is the same formulation used by Yeo [5] and is composed of 6 % w/w gelatine (Type A from porcine skin, Sigma Aldrich Ltd., Oakville, Canada) for the hydrogel matrix. Dissolved in this matrix are 3 % w/w N,N'-methylene-bis-acrylamide (Bis) and 3 % w/w acrylamide (Aam)(both from Sigma Aldrich Ltd.). The remaining mass is DI water, with 0.01 mM hydroquinone and 5 mM of Bis[tetrakis(hydroxymethyl-phosphonium)]sulphate (THPS) added to combat autopolymerisation and atmospheric oxygen diffusion respectively. Iodine contrast agent Isovue 300 was added in concentrations of 0.37 %, 0.73 %, 1.47 % and 2.93 % w/w, replacing an equivalent mass of water. These various preparations were then scanned using a Phillips Brilliance Big Bore X-ray CT to determine a correlation between Iodine concentration and HU.

### *2.2. Dosimetric impact of Iodine*

The impact on the dose response of the gel and how it has changed due to the addition of Isovue was investigated. The changes in the chemistry of the gel were examined as it relates to the measured dose response.

#### *2.2.1. Dose response Calibration Curve*

Dose response curves for 0.5 %, 0.7 % and 1.0 % InPAG have been recorded by producing gels at each concentration and irradiating them to known doses with known $D_{max}$ depths that could be investigated and measured from OCT. The 0.7 % samples were selected as the necessary concentration for a readily visible HU increase for planning purposes and was used to construct target region calibration samples. From a single batch of gel 5 samples were produced in 1 L PET containers. A cavity was set in these volumes that were then filled with InPAG and the containers flushed with $N_2$ gas to combat oxygen diffusion during gelation. The final samples were a 6cm depth of InPAG in the central axis of the container, surrounded by the same depth of nPAG, situating the expected depth of $D_{max}$ in the centre of the field of view of the Vista OCT scanner.

*2.2.2. Percentage Depth Dose.* Using the method explained above two more samples without central InPAG volumes were produced, along with a full 1L PET jar. The partially filled gel containers were used to investigate the build-up region behaviour in the gels and confirm the depth of $D_{max}$. The fully filled container would then provide the data for the depth dose in the gel and have a full build-up region and matched $D_{max}$ depth, which are usually obstructed by the sample holder and artefacts in the OCT reconstruction.

*2.2.3. Irradiation of samples.* All samples were placed in a water bath filled to a depth such that water surface was the same height as that of the gel. The surface of the gel was aligned at 100SSD and a machine output rate of 600 MU/min was used for all irradiations, from a Varian TrueBeam medical linear accelerator (linac) (Palo Alto, California). A 10x10 cm field of energy 6 MV was used for all irradiations overlaying the gel volume such that it fully contained the central InPAG volume. The target volume samples were irradiated to 0.8, 1.0, 1.2, 1.4 and 2.0 Gy, and the depth dose samples were irradiated to 1.2 Gy at $D_{max}$. All samples were then read out using a Modus Medical Devices (London, Ontario) VistaCT Optical CT scanner.

## 3. Results

The addition of Iodine contrast agent was expected to have an impact upon the chemical environment [6] and an impact upon the $Z_{eff}$ at imaging energies (~<140 kVp, ~50 kV mean energy), but a minimal physical interference for treatment energies of 6-10 MV. The changes to $Z_{eff}$ are illustrated in Figure 1.

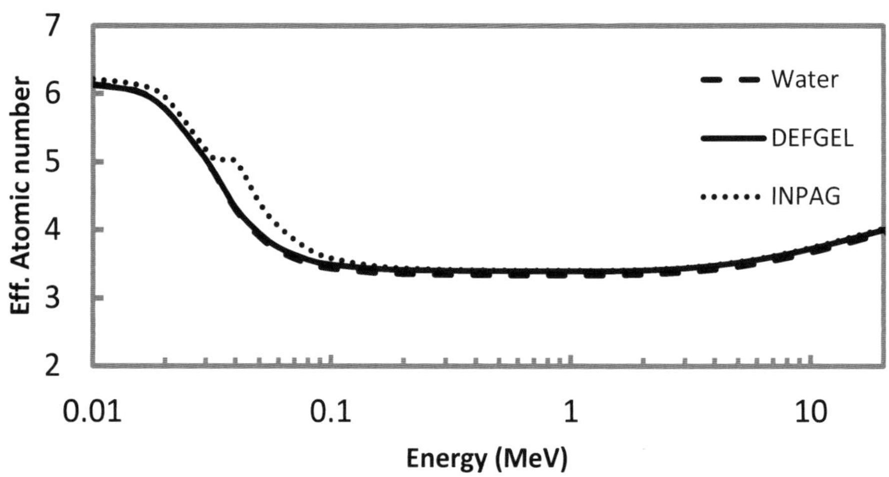


**Figure 1.** The effective atomic number of Water, DEFGEL(nPAG), and InPAG at energies from 0-20 MeV. The k-edge of Iodine is evident at 33 keV. $Z_{eff}$ calculations were performed using Auto-$Z_{eff}$ software [6].

### *3.1. Iodine Contrast Agent Concentration*

The sharp increase in $Z_{eff}$ of the InPAG at 33 keV visible in Figure 1is a primary reason why it is so valuable as medical imaging contrast material, the relationship between contrast agent concentration and HU is plotted in Figure 2. The various gel preparations were imaged inside a water bath in X-ray CT. The HU values are recorded as the mean of values along the central axis of the sample and off axis at three points throughout the height of the gel samples.

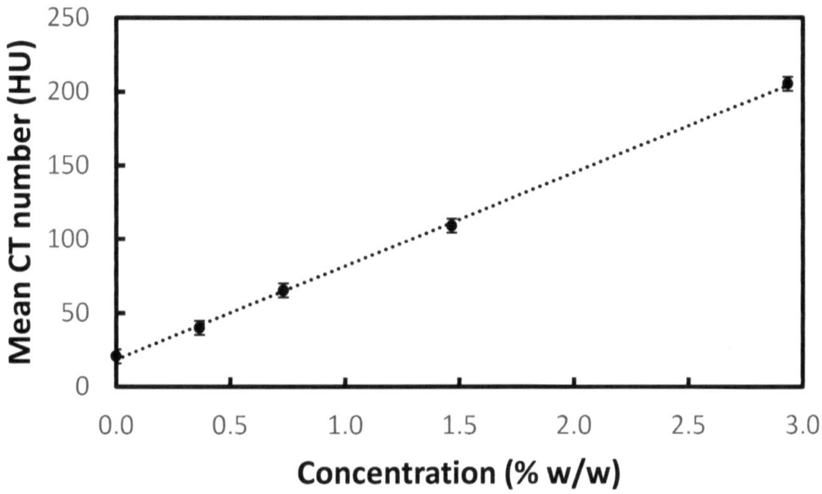


**Figure 2.** The mean increase in HU of a volume of nPAG containing Isovue as viewed in X-ray CT at 140 kVp. The variance of the signal is extremely low inside the dosimeter volume and as such produces very little uncertainty in the HU value. The error bars at each point are representative of the standard deviation in the HU within each sample.

From the curve displayed in Figure 2 it was determined that an Isovue concentration of 0.7% would provide a satisfactory contrast of InPAG(62 HU) surrounded by nPAG(16 HU).

*3.2. Dose Response*

The calibration points for the dose response curve of InPAG were taken from a region of interest in the slice that occurred at depth $D_{max}$ for a 6MV beam, 1.6cm below the surface. The results of this analysis are plotted in Figure 3.

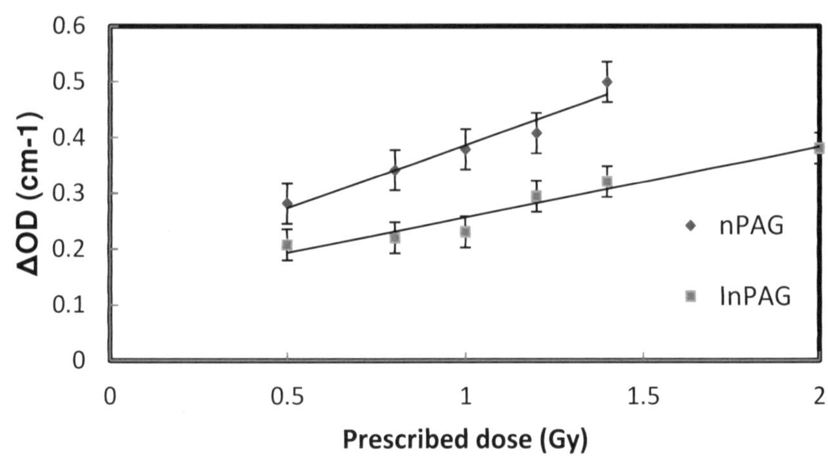


Figure 3: The change in optical density due to irradiation in a 6MV 10x10cm field at 100SSD for InPAG and nPAG in the same gel volume. The OD values are recorded for ROIs at $D_{max.}$ Uncertainty in the OD values is represented by the error bars present in this plot, being derived from the standard deviation in each of the ROIs examined. The larger variance in InPAG is likely due to tears caused by removing the test tube used to create a cavity.

The calibration curve plotted in Figure 3 was then used to calibrate the percentage depth dose curve that was extracted from the other containers. The PDD plotted in Figure 4 is a combination of 2 samples and the reasoning behind this approach is explained in the discussion.

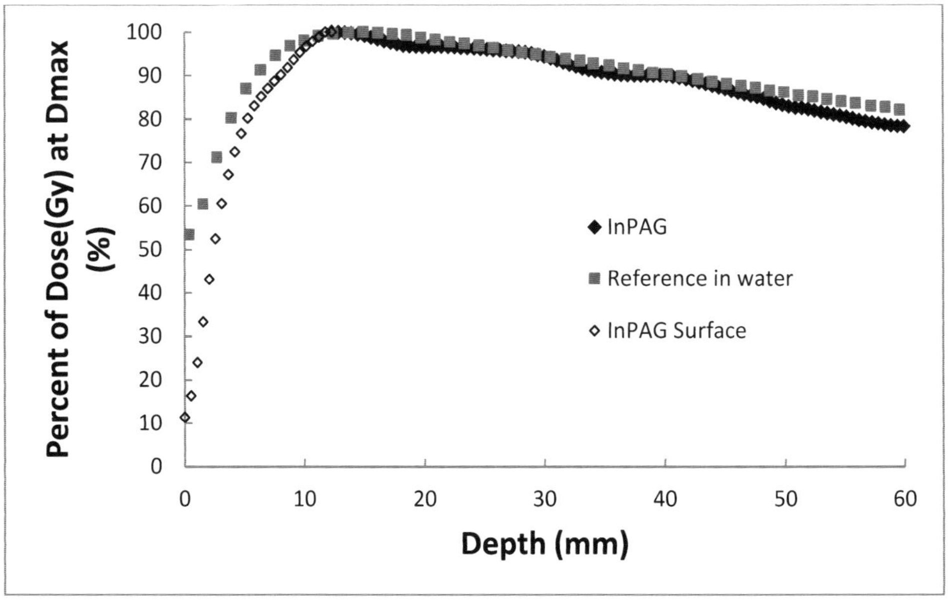


**Figure 4.** Percentage depth dose as derived from 3 gel volumes delivered a nominal dose of 1.2 Gy at $D_{max}$, with OD values calibrated using the functions presented in figure 3 then normalised to the maximum dose value at $D_{max}$. The reference PDD in water from the machine used to irradiate the samples is included in red. The InPAG surface data points are hollowed to differentiate the two samples that the PDD curve data is taken from.

The surface region of the gel shallower than $D_{max}$ is likely to have its dose response inhibited to some degree by atmospheric oxygen. The flushing of the sample container with $N_2$ gas was intended to combat this problem and appears it has been only partially effective.

## 4. Discussion

The mechanism by which Iodine alters the dose response has not been clearly identified as of yet but it is proposed that it is a chemical change rather than a physical change in the amount of dose deposited. The Iodine is likely to change the consumption rate of one or both of the co-monomer units (Bis and Aam) used in nPAG. As the polymerization reaction is exothermic there is also a significant amount of pre-irradiation auto-polymerisation that can be caused by the addition of the Iodine. The dose response curve observed in Figure 3 is likely to be dominated by the reaction of a single monomer ingredient, inhibiting the dynamic range of a dosimeter containing Iodine.

As has been the case of all polymer gel dosimeters oxygen continues to be a problematic factor in production of reliable dosimeters [7-9]. The method of production used in this study included several steps at which oxygen permeating the gel matrix would cause problems. Even with the dosimeter containers being flushed with an inert gas ($N_2$) it is expected that some damping of dose response occurred at the gel surfaces that were exposed and as such reduced the observed signal. This is most impactful in Figure 4 when examining the buildup region and the apparent depth of $D_{max}$ at 16 mm. More investigation into what is a permissible pre-irradiation gelation time for limiting the extent of atmospheric oxygen diffusion is necessary for complete quantification of the PDD and calibration curves.

## 5. Conclusion

The addition of Isovue to nPAG is shown to increase the HU as expected, to a degree that makes it easily visible and contourable on X-ray CT for the purposes of treatment planning. The concentration of Isovue needed for this HU was asserted as 0.7 % w/w and the dosimetric impact of this quantity was investigated. The Iodine is shown to alter the dose response curve compared to nPAG, with variances of the dose response noted. By careful combination of percentage depth dose curves from multiple samples a PDD of InPAG was produced, which has significant worth in understanding the dose response in the material when analyzing 3D dose distributions. The PDD exhibited some inhomogeneous dose response within the volume and the investigation of the causes of this will be the subject of further study [10-12].

## 6. References

[1] Baldock C *et al* 2010 *Phys. Med. Biol.* **55** R1-R63
[2] Mather M L *et al* 2003 *Phys. Med. Biol.* **48** N269-75
[3] Yeo U J *et al* 2013 *Med. Phys.* **40** 101701
[4] Vandecasteele J and De Deene Y 2013 *Phys. Med. Biol.* **58** 6241-62
[5] Yeo U J *et al* 2012 *Med. Phys.* **39** 2203-13
[6] Watson C J *et al* 2017 *J. Phys.: Conf. Ser.* **847** 012062
[7] Taylor M L *et al* 2012 *Med. Phys.* **39** 1769-78
[8] Hepworth S J *et al* 1999 *Phys. Med. Biol.* **44** 1875-84
[9] Deene Y De *et al* 2002 *Phys. Med. Biol.* **47** 3441-63
[10] Gastaldo J *et al* 2008 *Eur. J. Radiol.* **68** 118-20
[11] Meesat R *et al* 2009 *Phys. Med. Biol.* **54** 5909-17
[12] Chen M *et al* 2009 *Phys. Med. Biol.* **36** 4881-92

# 4D dosimetry and motion management in clinical radiotherapy

**S Å J Bäck[1], R D Franich[2], A Edvardsson[3] and S Ceberg[3]**
[1]Department of Hematology, Oncology and Radiation Physics, Skåne University Hospital, Lund, Sweden
[2]School of Science, RMIT University, Melbourne, Australia
[3]Department of Medical Radiation Physics, Clinical Sciences, Lund University, Lund, Sweden

E-mail: Sven.Back@skane.se

**Abstract**. Many novel modulated radiation treatment techniques are sensitive to patient motion which may degrade the dose distribution considerably. As there may be a simultaneous movement of the tumour and treatment machine, undesired heterogeneities in the dose distribution can be resulted. Methods to estimate the dosimetric effect of motion and treatment deliveries for both photons and protons are needed. We have recently studied Hodgkin´s lymphoma, liver and left sided breast cancer cases and developed tools to be able to simulate simultaneous organ movement and treatment delivery. Furthermore, it is of great importance to validate potential simulations in a realistic quality control set-up, ideally including a complete dosimetry volume and movement/deformation (4D). Radiation sensitive deformable gels have the potential to meet this dosimetry challenge owing to the unique 3D characteristic to form both phantom and detector in one volume. Multi-array detectors together with a moving platform and a realistic object trajectory is an alternative to evaluate the clinical setting. The evaluation could then in principle be done on-line. Gel/plastic 3D dosimeters have the potential to also be irradiated during motion in a similar matter but have to be read-out post irradiation.

## 1. Introduction

As radiotherapy techniques progress there are new challenges to assure that there is a homogeneous high dose to the tumour while minimizing the dose to the surrounding healthy tissue. There is a considerable risk that the timing of the treatment delivery and volume movement result in an inter-play effect that may affect the treatment outcome but this assumption has to be thoroughly validated. The inherent problem of volumes that also change shape within the patient due to movement can also be simulated but the results is very difficult to verify.

To minimize the uncertainties owing to motion in a clinical setting, it is first essential to use suitable immobilization devices and image guidance. Secondly, organ changes/motions have to be handled on different time scales depending on their inherent characteristics. A patient can lose weight or have a large change in bowel filling (interfraction), alternatively be temporarily affected by breathing, swallowing, and for example eye movements (intrafraction).

Most important is tumour motion due to breathing as the tumour can follow complex trajectories, and the breathing pattern might change regarding magnitude, period, regularity, and baseline position. For both photon and particle therapy, breathing motion might result in deviations between the planned and delivered dose distributions, in the form of dose blurring, interplay effects and range uncertainties

[1-4]. The combined effect might be difficult to predict and in clinical practice the treatment planning procedure often only include the static components of the incident beams (figure 1).

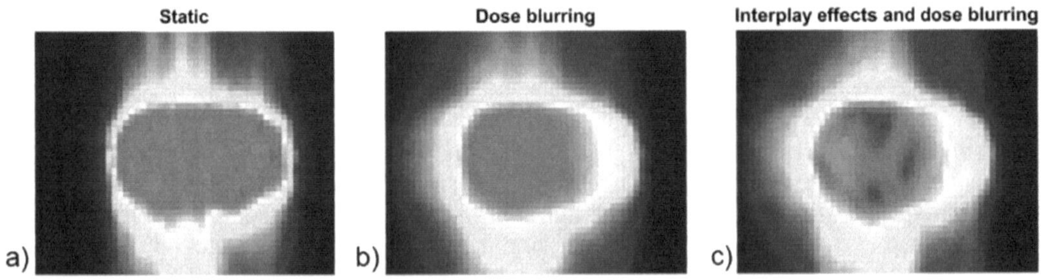


**Figure 1.** Example of a static dose distribution (a), a dose distribution including dose blurring (b), and a dose distribution including both interplay effects and dose blurring (c). The motion is in the left-right direction. Figure adopted from Edvardsson 2018 [5].

In this communication we will discuss the clinical challenges and significance of relevant dosimetry and quality assurance in conjunction with motion management in clinical radiotherapy.

## 2. Clinical motion management and treatment techniques

To handle the effect of motion the treatment margin can be increased, but larger volumes of healthy tissue will then be irradiated. Further, this will not account for interplay effects. Other techniques include respiratory gating, breath-hold and various way of tracking the tumour. Methods are needed to reproduce and monitor the patient during CT/MRI/PET and treatment, several techniques are available to accomplish this [6, 7].

One example is surface scanning systems that can provide both feed-back to the patient and trigger the treatment machine and thereby facilitate for example deep inspiration breath hold (figure 2).

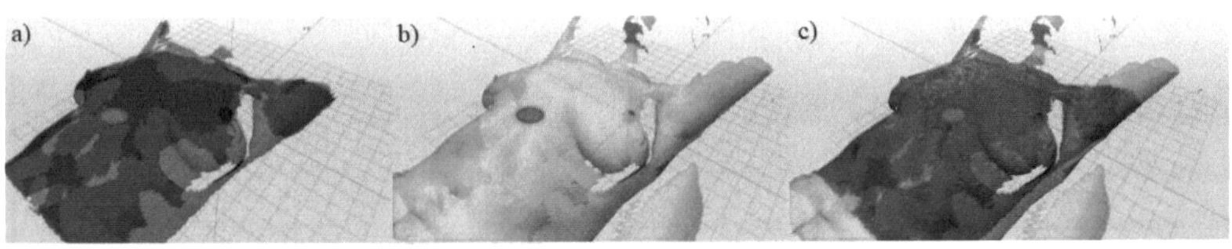


**Figure 2.** Example of a reference surface (a) and a real-time live surface (b) obtained by the Catalyst system during a deep inspiration breath hold (DIBH) left-sided breast cancer treatment. The treatment delivery is triggered when the surfaces within the red circle coincides (c). In addition, the live position of the iso-centre, calculated from the live surface using a deformable algorithm, is continuously compared to the planed iso-center position. Figure adopted from Kugele et al 2018 [8].

Another approach is to try to follow markers/edges in the tumour during treatment or as in the MRI-linacc concept, to have the potential to follow complete volumes of interest during treatment. Ideally this will also incur advanced and new methods of gating and motion management, at least for certain diagnosis.

However, there will always be inherent difficulties regarding how to handle anatomical deformations during acquisition/treatment. This problem may require theoretical methods including deformable image registration (DIR) algorithms. By the use of DIR, and the associated deformation vector field describing the motion for each voxel, the accumulation of dose from several different time points can be transferred into a common geometry [9-11]. This method was recently used and developed to evaluate the breathing-motion induced interplay effects for VMAT radiotherapy using relevant clinical data [5].

## 3. 4D verification dosimetry

To enable a reliable radiation treatment including motion management, control methods should be developed accordingly. On a patient by patient basis the treatment plans can now be optimized taking e.g. breathing into account. Consequently, it is of utmost importance to assure that the theoretical and practical methods are transferred to the patient in a safe way and that the dosimetric effects can be validated. There are also a number of basic motion control parameters that should be verified on a day-to-day/regular basis, including iso-center geometry, timing, latency, amplitude etc. and guidelines are needed as highlighted recently (eg. AAPM task group 302). This ideally include end-to-end tests with 3D and 4D dose measurements and several methods have been presented within the IC3DDose conference series (eg [12]).

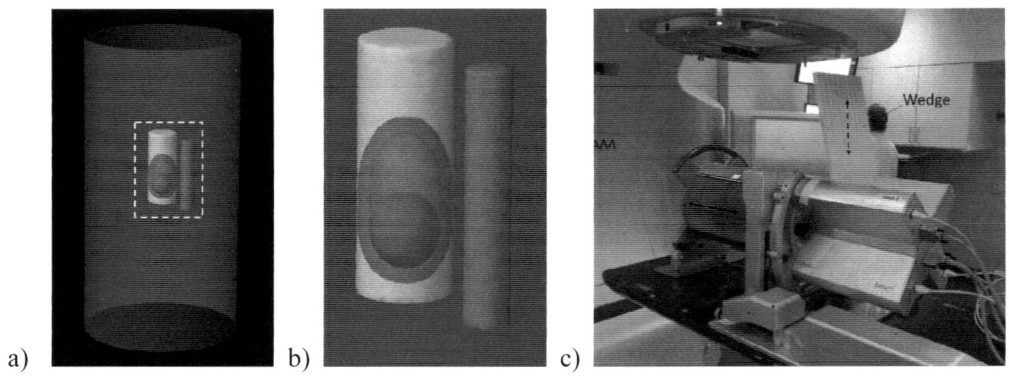


**Figure 3.** The virtual phantom of the Delta4 used for simulations of interplay effects, containing a CTV (purple), ITV (green), PTV (blue) and two OARs (yellow and red) (a), with a magification of the white dashed box in (a) displayed in (b). The actual Delta4 phantom, positioned on the HexaMotion platform in the setup used for the verification measurements, is displayed in (c). Figure adopted from Edvardsson 2018 [5].

Another approach to undertake a dose verification, including motion, inter-play effect, linacc output MLC movements etc, was to use a multi-array detector and a motion platform (figure 3, Delta4 and HexaMotion, Scandidos, Uppsala, Sweden). For the measurements during motion, the present system to synchronize the motion and the treatment delivery should be used. In our setting the results showed good agreement between the measured and simulated dose distributions during motion, comparable to the results obtained for the measured and simulated static dose distributions. Based on these results, the clinical simulation tool was considered validated, and could be used for further patient treatment optimisation. Gel/plastic 3D dosimeters [13, 14] have the potential to also be irradiated during motion in a similar matter but have to be read-out post irradiation [eg [15-17]].

To include verification of deformations, and other changes in the irradiated volumes of interest, new approaches are needed. One example is the use of microMOSFET dosimeters and an electromagnetic positioning system for real-time tracking purposes [18] which also could be used together with anthropomorphic and deformable phantoms [19]. Time-resolved in-vivo dosimetric verification techniques have recently also been presented using radio-luminescence detectors in Brachytherapy [20]. Further, deformable dosimeters forming both phantom and detector (e.g. gel dosimeters) have the potential to validate the complete treatment procedure. Also assumptions made in the treatment planning phase, regarding for instance image registration and inter-play effects for motion or deformation, will then be included in the validation. Achieving a valid gel-phantom model of any individual patient case

IOP Conf. Series: Journal of Physics: Conf. Series **1305** (2019) 012049 doi:10.1088/1742-6596/1305/1/012049

may be impractical, however mathematical approaches such as deformable dose accumulation based on DIR can be investigated [21, 22]. Potentially problematic cases such as those deficient in visible contrast or leading to interplay effects can be identified and the magnitude of those effects can be quantified [23]. Initial results have been presented for a deformable gel dosimeter which was dynamically deformed within a QUASAR respiratory phantom (figure 4) [24] along with other gel/readout approaches [25, 26].

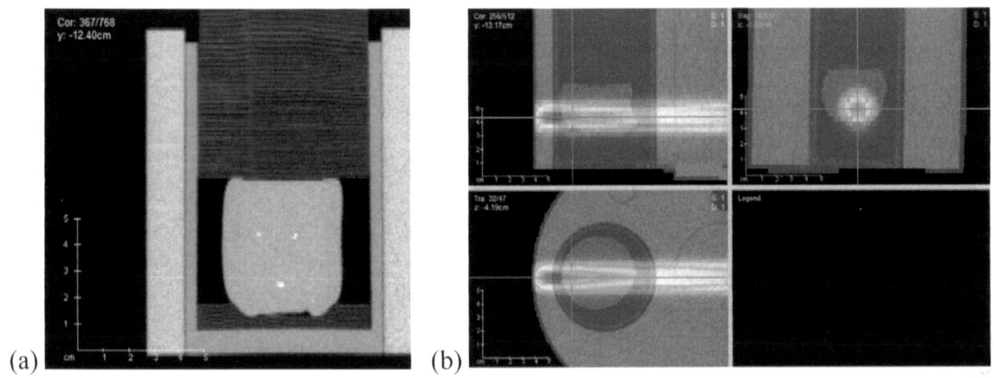

**Figure 4.** DEFGEL with dynamic deformation applied (a) and static planned dose distribution (b).

Ideally the dosimetric verification should be directly related to the actual patient irradiation in a complete 4D measurement (time resolved 3D dose distributions) for any treatment equipment and Cherenkov imaging is a promising technique (e.g. [27]). The patient geometry is clearly also in place for EPID based dosimetry and the method have also been developed in the direction of 3D/4D verification [28].

## 4. References


[1] Bortfeld T *et al* 2004 *Semin. Radiat. Oncol.* **14** 41-51
[2] Netherton T *et al* 2018 *Med. Phys.* **45** 2369-76
[3] Lomax A J 2008 *Med. Biol.* **53** 1043-56
[4] Bert C and Durante M 2011 *Phys. Med. Biol.* **56** 113-44
[5] Edvardsson A 2018 *Dosimetric effects of breathing motion in radiotherapy* (Lund: Lund University) PhD thesis ISBN 978-91-7753-804-2
[6] Boda-Heggemann J *et al* 2016 *Int. J. Radiat. Oncol. Biol. Phys.* **94** 478-92
[7] Korreman S S 2012 *Phys. Med. Biol.* **57** 161-91
[8] Kügele M *et al* 2018 *J. Appl. Clin. Med. Phys.* **19** 25-38
[9] Brock K K *et al* 2017 *Med. Phys.* **44** 43-76
[10] Rietzel E *et al* 2005 *Int. J. Radiat. Oncol. Biol. Phys.* **61** 1535-50
[11] Keall P 2004 *Semin. Radiat. Oncol.* **14** 81-90
[12] Ceberg S *et al* 2010 *J. Phys.: Conf. Ser.* **250** 012051
[13] Mather M L *et al* 2003 *Phys. Med. Biol.* **48** N269-75
[14] Hurley C *et al* 2006 *Nucl. Instrum. Meth. A* **565** 801-11
[15] Edvardsson A *et al* 2018 *Phys. Med. Biol.* **63** 085012
[16] Edvardsson A *et al* 2015 *J. Phys.: Conf. Ser.* **573** 012048
[17] Ceberg S *et al* 2013 *J. Phys.: Conf. Ser.* **444** 012098
[18] Gholampourkashi S *et al* 2017 *Med. Phys.* **44** 299-310
[19] Gholampourkashi S *et al* 2018 *Physica Medica* **51** 81-90
[20] Graversen Johansen J *et al* 2018 **17** 122-32
[21] Yeo U J *et al* 2012 *Med. Phys.* **39** 5065-5072
[22] Juang T *et al* 2013 *Rad. Oncol.* **97** 414-21
[23] Yeo U J *et al* 2013 *Med. Phys.* **40** 101701

[24] Franich R D *et al* 2015 *J. Phys.: Conf. Ser.* **573** 012024
[25] De Deene Y *et al* 2015 *Phys. Med. Biol.* **60** 1543-63
[26] Maynard E *et al* 2018 *Phys. Med. Biol.* **63** 075014
[27] Andreozzi J M *et al* 2018 *Med. Phys.* **45** 2647-59
[28] Spreeuw H *et al* 2016 *Med. Phys.* **43** 3969-74

IOP Conf. Series: Journal of Physics: Conf. Series **1305** (2019) 012050 doi:10.1088/1742-6596/1305/1/012050

# Validation of an ultrasound-guided prostate HDR brachytherapy dose delivery

**T Olding[1,2], KM Alexander[1], C Joshi[1,2] and LJ Schreiner[1,2]**
[1]Department of Physics, Queen's University, Kingston, Ontario, Canada, K7L3N6
[2]Cancer Centre of South Eastern Ontario at Kingston Health Sciences Centre, 76 Stuart Street, Kingston, ON, Canada K7L 2V7

E-Mail: tim.olding@kingstonhsc.ca

**Abstract**. As part of a recent commissioning process for an ultrasound-guided prostate HDR technique at our centre, a representative dose delivery was validated using Gafchromic EBT3 film dosimetry. Agreement between Oncentra Prostate-calculated plan dose and measured film dose was within 5-10% over most of the 2D film dose planes, except in the regions within 5 mm of the catheter axes. Given the uncertainties associated with the measurement, this result was deemed to be clinically acceptable.

## 1. Introduction

An ultrasound guidance-based high dose rate (HDR) brachytherapy prostate treatment program has recently been implemented at our cancer centre, with independent point dose calculations of simple geometries using the TG-43 formalism [1] included as part of the commissioning process. These calculations agreed to within 1-2% with the calculated values from the Oncentra Prostate v.4.2.2 treatment planning system (Nucletron BV, NL) throughout most of the treatment volume. Both calculations were based on the TG-43 formalism using the 2D anisotropy function data, so the good observed agreement was consistent with expectation. Greater disagreement (> 10%) was observed in the region within 5 mm of the catheter axes, again as expected [1].

In addition to independent dose calculations, two approaches were contemplated for providing a 3D dose measurement of a sample prostate HDR brachytherapy dose plan delivery in order to validate the planning system in a more clinical situation. The first approach would compare Oncentra Prostate plan dose to measured dose from radiochromic film fixed in coronal and sagittal planes with respect to the ultrasound probe/stepper system and associated treatment catheters immersed in a water tank. EBT3 film is well-suited for this type of validation, as it can be immersed in water for short periods of time with only minor edge perturbations, is effectively dose rate independent [2], has only a slight energy dependence at energies above 100 kV [3], and readily provides near tissue equivalent [4] high resolution 2D dose data for quantifying the dose delivery in representative planes.

The second approach would compare a CT-based Oncentra Brachy v.4.5.2 plan (Nucletron BV, NL, note difference in planning system) with a measured dose distribution in a suitable gel dosimeter to realize a full 3D measurement [5]. This approach requires the use of algebraic reconstruction techniques (ART) to remove the optical artifacts introduced to the 3D image volume by the treatment catheters. The effort to obtain full 3D dose data involves extra validation steps, and will be the subject of a separate report.

The purpose of this work is to assess the accuracy of the Oncentra Prostate treatment planning system dose calculations using ultrasound image tracked-catheters and Gafchromic EBT3 film (ISP, Wayne, NJ) film for a pseudo 3D evaluation of the dose delivery. This report is part of our ongoing efforts to show the application of 3D and pseudo-3D dosimetry to selected clinical problems and highlight the practical challenges associated with their use.

## 2. Materials and Methods

### *2.1. Experimental Setup*

The phantom used in this experiment was a 30x30x30cm water tank. Sixteen Proguide 5Fx240mm treatment catheters (Elekta, Atlanta, GA) were inserted into the stepper template in a standard pattern preferred by our clinic (figure 1a) and a typical depth of insertion (see figure 1b). These catheters were then locked in place. In addition to these treatment catheters, four other catheters were inserted in the template with the metal obdurators left in, and a metal template was inserted over the ends of the catheters to provide additional catheter position stability (figures 1b and c). The coronal films were fixed above and below the c3.5 (L3) and e3.5 (L4) needles (figures 1a and b) during plan delivery in the water tank (figure 1b). The sagittal films were attached to the two support catheters in the middle column of the template during delivery and exposed during a separate delivery of the same plan. The position of the sagittal and coronal film planes are as shown in figure 2 (upper right).

(a) (b) (c) (d)

**Figure 1.** (a) Reference ultrasound image plane showing the needle insertion pattern. (b) Experimental setup showing location of coronal films above and below the L3 and L4 catheters immersed in the water tank, and (c) sagittal films on either side of central axis catheters. (d) Illustration of the catheter tracking on a sagittal ultrasound image slice.

### *2.2. Delivery Validation*

The clinical system used in this experiment consists of a BK3000 ultrasound unit (BK Ultrasound, Peabody, MA) and an OncoSelect stepper system (Nucletron, NL) in combination with the Oncentra Prostate treatment planning system. The 3D ultrasound image used for treatment planning was acquired by immersing the described apparatus in the water tank and imaging according to the standard clinical procedure. Treatment catheters were tracked in Oncentra Prostate (figure 1d), a typical size and position of urethra and prostate were contoured on the ultrasound image, and inverse optimization used to generate a clinically acceptable treatment plan (figure 2). The 15 Gy clinical prescription dose was scaled down to 3 Gy to accommodate the preferred range for EBT3 film dosimetry, and the treatment plan was delivered using a Flexitron remote afterloader (Elekta, Atlanta, GA) with an Ir-192 radioactive source.

Two different calibration approaches were evaluated for their suitability. First, a calibration set of films were irradiated on a Varian TrueBeam linear accelerator (Varian Medical Systems, Palo Alto, CA) under machine reference conditions (6 MV, 10x10 $cm^2$ field size, 2 cm depth in solid water, 100 cm SAD) to doses spanning the range of 0-9 Gy. Second, three different well-characterized Oncentra Brachy

IOP Conf. Series: Journal of Physics: Conf. Series **1305** (2019) 012050 doi:10.1088/1742-6596/1305/1/012050

CT-planned five catheter HDR brachytherapy plans (unpublished work) were delivered to a fixed geometry solid water phantom using the Flexitron remote afterloader unit. Calibration EBT3 films were located 1.0 cm below the central axis plane of the catheters in the solid water phantom (see figure 3a). These plans were developed at our centre specifically for film calibrations in HDR brachytherapy applications (unpublished work). EBT3 film dose readout was accomplished using an Epson 10000XL flatbed scanner (Epson Canada, Markham, ON). Calibrated film dose was compared to Oncentra Brachy dose in 3D Slicer (www.slicer.org) [6].

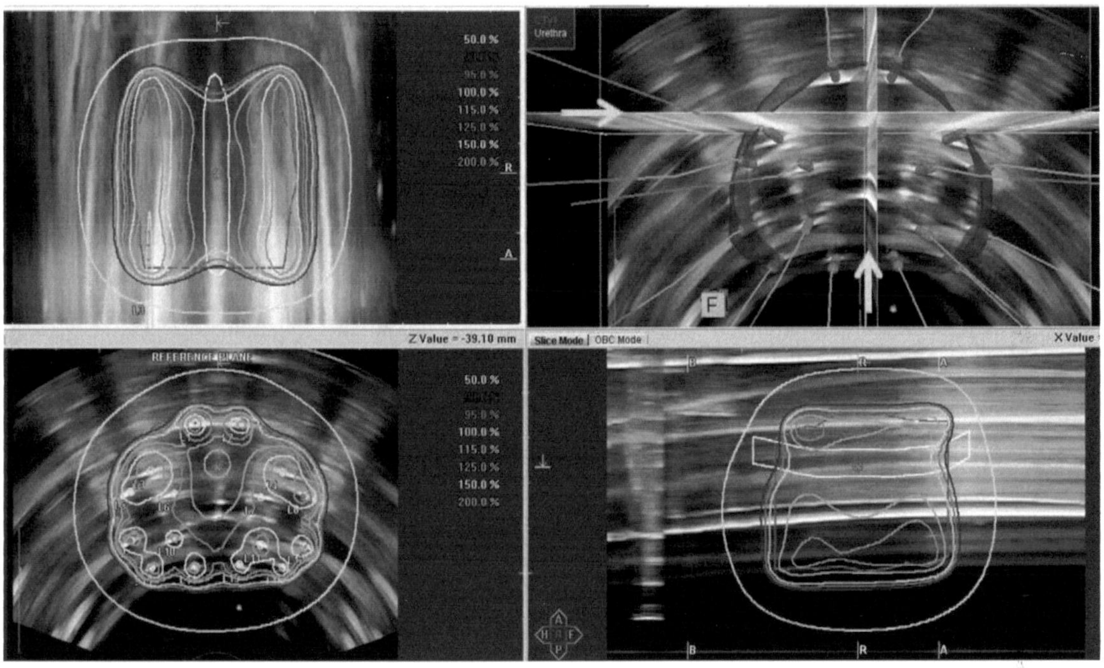

**Figure 2.** Views of the Oncentra Prostate-calculated dose distribution, with prostate contour drawn in red and urethra in yellow. The 3D image in the upper right shows the planar locations of the coronal and sagittal films. Arrows highlight the 2D slice locations of the film planes.

## 3. Results

Figure 3a shows a CT image of the HDR brachytherapy catheter geometry in the solid water phantom, along with calculated plan dose. The catheters were fixed in place in machined channels on the surface of a 2 cm thick solid water slab. Figure 3b shows the corresponding overlaid optical density-to dose polynomial fits generated from three simple geometry HDR brachytherapy plan deliveries. The calibrations are labelled as hut, wedge, and circular calibrations in reference to the general shape of the plan dose distributions at the film plane. Overlaid optical density-to-dose calibration curves are shown in figure 3b, with only the data points from the external beam calibration included in the figure, for clarity of comparison.

Figures 4a and 4b show calibrated film dose from the plan delivery. The two coronal and two sagittal films yielded similar results as compared to the calculated plan dose, so only one of each is shown. Figures 4c-4e show representative profile comparisons between calculated plan dose and measured film dose along the 3 image axes, as illustrated by dotted lines in figures 4a and 4b. A gamma evaluation is not shown, as it was deemed to be of limited utility in this application. As seen in the comparisons, the film dose was typically in the range of 5-10% greater than the plan dose in the near-flat regions of dose delivery to the target (i.e. not in close proximity to the treatment catheters).

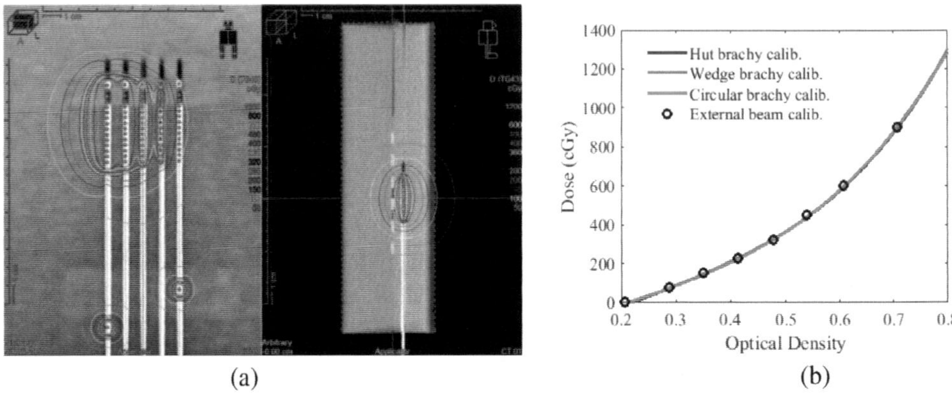


**Figure 3.** (a) Oncentra Brachy planning system visualization of a five catheter wedge dose distribution plan calculated to a CT-imaged solid water phantom, with EBT3 film located 1.0 cm posterior to the central axis plane of the five catheters (as shown by the green dotted line). (b) EBT3 film optical density-to-dose calibration curves generated from (i) three simple geometry HDR brachytherapy plans, and (ii) a set of varied dose 6 MV external beam single fields delivered to EBT3 film in solid water. Error bars are too small to show.

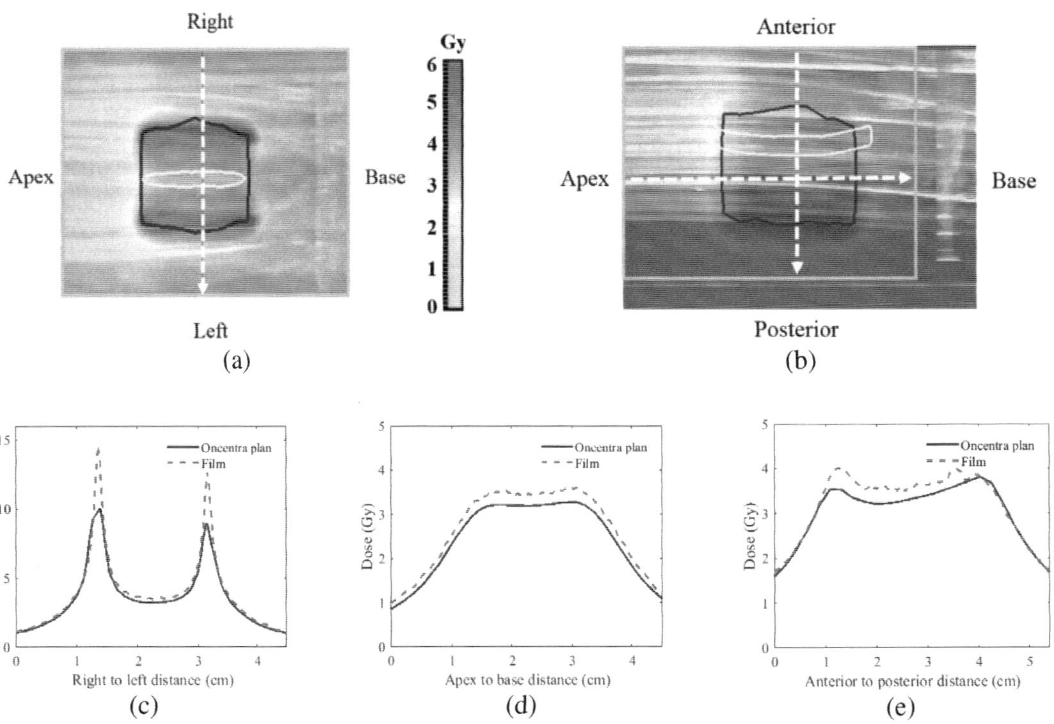


**Figure 4.** Calibrated film dose from one each of the (a) coronal and (b) sagittal films overlaid on the ultrasound image. The film location is shown in orange, the prostate contour in blue and the urethra contour in yellow. (c) – (e) Dose profile comparisons along the dotted lines shown in (a) and (b).

## 4. Discussion & Conclusions

External beam radiotherapy-based EBT3 film calibrations have been well-investigated for use in the literature e.g. [7, 8], and our clinical experience with this film dosimeter has yielded similar results. The good agreement between calibration fits shown in figure 3b therefore illustrates that simple geometry HDR brachytherapy plans can also be used for calibration of EBT3 film as long as a proper amount of care is taken to ensuring geometric accuracy in film and catheter placement. This is in agreement with the conclusions reached by Palmer *et al* [2].

The disagreement between measured film dose and calculated plan dose was larger than expected and warrants further investigation, even factoring in the uncertainties inherent to film dosimetry [2, 7, 8]. It is likely that the discrepancy is primarily due to a combination of (a) uncertainty in the film dosimetry, (b) some form of systematic uncertainty in catheter position in the water tank, leading to a 5-10% average increase in dose to the target region, and (c) uncertainty in the treatment planning system calculation. There is a known uncertainty in the template position calibration leading to mis-reporting of catheter position with increasing distance away from the probe surface in the ultrasound image. If a majority of the nearest (to film plane) treatment catheters were located closer to the sagittal and coronal film planes than
reported in the ultrasound image, then an overall increase in delivered dose is possible. The concern is how much of the discrepancy is due to inaccuracy in the planning system dose calculation algorithm.

A few other sources of error have been contemplated. The location of the film in the water tank was not considered to be an issue, as the films were not placed in high dose gradient regions except near the L3 and L4 catheters (figures 3a and 4a). Adjacent plan dose planes were assessed to arrive at this conclusion. Roundoff in treatment delivery time accumulated over multiple dwell positions was assessed as a possible source of error and found to be insignificant. The Ir-192 source strength was well-determined to within 1% via well chamber calibration (corroborated with the calibration report from the manufacturer), and is not deemed to be a significant contributing factor. Transit dose during the delivery could also add an additional source of error, but this was not evident in the comparison between the brachytherapy and external beam calibrations reported in figure 3b, and therefore is not likely to have a significant contributing effect.

Given the practical challenges and inherent uncertainties associated with the validation of ultrasound-guided prostate HDR brachytherapy dose deliveries, and against the backdrop of the reality that this technique is used widely with excellent clinical effectiveness, the results from this work were not deemed to be a deterrent for clinical use of this technique, but rather as an impetus for further investigation. Future work will include a focus on validating the ultrasound imaged-derived catheter positions against an electromagnetic tracking system under development [9].

## 5. Acknowledgements

This work is funded by the Canadian CIHR and NSERC funding agencies, projects MOP-115101 and CPG 151964. The authors would like to thank Miranda Nixon for her work in characterizing the film calibration plans for use (undergraduate thesis).

## 6. References

[1] Rivard MJ *et al* 2004 *Med. Phys.* **31** 633
[2] Palmer A L *et al* 2013 *Phys. Med. Biol.* **58** 497
[3] Arjomandy B *et al* 2010 *Med. Phys.* **37** 1942
[4] Sutherland J G H and Rogers D W 2010 *Med. Phys.* **37** 1110
[5] Baldock C *et al* 2010 *Phys. Med. Biol.* **55** R1-63
[6] Alexander K M *et al* 2017 *J. Phys.: Conf. Ser.* **847** 012061
[7] Marroquin E Y L *et al* 2016 *J. Appl. Clin. Med. Phys.* **17** 466
[8] Borca C *et al* 2013 *J. Appl. Clin. Med. Phys.* **14** 158-71
[9] Lugez E *et al* 2017 *Int. J. Comput. Assist. Radiol. Surg.* **12** 681

# High dose rate brachytherapy three-dimensional gel dosimetry using optical computed tomography readout

**DA DeVries[1], C Joshi[1,2] and LJ Schreiner[1,2]**
[1] Department of Physics, Queen's University, Kingston, Ontario, Canada K7L 3N6
[2] Department of Medical Physics, Cancer Centre of Southeastern Ontario at Kingston Health Sciences Centre, Kingston, Ontario, Canada, K7L 5P9

Email: david.devries@queensu.ca

**Abstract**. The cause and removal of streaking artifacts from catheters within radiochromic 3D dosimeters was investigated to allow the use of these dosimeters for high dose rate (HDR) brachytherapy applications. The OSC-TV iterative reconstruction algorithm's ability to remove the streaking artifacts from one or more catheters was validated using external beam irradiations and leuco crystal violet (LCV) micelle gels. One and two catheter HDR plans were then delivered to LCV gels. Dose volumes from the gels were reconstructed with the OSC-TV algorithm, calibrated, and registered to treatment planning system dose volumes, and were shown to be artifact free and in agreement with the expected values.

## 1. Introduction

Three-dimensional (3D) dosimetry has become more easily accessible because of the development of optical computed tomography (optCT) gel scanners, which are less costly when compared to x-ray CT or magnetic resonance imaging (MRI) scanners [1]. The use of radiochromic dosimeters with optCT readout for quality assurance (QA) of external beam radiation therapy has become commonplace, with developments such as improved scanners, reconstruction algorithms, radiochromic gels and plastics, and analysis software, making the QA process with gel dosimetry faster and more accurate.

Despite the progress in the field of 3D dosimetry, most applications remain within external beam radiotherapy (EBRT), while applications in high dose rate (HDR) brachytherapy are limited. Work with polymer-based and Fricke gels with MRI read-out for brachytherapy dosimetry has shown promising results [2-5], but the barrier to entry for clinical use is high [6]. Work using radiochromic 3D dosimeters with optCT readout has also shown promise [7-9], but streaking artifacts radiating from the inserted catheters created during the dose volume reconstruction can leave sections of the volume unusable for dosimetry. In this work we report on efforts to use iterative reconstruction algorithms to extend the readable volume of the dosimeter when catheters are present. It is anticipated that this work will allow for patient specific QA for HDR brachytherapy and eventually possible HDR source characterization. Patient specific QA has been adopted for EBRT as a standard of care in many clinics, but has not been adopted for HDR brachytherapy. This work seeks to improve 3D gel dosimetry for HDR brachytherapy to provide a dosimetry technique for patient specific QA for HDR brachytherapy.

In 3D dosimetry optCT, change in attenuation ($\Delta\mu$) from pre to post-irradiation is found at each voxel rather than optical attenuation values ($\mu$). This is accomplished by capturing projection images before and after irradiation, and performing the reconstruction using the ratios of the pre and post-irradiation images. With this is mind, it would stand to reason that any catheters within an imaged gel would not change in attenuation, and so should "disappear" when the ratio of pre and post-irradiation images are taken, and reconstruction results should not be effected. This is not the case in practice however, and so the cause of this effect will first be investigated.

The reduction or removal of streaking artifacts in CT reconstructions with the use of iterative reconstruction algorithms instead of the traditional filtered-backprojection type algorithms has been shown in the field of x-ray CT [10]. These iterative algorithms are more computationally intensive than filtered-backprojection algorithms, but the use of graphics processing units (GPUs) to cut computation times makes it possible to now use the algorithms practically. Such iterative algorithms, in particular the Ordered Subsets Convex algorithm with Total Variation minimization regularization (OSC-TV), have been applied to optCT [11, 12]. The OSC-TV algorithm has been used to reduce noise and jar wall artifacts, but has not yet been applied to removing streaking artifacts from catheters. It is investigated here to characterize its ability to remove the streaking artifacts found in brachytherapy applications.

**Figure 1.** LCV gel with catheter inserted with catheter not yet trimmed.

**Figure 2.** X-ray CT imaging of gels with catheter guidewire in place.

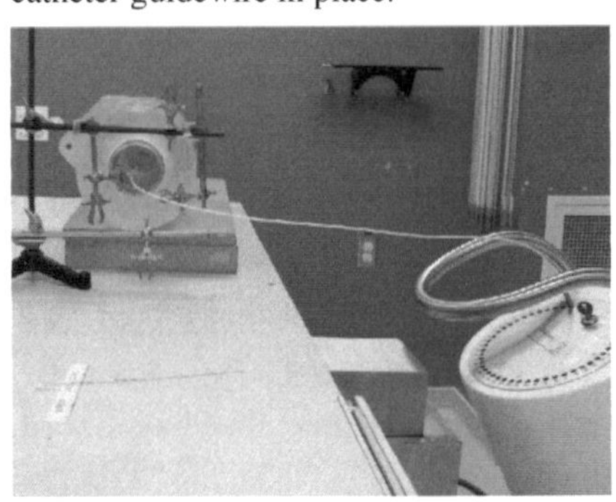

**Figure 3.** Gel HDR brachytherapy irradiation. To prevent damaging the gel, the transfer cables from the afterloader were supported with a retort stand.

## 2. Materials and Methods

All the gels used in this study were leuco crystal violet (LCV) micelle gels prepared according to the recipe of Babic *et al* [13], which respond linearly to dose, are free from dose-rate or energy dependence, and do not suffer from spatially instability resulting from diffusion [13]. The LCV micelle gel recipe has low dose sensitivity, and so the irradiations delivered were above the clinical range (80Gy at 1cm from the HDR source axis). The results presented here are applicable not only for the gels used, but for all types of 3D radiochromic dosimeters. Gels that required catheters to be inserted were prepared by suspending the catheters within the gels as they set using special lids or plastic films over the jar that held the catheters in place (Figure 1). Once the gel was set, the catheters would be cut such that the jar lid could be fastened, allowing the gel to be optically scanned.

In order to investigate the use of the OSC-TV algorithm for artifact removal, the algorithm was first validated using irradiations that have been well characterized (e.g. external beam radiation). To characterize the artifact removal, two gels received an identical irradiation (two perpendicular beams); one gel containing an inserted catheter and the other no catheter. The gel with the catheter had its dose volume reconstructed via the OSC-TV algorithm, while the gel without the catheter was reconstructed via a gold-standard filtered-backprojection algorithm. The reconstructed dose volumes were then compared to characterize the OSC-TV algorithm's artifact removal capabilities. This also allowed the strength of the noise reduction of the OSC-TV algorithm to be "tuned" to ensure it did not blur sharp dose gradients.

IOP Conf. Series: Journal of Physics: Conf. Series **1305** (2019) 012051 doi:10.1088/1742-6596/1305/1/012051

With the OSC-TV algorithm verified, gels were prepared with one or two catheters inserted. The gels were then imaged with x-ray CT (1mm slice thickness, 0.3mm spacing) using a Phillips Brilliance Big Bore CT scanner (Philips Medical Systems, Cleveland, OH). Guidewires were inserted into each catheter during imaging to allow for tracking of the catheters during planning (figure 2). X-ray CT and matching optCT fiducials (metal BBs and black marker spots) were also placed on the jar to allow for the registration of treatment planning and gel dose volumes. Once imaged, catheter tracking, treatment planning, and dose calculation were performed within the Oncentra Brachytherapy treatment planning system (TPS) (Elekta Ltd., Stockholm, Sweden). During irradiation, the gels were placed within a custom PMMA phantom and then the catheters were connected to a Flexitron Remote Afterloader and were irradiated with a Flexisource $Ir^{192}$ source (Elekta Ltd., Stockholm, Sweden) (figure 3).

In order to calibrate the gels for dose comparisons, the gels were made in batches, with one gel reserved for calibration. This gel received an electron-beam irradiation following the protocol from Alexander *et al* [14], and was analyzed in 3DSlicer (www.slicer.org) using the 3DSlicer Gel Dosimetry Analysis slicelet. This slicet was also used for dose volume registration and comparison.

The gels were scanned pre and post-irradiation with the Vista 15 cone-beam optCT scanner (Modus Medical Devices Inc., London, ON) in a refractive index matched solution under amber light, with 410 projections recorded per scan. Flood field projections were also captured by filling the scanner tank fully with the matched solution. Filtered-backprojection reconstructions were performed using the VistaRecon software provided with the Vista 15 scanner. OSC-TV reconstructions were performed by using an in-house open-source computed tomography reconstruction toolkit (https://github.com /DavidADeVries/Gyrfalcon-CT-Simulation-And-Reconstruction) programmed in MATLAB (The Mathworks, Inc., Natick, MA). GPU acceleration of the OSC-TV algorithm was achieved using functions from the open-source TIGRE toolkit [15]. All reconstructions were computed at the same resolution of a 256 × 256 × 256 voxel data set with 0.5 × 0.5 × 0.5mm voxels.

## 3. Results and Conclusions

### *3.1. Cause of streaking artifacts*

The cause of the streaking artifacts was found to be partly from a misalignment of the imaged catheter in the pre and post-irradiation projection images. As shown in figure 4a-b, the pre and post irradiation projections clearly show the catheter, but when the ratio of the projections is taken (figure 4c) the catheter does not "disappear" as expected. Instead two lines, one dark and one light are shown, pointing to a shift between the projections. This is due to the limitation of being able to reproducibly position the gel within the optCT scanner, since the gel is aligned within the scanner by lining up a visual marker with the jar clamp. By comparing figure 5b and 5d it is clear that taking the ratio of the projections reduces the streaking artifacts, but does not remove them, motivating the need to use the OSC-TV algorithm. Perfect alignment of the projection images may not necessarily eliminate the artifacts either, as photon starvation or scattering effects may also produce artifacts.

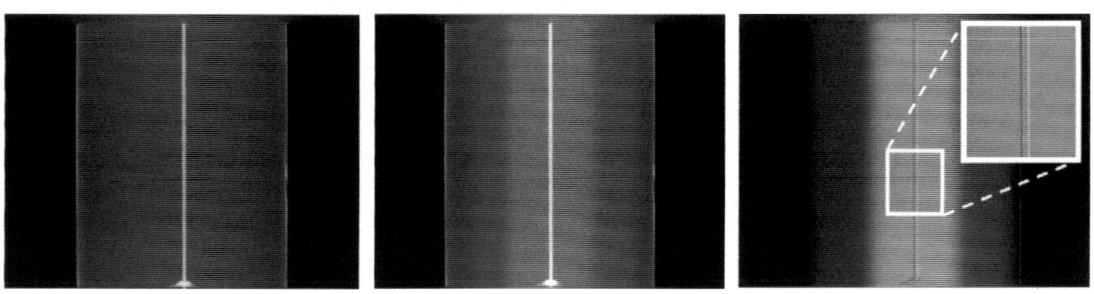

(a) (b) (c)

**Figure 4.** Natural logarithm of the ratio of (a) flood field and pre-irradiation, (b) flood field and post-irradiation, and (c) pre and post-irradiation projections from an external beam irradiation. The inset in (c) shows the misalginment of the catheter between the pre and post-irradiation projections.

### 3.2. *OSC-TV reconstruction algorithm validation*

The OSC-TV algorithm was validated for removing the streaking artifacts by using the algorithm to reconstruct dose volumes for gels with inserted catheters that had been irradiated with external beam radiation, and comparing these results against a gel with no inserted catheters that had received the identical irradiation. The results of one such trial are shown in figure 5, demonstrating the ability of the OSC-TV algorithm to remove the streaking artifacts, as well as reduce noise, without compromising the accuracy of the dosimetry results. The trials also found that within a 5mm radius of any catheter, the results of the OSC-TV algorithm has a high uncertainty, and so dosimetry within this range is not yet possible.

These validation trials also motivated modifying of the OSC-TV algorithm from its original implementation following the results of trials with multiple catheters in the gels. During these trials new artifacts appeared that joined the catheters (figure 7c), due likely to the inability of the optical attenuation model of the reconstruction algorithm to model the joint attenuation of multiple catheters. To combat this effect, the voxels containing the catheters were first found by thresholding the pre-irradiation projection images, such that only the highly attenuating catheters were set to "1" and all other pixels were set to "0", and preforming a single reconstruction iteration. All rays passing through these catheter voxels would then be rejected from the reconstruction process, which is allowable in iterative reconstructions which do not require symmetric projection data sets. This modification of the algorithm effectively removed the new artifact and was used for all results presented.

### 3.3. *Brachytherapy dosimetry results*

Two HDR brachytherapy deliveries were measured using the LCV micelle gels and OSC-TV reconstruction algorithm. The first consisted of a single catheter with a single dwell position. The comparison of the expected dose volume from the treatment planning system to the calibrated gel dose volumes reconstructed by the filtered-backprojection and OSC-TV algorithms are shown in figure 6.

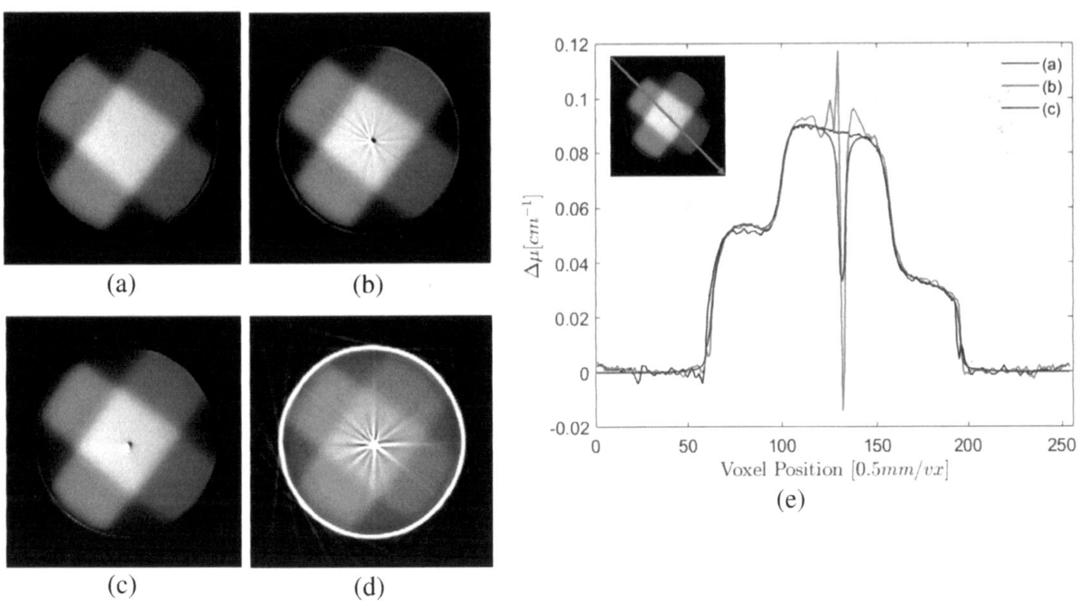


**Figure 5.** Axial slices of a dose volume reconstruction of an identical external beam irradiation delivered to two gels. The first gel (a) had no catheter inserted and was reconstructed using filtered-backprojection. The second gel had one catheter inserted and was reconstructed using filtered-backprojection (b), the OSC-TV algorithm (c), and filtered-backprojection with only using post-irradiation projection images (d). (e) shows a line profile across the catheter, demonstrating the artifact removal and noise reduction in (c) without compromising accurate dosimetry.

The modified OSC-TV algorithm is shown to remove all the artifacts present in the filtered-backprojection reconstruction, as well as to have close agreement to the TPS.

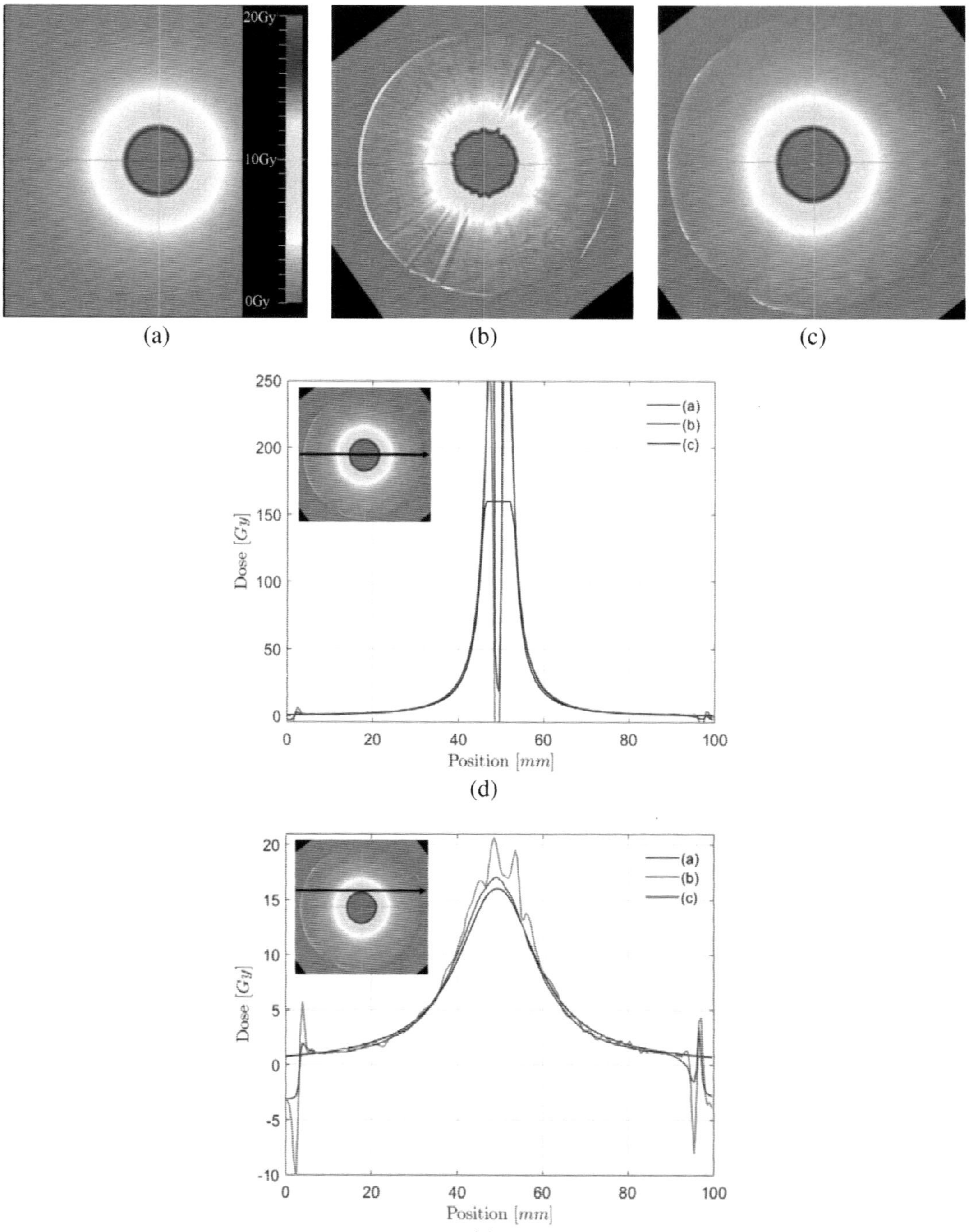


**Figure 6.** Axial slices of dose volumes of a single dwell position irradiation from the TPS (a), and from volumes reconstructed by filtered-backprojection (b), and OSC-TV (c). The line profiles through the catheter (d) and near the catheter (e) show the agreement between (a) and (c) and the effect of the streaking artifacts in (b). The TPS dose calculation cuts-off within 3mm of the source, causing the flattened profile in (d).

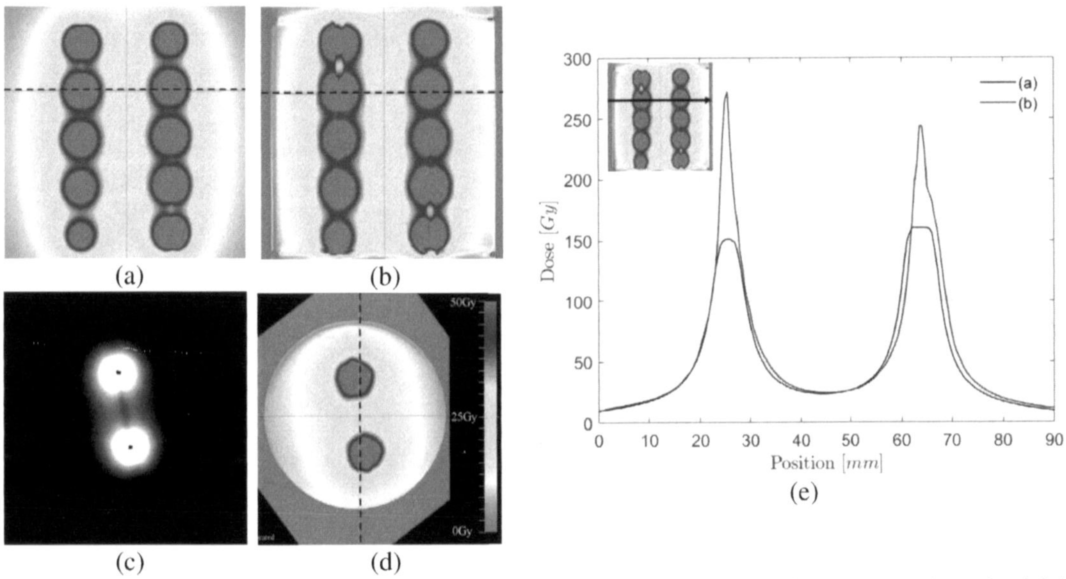


**Figure 7.** Transverse slice of the TPS dose volume for a two catheter plan (a) and a transverse (b) and axial (d) slice from the calibrated gel dose volume reconstructed with the OSC-TV algorithm. Agreement between the two is shown by the line profile (e), though a shift in the right-most peak also shows a probable shift in the catheter. The artifact caused by not removing rays passing through the catheters during reconstruction is shown in (c) (window/level set to highlight the artifact). The TPS dose calculation cuts-off within 3mm of the source, causing the flattened profile in (e).

The second brachytherapy delivery consisted of two catheters with five equally spaced and timed dwell positions along each catheter. As shown in figure 7, the comparison between the TPS and OSC-TV reconstructed dose volumes shows close agreement. However, the apparent shift of the right-most dose peak in figure 7e that is not present in the left-most peak points to one of the catheters shifting between x-ray CT imaging and plan delivery. This is most likely due to the strain put on the catheters will being secured to the afterloader.

Planned future studies using the OncoSmart Catheter System (Elekta Ltd., Stockholm, Sweden) will aim to avoid placing strain on the catheters during irradiation. The OncoSmart Catheter System features a nested catheter system, allowing for an outer catheter to remain within the gel, while an inner catheter can be connected to the afterloader before being inserted into the outer catheter, avoiding any strain on the catheter in the gel.

## 4. Acknowledgements

This work is funded by the Canadian CIHR and NSERC funding agencies, projects MOP-115101 and CPG 151964, and the NSERC CGS-M program.

The authors would also like to acknowledge the assistance provided by K. Alexander in gel preparation and analysis and Dr. S. Peca in the brachytherapy irradiation planning and delivery.

## 5. References

[1] Baldock C *et al* 2010 *Phys. Med. Biol.* **55** R1-63
[2] Hurley C *et al* 2006 *Nucl. Instrum. Meth.* A. **565** 801
[3] Pantelis E *et al* 2005 *Phys. Med. Biol.* **50** 4371
[4] Senkesen O *et al* 2014 *Med. Dosim.* **39** 266
[5] Schreiner L *et al* 1994 *Magn. Reson. Imaging* **12** 901
[6] Mather M L *et al* 2003 *Phys. Med. Biol.* **48** N269-75
[7] Massillon G *et al* 2009 *Phys. Med. Biol.* **54** 1661
[8] Šolc J and Sochor V 2012 *Metrologia* **49** 231
[9] Yeo U *et al* 2013 *J. Phys. Conf. Ser.*, **444** 1
[10] Wang G *et al* 1996 *IEEE Trans. Med. Imaging* **15** 657

[11] Dekker K *et al* 2017 *Med. Phys.* **44** 6678
[12] Matenine D *et al* 2015 *Med. Phys.* **42** 6376
[13] Babic S *et al* 2009 *Phys. Med. Biol.* **54** 6791
[14] Alexander K *et al J. Phys. Conf. Ser.* **573** 6
[15] Biguri A *et al* 2016 *Biomed. Phys. Eng. Express.* **2** 1

# Spatial dose distribution analysis of Co-60 HDR brachytherapy of cervical cancer using an AQUAJOINT®-based VIPET polymer gel dosimeter

**Ayumi Oe[1], Mikio Nemoto[1,2], Masanori Miyazawa[3], Daniel Antonio Sahade[2,4] and Toshimasa Hamada[2,5]**
[1]Department of Radiology, Jichi Medical University Hospital, 3311-1, Yakushiji, Shimotsuke, Tochigi 329-0498, Japan
[2]3D Gel Dosimeter Research Laboratory, Cluster for Science, Technology and Innovation Hub, RIKEN, 2-1, Hirosawa, Wako, Saitama 351-0198, Japan
[3]R-TECH INC., 1-2-9, Akatsuka, Itabashi-ku, Tokyo 175-0092, Japan
[4]Frontier Materials Research Department, Materials Research Laboratories, Nissan Chemical Corporation, 488-6 Suzumi-cho, Funabashi, Chiba 274-0052, Japan
[5]Advanced Materials & Planning Department, Nissan Chemical Corporation, 5-1, Nihonbashi 2-Chome, Chuo-ku, Tokyo 103-6119, Japan

E-mail: toshimasa.hamada@riken.jp

**Abstract.** High-dose rate brachytherapy treatment has become more complicated with the use of three-dimensional image-guided brachytherapy (3D-IGBT) to prepare treatment plans using 3D images, such as those obtained from computed tomography (CT). In planning a 3D-IGBT, spatial measurement of dose distribution verification is recommended. In this study, the spatial dose distribution of intrauterine cavity irradiation by a Co-60 sealed brachytherapy source was acquired using an AQUAJOINT® polymer gel dosimeter. A CT/magnetic resonance imaging (MRI) compatible applicator was inserted into the gel in a bottle. It was irradiated and visualized using MRI. The image was converted to an R2 image using a DD-System (R-TECH INC, Japan), and dose distribution was evaluated using the dose–R2 response curve. The obtained dose distribution, dose profile of the radiotherapy treatment planning system, isodose curve, and gradient passing were calculated. In the dose profile, the dose difference was large near the applicator (high-dose region) and was less than ±2% in the middle-dose region. The isodose curves showed good agreement in the region of 2–6 Gy near the prescription point. The gamma analysis was 87.594% on the sagittal cross-section and 93.711% on the coronal cross-section based on the dose difference (DD)/distance to agreement (DTA) of 3%/3 mm.

## 1. Introduction

For external radiation therapy, dose distribution verification is performed using various film and semiconductor detectors, and the use of three-dimensional (3D) detectors is also increasing [1-3]. In contrast, verifying the spatial dose distribution in brachytherapy is difficult, and verification by point measurement is a realistic method. In clinical practice, we are moving to 3D-image-guided brachytherapy (IGBT), which plans and treats three-dimensionally, using computed tomography (CT) and magnetic resonance imaging (MRI) for the treatment of cervical cancer. 3D-IGBT leads not only to general intraluminal irradiation but also hybrid irradiation combined with interstitial irradiation that

IOP Conf. Series: Journal of Physics: Conf. Series **1305** (2019) 012052 doi:10.1088/1742-6596/1305/1/012052

needs to be performed in complicated treatment plans. Shifting to 3D-IGBT, spatial measurement of dose distribution verification is recommended. However, no practical individual measurement devices exist, and the validation of individual patient's plans has not yet been conducted. A previous study using Fricke gel dosimeter for Ir-192 intracavitary brachytherapy for cervical cancer has been reported [4]. To our best knowledge, there is no measurement report of Co-60 using a polymer gel dosimeter, and gradient passing rate similar to external irradiation has not been evaluated.

In this study, an AQUAJOINT®-based polymer gel dosimeter developed by RIKEN and Nissan Chemical Corporation was used [5-7]. AQUAJOINT® is a physical hydrogel formed by mixing two liquids at room temperature. The gel is stretchable, and the elastic property enables an applicator to be inserted into the gel. We aimed to obtain and analyze spatial dose distribution of Co-60 sealed brachytherapy for cervical therapy using an AQUAJOINT®-based VIPET polymer gel dosimeter.

## 2. Materials and Methods

### *2.1. Gel dosimeter fabrication*

We prepared a dosimeter mixture by first dissolving *N*-vinyl-2-pyrrolidone (NVP), *N,N'*-methylene-bis-acrylamide (BIS), tetrakis-(hydroxymethyl)phosphonium chloride (THPC), and AQUAJOINT® B-5 (Nissan Chemical Corporation, water content: 86%) in deionized water. The addition of AQUAJOINT® A-5 (Nissan Chemical Corporation, water content: 81%) to the aqueous solution yields a homogeneous system comprising of a gel matrix with monomers. The resultant mixture contained 8% NVP, 4% BIS, 50 mM THPC, 11% AQUAJOINT® A-5, and 11% AQUAJOINT® B-5. The gel precursor was poured into eleven 50 mL PET vials to prepare the response curves for the dose–R2. One 2-L gel dosimeter was prepared to insert an applicator. The bottle was a fluorine-treated polypropylene cylindrical container. Three holes adjusted to the applicator were drilled in the lid of the bottle and the CT/MRI compatible applicator was inserted and fixed. After preparation, it was sealed in a nitrogen-purged aluminum laminated bag.

### *2.2. Irradiation Planning*

In order to prepare the response curve of the dose–R2, a radiation treatment planning system (RTPS, SagiPlan®, Eckert & Ziegler BEBIG, Germany) calculation was performed to irradiate stepwise from 0 Gy to 56 Gy on a slice plane 3.5 mm from the center of the radiation source. We planned to shoot the sample with CT (Optima 580 W, GE, imaging conditions: 120 kV, 335 mA, slice thickness 2.5 mm) and irradiate point A, which is the prescription reference point for intrauterine cavity irradiation, at 4 Gy.

### *2.3. Irradiation*

Vials to prepare the response curve were placed in water with the plastic applicator fixed with adhesive (Figure 1 (a)). Then, irradiation was performed with a small source irradiation apparatus, MultiSource® (Eckert & Ziegler BEBIG, Germany). A 2-L gel sample was connected to the CT/MRI compatible applicator and irradiated based on the treatment plan (Figure 1 (b)).

### *2.4 MRI scan and analysis*

Imaging was performed using MRI (MAGNETOM Avant, Siemens Healthcare) 24 h after irradiation. The imaging conditions were TR = 4000 ms, TE 1 = 30 ms, TE 2 = 136 ms, slice thickness = 3 mm, and FOV = 256 × 256 pixels using multi-SE sequence.

Depending on the direction of frequency and phase, whether metal artifacts appeared were examined in advance. A CT/MRI compatible applicator was placed in the water, and images were compared by changing the direction of frequency and phase parallel or perpendicular to the applicator.

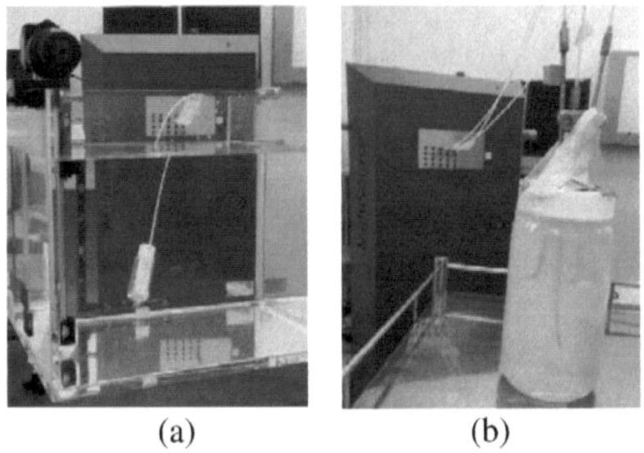

**Figure 1.** Measurement arrangement (a) Response curve vial, (b) 2-L sample with an applicator.

*2.5 Profile analysis*

Using Image J, the R2 (R2 = 1/T2, $s^{-1}$) image was created from the MRI image. A region of interest (ROI) was set on the R2 image of the vial, and the R2 value was read to create a response curve (Figure 2). In the DD-System (R-TECH INC, Japan), the R2 image of the sample (Figure 3) was converted into a dose-distribution map. In addition, the dose profile was created by converting the R2 value of the sample using the response curve. We obtained the planned dose distribution from RTPS and compared and analyzed the same section and profile of the AQUAJOINT®-based VIPET polymer gel dosimeter. The profile to be analyzed was the axial section including point A, which was defined as the prescription point. Since the value was saturated in the high dose range after 30 Gy of the response curve, the analysis that results up to a 1 cm distance from the center of the radiation source was excluded because the R2 value has no dose dependency.

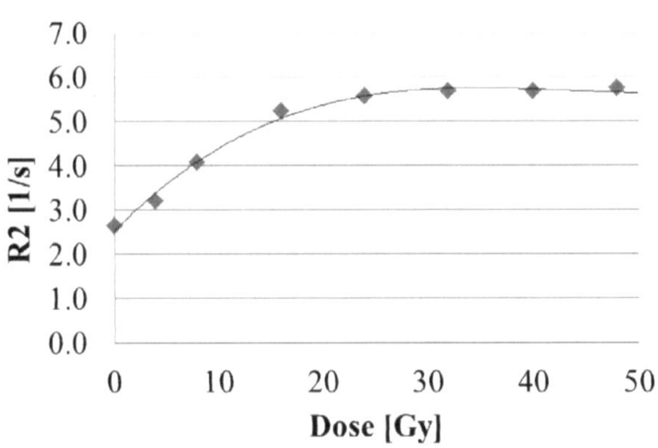


**Figure 2.** Dose-response curve.

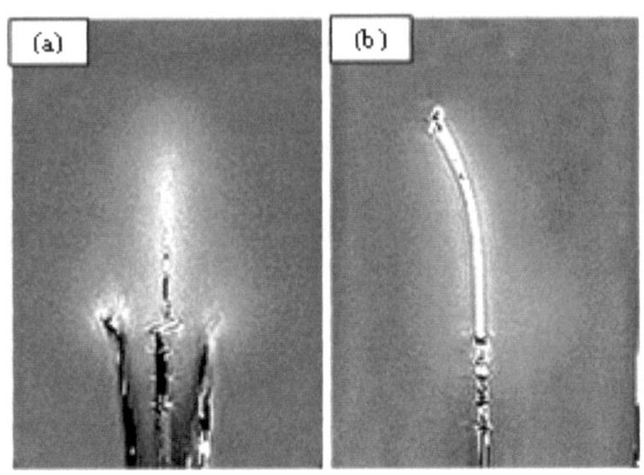


**Figure 3.** The R2 image (in container) of 2L sample with an applicator: (a) Coronal, (b) Sagittal.

*2.6 Gradient and gamma analyses*

The data were transferred from RTPS to the DD-System through the radiation treatment planning support device, MIM Maestro™ (MIM Software, USA). MIM Maestro™ extracts the dose distribution of an arbitrary cross-sectional data. Sagittal and coronal cross-sections, which are characteristic for intrauterine cavity irradiation, were prepared with the same cross-section as the MRI image. The isodose curve and gradient analysis [8] were calculated for the dose distribution of RTPS and gels as converted by the DD-System.

IOP Conf. Series: Journal of Physics: Conf. Series **1305** (2019) 012052 doi:10.1088/1742-6596/1305/1/012052

## 3. Results and Discussion

### *3.1 MRI artifacts*

Despite a CT/MRI compatible applicator, metal artifacts were apparent (Figure 4). The image with the phase direction set in the axial direction (AP) of the applicator indicated an influence of artifacts of 10-15 mm. By contrast, the artifact was within 10 mm in the orthogonal direction (RL) (Figure 5). In the RL, the metal area of the applicator was thought to be small. In both directions, an increase in the R2 value was observed due to artifacts up to a distance of 10 - 15 mm.

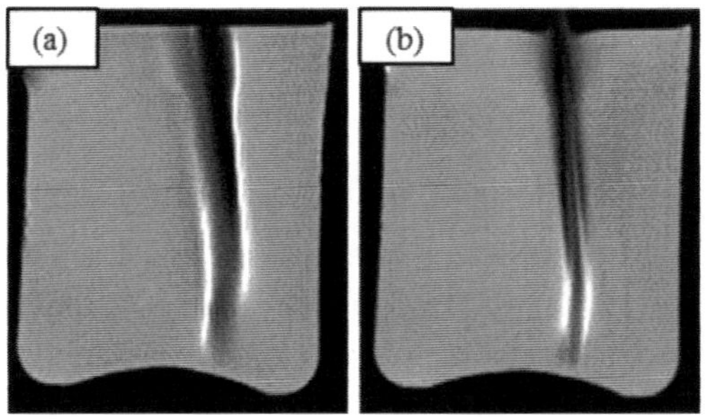


**Figure 4.** Influence of metal artifacts due to differences in frequency and phase direction: (a) The AP phase direction, (b) The RL phase direction.

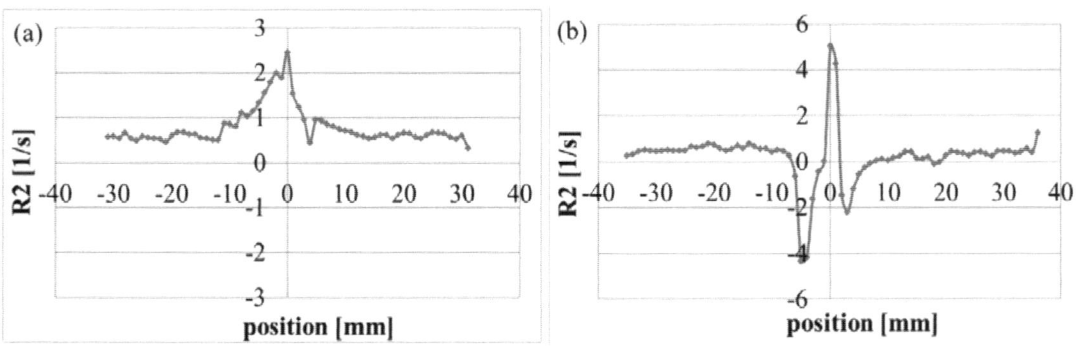


**Figure 5.** Profile of metal artifacts due to differences in the frequency and phase direction. (a) The AP phase direction. (b) The RL phase direction.

### *3.2 Profile analysis*

The R2 image was converted to a dose distribution of Co-60 (Figure 6). Measurement and planned dose profile on the cross-section passing through point A of the sagittal image is shown in Figure 7. When the treatment plan data and profile of the AQUAJOINT®-based VIPET polymer gel dosimeter were normalized at point A and compared with RTPS, the dose difference tended to be large in the region within approximately 1.5 cm from the center of the applicator (high-dose range). On the contrary, the dose difference in the middle-dose range was less than ± 2%.

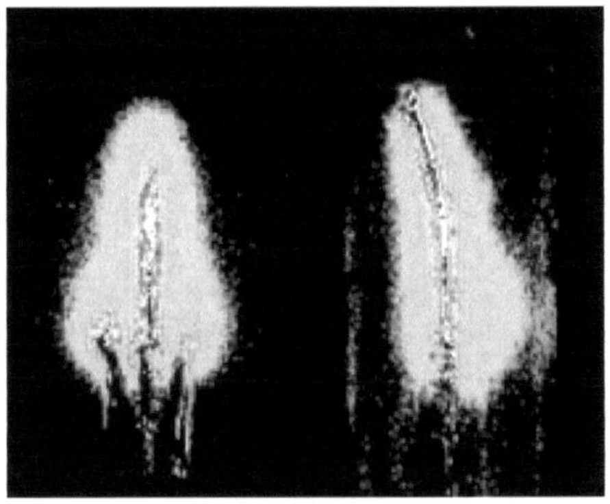


**Figure 6.** Image converted to dose distribution: (a) Coronal, (b) Sagittal.

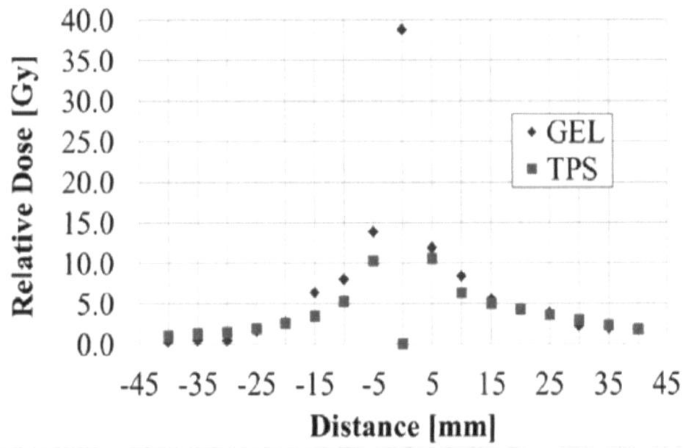


**Figure 7.** Comparison of measurement and planned dose profile at axial slice of point A. Because the position at the distance of 0 mm was as high as ≥1,000 Gy, the value is converted to zero.

At the high-dose range, the large dose difference in the vicinity of the applicator (within 5 mm) was due to the inability of the AQUAJOINT®-based VIPET polymer gel dosimeter dose to exhibit dose dependency over 30 Gy. In the dose range of ≥20 Gy, the response curve tends to gently saturate; therefore, the planned dose of RTPS in the vicinity of the applicator is underestimated. In order to measure this range, improving the dynamic range of the AQUAJOINT®-based VIPET polymer gel dosimeter in high-dose range is required.

In contrast, within 5–15 mm of the applicator, the AQUAJOINT®-based VIPET polymer gel dosimeter showed higher doses than that of RTPS. The range in which this dose difference occurred was close to the region of the metal artifact. Imaging was performed to minimize the artifacts as much as possible; however, it was difficult to completely remove them even with CT/MRI compatible applicators. In order to improve the measurement accuracy, an applicator made of a material, such as plastic, that does not generate artifacts should be used.

*3.3 Gradient and gamma analyses*

In order to compare the spatial dose distribution, the isodose curves and calculation of the gradient analysis are shown (Figure 8). Good agreements were confirmed in the region of 2 to 6 Gy near the prescription point.

As a result of the gradient analysis excluding the pixel data of ≥20 Gy, the gradient passing rate was 62.38% in the sagittal plane and 67.81% in the coronal plane (Figure 9). In the brachytherapy source, the 3%/3 mm of dose difference (DD)/ distance to agreement (DTA) generally used for IMRT (intensity-modulated radiation therapy) was adopted because no evaluation condition acts as a reference for the gradient analysis.

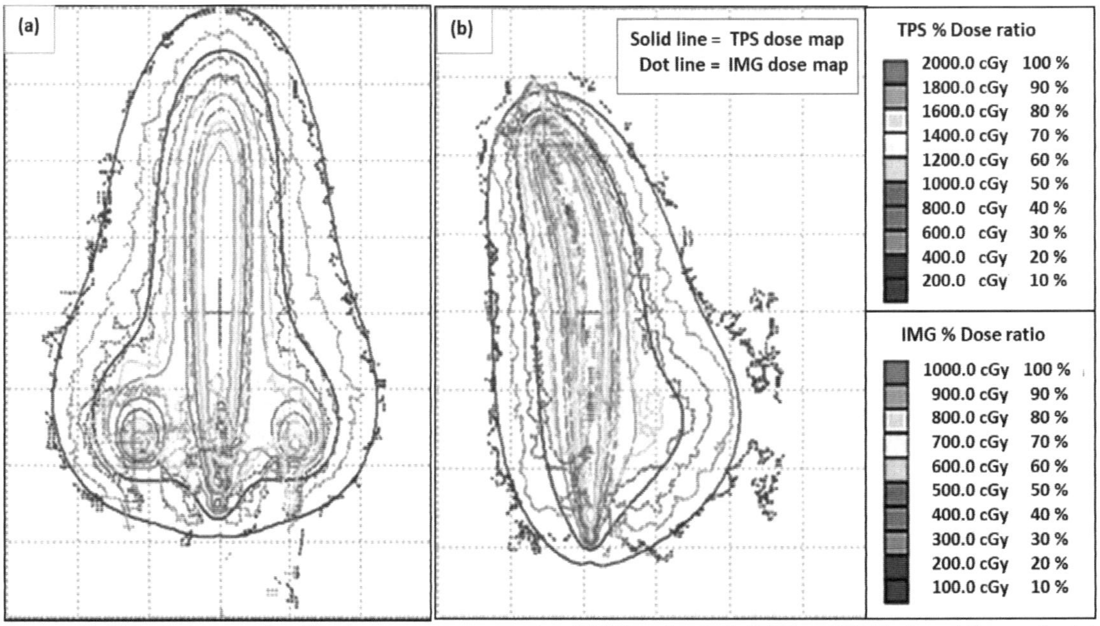


**Figure 8.** Calculation result of isodose curves comparing RTPS and gel. Solid line: RTPS, dotted line: gel. (a) Coronal plane. (b) Sagittal plane.

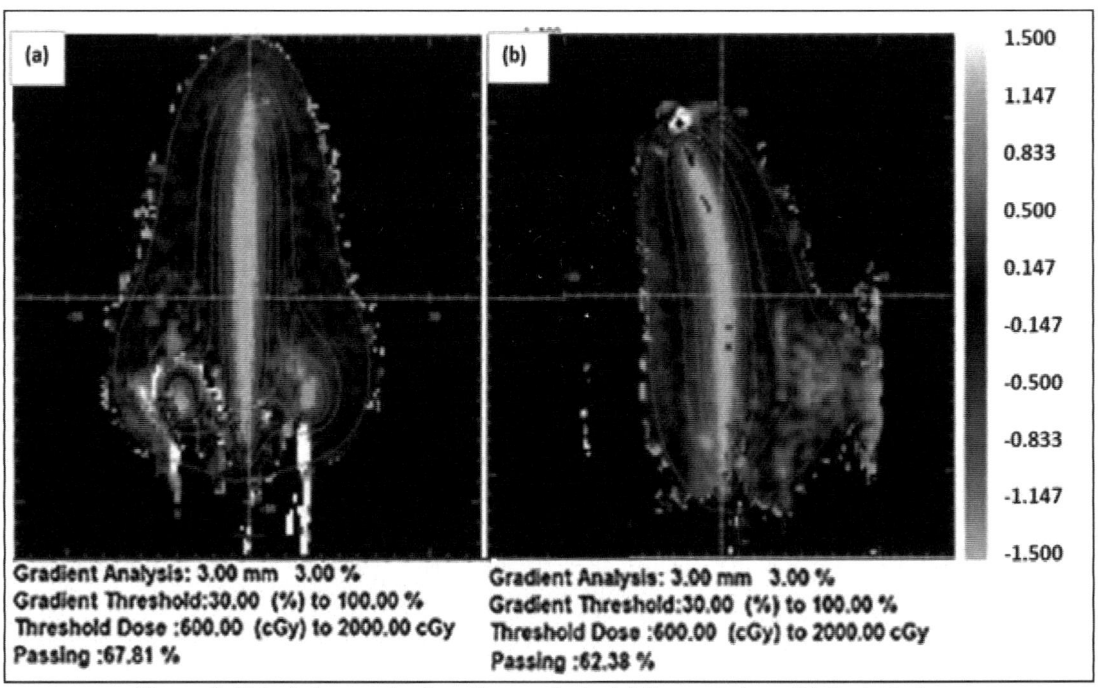


**Figure 9.** Calculation result of gradient analysis. (a) Coronal plane. (b) Sagittal plane.

Furthermore, we improved the DD-System and evaluated by gamma analysis. The result of gamma analysis was 87.594% in the sagittal plane and 93.711% in the coronal plane (Figure 10). Similarly, the evaluation condition was 3%/3mm of DD/DTA. However, since the DD-System used in this study is in the development stage of the program, no reference coordinates were available to match the MRI and dose distribution. The accuracy of the spatial position of MRI images remains uncertain [9]. The decrease in pass rate was mainly due to the uncertain accuracy of this spatial position. Moreover, a phenomenon with an overestimated dose distribution due to the strong occurrence of MRI artifacts was found in the vicinity of the uterus mouth, near each applicator, especially the connecting parts.

In this study, analyzing the gradient was possible because the MRI slices were 3 mm thick at 256 pixels. In order to improve the measurement accuracy, the following improvements should be considered: (1) expansion of the dynamic range of gel dosimeters to be sensitive at high doses of ≥30 Gy, (2) improvement of spatial position accuracy, and (3) removal of artifacts by changing the applicator.

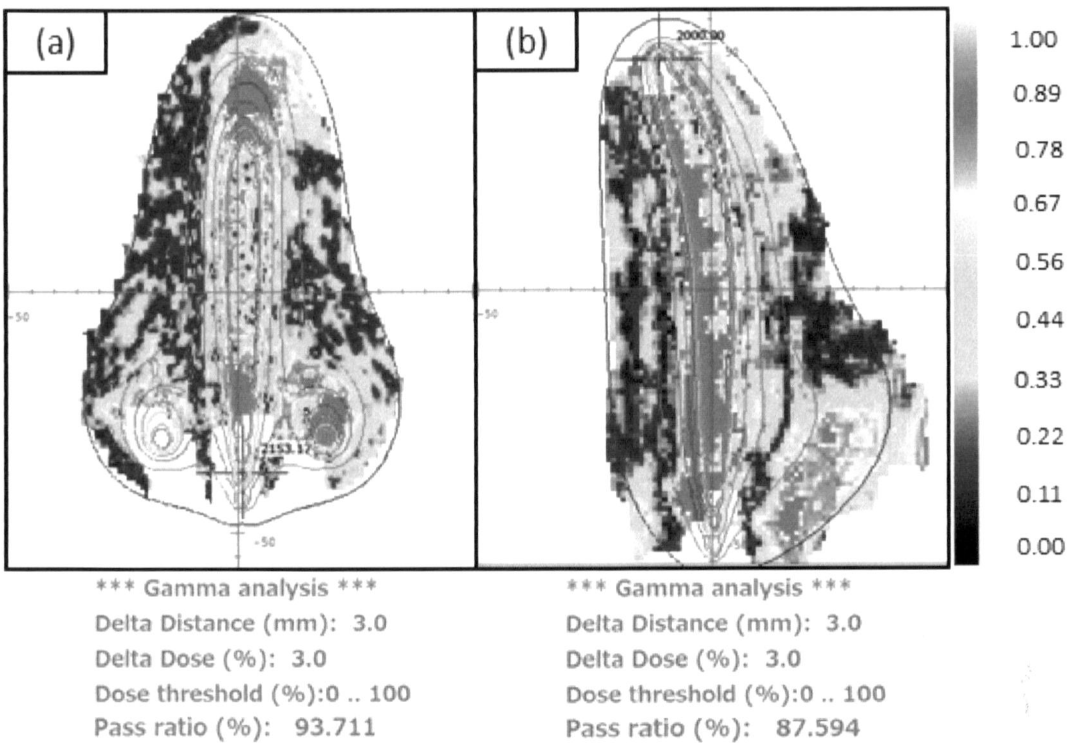


**Figure 10.** Calculation result of gamma analysis. (a) Coronal plane. (b) Sagittal plane.

## 4. Conclusions

We were able to obtain and evaluate 3D dose distribution with pass rate of the intracavitary brachytherapy by Co-60 radiation source using an AQUAJOINT®-based VIPET polymer gel dosimeter.

## 5. References

[1] Baldock C *et al* 2010 *Phys. Med. Biol.* **55** R1-63
[2] Mather M L *et al* 2003 *Phys. Med. Biol.* **48** N269-75
[3] Vial P *et al* 2008 *Med. Phys.* **35** 4362-74
[4] Knutsen B H *et al* 1997 *Radiother. Oncol.* **43** 219-27
[5] Maeyama T *et al* 2015 *Fourth 3D Gel Dosimeter Research Meeting, Teikyo University*
[6] Takanashi T *et al* 2017 *J. Phys.: Conf. Series* **847** 012013
[7] Maeyama T *et al* 2018 *Radiat. Phys. Chem.* **146** 121-5
[8] Moran J M *et al* 2005 *J. Appl. Clin. Med. Phys.* **6** 62-73
[9] De Deene Y *et al* 2001 *Phys. Med. Biol.* **46** 2801-25

IOP Conf. Series: Journal of Physics: Conf. Series **1305** (2019) 012053 doi:10.1088/1742-6596/1305/1/012053

# Dose reconstruction including dynamic six-degree of freedom motion during prostate radiotherapy

**C G Muurholm[1], T Ravkilde[2], S Skouboe[1], T Eade[3], D T Nguyen[4], J Booth[3], P J Keall[4] and P R Poulsen[1,5]**
[1]Department of Oncology, Aarhus University Hospital, Denmark
[2]Department of Medical Physics, Aarhus University Hospital, Denmark
[3]Northern Sydney Cancer Centre, Royal North Shore Hospital, Australia
[4]Radiation Physics Laboratory, Sydney Medical School, The University of Sydney, Australia
[5]Danish Center for Particle Therapy, Aarhus University Hospital, Denmark

E-mail: per.poulsen@rm.dk

**Abstract**. An in-house developed program for real-time reconstruction of motion-induced dose errors, DoseTracker, was extended to handle rotational target motion in addition to the previously implemented translational motion, and applied offline for prostate VMAT treatments. For translational motion, the motion-induced errors of DoseTracker were in good agreement with ground truth dose reconstructions performed in a commercial treatment planning system. For rotational motion, no ground truth was available, but DoseTracker showed that the VMAT dose is highly robust against static interfractional rotations but quite sensitive to dynamic intrafraction rotations due to interplay effects between target motion and machine motion.

## 1. Introduction

Image guidance is commonly used in prostate radiotherapy for daily pre-treatment alignment of implanted markers with the planned marker positions [1]. The alignment will typically only include translational motion although both inter- and intrafractional prostate rotations may be substantial [2]. Real-time monitoring of six-degree of freedom (6DoF) prostate motion during radiotherapy delivery was recently demonstrated using kilovoltage intrafraction monitoring (KIM) [3]. However, real-time motion adaptation by gating or tracking so far only accounts for the translational component of the motion [4-5]. The impact of the uncompensated rotational motion is unknown since commercial treatment planning systems (TPSs) lack the ability to include rotations. In this study, we extend an in-house developed fast motion-including dose reconstruction program to 6DoF motion and demonstrate real-time use of the program to evaluate the dosimetric effects of dynamic prostate translations and rotations during volumetric modulated arc therapy (VMAT) treatments.

## 2. Material and methods

A software program, DoseTracker, was developed for real-time reconstruction of the dose delivered to any set of moving points within a body contour during radiotherapy delivery [6]. DoseTracker first calculates the 2D dose distribution in the isocenter plane by convolving the beam aperture with a dose kernel and then determines the dose in any point by depth scaling of the isocenter plane dose to the point by use of depth dose curves [7]. DoseTracker currently assumes water density.

In this study, DoseTracker was extended to pelvic applications and handling of 6DoF internal motion. The extended program was applied offline for a prostate cancer patient treated with a dual arc VMAT plan created in the Eclipse TPS (Varian Medical Systems). Connection to the accelerator was simulated by sending a data stream at 21 Hz to DoseTracker with 6DoF prostate positions and accelerator parameters (gantry angle, MLC positions, MU increment, etc.) corresponding to delivery of the VMAT plan. In a repeated loop, DoseTracker continuously calculated the dose increment to a set of calculation points twice: once with motion and once without motion. These where then used to obtain the time-resolved motion-induced dose error for each calculation point. The dose calculations were performed for 50,000 calculation points that included the PTV and constituted a subset of the original TPS calculation points. The same rigid 6DoF motion was applied to all calculation points while the outer body contour was assumed to be static.

A treatment simulation was first performed with dynamic prostate translations measured with implanted electromagnetic transponders [8] to investigate the accuracy of DoseTracker for prostate VMAT. A ground truth motion-including dose reconstruction was obtained by a previously validated dose reconstruction method that emulates rigid target translations as multiple isocenter shifts in the TPS [9] and has been extended for time resolved calculations [10]. A time resolution of 3Hz was used corresponding to a dose calculation for every 2° gantry rotation during the simulated VMAT delivery.

Next, simulations were performed with static prostate rotations about the left-right (LR) axis and with dynamic prostate rotations measured with KIM [2]. The dosimetric impact of the rotations was quantified as the CTV volume covered by 95% of planned mean CTV dose (V95%).

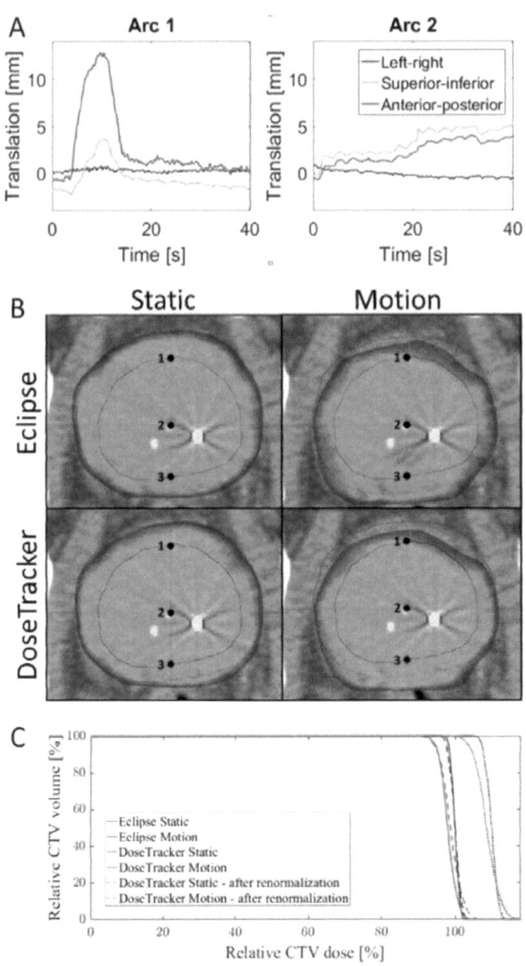


**Figure 1.** (A) Translational prostate motion during the two VMAT arcs. (B) Reconstructed dose distributions in an axial plane through the prostate CTV (green structure) and PTV (red) calculated by the Eclipse TPS and DoseTracker with and without the motion. The dose color wash shows doses in the range 90-110%. The illustrated DoseTracker doses were renormalized to the same static mean CTV dose as in the TPS. Time resolved doses in the numbered dose points are presented in figure 2. (C) CTV dose volume histograms for reconstructed doses in the Eclipse TPS and in DoseTracker with and without translational motion. The DoseTracker dose is shown with and without renomalization to the same static mean CTV dose as in the TPS.

## 3. Results and discussion

DoseTracker performed the motion-including and static dose reconstructions with a mean frequency of 1.7 Hz, demonstrating that real-time application is feasible. The translational motion included a large transient excursion in mainly the anterior direction during the first VMAT arc and a more modest continuous drift in the superior-anterior direction during the second arc (figure 1A). DoseTracker overestimated the absolute dose level for the prostate with ~10% (figure 1C) due to its simplified modelling of the output factor and phantom scatter. Still, the distortion of the dose distribution caused by prostate motion during dynamic VMAT delivery (i.e. interplay effects, figure 1B) and the overall impact of motion on the CTV dose-volume histogram (figure 1C) were in good agreement with the ground truth dose reconstruction by the TPS. The motion-induced reduction in CTV D95 relative to the planned dose was 3.6 percent-point for DoseTracker and 3.7 percent-point for the TPS dose reconstruction. The time-resolved dose rate and motion-induced dose rate error of DoseTracker in selected points in the prostate (figure 1B) were also in good agreement with the TPS ground truth as shown in figure 2. Note in particular the large dose deficit in the anterior part of the prostate (Point 1) caused by the large anterior excursion of the prostate 6-12 s into the treatment delivery.

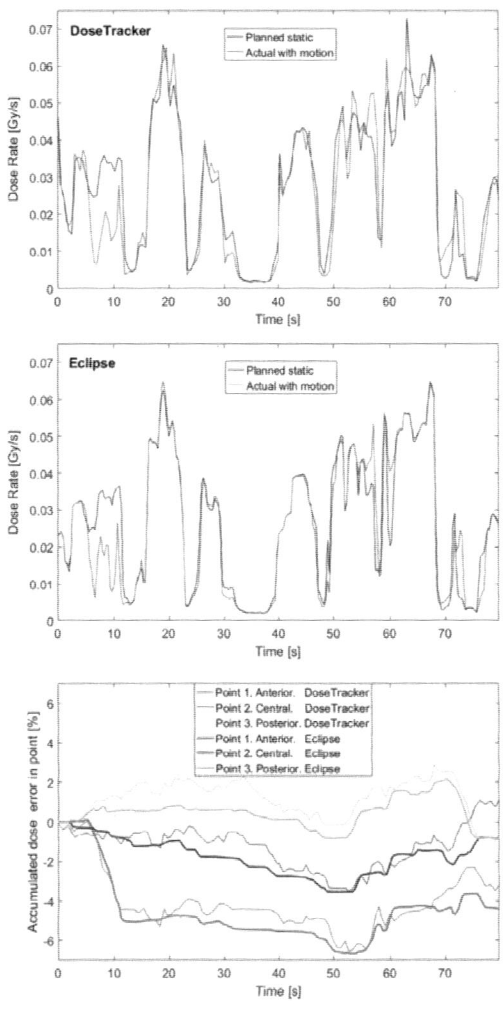


**Figure 2.** Top and middle: Dose rate during dual arc VMAT delivery as a function of time in Point 1 in figure 1B (anterior part of the prostate) as reconstructed by DoseTracker (top) and the Eclipse TPS (middle) with and without the translational motion of figure 1A. Bottom: Cumulative motion-induced dose error as a function of time (i.e. cumulative difference between the red and black curves in the top and middle figures) in the three prostate calculation points shown in figure 1B as reconstructed by DoseTracker and Eclipse. The dose error is shown in percent of the prescribed fraction dose of 2 Gy.

Static interfractional prostate rotations caused a rotation of the high dose volume relative to the prostate, but otherwise kept its shape almost unchanged (figure 3). For the investigated VMAT plan, CTV V95% remained high for rotations up to ~10° and then dropped steeply at larger rotations as the prostate moved out of the high dose region (figure 3). The patient-measured dynamic prostate rotation was in mean 4.4° (LR axis), 0.5° (superior-inferior axis), and -1.6° (anterior-posterior axis), but it was highly dynamic with a maximum of 19° (figure 4A). Interplay effects between the dynamic rotation and VMAT delivery markedly distorted the dose distribution and reduced the CTV V95% from 100% to 94.9% (figure 4B-C) because of mismatches between the modulated doses received by the prostate from different directions. This was in stark contrast to the small dosimetric impact of static rotations, which can more easily be mitigated by application of safety margins.

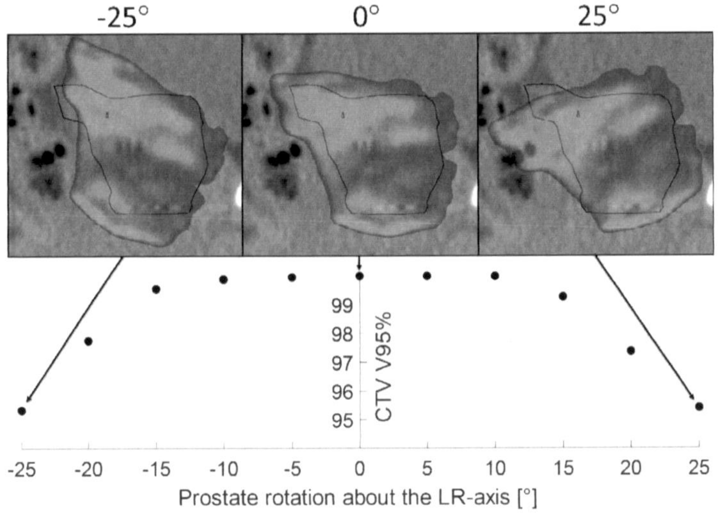


**Figure 3.** Top: Examples of dose distributions in the sagittal plane with static prostate rotations of -25°, 0°, and 25° about the left-right axis (doses above 95% are shown). Note that only doses inside the CTV (black contour) are relevant since the remaining anatomy outside the prostate (e.g. bones) does not rotate in the same way as the prostate. Bottom: CTV V95% as function of static left-right prostate rotation.

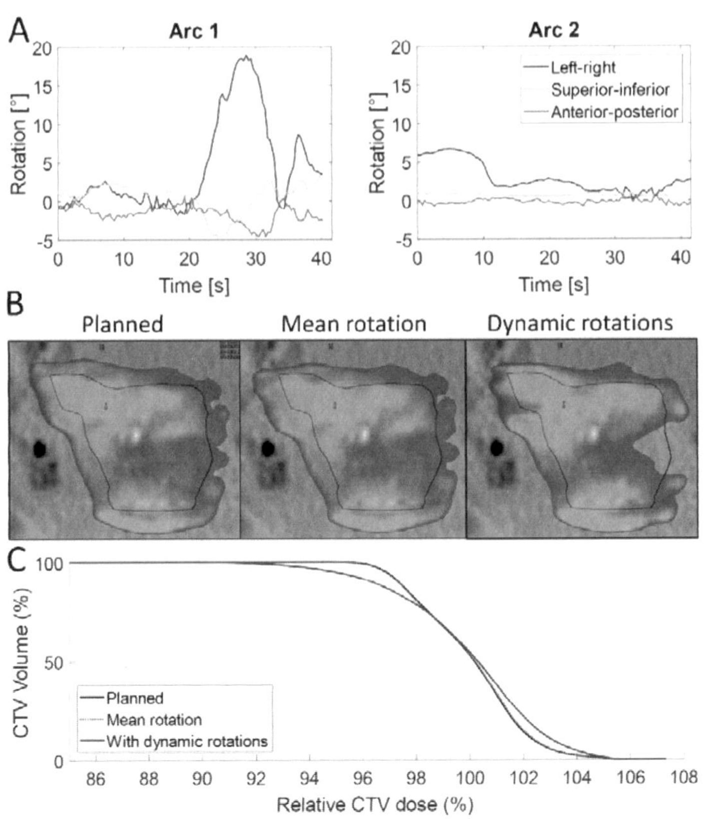


**Figure 4.**
(A) Patient-measured dynamic prostate rotation applied during the simulated VMAT treatment.
(B) Reconstructed dose distribution (above 95%) in the sagittal plane as planned, with a static rotation equal to the mean prostate rotation, and with the dynamic rotation of figure 4A.
(C) CTV dose volume histogram as planned and delivered using both the mean prostate rotation and the dynamic rotation of figure 4A.

## 4. Conclusion

Real-time dose reconstruction including the effects of dynamic 6DoF motion was demonstrated during simulated VMAT delivery to the prostate with the in-house developed DoseTracker program. For translational motion, a ground truth dose reconstruction could be created in a commercial TPS demonstrating ~10% overestimation of the absolute dose level by DoseTracker, but good estimation of the motion induced dose errors. Errors in the absolute dose level tend to cancel out when DoseTracker calculations with and without motion are compared. For rotational motion, the CTV dose was much more susceptible to dynamic prostate rotations than to static rotations. The calculations can be integrated with online real-time 6DoF motion monitoring [3] for on-the-fly 6DoF dose reconstruction during treatment delivery, e.g. to estimate the residual dose errors after the translational real-time corrections currently applied in clinical protocols with prostate gating or MLC tracking [5].

## 5. Acknowledgements

This work was supported by The Danish Cancer Society and Varian Medical Systems, Inc., Palo Alto, CA. We gratefully thank Katja Langen, Maryland Proton Treatment Center, for the translational prostate motion data.

## 6. References

[1] Poulsen P R *et al* 2007 *Radiother. Oncol.* **85** 201
[2] Kim J-H *et al* 2018 *Radiother. Oncol.* **126** 236
[3] Nguyen D T *et al* 2017 *Radiother. Oncol.* **123** 37
[4] Keall P *et al* 2018 *Radiother. Oncol.* **127** 6
[5] Keall P *et al* 2017 *BMC Cancer* **17** 180
[6] Ravkilde T *et al* 2018 *Med. Phys.* **45** 3893
[7] Ravkilde T *et al* 2014 *Phys. Med. Biol.* **59** 7279
[8] Langen K M *et al* 2008 *Int. J. Radiat. Oncol. Biol. Phys.* **71** 1084
[9] Poulsen P R *et al* 2012 *Med. Phys.* **39** 6237
[10] Ravkilde T *et al* 2013 *Acta. Oncol.* **52** 1497

IOP Conf. Series: Journal of Physics: Conf. Series **1305** (2019) 012054 doi:10.1088/1742-6596/1305/1/012054

# Development of an experimental 3-D tool based on radiochromic films to determine normal tissue doses in external radiotherapy

**J Colnot[1], G Garnier[1], S Zefkili[2], J-L Dumas[2], R Gschwind[3] and C Huet[1]**
[1]Institut de Radioprotection et de Sûreté Nucléaire (IRSN), Service de Recherche en Dosimétrie, Laboratoire de Dosimétrie des Rayonnements Ionisants, Fontenay-aux-Roses, France
[2]Institut Curie, Service de Physique, Paris, France
[3]LCE, Université de Bourgogne-Franche-Comté, Montbéliard, France

E-mail : julie.colnot@irsn.fr

**Abstract**. External radiotherapy largely proved its efficacy to treat cancer. However, this technique leads to unavoidable exposure of normal tissues that may results in adverse effects. One of the main concerns is the induction of second cancers that may appear at the periphery or even away from the treated area. Risk estimation models need accurate 3-D dosimetric data on healthy organs. To this aim, we developed an experimental tool based on radiochromic films measurements. In this work, we present a study performed with a heterogeneous phantom. EBT3 films were positioned in between the phantom slices and irradiated according to a VMAT plan. The dose was reconstructed in 3-D by interpolation thanks to an in-house Matlab tool. Two interpolation methods are studied and the reconstructed dose is compared to independent film measurements for validation. Finally, dosimetric data such as DVH are compared to the TPS evaluation. Continuous interpolation proves to better reconstruct the doses in high gradient areas whereas linear interpolation seems to be a better option to evaluate doses away from the field edge. Besides, TPS and 3-D measurements give comparable results for organs placed inside the beams (maximum 6.3% difference on organs mean doses). However, the TPS tends to underestimate doses to organs placed at least partially outside the fields (from 29 to 59% difference on mean doses). Optimization of the dose reconstruction tool by using a combination of the two interpolation methods is ongoing. Comparison with gel dosimetry will also be investigated.

## 1. Introduction

External radiotherapy efficacy is widely acknowledged for its curative benefits to treat cancer. Therefore, this technique takes part in more than 50% of current treatments [1]. However, its most important drawback is the unavoidable exposure of healthy tissues to dose that may lead to secondary effects [2-3]. Some of those effects such as second cancers are stochastics and can appear at the periphery or even away from the treated target [1, 4]. The risk of second cancer induction raises a growing concern especially since modern techniques tend to irradiate a more important volume of tissues and since the better life expectancy of survivors increases the probability of second cancer appearance. Thus, evaluating the risk of adverse effects induction represents an important challenge. This estimation is based on risk models that correlate volumetric dosimetry and occurrence of adverse effects. Accurate 3-D dosimetric data inside and away from the treatment field are needed. For this purpose, the treatment

IOP Conf. Series: Journal of Physics: Conf. Series **1305** (2019) 012054 doi:10.1088/1742-6596/1305/1/012054

planning system (TPS) can only be used to evaluate in-field doses as its lack of precision for out-of-field dose evaluation is well documented in the literature [5]. Over the years, other techniques have been studied to estimate doses away from field edge: water or anthropomorphic phantom measurements and Monte-Carlo simulations. However, most of the experimental studies provide point measurements or unrealistic phantom geometries [6]. Within this context, we developed a new numerical tool on Matlab based on EBT3 films measurements to evaluate 3-D doses to normal tissues. The procedure consists in irradiating films placed in a sliced phantom. After a rigorous analysis of the films, 2-D dose distributions are interpolated to reconstruct the 3-D dose distribution. First, a preliminary study has been performed in a homogeneous phantom to evaluate the feasibility of the method. In this work, we present the results obtained with a heterogeneous configuration. The first part of this study is the validation and optimization of the tool. Finally, the reconstructed dose is compared to the evaluation of the Eclipse™ TPS (Varian Medical Systems) performed with two dose calculation algorithms: AAA and Acuros®.

## 2. Material and methods

### *2.1. EBT3 film calibration*
In this study, EBT3 radiochromic films (Ashland) were prepared according to a rigorous protocol (cutting, calibration, readout) [7]. The irradiations were performed at the Institut Curie with a Clinac 2100CS linear accelerator (Varian Medical Systems), all the films were irradiated the same day. The films were calibrated from 1.35 cGy to 37 Gy with 24 points (2 films per dose) between tissue-equivalent slabs at 10 cm in depth with a 10×10 cm² field (SSD = 100 cm).

### *2.2. Phantom preparation*
The EasyCube modular phantom (Euromechanics Medical) was used in this work in a heterogeneous configuration. The tissue-equivalent cube measuring 18 cm of side was filled with soft tissue, bone and lung inserts (Figure 1). Nine EBT3 films were cut to perfectly fit in the phantom (16×16 cm²), seven of them were spaced 2 from 2.5 cm apart in the phantom. The two leaving films were placed in between for validation (red arrows on Figure 1): one is dedicated to be placed at the target level and the other outside the beam path. The phantom was scanned (Siemens Definition AS OpenRT) with substitute films in place. A target volume of 3.5 cm diameter and the five heterogeneities were then contoured on the Eclipse™ TPS: the PTV is represented in orange on Figure 2, two lung volumes in blue and three bones in yellow. Two other organs were added in the soft tissue: the red organ is intended to be partially in the directly irradiated area (named OAR in this paper) whereas the brown volume is completely outside the beams (referred as OOF organ in this paper). A VMAT plan has been carried out on this geometry. 34 Gy was prescribed to the target volume in 17 fractions (one complete arc per fraction) for approximatively 600 MU per fraction. The plan was calculated using two dose calculation algorithms available on the TPS: AAA (Analytical Anisotropic Algorithm) and Acuros®.

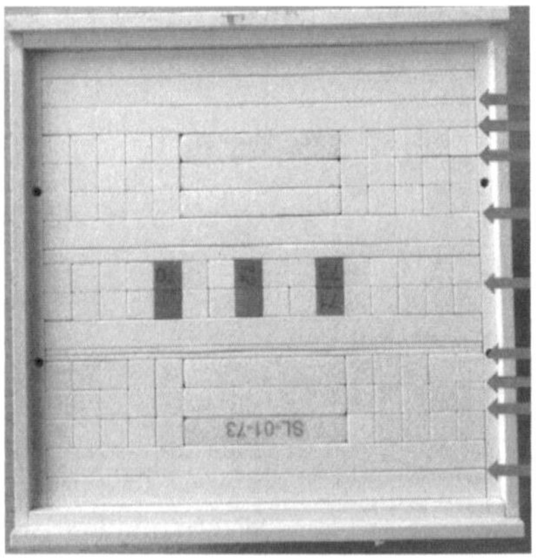


**Figure 1.** The EasyCube configuration and films location: head direction at the top (lungs in yellow, bone in grey, soft tissue in white).

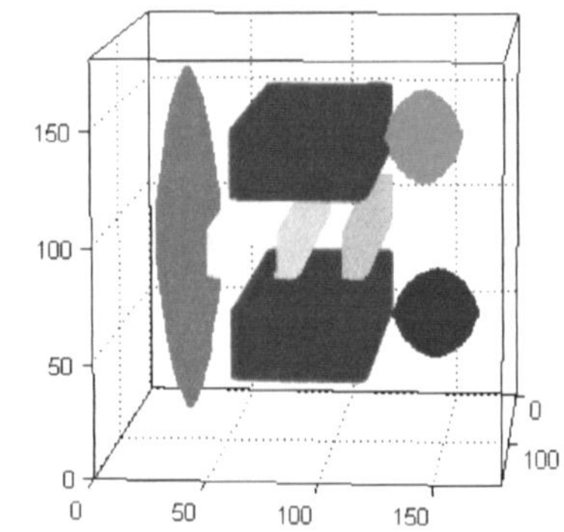


**Figure 2.** Contoured organs present in the phantom (orange: PTV, blue: lungs, yellow: bones, red and brown: organs placed in soft tissue (OAR and OOF respectively)).

### *2.3. 3-D dose reconstruction tool*

After a first step of film processing (subtraction of background signal, denoising and conversion to dose), the tool developed on Matlab (Matlab 2013 and Image Processing Toolbox, The MathWorks, Inc.) proceeds as following:

- Slices corresponding to the films location are identified on the CT scanner images.
- A precise intensity-based registration of the films on the Acuros® dose distribution is performed (for the four films placed on the beam path). For the others, a manual registration of the films has been done.
- Films dose distributions are rescaled according to the TPS resolution.
- Film doses are interpolated to reconstruct the 3-D dose map contained between the two extreme dosimeters: only the seven films (represented by the blue arrows on Figure 1) are used in this process. Two interpolation methods available on Matlab are studied in this work: a linear interpolation and a shape-conservative continuous method.
- The reconstructed dose is compared to the dose obtained on the two intermediate films (red arrows on Figure 1) to validate the interpolation.
- 3-D dosimetric data obtained on the reconstructed distributions (DVH, mean doses) are compared with those evaluated by the TPS.

## 3. Results

### *3.1. Validation of the reconstruction tool*

First, the comparison between the intermediate film placed at the target level and the reconstructed dose allows to evaluate the tool precision on important dose gradient areas. In this case, the two interpolation methods used in the reconstruction underestimate the dose in the gradient. However, the continuous interpolation reproduces better the gradient shape (Figure 3). Away from the field edge, the linear method gives a better estimation of the dose in comparison with continuous interpolation. In both cases, the mean dose differences with the film measurement are around 24%, this value is representative of the discrepancies obtained in the gradient (from 90 to 110 mm on Figure 3).

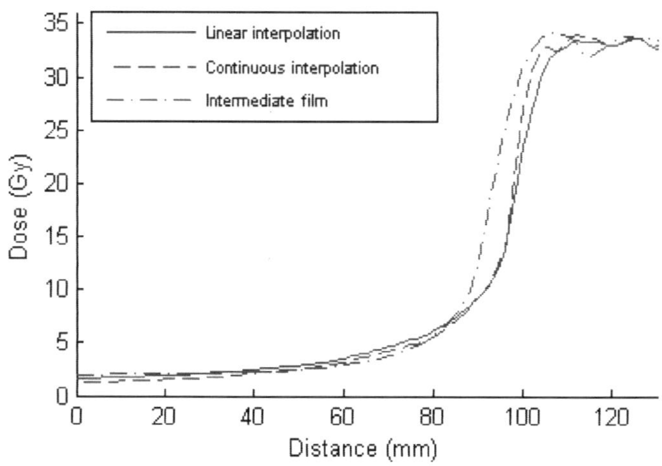


**Figure 3.** Dose profiles of reconstructed dose using two different interpolation methods and film measurement on the intermediate film slice placed within the beam path.

However, when comparing the two interpolations with the intermediate film placed away from the beams, the linear interpolation proves to be the best option (Figure 4). The mean dose difference between reconstructed dose with linear interpolation on this slice and the intermediate film is of 7.0% against 9.4% for the continuous interpolation and 95.5% of the points pass the 10% / 3 mm gamma test against 87.6%. The agreement between intermediate film measurement and interpolations is much better than the comparison with the TPS evaluation. In fact, the mean dose difference is of 30.2% when comparing the AAA dose distribution to the film (13.1% of points passing the 10% / 3 mm gamma test) and 35.2% when comparing the Acuros® slice (2.8% of points passing the 10% / 3 mm gamma test). Figure 5, representing a perpendicular dose profile, shows that continuous interpolation better reconstructs the gradient shape. Thus, this technique allows to slightly reduce the reconstruction errors in this area.

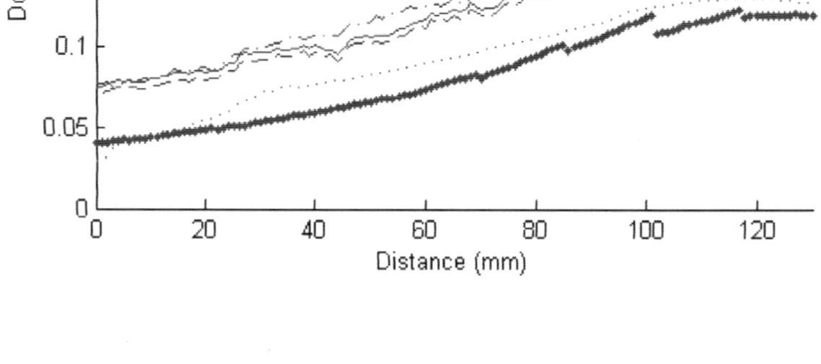


**Figure 4.** Dose profiles of reconstructed dose, film measurement and TPS on the intermediate film slice placed away from the beam path.

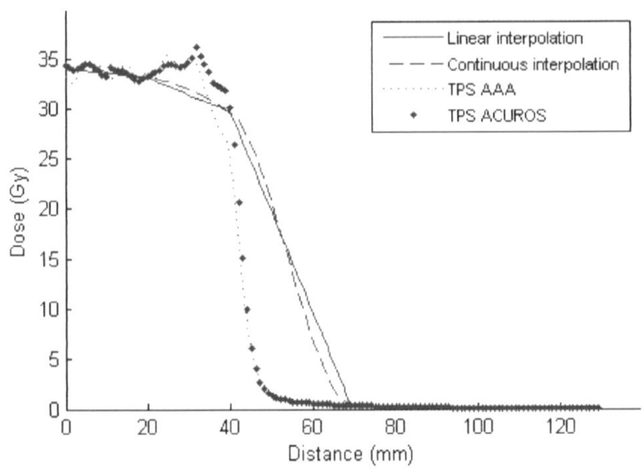


**Figure 5.** Dose profiles perpendicular to the films plane obtained with the TPS and the two different interpolation methods.

### *3.2. Normal tissue doses evaluation*

Doses delivered to contoured organs are compared between reconstructed dose (with the two interpolations) and TPS. Table 1 summarizes dose difference between mean organ doses. The tool underestimates doses to the PTV as high gradient cannot be perfectly reproduced. Linear interpolation overestimates doses to the three bones, which are placed in the gradient between in-field and out-of-field areas. For the other organs the difference between the two interpolation methods are small. AAA and Acuros® give similar results for the PTV and for organs placed inside the beams (maximum difference in mean organ doses 6.3%), but for organs placed outside the field Acuros® underestimates organ doses by up to 17% compared to AAA. Finally, the TPS (both AAA and Acuros®) largely underestimates the doses delivered to out-of-field organs (from 29% for the OAR to 59%) compared to the reconstructed dose. Figure 6 represents the DVH obtained with the experimental tool (two interpolation methods) and with the TPS AAA algorithm.

**Table 1.** Difference between mean organ doses obtained with the reconstructed film doses and the TPS.

| Organs | Difference between mean organ doses (%). | | |
|---|---|---|---|
| | Linear / continuous interpolation | Linear interpolation / AAA | Continuous interpolation / AAA |
| PTV | -2.3 | -7.3 | -4.8 |
| OOF organ | 3.6 | 37 | 35 |
| OAR | 6.3 | 29 | 25 |
| Bones | From 15% to 29% | From 52% to 59% | From 43% to 47% |
| Lung (in-field) | 2.0 | -1.1 | -2.4 |
| Lung (out-of-field) | 1.3 | 24 | 22 |

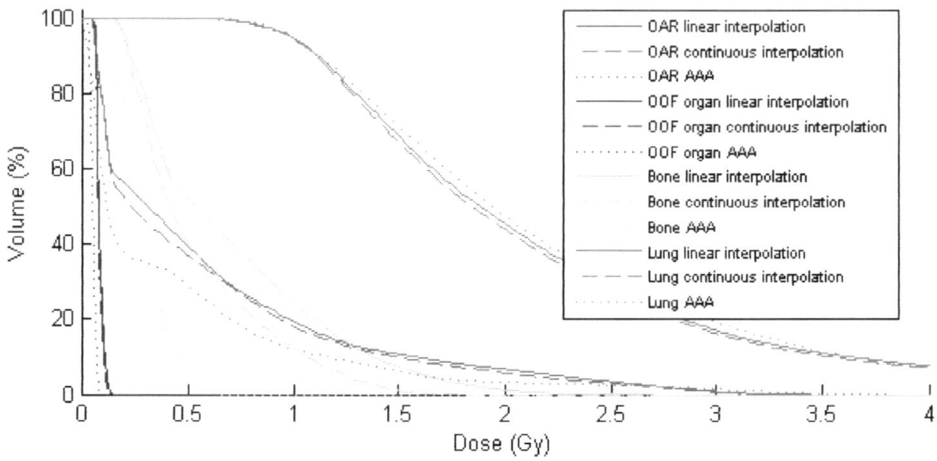


**Figure 6.** Dose-Volume histogram for the in-field lung, one bone, the OAR and the OOF organ obtained with two different interpolation methods and with the AAA algorithm (TPS).

## 4. Discussion

### *4.1. Validation of the reconstruction tool*

Results obtained when comparing reconstructed dose distribution and intermediate film measurements show that our method can be used to evaluate out-of-field doses or doses in low doses areas within the field. In fact, the reconstructed dose proves to be more accurate than the TPS away from the field edge (see 3.1.). Linear interpolation seems to better reproduce the dose evolution away from the target (Figure 2 and 3). However, even though continuous interpolation reduces errors in high gradients, they cannot be precisely reconstructed. A combination of the two interpolation methods seems to be a good alternative.

### *4.2. Normal tissue doses evaluation*

The TPS normal tissue doses underestimation observed in our work is consistent with the results of the literature [6-8]. This is mainly due to head scatter and head leakage that are not modelled in the TPS, in fact this radiation contributes to more than 40% of the out-of-field dose: Joosten *et al.* [5] report that TPS and Monte-Carlo simulation show major discrepancies, even for organs located near the treated area, that can reach 70%. In this work, the dose reconstruction tool was used in a narrow volume, but it enables dose determination for organs away from the target even in area where TPS doses are often not calculated (outside the scanned volume).

## 5. Conclusion

In this work, we present an original in-house tool that enable to obtain reconstructed 3-D doses to normal tissues by experimental means. This tool developed on Matlab and based on radiochromic films has been validated against film measurements and the reconstructed doses compared to the TPS evaluation. The results show major TPS underestimations of the out-of-field doses. In order to improve the accuracy of the developed tool, a combination of linear and continuous shape-preservative interpolations is currently investigated. Finally, a comparison with 3-D gel dosimetry is also considered. This tool will be used in future work to estimate normal tissue doses delivered by advanced radiotherapy techniques on paediatric anthropomorphic phantoms.

## 6. References

[1] Brenner D J *et al* 2003 *PNAS* **100** 13761-66
[2] Xu X G *et al* 2008 *Phys. Med. Biol.* **53** 193-241

[3] Kang K H *et al* 2015 *Cancers* **7** 981-1004
[4] Diallo I *et al* 2009 *Int. J. Radiation Oncology Biol. Phys.* **74** 876-83
[5] Joosten A *et al* 2013 *Radiot. and Onc.* **106** 33-41
[6] Kry S *et al* 2017 *Med. Phys.* **44** 391-429
[7] Moignier C *et al* 2016 *Rad. Meas.* **89** 107-18
[8] Qi Y *et al* 2017 *Rad. Prot. Dosim.* **177** 389-96

# Preliminary investigations of a dynamically deforming target-in-gel dosimeter

**C J Watson[1], A U Yeo[2], J R Supple[3], M Geso[4], T Kron[2] and R D Franich[1]**
[1]School of Science, RMIT University, Melbourne, Australia
[2]Physical Sciences, Peter MacCallum Cancer Centre, Melbourne, Australia
[3]Australian Clinical Dosimetry Service, ARPANSA, Yallambie, Australia
[4]School of Health & Biomedical Sciences, RMIT University, Bundoora, Australia

E-mail: callum.watson@rmit.edu.au

**Abstract**. To increase the clinical utility of deformable gel dosimeters an X-ray CT visible target region is introduced into the DEFGEL dosimeter, with the goal to enable the determination of the change in marginal doses around planned boundaries. The production of the dosimeter components is detailed and the requirements in making, deforming, irradiating and analysing the dosimeter are explored with several areas of required improvement identified. A single degree of motion is used to deform the dosimeter with the intent of identifying the volumetric dose changes caused by the deformation. The planned target volume dose structure was registered to the physical target region present in an optical CT data array and the dose volume histograms of the target volume while static and undergoing deformation are investigated and reported. The total dose to the target volume was shown to decrease by 9.4% due to a 1cm compression of the entire dosimeter volume. Further ways to improve the use of the target region are suggested with a view to improving the dose map information that can be obtained from target-in-gel dosimeters.

## 1. Introduction

Polymer gel dosimeters [1] are uniquely suited for the validation of the effects of motion and deformation [2] in radiotherapy treatment planning systems and treatment delivery [3]. The registration of points is usually achieved by tracking the movement of fiducial markers which provide details of motion within the entire volume of the dosimeter [4]. The lack of internal features makes it difficult to assess the accuracy of a volumetrically targeted treatment plan, as the only structure observable is the outside of the dosimeter volume, which would assumedly contain all treatment volume margins.

It may be that the addition of an internal feature to define a target region can improve the assessment of dose coverage in a planning target volume (PTV) [5], or indeed any volume that changes as a result of motion and deformation. There is also the complicated issue of quantifying the dosimetric impacts of motion, positioning and deformation, which may be approached with more rigour by 3D dosimeters that can measure whilst undergoing dynamic changes. This internal feature is implemented as gel dosimeter material nPAG with the addition of an X-ray contrast agent in the form of Isovue. The resulting dosimeter configuration is termed the IDEFGEL (Isovue +DEFGEL [6]). The ability to discriminate between in and out of volume dose is performed through the investigation of the resulting dose volume histograms in each region. By comparing a static and deforming case it is expected that the amount of dose that is deposited in material that moves into or out of target region can be volumetrically assessed and provide a measurement of the degree to which a margin should be extended to deal with a given range of motion or deformation.

The following study was conducted using a concentration of 0.7% w/w Isovue that provided a contourable target region within the dosimeter volume. A simple breathing motion is used to provide the deformation of the dosimeter that is to be examined, with the intent being that a direct correlation between the volumetric deformations can be drawn to the change in the total dose delivered to the two volumes. The impact of several external factors as well as those directly involved in the delivery and measurement of the dose distributions in the target volume are discussed with several proposed causative relationships detailed, with the intent that addressing these factors will produce a clinically viable gel dosimeter containing a dose sensitive target region that is visible on X-ray CT.

## 2. Method

The feasibility of the gel dosimeter containing an Iodine doped nPAG (InPAG) target region were investigated through analysis of the change in dose response localized within and around the target volume as it underwent a repeatable deformation. The impact of Isovue on the nPAG dosimetric properties is outlined in a previous article and the production process follows closely that of Yeo *et al* [6], with details of inserting the target region outlined below.

### *2.1. Gel formulation, production and target region insertion*

The nPAG used in this study is the same formulation used by Yeo *et al* [5] and is composed of 6 % w/w gelatine (Type A from porcine skin, Sigma Aldrich Ltd., Oakville, Canada) for the hydrogel matrix. Dissolved in this matrix are 3 % w/w N,N'-methylene-bis-acrylamide (Bis) and 3 % w/w acrylamide (Aam)(both from Sigma Aldrich Ltd.). The remaining mass is DI water, with 0.01 mM hydroquinone and 5 mM of Bis[tetrakis(hydroxymethyl-phosphonium)]sulphate (THPS) added to combat autopolymerisation and atmospheric oxygen diffusion respectively with the target region containing 0.7% w/w Isovue as detailed by Watson *et al* [7]. The standard nPAG formulation is first poured into a plastic PET sample container lined with a latex membrane and a hollow plastic cavity is inserted into the center of the gel with a foil lining used to seal the top of the vessel. A portion of the same batch of nPAG mixed with 0.7% w/w Isovue and set inside the cavity. This concentration was used as it was the smallest amount that still provided enough contrast in X-ray CT. Using 0.7% w/w of Isovue provides more definition for the edge of the target region but to minimize the impact on the dosimetric properties a lower concentration was used. With further experimentation and better knowledge of the dosimetry and chemistry changes stemming from the addition of Isovue 0.7% w/w will be used. A silicone mold the same shape and dimension of the plastic pipette bulb used to create the cavity in the dosimeter was manufactured and the target gel is set in this material. The silicone is well sealed to prevent oxygen diffusion during setting and has the added advantage of not adhering to the gel surface, allowing the gel to be easily removed. For this study 3 samples of this configuration were fabricated with one sample containing several internal fiducial markers at the target boundary to track the magnitude of motion.

Both gel containers were placed in a refrigerator at 4°C and set for 12 hours before the target region was inserted into the larger gel volume and standard nPAG poured into the remaining gap at the top of the dosimeter. The latex membrane was sealed and remained in the form fitting sample holder and set for 10 more hours before imaging and irradiation.

### *2.2. Motion phantom setup*

The IDEFGEL samples were housed in an acrylic cylinder holder lined with soft foam to centre the dosimeter in the holder without impeding deformation and ensuring the "relaxed" position was repeatable for all 3 samples. The cylinder holder was placed in a Modus Medical Devices Quasar™ respiratory motion phantom and fixed in place using a phantom setup board built at RMIT University (Melbourne, Victoria). The phantom setup board prevented the compressed gel from shifting the position of moving components when relaxing and made repeating setups on multiple devices with several samples an easier process. The imaging and irradiation setup is pictured in Figure 1.

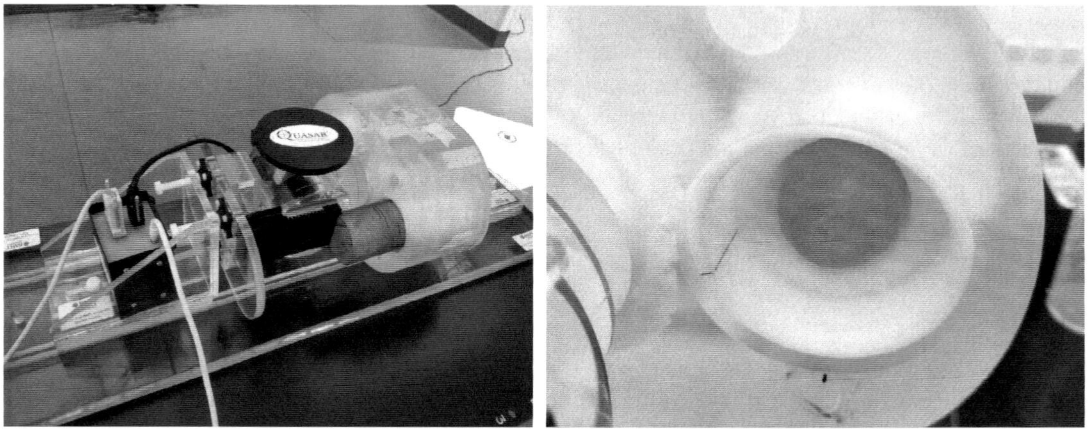

**Figure 1.** (Left) Motion phantom setup on a linac couch with the gel housed in the thorax phantom where the wood motion insert is located. The motion setup is held in place by a phantom setup board. (Right) the IDEFGEL housed inside the acrylic holder surrounded by a layer of foam.

The Quasar phantom was used to simulate a breathing motion of 1 cm in magnitude at a rate of 20 breaths per minute. It was noted that there seemed to be some damping of the oscillation due to over compression of the gel and some possible slippage of components in the motor.

### *2.3. Imaging and treatment planning*

A Phillips Brilliance Big Bore X-ray CT was used for imaging the dosimeters housed in the motion phantom as pictured in Figure 1. A 4D CT of the dosimeter containing multiple fiducials was captured with a set of bellows attached to the motion phantom to track the breathing pattern. Standard CT of the other two samples were imaged to investigate the uniformity the target regions between samples and for planning treatments on.

A static CT plan was created in Eclipse and the visible Iodinated nPAG (InPAG) target region was denoted as the PTV. A dose of 1.5 Gy was prescribed to the target volume to ensure irradiation corresponded with the full dynamic range of the IDEFGEL [7].

*2.3.1. Irradiation of IDEFGEL.* The treatment plan was delivered using a Varian TrueBeam medical linear accelerator (Varian Medical Systems, Palo Alto). The plan was delivered to one sample that remained static and one sample that underwent the same motion during delivery as imaged on the 4D CT. The treatment plan was a VMAT protocol that delivered 208 MU through two 180° arcs.

The samples were all then read out using a Modus Medical Devices (London, Ontario) VistaCT Optical CT scanner.

## 3. Results

Dose coverage within the target volume is a metric that can be used for an assessment of how large the variation introduced by motion is, with several tools available for analyzing the extent of dose coverage and agreement between planned and delivered dose. The plan DVH for the target region was reported using the CERR MATLAB extension and a binary mask of the structure volume was overlaid on the gel volume array using the target region boundaries as co-location points. The comparison of DVH's for the PTV is plotted in Figure 2.

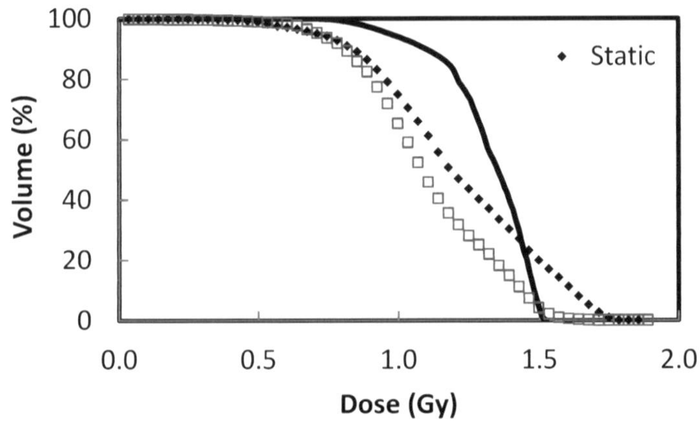


**Figure 2.** Cumulative DVH for the planned treatment volume compared to the doses measured in the target volume of the static and deforming IDEFGEL samples. The doses have been calibrated using the calibration factors in Watson *et al* [7].

A second structure was delineated in eclipse that encapsulated the gel dosimeter volume with the exclusion of the target region. This region includes the peripheral doses at the boundaries of the PTV and dose wash due to the VMAT protocol. The boundaries of this dose structure in eclipse extend beyond that of the ITV and include some volume that was expected to receive no dose which could artificially inflate the perceived effect of excess dose deposited due to dose wash. However, including the entire gel volume was considered a more sensible option rather than defining an arbitrary boundary within the dosimeter that had no correlation to a physical feature, such as the boundary created by the Isovue in the target volume. The DVH of the remaining IDEFGEL volume in the static and deforming samples is displayed in Figure 3.

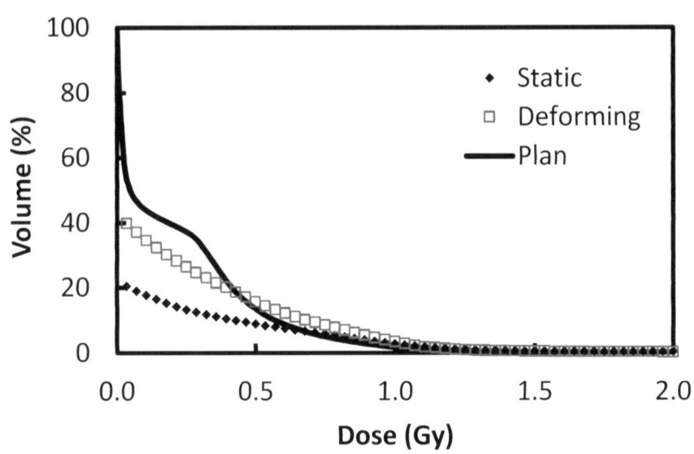


**Figure 3.** Cumulative DVH of the dosimeter volume excluding the target volume for the static and deforming IDEFGEL compared to the plan DVH. A scaled calibration factor is used to account for a small amount of Isovue diffusion prior to irradiation, this factor is explained in more detail by Watson *et al* [7].

Some key dose metrics from the DVHs above are collated in Table 1.

**Table 1.** The significant dose values of the two regions present in the IDEFGEL for both the static and DEF samples compared to those in the delivered treatment plan. The mean values are weighted to represent the volume average of the dose structure. The minimum dose values in the greater dosimeter volume are subject to more uncertainty than the other gel results due to RI mismatch artefacts.

| | **PTV** | | | **Dosimeter Volume** | | |
|---|---|---|---|---|---|---|
| | Max(Gy) | Mean(Gy) | Min(Gy) | Max(Gy) | Mean(Gy) | Min(Gy) |
| **Plan** | 1.52 | 1.32 | 0.68 | 1.29 | 0.21 | 0.01 |
| **Static** | 1.78 | 1.16 | 0.18 | 1.36 | 0.51 | 0.03 |
| **DEF** | 1.89 | 1.05 | 0.14 | 1.14 | 0.55 | 0.03 |

By being able to isolate the regions of the dosimeter and analyse the dose metrics of the PTV and the greater dosimeter volume independently the impact of the DVH analysis should be greater and inform more rigorous absolute dose measurements in conjunction with other dose metrics that are more suitable to conformation analysis of the 3D dosimeter volume.

## 4. Discussion

The analysis of the target region in this study was focused upon the dosimetric features of the plan comparison, which led to a focus upon the dose volume histogram of the target region rather than a gamma test. Given the simple object shapes that were present in this study, both in the dosimeter and planned dose delivery, the comparison of DVHs for the static and deforming cases provides a more meaningful study of a single feature. This feature is of course the deformation of the target region as the applied breathing motion forces it to depart from the conformation expected by the plan, quantifying the impact of this motion directly in relative dose terms.

In the comparison of the dose coverage of the target region displayed in Figure 1 it is evident that the deformation of the target region has reduced the dose delivered to the target region, decreasing the mean dose delivered to the volume as documented in Table 1. The decrease in mean dose to the volume is likely to stem from the plan assuming a static volume as expected. The change in dosimeter volume introduces peripheral gel material into the target volume receiving a higher dose, while some mass from the target region is forced outside the primary beam decreasing the dose readout when the gel is once again in the static state. This effect can also be seen in Figure 3 where the DEF sample DVH receives on average a much higher dose compared to the static sample and has a higher mean dose. A large portion of the dosimeter volume also exceeds the dose expected according to the plan DVH indicating that these parts of the dosimeter volume spent significant portions of time in the target region during treatment delivery.

The static gel irradiation does not appear to correlate well with the planned dose distribution, to a significant degree when inspecting the absolute dose values present in Table 1. This discrepancy points to an error in the calibration function used to convert the transmission values from OCT to dose, which may be an effect of several factors. As has been stated previously the diffusion of Isovue out of the target region presents an issue as there is a spatial and temporal dependence on dose response due to the change in the consumption rates of the monomers proposed to be caused by Iodine. The oxygen permeability of the latex membrane is also higher than would be considered ideal and while housing it in a less oxygen permeable container was done, it seems to have had a limited buffering effect. Combating these issues is an ongoing process and subject to immediate investigation.

The relative comparison of the static dosimeter to the IDEFGEL provides a quantification of the degree to which the deformation has blurred the expected dose distribution. The amount of outer dosimeter gel pushed into the target region appears to have collected 9.4% of the dose delivered to the target region, derived from the total dose deposited in both regions for both samples, a reasonably significant amount for a 1 cm compression of the material. The spreading of the maximum and minimum dose values is also interesting given the expected dose blurring, but with the suspected dose calibration inconsistencies it is difficult to derive a solid conclusion from these values.

## 5. Conclusion

A target region containing 0.7% w/w Isovue X-ray contrast agent was successfully fabricated from the same gel batch as the outer dosimeter and used to measure a planned dose distribution to the contoured target region in both a static and deforming configuration. By performing comparisons of the DVHs of the two samples to one another, it can be seen that the degree of motion was enough to significantly impact the dose coverage of the target region. The IDEFGEL measurements are precise enough indicate that the addition of the target region as a physical feature that is dose sensitive can provide meaningful results in terms of assessing the dosimetric impact of the margins included in treatment planning, most pertinently in this study margins for motion and deformation. The IDEFGEL dosimeter is currently restricted to simple shape configurations and equally simple dose structures in planning. These restrictions do provide a focus on single phenomena for the conducted test and can be probed further for dose 3D distribution shape analysis. The calibration of the dose response function appears to have been severely impacted by external environmental factors and the change in chemistry caused by diffusion of the Isovue throughout the dosimeter volume, necessitating extremely strict controls on the time scale at which the experiments are conducted for not only the deformation study but also the calibration studies. With further optimization of the contrast agent concentration and the dosimeter configuration a IDEFGEL that can reliably assess the impact of motion and deformation on 3D dose distributions is achievable and can inform treatment planning that involves margins to account for motion and deformation.

## 6. References

[1] Mather M L *et al* 2003 *Phys. Med. Biol.* **48** N269-75
[2] Yeo U J *et al* 2013 *J. Phys.: Conf. Ser.* **444** 012107
[3] Yeo U J *et al* 2013 *Med. Phys.* **40** 101701
[4] Juang T *et al* 2013 *Int. J. Radiat. Oncol. Biol. Phys.* **87** 414-21
[5] Burnet N G *et al* 2004 *Cancer Imaging* **4** 53-61
[6] Yeo U J *et al* 2012 *Med. Phys.* **39** 2203-13
[7] Watson C J *et al* 2018 In: *10th International Conference on 3D Radiation Dosimetry.*

IOP Conf. Series: Journal of Physics: Conf. Series **1305** (2019) 012056 doi:10.1088/1742-6596/1305/1/012056

# Development of fast Monte Carlo code for high dose rate brachytherapy

**Ankang Hu[1, 2], Rui Qiu[1, 2], Zhen Wu[1, 3], Chunyan Li[1, 3], Hui Zhang[1, 2] and Junli Li[1, 2]**

[1]Department of Engineering Physics, Tsinghua University, Beijing, China
[2]Key Laboratory of Particle & Radiation Imaging (Tsinghua University), Ministry of Education, Beijing, China
[3]Nuctech Company Limited, Beijing, China

E-mail: qiurui@mail.tsinghua.edu.cn

**Abstract**. Brachytherapy plays an important role in the treatment of various cancers. The TG-43 formalism is widely used in dose calculation for brachytherapy, but it is not accurate as it assumes the patient as homogeneously constituted of water. Monte Carlo simulation can obtain accurate result, but computing time using traditional code is too long to be accepted in clinical application. This work developed a fast Monte Carlo code THUBrachy for high dose rate brachytherapy dose calculation. THUBrachy can simulate photon with energy up to 3MeV for heterogeneous materials and obtain dose, fluence and dose distribution. It applies several accelerators such as Graphics Processing Unit (GPU), Intel Xeon Phi to perform dose calculation with good accuracy and high speed. THUBrachy adopts several parallel programming models to transplant the code onto different hardware accelerators. Voxel geometry based on CT images is used to describe the tissues of patient. A linear track-length estimator is used for dose calculation. In the preliminary accuracy test, a cuboid water model and a heterogeneous-material model were selected as the phantoms. Dose and fluence distributions were calculated by both THUBrachy and FLUKA which is a well-benchmarked Monte Carlo code. A total of 5×108 photons were simulated to make the uncertainty in all area less than 5%. The maxium difference of the results in low dose area is less than 3% and in high dose area is less than 1%, indicating a good agreement. In the performance test, a phantom consisted by 24 types of human tissue materials was chosen to estimate the performance in practical application. Dose distributions with an uncertainty of 2% or less for the target volume were obtained in less than 1 min by simulating 5×107 photons with a GTX 1080Ti, while FLUKA taking several hours to simulate the same number of particles. There is great potential for clinical application in brachytherapy dose planning using this fast Monte Carlo code.

## 1. Introduction

Brachytherapy plays an important role in the treatment of various cancers by placing radiation sources inside or next to the area requiring treatment. The radiation field of brachytherapy is extremely nonuniform [1]. Dose rate in some area is very high, especially for high dose rate brachytherapy which is up to 300cGy/min. To ensure curative effect and reduce the damage on healthy tissue as possible, medical physicist should formulate therapy plans carefully based on accurate dose distributions. Brachytherapy dose calculation is a key process of therapy planning. Nowadays, the TG-43 formalism is widely used in dose calculation for brachytherapy. This formalism depends on a set of predefined parameters for each radiation source seed model. These parameters are obtained from Monte Carlo simulations and experimental measurements in water phantom [2]. For complex human tissue especially for bone, TG-43 formalism would cause large deviation. More accurate dose calculation methods are required for brachytherapy.

Monte Carlo methods could describe the medium and interaction of particles accurately to obtain accurate dose distribution. It is often regarded as the gold standard for dose distribution calculations in radiation therapy [3]. But it takes long computation time to track a large number of particles to obtain a result with low uncertainty. The computation time using traditional Monte Carlo code in brachytherapy dose calculation is not acceptable in clinical application.

Hardware accelerators such as Graphics Processing Unit (GPU) and Intel Xeon Phi have been developed greatly in recent years. These hardware accelerators provide high calculation speed and reduce the power consumption by simplifying the complex function of processors and strengthening the function of calculation. Hardware accelerators have been adopted to accelerate Monte Carlo simulation. Jia et al developed a GPU-based fast Monte Carlo code for brachytherapy [4]. Parallel computing with hardware accelerators is considered as an effective method to accelerate Monte Carlo simulation [5].

This paper describes the development of a fast Monte Carlo code THUBrachy which can be accelerated by various hardware accelerators. This code is designed for dose calculation and image simulation for low energy photon. Preliminary accuracy test and performance test were performed to verify that this code could complete dose calculation and get accurate result for high dose rate brachytherapy in clinically reasonable time.

## 2. Material and methods

### *2.1. Physics models*

THUBrachy takes transportation and interaction of low energy photon into account. Photoelectric effect, Compton scattering, Rayleigh scattering and pair production were modeled in this code. Because the continuous slowing down approximation (CSDA) range of secondary electrons inside the human body at the considered energy range is generally much smaller than voxel size. The electron are not simulated in this code.

The cross-section values used in THUBrachy come from the NIST XCOM database [6]. A pre-calculated lookup table is used to speed up the calculation of photon transportation. Accurate cross section value is obtained by interpolating from the value of lookup table based on the energy of photon. For dose calculation, a linear track-length estimator is used to reduce the number of simulated photons while keeping a low statistical uncertainty. This method calculates the tack-length in each voxel to obtain fluence and covert fluence to dose by multiplying the mass attenuation coefficient.

$$D = \int \varphi(x,E) \frac{\mu_{en}(E)}{\rho} E dE \qquad (1)$$

$\varphi$-photon fluence; $D$-dose, $\mu_{en}/\rho$-mass attuation coefficient; $E$-photon energy.

### *2.2. Random generators*

Accurate dose calculation requires the simulation of a large number of primary particles. And for codes executes on accelerators have many threads. To obtain high quality random, Tiny Mersenne Twister (TinyMT) random generated were used to provide a random series with period up to $2^{127}$. A jump function making the internal state of TinyMT $N^{th}$ forward state was used for each parallel thread to avoid repetition.

### *2.3. Parallel programming*

This set of fast are consisted of 5 versions. Every version of code is programmed using a parallel programming model used in the selected accelerators. This table shows the programming model and accelerators used in each version of THUBrachy.

**Table 1.** Programming model and accelerators for each version of THUBrachy.

| Version | Programming model | Accelerator |
|---|---|---|
| 1 | OpenMP | Multicore CPU |
| 2 | OpenMP | Intel Xeon Phi |
| 3 | OpenACC | GPU |
| 4 | CUDA | NVIDIA GPU |
| 5 | OpenACC | SW26010(used in Sunway Taihu Light) |

### *2.4. Preliminary accuracy and performance test*

This set of fast are consisted of 5 versions. Every version of code is programmed using a parallel programming model used in the selected accelerators. This table shows the programming model and accelerators used in each version of THUBrachy. To validate the code, a water model was selected as the test phantom. The size of water model is 20×20×20cm$^3$. And the model was devided to 40×40×40 voxels. A 2×2×2cm$^3$ isotopic 350keV photon source is placed in the center of the model.

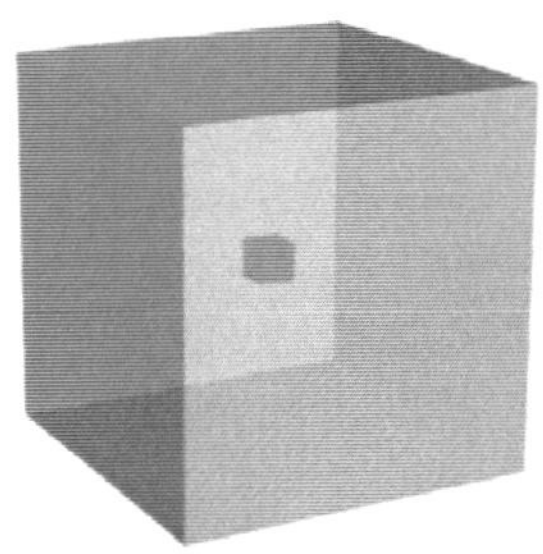

**Figure 1.** Diagram of source and model.

Both dose and fluence in each voxel were calculated when performing simulation. A general purpose Monte Carlo code FLUKA was selected as the benchmark. 5×10$^8$ primary photons were simulated to make the uncertainty of dose in fringe less than 5%. Results obtained by the fast Monte Carlo code were compared with the benchmark.

For performance test, a heterogeneous-material model containing main human tissue were selected as the phantom. The phantom is consisted by 24 materials differing in composition and density. The size of model is 512×512×200mm$^3$, which is devided to 512×512×200 voxels. To compare the acceleration effect of different version of codes, a server equipped with 2 Intel Xeon E5-2630 v4 CPUs, 64GB RAM and a Nvidia GTX 1080Ti was chosen as the test platform. Intel Parallel Studio XE 18.0 was used to compile code executing on CPU. The GPU code was compiled with gcc 5.0 and CUDA toolkit 8.0. A total of 5×10$^7$ primary photons were simulated using each tested version of codes. The model and number of simulated particles are both similar to clinical application. So, the performance of code in this test could represent the true performance in clinical application.

## 3. Results

### *3.1. Preliminary accuracy test*

To compare the result obtained from THUBrachy in this work and benchmark intuitively, dose distributions in cross section are shown below. The difference between the two results can hardly be observed.

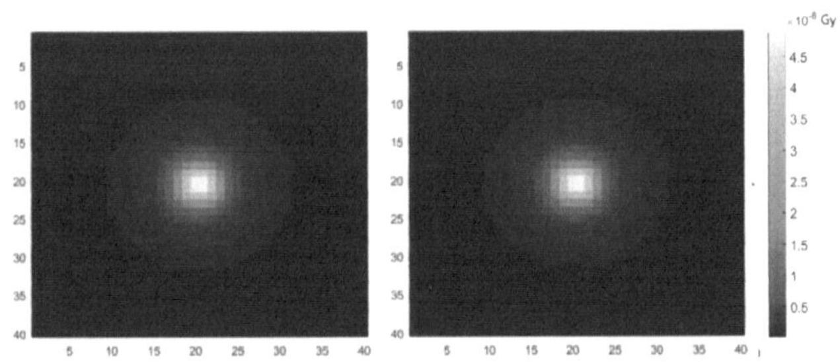

**Figure 2.** Dose distribution in cross section obtained by THUBrachy (left) and FLUKA (right).

For quantitative comparison, dose distributions on x-axis were shown in Figure 3.

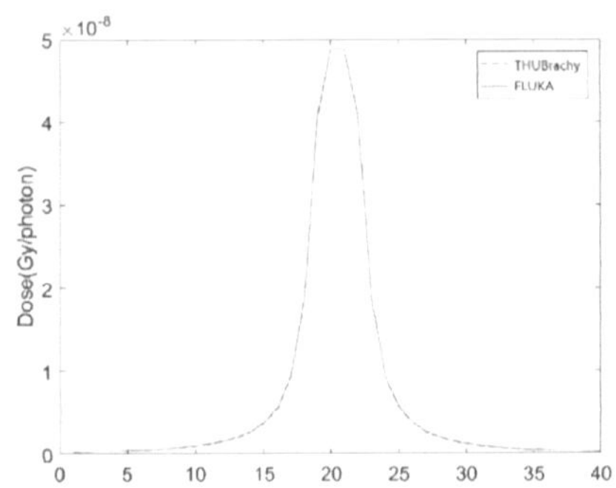


**Figure 3.** Dose distribution in x-axis obtained by THUBrachy and FLUKA.

Two curve is almost coincident with each other. For low dose area, the maxium difference between two codes is less than 3%. And in high dose area, the maxium difference is below 1%. This result verifies the accuracy of the fast Monte Carlo code preliminarily. Further accuracy test is ongoing.

### *3.2. Preliminary accuracy test*

THUBrachy executed on different accelerators to test the performance. Limited by the test platform, GPU code and multicore CPU code were tested. The results were shown below.

**Table 2.** Performance comparison between different accelerators.

| Code | Number of threads | Compiling option | Executing speed |
|---|---|---|---|
| Single core | 1 | -O3 -ipo | 946s/$10^7$ photons |
| Multicore | 32 | -O3 -ipo -qopenmp | 55s/$10^7$ photons |
| GPU | - | default | 15s/$10^7$ photons |
| FLUKA | 1 | - | 4h/$10^7$ photons |

The result shows that multicore CPU and GPU could speed up the code effectively. It takes 4 hours to simulate $10^7$ particles using FLUKA on a single core. THUBrachy is much faster than general purpose Monte Carlo codes. And the computational speed of GPU code reaches the demand of clinical application.

## 4. Conclusion

This work develops a fast Monte Carlo code THUBrachy for high dose rate dose calculation. The code takes main physics interactions of photon in considered energy into account. Preliminary accuracy test validated the code by comparing the result generated by this code and general-purpose Monte Carlo code. Performance test indicated that accelerators could accelerate the dose calculation effectively. The speed of this code is significantly faster than general purpose code. And there is great potential for clinical application in brachytherapy dose planning using this fast Monte Carlo code.

For future work, the code will be used in clinical case to perform stricter test. This code is expected to execute on other accelerators such as Intel Xeon Phi, SW2601 and future accelerators.

## 5. References

[1] Hurley C *et al* 2006 *Nucl. Instrum. Meth. A* **565** 801-11
[2] Nath R *et al* 1999 *Med. Phys.* **26** 2514
[3] Taylor R E *et al* 2007 *Med. Phys.* **34** 445-57
[4] Tian Z *et al* 2016 *Brachytherapy* **15** 387-98
[5] Xu X *et al* 2015 *Annals of Nuclear Energy* **82** 2-9
[6] Hubbell J H and Seltzer S M B 2005 *XCOM: Photon Cross Sections Database*

# Innovation in education: computer simulation in physics training

**Andrew W Beavis[1,2,4,5] and James W Ward[3,4]**
[1]Department of Medical Physics, Queens Centre for Oncology, Cottingham, UK.
[2]School of Life Sciences, University of Hull, Hull, UK
[3] Department of Computer Science, University of Hull, Hull, UK
[4]Vertual Ltd, Logistics Institute, University of Hull, Hull, UK
[5]Faculty of Health and Well Being, Sheffield-Hallamshire University, Sheffield, UK

E-mail: andy.beavis@nhs.net

**Abstract**. Interest in the application of Computer Simulation within Medical training continues to grow and the use of the VERT (Virtual Environment for Radiotherapy Training) system in radiotherapy training equally so. Herein we explore the use of the VERT system to understand, illustrate and experience the process required to calibrate an ion chamber and use it to obtain an accurate output measurement on a high energy Linac. The VERT system provides a detailed and accurate simulation of conventional Linacs and has a variety of physics activities that mirror the tasks medical physicists are required to become familiar with during their training. One such activity has been implemented to follow the steps defined by the IAEA TRS398 protocol. All activities on the virtual Linac can be undertaken with deliberate errors having been introduced to explore their consequences. The system was used to make the necessary 'measurements' to determine the various calibration factors required to correct the raw electrometer reading for ion recombination, polarity, temperature/pressure. Likewise, the Quality Index of the Linac's 6MV beam was determined to enable the correction of the Absorbed Dose to Water factor measured in the reference Quality. The VERT system provides a simulation environment for training and enhancing the trainees understanding of radiotherapy concepts and practices. Dosimetry practice can be experienced and explored in the classroom during standard working hours rather than waiting until a clinical Linac is available. Random and systematic errors can be introduced, however, the virtual machine can be used without risk of leaving it in a mis-calibrated or unsafe condition.

## 1.Introduction

We have developed an education tool for Radiotherapy that uses virtual reality to provide a computer simulation environment that enables students and trainees to gain experiential learning or practical experience in the classroom. Over the ten years that it has been commercially available the use of the VERT system (Virtual Environment for Radiotherapy Training) has become established in 28 countries with some 140 installations, in Universities, Community Colleges, Hospitals and Standards Labs; eight of which are primarily used for Physics training programmes.

The VERT system has previously been described elsewhere [1-3] as has its origins [4, 5]. In summary, it was developed as a training tool for radiographers with a philosophy to bring an interactive, 'fully' functional, Linac into the classroom. Whereas the virtual Linac has full articulation and is faithfully and accurately modelled to resemble its real-world counterpart, it does not (of course) produce ionizing radiation. The immediate benefit to trainee and trainer being that the learning environment is therefore safe, without electrical, radiation, mechanical crush risk and the virtual patients cannot be harmed in any manner, nor can the career of the trainee! The similarity to the flight simulators used by trainee pilots and astronauts has led to the VERT system being described as 'the flight simulator for radiotherapy'.

In recent years there has been an increasing interest in the use of computer simulation training for medical physics applications. The concept and rationale being the same as for radiographer or therapist training, that practical experience can be gained, safely, at any time of the day without having to wait for a clinical unit to be taken out of clinical service. A common theme in the use of virtual reality is that we can learn from the experience of making mistakes. Therefore, not only can procedures and scenarios be repeatedly explored without delaying the return to use of clinical equipment, it can be done without impacting the integrity of its safety.

To explore this opportunity, the 'physics tools' in VERT have been expanded in order to make them more applicable to a wider audience and to add functionality. Previous implementations have been reported [6-8], however here we present a new feature aimed at expanding the understanding of dosimetric concepts and to broaden the systems applicability.

## 2.Materials

A variety of conventional Linacs produced by the two major manufacturers are available in VERT. In addition a Varian ProBeam Proton therapy system has recently been introduced and plans exist for similar systems from other manufacturers. They are controlled using the authentic hand control pendants or simple menu controls. Virtual patients are created using information taken from DICOM descriptions of treatment plans which can be sourced from any commercial treatment planning system and the radiation characteristics of the Linac are modelled via comprehensive beam models. All aspects of the treatment delivery and image verification are simulated within the system allowing trainees to explore concepts, techniques and related processes.

A Physics module is available that has previously been reported [8] which was essentially designed to make some of the 'physics concepts' more accessible to 'other' staff groups. However, one of the activities therein allowed the process of calibrating an ion chamber against a reference class chamber to be simulated was implemented. This was designed with the IPEM code of practice [9] in mind. To make this activity more universal a workflow that follows the IAEA TRS398 [10] has since been introduced.

The IAEA TRS398 activity guides the user through using the published protocol by simulation of the relevant equipment and process. The simulation provides a scanning water phantom with which to make the necessary measurements, see Figure 1. The system has a touchscreen interface that hosts a worksheet which is used to work through the process in conjunction with the published protocol. The worksheet also contains a virtual electrometer which is used to measure output for a configurable bias voltage.

The philosophy behind the simulation is that the protocol can be followed and measurements can be made using VERT as if using a clinical system. The trainee is required to understand the protocol which can be achieved through self-study or via traditional didactic class teaching. The VERT system can then be used, again individually or as a demonstration by the instructor, to follow the process of taking appropriate measurements and calculating component correction factors of the chamber calibration factor, in a manner that mirrors the actual workflow. The true benefit is gained by the trainee using the system themselves whilst systematically following the protocol.

As required by the protocol, measurements can be made for the appropriate field size with the chamber at the appropriate depth of water and distance from the target, and with selection of an appropriate bias voltage for the chamber. Once the chamber has set up readings can be captured and recorded via interaction with a worksheet Fig 2. Repeated acquisitions yield values that demonstrate random measurement uncertainty, the magnitude of which can be configured. The temperature and pressure within the virtual treatment room is varied each time the software is used to ensure the trainee is required to calculate and correctly apply the associated correction factors.

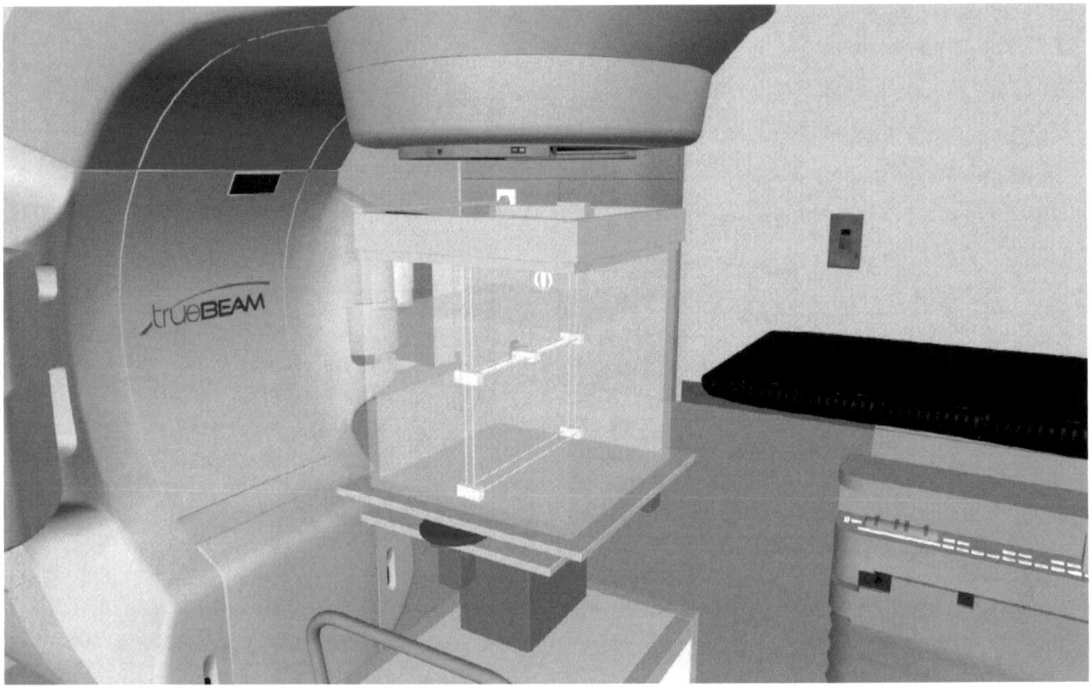


**Figure 1.** Screenshot from VERT showing a scanning water phantom in position under a Varian TrueBeam.

## 3. User Case scenario: TRS398

In this section we illustrate the use of the simulation. Fig 2 shows the worksheet associated with the TRS398 ion chamber calibration activity which is presented as a touchscreen interface. On the left side of the panel we note that it displays the beam quality, field size, depth of chamber and SSD. The measurement configuration and conditions may be altered by the user; the Linac settings may be altered using the Pendant or via menus in the software GUI and the depth of the chamber using a menu in the GUI. When the 'Measure' button is depressed a reading commensurate to the current configuration, number of monitor units delivered and beam quality selected, is captured. The value produced is a function of the 'assigned' chamber calibration factor, the expected dose at the measurement point, the defined output (systematic) uncertainty and the measurement (random) uncertainty.

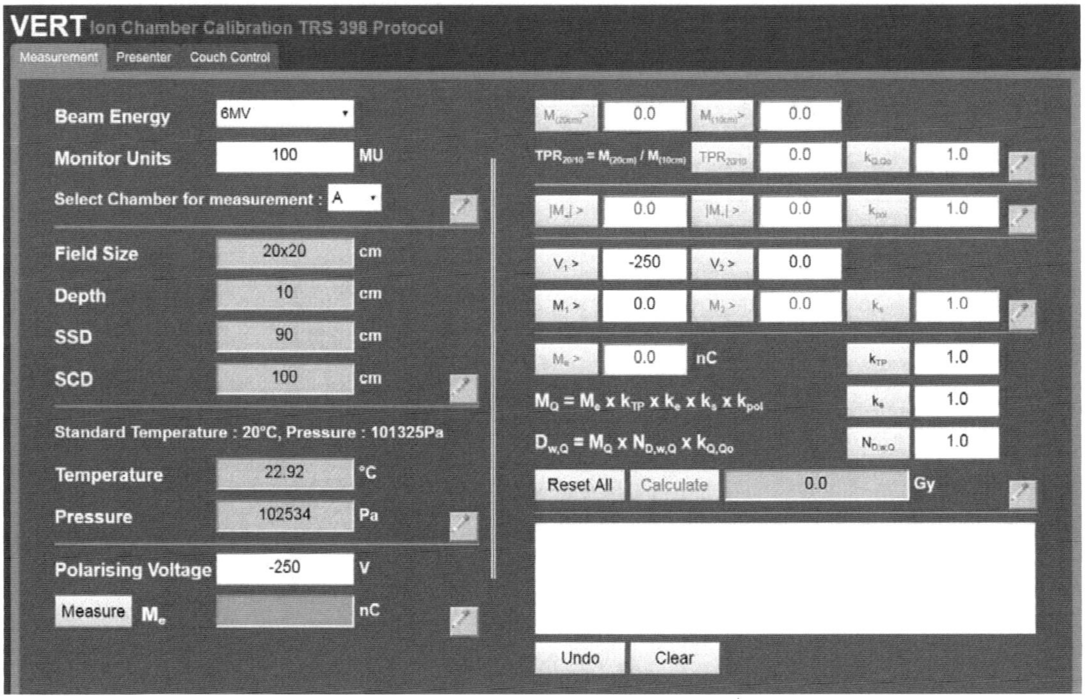


**Figure 2.** The worksheet associated with the TRS398 activity.

Assuming the trainee is familiar with the TRS398 protocol, the worksheet is intended to make the workflow of the activity / exercise very intuitive. The layout of the righthand side of the worksheet groups measurement steps and factor generation/ look up and uses nomenclature from the protocol. The equations are presented as a reminder for which factors are used to correct the raw electrometer measurement and then how the calibration factor is converted to the clinical quality and deployed.

| *Measurements* | |
|---|---|
| $M_e$ | Electrometer reading |
| $M_{20cm}$ / $M_{10cm}$ | Reading at (20cm/ 10cm) depth |
| $TPR_{20/10}$ | Tissue Phantom Ratio for 20cm to 20 cm depth |
| $\|M_-\|$ / $\|M_+\|$ | Reading with negative/ positive polarising voltage |
| $V_1$/ $V_2$ | Polarising Voltage (1/2) |
| *Individual (correction) factors* | |
| $k_{Q,Qo}$ | Quality correction factor |
| $k_{pol}$ | Polarity correction factor |
| $k_s$ | Ion recombination factor |
| $k_{T,P}$ | Pressure and Temperature correction factor |
| $k_e$ | Electrometer factor |
| $N_{D,w,Q}$ | Absorbed dose to water factor calibrated in reference Quality |
| *Calculation terms* | |
| $M_Q$ | Corrected electrometer reading made in Quality Q |
| $D_{w,Q}$ | Calibrated dose to water for Quality Q |

**Table 1.** Summary of the nomenclature used in the worksheet.

Where individual factors are assumed ($k_e$ and $N_{D,w,Q}$) then default values are assigned in the simulation software. Whereas these are 'blind' to the trainee, the system administrator is able to update the default values should they have preferred 'local' values. When the corresponding button is depressed the value box to its right is populated by the assigned value.

Where individual factors are taken from the protocol ($k_{Q,Qo}$), depressing the appropriate 'factor' button populates the value box with the value looked up from the protocol for an appropriate Chamber type. This value is a function of the $TPR_{20/10}$ ratio which is calculated when the button is depressed.

Where individual factors are calculated ($k_{pol}$, $k_s$, $k_{T,P}$) then depressing the appropriate factor button will populate the associated value box with a value derived using the appropriate formulae (and any required parameters) from the TRS398 protocol.

Alternatively, for all classes of individual factors, the user can type values into the 'value box' to the right of each factor button and those values will be used in the subsequent calculation of dose.

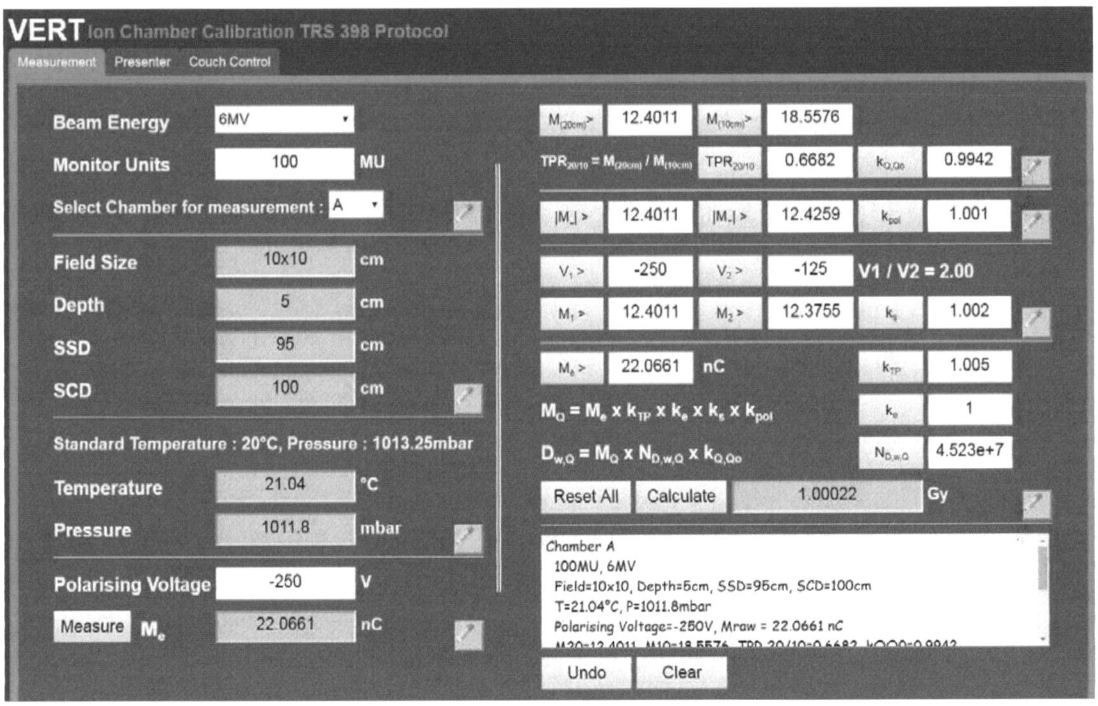


**Figure 3.** Worked example, having made all appropriate measurements and created the individual correction factors. The Dose at 5 cm in water for a 10 x 10 field for a measurement of 95cm SSD is calculated from the uncorrected electrometer reading and the various factors. Note: a variety of information was captured by the user for evidence to be submitted for their training records.

Once a measurement is made and captured it is transferred to the 'work area' on the righthand side of the worksheet. For example, the measurement made under the conditions illustrated in figure 2 may be used in the calculation of the polarity factor. By depressing the button labelled with $|M_-|$ the current reading would be used to populate the measurement value box to the right of that button. Next, the polarising voltage would be reset to +250V, the measurement captured and that reading be transferred to the appropriate field by depressing the button labelled $|M_+|$. The measurement reading obtained for the 'reversed' polarity bias voltage is reversed engineered using the default polarity correction factor defined in the user preferences of the system. Therefore, when the $k_{pol}$ button is depressed, the correct value is recovered from a 'forward' calculation.

Figure 3 illustrates the completed activity. The various individual correction factors were generated as per the TRS398 protocol document. The Linac used in the simulation was calibrated to give 1 Gy per

100 MU at 5cm depth at isocentre for a 10 x 10 field. The uncorrected chamber/ electrometer reading was acquired under these conditions. This value was transferred to the 'uncorrected reading' box in the work area on the righthand side by depressing the button labelled $M_Q$. When the button labelled 'Calculate' is depressed the currently defined correction factors are applied to that value as described by the appropriate formulae in the protocol and reproduced on the worksheet to produce a corrected reading. In the example shown in figure 3 we see the expected (calibration) dose value is recovered.

The pencil icons on the right of the two columns capture pertinent information when depressed. This information is shown in the 'scratch pad' box on the lower right of the screen (as seen in fig 3) and can be 'copy/pasted' into word pad or similar to make a record of the trainee's experience.

## 4. Discussion

There is a common theme of discussion across all professions that training and education is moving, or has moved, away from the 'talk and chalk' or purely didactic methods of the previous centuries. Computer simulation technologies have been adopted in various medical fields; within radiotherapy it is beginning to become a standard in Radiographer/ Therapist/ RTT/ Technologists training, along with emerging published evidence [11-13]. The VERT system has been an enabler of this approach and is becoming used for Physics training [14-16] of Radiographers, Dosimetrists and Physicists. The VERT system has proven useful for the Group learning or inverted classroom approaches that are becoming more common in contemporary education.

In this manuscript an activity within VERT to simulate the process for calibrating an ion chamber and subsequently making a calibrated measurement of the output of a Linac, following the IAEA TRS398 protocol is described. It has been kept very simple in order to demonstrate how the VERT system is used as a platform to illustrate or practice skills and methodologies in the safe and relaxed classroom environment. Computer simulation platforms are intended as additional tools to enhance the educational experience. For the trainee they are safe in the knowledge that, no matter how long they take, they are not holding up the return to service of any equipment nor could it be handed over incorrectly calibrated. For the trainer, they can concentrate on passing on their knowledge and experience without any distractions [17] that could lead to failing to catch an incorrectly calibrated piece of equipment [18], or by ensuring this is not the case, failing to provide a high quality of training.

A key benefit that computer simulation brings to training is the ability to explore errors and process failures in a safe and instructive environment [5]. In the training of Physicists or Physics technicians this concept can be expanded to fault finding [8]. The VERT system has a facility (known as Virtual Presenter) for the trainer to set up scenarios for the trainee to explore and therefore if fault conditions are incorporated then failed QC tests can be illustrated and the appropriate course of action followed to rectify the experienced problem. Anecdotally, it might take several years of traditional post-training clinical practice to experience all the typical mis-calibration errors that a medical physicist is expected to know how to identify or 'capture'. Computer simulation allows them to be demonstrated or experienced during the training phase.

This scenario setting facility can be taken one step further and used for competency assessment or sign off of the trainee. Since the scenario can be reset countless times without restriction or deviation, every trainee in every training cohort can be given the exact same test. Clearly such an approach offers benefit from an educational ethics perspective. Furthermore, if the note/ record generation is used (as illustrated in figure 3) then evidence can be collected and provided to demonstrate competency.

## 5.Conclusion

The VERT system is a computer simulation training system that is offered commercially and at the time of writing is used at approximately 140 installations in 28 countries across the world. It was initially designed and promoted as a Radiographer training tool, however it has developed more broadly and has modules designed to assist 'physics' training. The physics tools enable activities such as calibration of ion chambers and measurement of Linac output to be introduced to trainees in the classroom environment with a blended learning approach that includes discussion of process and 'hands on' experiential learning. The physics activities in the VERT system are not limited to those discussed here and the scope of this computer simulation training environment continues to be developed.

## 6. References

[1] Phillips R *et al* 2005 *Studies in Health Technology and Informatics* **111** 390-6
[2] Bridge P *et al* 2007 *Computers and Education* **49** 481-94
[3] Boejen A and Grau C 2011 *Surgical Oncology* **20** 185-8
[4] A major contribution to practice and to the education and training of practitioners in radiotherapy; an interview with Professor Andy Beavis, Consultant Clinical Scientist and Head of Radiation Physics at the Department of Radiation Physics, Queen's Centre for Oncology and Haematology, Castle Hill Hospital, Hull, UK. (2016). *Journal of Radiotherapy in Practice* **15** 66-75
[5] Beavis A W 2018 *Journal of Medical Radiation Sciences* **65** 77-9
[6] Beavis A W *et al* 2012 *Med. Phys.* **39** 3969
[7] Beavis A W and Ward J W 2012 *Med. Phys.* **39** 3970
[8] Beavis A W and Ward J W 2014 *J. Phys. Conf.: Ser.* **489** 6
[9] Lillicrap S C *et al* 1990 *Phys. Med. Biol.* **35** 1355
[10] Andreo P *et al* 2000 IAEA Technical Report Series No 398, International Atomic Energy Agency, Vienna
[11] James S and Dumbleton 2013 *Radiography* **19** 142-50
[12] Bridge P *et al* 2017 *Journal of Radiotherapy in Practice* **16** 375-82
[13] Kane P 2018 *J. Med. Rad. Sci.* **65** 131-6
[14] Kirby M C 2015 *Medical Physics International* **3** 87-93
[15] Leong A *et al* 2018 J. Med. Rad. Sci. **65** 97-105
[16] Jimenez J A *et al* 2018 *J. Med. Rad. Sci.* **65** 106-13
[17] Lane J L *et al* 2001*Simul. Gaming* **32** 297-314
[18] Vial P *et al* 2008 *Med. Phys.* **35** 4362-74

IOP Conf. Series: Journal of Physics: Conf. Series **1305** (2019) 012058 doi:10.1088/1742-6596/1305/1/012058

# Teaching the principles of X-ray CT and SPECT using optical CT, glowsticks and a scaled anthropomorphic phantom

**Yves De Deene**
School of Engineering, Macquarie University, North Ryde, Sydney, Australia

E-mail: Yves.DeDeene@mq.edu.au

**Abstract**. Teaching demonstrations of computerized tomography (CT) and Single-Photon Emission Tomography (SPECT) to biomedical engineering and medical physics students is hampered by a limited accessibility to clinical scanners, especially during day time. The use of ionizing radiation and radioactive sources in X-ray CT and SPECT further complicates the design of a teaching laboratory session. We here propose an inexpensive and safe educational demonstration of CT and SPECT on an anthropomorphic phantom whereby a visible light source serves as source and a CCD camera serves as detector. The equivalent of a SPECT radionuclide in optical CT scanning is a chemiluminescent material which can be obtained relatively inexpensive in the form of party glow sticks. The proposed teaching tool comprises several learning outcomes such as hands-on construction of the scanner, the acquisition of images and image reconstruction. Also, different imaging artefacts can be simulated and investigated.

## 1. Introduction

'Medical imaging' is gaining a lot of interest in the medical physics curriculum as there is an increasing importance on accurate tumour volume definition and the non-invasive assessment of tumour biology. Practical laboratory teaching modules have been developed for many medical imaging technologies such as for ultrasonic imaging [1] and MRI [2,3]. Commercial MRI benchtop scanners for teaching the principles can be purchased from companies such as Pure Devices, SpinCore Technologies, Magritek and NiuMag. Clinical laptop ultrasound scanners are ideal for in class demonstration of ultrasound. Laboratory models for X-ray CT and nuclear imaging (SPECT and PET) are not so widely available because of the involvement of ionizing radiation. A demonstration experiment to illustrate the principle of back projection, the underlying principle in Computed Tomography has been proposed previously using optical components [4,5]. While these demonstration experiments are useful in helping students to understand that an internal image can be constructed from a combination of projections, they do not provide a realistic picture of current X-ray CT scanning and do not illustrate SPECT scanning. These designs need to be extended to demonstrate the principle of cone beam CT scanning of realistic humanoid geometries.

Most modern X-ray CT scanners and dual modality SPECT-CT scanners are cone beam CT scanners in which an X-ray beam is directed to the patient and transmission projections are acquired in a plane where the beam leaves the patient. Cone beam CT scanning is also used in dental imaging, radiotherapy planning and interventional radiology. The 3D reconstruction from the obtained set of planar projection data in cone beam scanning is performed using an extension to the fan beam reconstruction algorithm [6]. A commercial optical CT scanner for teaching the principles of conebeam CT scanning has recently been marketed by Modus QA under the name DeskCAT™. In this benchtop system, the X-ray source is replaced by a diffuse light source and a CCD camera is used to capture projection images of transmitted light. Where in X-ray CT, the source and detector are rotated around the patient, in the optical cone beam CT scanner, the object is rotated while the source and detector are kept fixed in the laboratory frame of reference. As light in the visible spectrum is not transmitted through opaque objects, optically transparent phantoms are used instead. Where in X-ray CT the image contrast relies on differences in absorption of high-energy photons, the image contrast in optical CT relies on differences in optical absorption at the operating wavelength of the scanner. The absorption of visible light in an optical CT scanner can be varied by adding different amounts of color dye to a crystal clear optically transparent phantom.

A recent breakthrough in X-ray CT has been the implementation of dual energy CT (DECT) where two sets of CT images are acquired with different X-ray tube voltages [7]. Dual energy X-ray CT scanning can be simulated by a dual wavelength optical CT scanner [8]. Until now, no laboratory demonstration has been proposed to illustrate Single Photon Emission Computed Tomography (SPECT). We here propose an optical conebeam CT scanner and the first miniaturized humanoid phantom for illustrating the principles of dual energy cone beam CT and SPECT scanning. For the principles of optical CT and back projection reconstruction, the reader is referred to a comprehensive published in one of the earlier conference proceedings [9].

## 2. Methods and Materials

### *2.1. Test phantoms*

Three phantoms were developed to illustrate the principles of dual energy CT and SPECT:

(1) A first phantom was given a humanoid shape which was based on a male fashion doll (Ken). Based on the formulation of a 3D radiation dosimeter developed in our group previously [10], the phantom was composed of a transparent silicone elastomer (Sylgard™ 184) and a solution of leucomalachite green (LMG) in chloroform, which gave the phantom a green color. A negative cast was made of another silicone elastomer, named Pinkysil™ that was obtained from a local handcraft store (Barnes). The negative cast was coated with a thin film of petroleum jelly that acted as a releasing agent. The phantom was suspended on the lid of a jar that acted as a tray which could be easily mounted on the rotation stage of the optical CT scanner. A 1 mm diameter hole was drilled from the left shoulder of the phantom in an oblique direction and filled with a chemiluminescent dye to emulate a radionuclide in SPECT imaging.
(2) A second phantom consisted of a 3D printed skeleton in translucent red poly-lactic acid (PLA).
(3) A third phantom consisted of transparent silicone elastomer in which a PLA skeleton was immersed. The phantom also contained a cavity in the shape of the myocardium in which a chemiluminescent dye was inserted. A chemiluminescent dye was extracted from party glow sticks that were purchased from a local store.

To mimic a SPECT radionuclide in the visible spectrum, a chemiluminescent material was purchased in the form of glow sticks which were obtained from a local supermarket. Glow sticks contain diphenyloxalate with a catalyst (sodium salicilate), a colour dye and hydrogenperoxyde which are separated by a thin glass wall.

When the glow stick is squeezed, the glass wall that separates the hydrogen peroxide from the other chemicals is broken and the chemicals mix, upon which a chemical reaction occurs that oxidizes diphenyl oxalate to phenol and 1,2-dioxetanedione. The peroxyacid 1,2-

dioxetanedione spontaneously decomposes to carbon dioxide, releasing energy that excites the dye, which then relaxes by releasing a photon.

*2.2. Optical scanning*

Our multi-colour conebeam optical CT scanner was based on the Modus Vista™ scanner in which both the light source and camera were replaced. The light source was developed in house using 10 strips of 12 tri-color RGB Light Emitting Diodes (LEDs). The camera was replaced with a higher resolution scientific grade CCD camera (AVT Stingray F-145B). A MATLAB program was developed in house to control the scanner and to reconstruct the images. A photograph of the optical conebeam CT scanner is shown in figure 1.

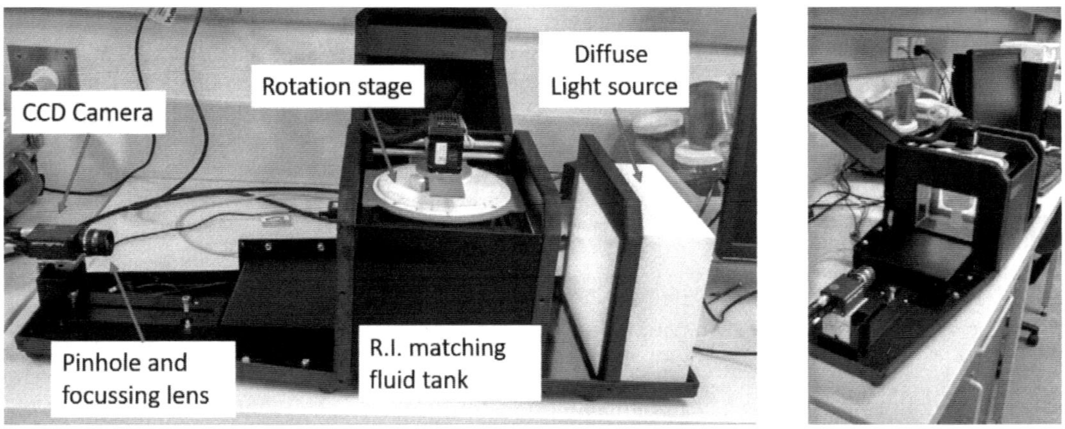


**Figure 1.** The optical CT scanner used in this study. The enclosure and fluid tank of the Modus Vista™ is retained but the camera and light source were replaced with an in house developed tri-color RGB light source, and the control program was rewritten in MATLAB code.

For each scanning experiment, 256 projection images were recorded at equi-angular increments of 1.40625 degrees which corresponds to a total revolution of the phantom over 360 degrees. Projection images can be recorded with the light source switched to different colors or with all light sources switched off for luminescence imaging. For dual wavelength scanning, projection images were acquired with the blue-light source (reference) and with the red-light source (data). A similar technique to luminescence imaging has been applied using fluorescence tracers in optical emission tomography of xenografts [11].

## 3. Results and discussion

*3.1. Humanoid shaped phantom with luminescent needle insert*

The humanoid shaped phantom containing the chemiluminescent dye is shown in figure 2. A structural scan is acquired with the red-light source switched on and a luminescent activity scan is acquired with the light source switched off. After filtered back projection, 3D data sets of the humanoid phantom and the luminescent region are obtained. As both scans are obtained without repositioning the phantom in the optical CT scanner, both data sets can be easily co-registered. A sharp spatial resolution in the luminescent scan can be appreciated. The contours of the phantom show hyperintense in the image which can be attributed to light scattering and a small refractive index mismatch at the surface of the phantom.

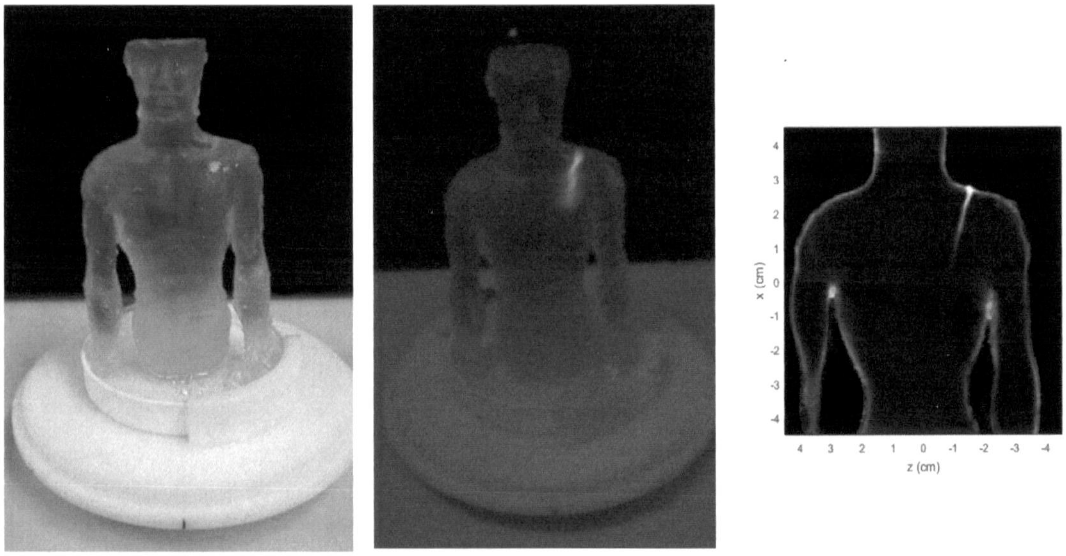


**(a)** **(b)** **(c)**

**Figure 2.** Humanoid shaped silicone phantom with chemiluminescent dye injected in the left shoulder. Photograph in ambient light (a) and in a darkened room (b). A coronal reconstructed image through the 1 mm wide cavity shows the luminescent signal.

### *3.2.* 3D printed skeleton

The 3D printed PLA phantom was scanned in the optical CT scanner with the red-light source. As the PLA phantom is translucent red, some of the light will still be transmitted through the phantom. The reconstructed CT images and the volume rendered image match remarkably well with clinical X-ray CT images of the human skeleton.

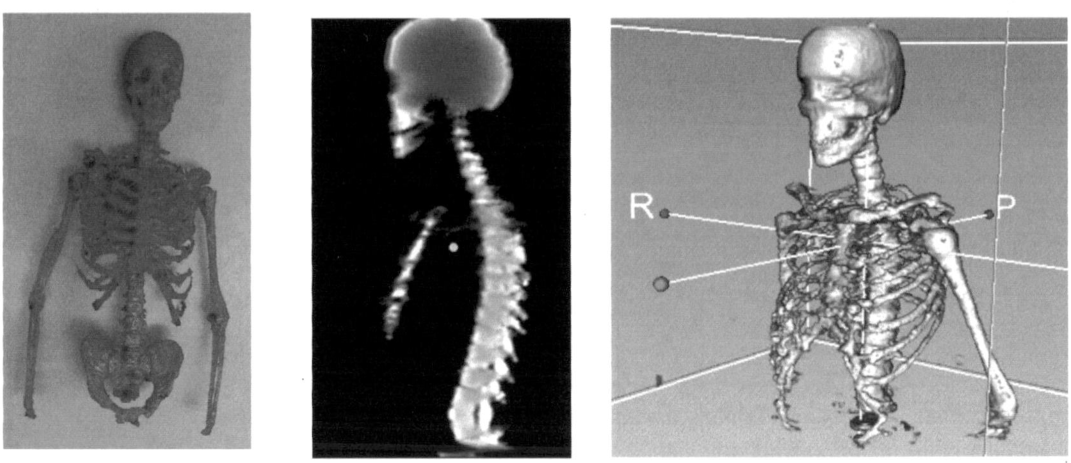


**(a)** **(b)** **(c)**

**Figure 3.** 3D printed PLA phantom (a) and corresponding reconstructed sagittal image (b) and 3D volume rendered image (c).

*3.3. Anthropomorphic phantom with skeleton and luminescent myocardium cavity*

The luminescent dye is clearly visible with the naked eye in a darkened room (Figure 4). The luminescent light is significantly attenuated by the ribs which is also apparent in the luminescence projection images. Unfortunately, several air bubbles in the silicone phantom create significant image artefacts. It is concluded that regardless these artefacts, the phantom is useful as a qualitative impression of SPECT-CT imaging.

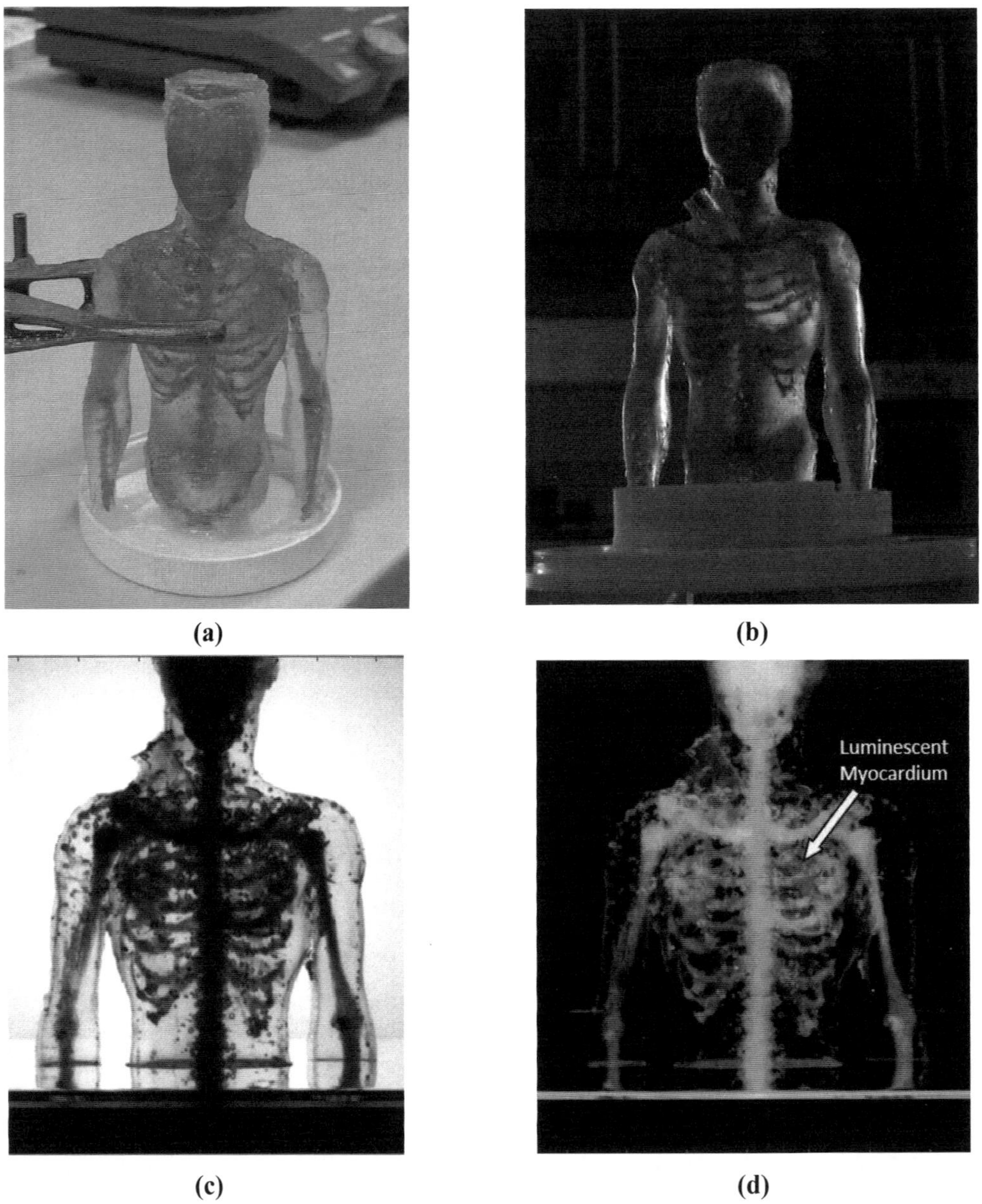


**Figure 4.** Anthropomorphic phantom with PLA skeleton and myocardium cavity containing a chemiluminescent dye. Photograph in ambient light (a) and in a darkened room showing chemi-luminescence (b). A projection image acquired with the red light source (c) and a reconstructed coronal slice with the luminescent signal superimposed in a colourwash image (d).

IOP Conf. Series: Journal of Physics: Conf. Series **1305** (2019) 012058 doi:10.1088/1742-6596/1305/1/012058

## 4. Conclusions

A realistic laboratory demonstration to illustrate the principles and concepts of both conebeam X-ray CT and SPECT imaging is proposed. A chemiluminescent dye can be injected in the phantom that simulates the radionuclide in SPECT imaging. Like the radioactive decay of a radionuclide, the activity of the chemiluminescent dye decays exponentially. Several realistic demonstration experiments and hands-on exercises can be designed. Using multi-colour LEDs, the principle of dual energy scanning can be demonstrated. While in this study, a scientific grade scanner was used, for demonstration purposes, an inexpensive optical cone beam scanner can be constructed relatively easy using a webcam, an Arduino controlled stepper motor and a diffuse light source built with off-the-shelf LED strips. Alternatively, optical CT scanners for teaching have also been marketed as commercial products.

The creative component of the learning experience can be enhanced as students can fabricate their own phantoms with either mould casting or 3D printing and can import the acquired optical CT scans in image processing software.

## 5. References

[1] Andrews D G H 2000 *Phys. Educ.* **35** 350
[2] Benli K P *et al* 2015 *Eur. J. Phys.* **36** 035032
[3] Taghizadeh S and Lincoln J 2018 *The Physics Teacher* **56** 266-8
[4] Ennemoser O *et al* 1986 *Eur. J. Phys.* **7** 88
[5] Mylott E *et al* 2011 *Eur. J. Phys.* **32** 1227
[6] Feldkamp L A *et al* 1984 *J. Opt. Soc. Am. A* **1** 612-9
[7] Johnson T R C 2012 *A. J. Radiology.* **199** S3-S8
[8] De Deene Y 2015 *J. Phys.: Conf. Ser.* **573** 012058
[9] Doran S J 2009 *J. Phys.: Conf. Ser.* **164** 012020
[10] De Deene Y *et al* 2015*Phys. Med. Biol.* **60** 1543-63
[11] Oldham M *et al* 2007 *J. Biomed. Optics* **12** 014009

IOP Conf. Series: Journal of Physics: Conf. Series **1305** (2019) 012059 doi:10.1088/1742-6596/1305/1/012059

# How important is the dose rate sensitivity of 2D and 3D radiation dosimeters?

**Yves De Deene**
School of Engineering, Macquarie University, North Ryde, Sydney, Australia

E-mail: Yves.DeDeene@mq.edu.au

**Abstract**. Several kinds of chemical 3D radiation dosimeters have been fabricated to acquire the dose distribution in clinical radiotherapy. A legitimate concern with every new dosimeter is its dose rate dependent response because, the dose rate in each point of the phantom during a clinical treatment is unknown. Unfortunately, many 3D dosimeters have shown some degree of dose rate dependence. In practice, radiation dosimeters are mostly calibrated using a calibration curve that is obtained by irradiating calibration vials with different doses but using a fixed dose rate. Therefore, when applying a dose calibration to an experimentally obtained dose distribution, a deviation between the measured dose distribution and the actual dose distribution can be expected. In this computational study, the effect of a dose rate dependent dosimeter response on a theoretical dose distribution has been investigated. In order to compare the effect of dose rate, a gamma evaluation is performed between the predefined dose distribution and the dose distribution that is affected by a dose rate dependent radiation response. It is found that a dose rate difference of -10%, results in a gamma pass rate of 100% in the 3D dose distribution.

## 1. Introduction

Experimental 3D dosimeters show some degree of dose rate dependence, but the dose rate dependence varies significantly between the different dosimeters. The dose rate dependence of 3D radiation dosimeters was extensively studied for some polymer gel dosimeters [1], PRESAGE [2] and Flexydos3D [3] and for different imaging systems including MRI, optical CT and X-ray CT. However, the dose rate dependence has not been documented for all gel systems.

Dose rate and temperature effects on a dose distribution need to be assessed relative to the dose range in which the dosimeter is used. If the imaged property is labeled $\psi$, the dose sensitivity of the dosimeter is $\frac{\partial \psi}{\partial D}$ and the change of the dose sensitivity with dose rate is $\frac{\partial}{\partial \dot{D}}\left(\frac{\partial \psi}{\partial D}\right)$ where $\dot{D}$ is the dose rate mostly expressed in units of (cGy/min). Temperature may affect both the reading of a non-irradiated sample ($\psi_0 = \psi(D = 0\ Gy)$) and the dose sensitivity ($\frac{\partial \psi}{\partial D}$). The imaged property $\psi$ can be the NMR spin-lattice relaxation rate ($R_1$ ($s^{-1}$)), the NMR spin-spin relaxation rate ($R_2$ ($s^{-1}$)), optical density (OD), X-ray CT number ($CT_{Number}$ (HU)). From figure 1, it can be easily seen, that if a dosimeter is calibrated using samples at a fixed dose rate ($\dot{D}_{cal}$), the derived dose ($D_{meas}$) in any point of the dosimeter will be:

$$D_{meas} = \left( \frac{\left(\frac{\partial \psi}{\partial D}\right)_{\dot{D}}}{\left(\frac{\partial \psi}{\partial D}\right)_{\dot{D}_{cal}}} \right) \cdot D \qquad (1)$$

where $\left(\frac{\partial\psi}{\partial D}\right)_{\dot{D}}$ is the dose sensitivity for a dose rate $\dot{D}$ received in the corresponding pixel of the dosimeter, $\left(\frac{\partial\psi}{\partial D}\right)_{\dot{D}_{cal}}$ is the dose sensitivity for the dose rate at which the calibration was performed $\dot{D}_{cal}$.

In a first order approximation, the dose response curves can be considered as linear and the dependence on dose rate can also be considered as linear:

$$\left(\frac{\partial\psi}{\partial D}\right)_{\dot{D}} = \left(\frac{\partial\psi}{\partial D}\right)_{\dot{D}_{cal}} + \frac{\partial}{\partial\dot{D}}\left(\frac{\partial\psi}{\partial D}\right)\cdot(\dot{D}-\dot{D}_{cal})$$

Note that in an actual dosimetry experiment, every spatial point in the dosimeter phantom will have received a total dose that will be a combination of doses delivered at several different dose rates as every point consists of a superposition of several beams.

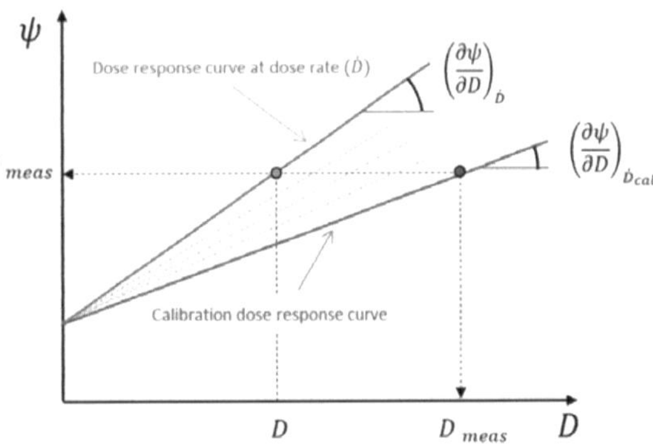


**Figure 1.** Illustration of the effect of calibration using calibration samples irradiated at dose rate $\dot{D}_{cal}$. With any given spatial point in the dosimeter corresponds a dose response curve (blue line) that results in a measurement $\psi_{meas}$. This value is converted in a 'measured' dose value $D_{meas}$ using the calibration curve (red).

Using equation 1, the percentage dose error as a result of dose rate dependent dose response can be written as:

$$\epsilon_{\dot{D}} = \left(\frac{D_{meas}-D}{D}\right)\cdot 100\% = \left(\frac{\left(\frac{\partial\psi}{\partial D}\right)_{\dot{D}} - \left(\frac{\partial\psi}{\partial D}\right)_{\dot{D}_{cal}}}{\left(\frac{\partial\psi}{\partial D}\right)_{\dot{D}_{cal}}}\right)\cdot 100\% \quad (2)$$

The dose rate related dose error $\epsilon_{\dot{D}}$ for a dose rate range between [50 cGy/min, 400 cGy/min] is summarized in table 1.

**Table 1.** Dose rate dependency and temperature dependency of some potential 3D radiation dosimeters. Literature values for dose rate dependence are (re)calculated for the dose rate range of [50 cGy/min, 400 cGy/min] by linear interpolation/extrapolation when not available. Note the big difference in dose rate dependence between the different dosimeters.

| Dosimeter | Read out | $\epsilon_{\dot{D}}\vert_{50\ cGy/min}^{400\ cGy/min}$ (%) | $\frac{\partial\psi_0}{\partial T}$ | $\frac{\partial}{\partial T}\left(\frac{\partial\psi}{\partial D}\right)/\left(\frac{\partial\psi}{\partial D}\right)_{20\ °C}$ |
|---|---|---|---|---|
| PAG [1] | MRI (R2) | -5 % | Nil | 0.2 % $K^{-1}$ |
| MAGAT [1] | MRI (R2) | -45 % | Nil | $-2.1$ % $K^{-1}$ |
| PAGAT [1] | MRI (R2) | -2 % | Nil | 0.4 % $K^{-1}$ |
| Micelle Gel [2] | Optical (OD) | -32 % | | 0.2 % $K^{-1}$ |
| PRESAGE™ [3] | Optical (OD) | -16 % | - | - |
| Flexydos3D [4] | Optical (OD) | -21 % | Nil | 0.3 % $K^{-1}$ |
| NIPAM PG [5] | X-ray CT | -12 % | - | - |

Important to note is that the dose rate dependence of most 3D dosimeters is not linear, with higher dose rate dependence at lower dose rates. The values provided in table 1 are therefore only indicative.

Despite the many attempts to fabricate radiation dosimeters that exhibit minimal dose rate dependency, there is no study on how a dose rate dependent radiation response affects a clinically realistic dose distribution. In this study, we investigate the effect of a dose rate dependency of the 3D radiation dosimeter on the measured dose distribution.

## 2. Methods and Materials

### *2.1. Construction of the dose distribution*

A hypothetical dose distribution is obtained by superimposing several beams from different directions on a cylindrical phantom using a coplanar beam set-up. In one part of the simulation study, the dose rate effect on the dose distributions was studied for various phantom diameters. The dose distribution was calculated by using a superposition of theoretical square radiation beams. The theoretical radiation beams are based on dose measurements acquired with an ionization chamber. Depth dose and lateral dose profiles were fitted against an in house developed physical model:

$$D\,(x,y,z) = D_0 \frac{sinh(\xi a(z))}{cosh(\xi x)+cosh(\xi a(z))} \cdot \frac{\sinh(\xi a(z))}{cosh(\xi y)+\cosh(\xi a(z))} \cdot \frac{SSD^2}{(z-SSD)^2}\left(e^{-\psi_p z} - e^{-\psi_e z}\right) \quad (3)$$

where x and y are lateral directions and z is the beam direction. The origin of the coordinate system $(x, y, z)$ is considered to coincide with the entrance point of the beam in the phantom. The first two hyperbolic factors in the equation define the lateral profile with field size $a$. The field size varies with depth according to $a(z) = \frac{(SSD+z)\cdot a(0)}{SSD}$. The parameter $\xi$ accounts for the penumbra width and is derived by fitting to the ion chamber measured beam data, $SSD$ is the source-to-surface distance and $\psi_p$ and $\psi_e$ are parameters that are linked to the photon and electron mass attenuation coefficients. However, they deviate from the true mass attenuation coefficients as a result of scatter contributions in the actual beam. $D_0$ is the dose defined at the surface of the phantom. The equation can be easily rewritten using the Tissue-Maximum Ratio (TMR). Regardless the simplified radiation model which doesn't account for any tissue inhomogeneities, we found however that an excellent fit of beam dose data was obtained against equation 1 for a range of square field sizes between 2 and 10 cm. This simplified model allowed us to quickly generate a realistic dose distribution in a cylindrical geometry with a multitude of beam configurations and phantom sizes. In this example, 2 cm × 2 cm fields were used and the output factor was taken into account in the scaling factor $D_0$ of equation 3. The off-axis-factor (OAF) was taken equal to 1 for simplicity.

The hyperbolic factors in equation 3 describing the lateral profiles of the beam, satisfy the property of superposition [6]:

$$\begin{aligned} P(x,a) &= \frac{\sinh(\xi a(z))}{cosh(\xi x) + \cosh(\xi a(z))} \\ &= \sum_{i=0}^{n-1} p[x + (2i+1)\cdot\delta, \delta] + p[x - (2i+1)\cdot\delta, \delta]\ , \quad \forall n \in \mathbb{N} \end{aligned} \quad (4)$$

Because of this particular property of superposition, the beam was split up in $n$ small segments with half width $\delta$ in the calculation algorithm to accommodate for the curved surface of the cylindrical phantom in the x-direction.

While this simulation comprises a simplified and fast algorithm written in MATLAB, as compared to the use of more sophisticated convolution-superposition or Monte Carlo based treatment planning systems, it is considered not to make a significant difference on the assessment of the dose rate effect. It can be assumed that the distribution of dose values above a certain minimum dose value threshold is not significantly affected by the geometry.

### *2.2. The effect of dose rate dependent dose response*

The effect of the dose rate dependent dose response was taken into account by considering equation 2 which can be rewritten as:

$$D_{meas} = D\,(1 + \epsilon_{\dot{D}}) \tag{5}$$

where $\epsilon_{\dot{D}}$ is dependent on the dose rate $\dot{D}(x, y, z)$ and on the dose rate $\dot{D}_{cal}$ at which the calibration is performed according to equation 2. Combining equation 1 and 2 and using the values from table 1, $\epsilon_{\dot{D}}$ can be linearized as:

$$\begin{aligned} \epsilon_{\dot{D}} &= \left( \frac{\dot{D}_{cal}}{(400\,cGy/min - 50\,cGy/min)} \cdot \epsilon_{\dot{D}}\Big|_{50\,cGy/min}^{400\,cGy/min} \right) \left( \frac{\dot{D}}{\dot{D}_{cal}} - 1 \right) \\ &= \epsilon_{\dot{D}}\Big|_{0}^{\dot{D}_{cal}} \left( \frac{\dot{D}}{\dot{D}_{cal}} - 1 \right) \end{aligned} \tag{6}$$

In this equation, $\epsilon_{\dot{D}}\big|_{0}^{\dot{D}_{cal}}$ is the dose rate dependency factor for calibrations performed at a dose rate $\dot{D}_{cal}$ for actual dose delivery at infinitely small dose rates ($\dot{D} \rightarrow 0\ cGy/\mathrm{min}$). For every single beam, the dose rate can be considered proportional to the dose. During the calculation of the dose distribution, the dose of each individual beam is stored. After the calculation of the actual dose distribution for every beam, the simulated dose incorporating a dose rate dependence is calculated for every beam using equation 6. After summing the dose rate affected dose for every beam, the simulated dose rate affected dose distribution is obtained.

After calculating the actual and the dose rate affected dose distribution, a gamma comparison is performed between the two dose distributions. The gamma pass rate is calculated for all dose values above a dose threshold of 5%.

## 3. Results and Discussion

The Matlab program using the beam superposition algorithm allowed to calculate dose distributions very quickly for different levels of dose rate dependence and for a multitude of different scenarios (different phantom size, number of beams and beam directions). 3D gamma evaluations were performed between the actual (theoretical) dose distribution and the dose distributions affected by the dose rate dependence.

Figure 1 shows a typical simulation of a 12 beam coplanar irradiation with square 6 MV photon beams within an arc of 250 degrees. The dose distribution of a single beam is also shown (figure 1c) that was obtained by splitting the beam in 10 beam segments along the x-direction.

A comparison between the actual and dose rate affected ('measured') dose distribution is shown in figure 2. Some groups have been using 3D dosimetry in a relative way, where the radiation dose distribution is scaled towards the maximum dose. The effect of this renormalization is illustrated in figure 2, d-f.

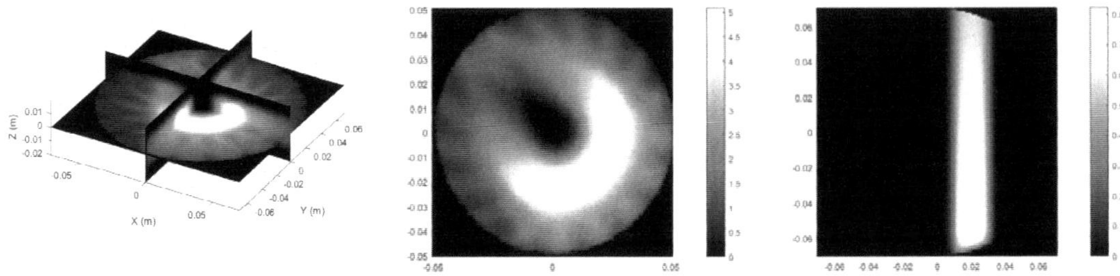


**(a)** **(b)** **(c)**

**Figure 2.** A typical simulated dose distribution in 3D (a) and in a central plane (b), for a cylindrical phantom (R = 5 cm) irradiated with 12 beams in an arc of 250 degrees. The calculated dose distribution of a single beam (2 cm × 2 cm) is also shown (c).

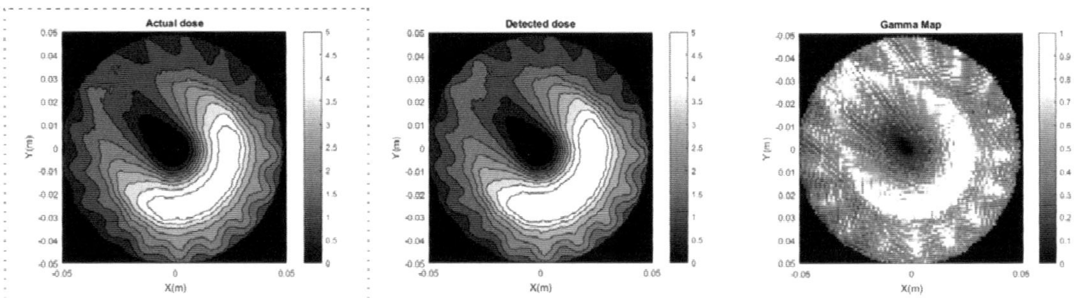


(a) (b) (c)

(d) (e) (f) (g)

**Figure 3.** Contour isodose plot of the computed dose distribution in the isocentric transverse plane with a 12 beam 250 degree arc (a) and corresponding computed dose distribution assuming a dose rate dependence $\epsilon_{\dot{D}}|_0^{\dot{D}cal} = -20\%$ (b). The corresponding gamma map is shown in c. After renormalization of the detected dose distribution (d), the gamma map shows no values above 1 (e). Corresponding gamma value histograms are shown in f and g for absolute calibration and relative dose calibration (renormalization towards the maximum dose) respectively.

Most dosimeters exhibit a decrease in dose sensitivity with increasing dose rate. As a result, the detected dose will be overestimated if the calibration samples are irradiated at a higher dose rate. The overestimation of dose is also apparent when comparing the detected dose distribution (figure 2b) with the reference dose distribution (figure 2a). After normalization of the detected dose distribution towards the maximum dose (figure 2d), a closer agreement is found in the gamma evaluation (figure 2e, g).

Figure 3a shows the trend in gamma-pass rates as a function of dose rate dependency coefficient ($\epsilon_{\dot{D}}|_0^{\dot{D}_{cal}}$) for three different phantom radii. It is clear that the phantom radius and as a result the dose distribution has only a small effect on the gamma pass rates. Figure 3b shows the trend in gamma-pass rates as a function of dose rate dependency coefficient ($\epsilon_{\dot{D}}|_0^{\dot{D}_{cal}}$) for a different number of beams.

It is clear from figures 2 and 3 that renormalization of the dose towards the maximum dose in the dose distribution has an enormous effect on the dose rate sensitivity. While one could conclude that it is better to use a renormalization of the dose distribution, it is important to realize that such a renormalization may also render the dosimetry insensitive for dose errors originating from the treatment.

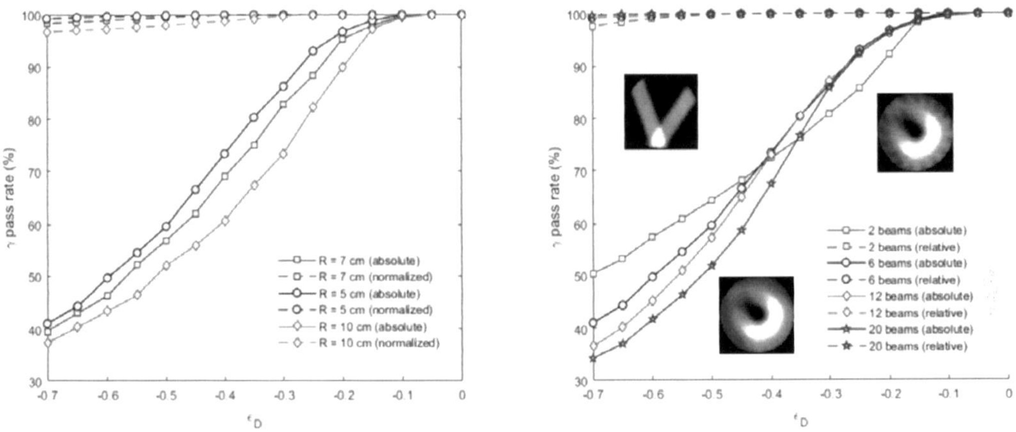


**(a)** **(b)**

**Figure 4.** Dependence of the gamma pass rate on the dose rate dependency coefficient for various phantom radii (a) and for different number of beams (b). The inset pictures are dose distributions for 2 beams, 12 beams and 20 beams.

Figure 3a and b demonstrate that the conclusions drawn from this study with respect to the dose rate dependency are relatively general irrespective of the phantom size and number of beams.

## 4. Conclusions

A fast dose calculation engine for homogenous irradiations was developed in Matlab to study the effect of a dose rate dependent dose response of radiation dosimeters.

Dose rate dependency of -10% does not affect the dose distribution significantly (gamma pass rate = 100%) and a dose rate dependency up to -15% results in a gamma pass rate of more than 98%. When the dose distribution is renormalized towards the maximum dose, a gamma pass rate of 99% is found for dose rate dependencies up to -40%. However, we do not recommend renormalization of the dose distribution as it may also obscure real treatment dose errors.

## 5. References

[1] De Deene Y *et al* 2006 *Phys. Med. Biol.* **51** 653-73
[2] Vandecasteele J *et al* 2011 *Phys. Med. Biol.* **56** 627-51
[3] Pappas E P *et al* 2017 *Radiother. Oncol* **123** S410
[4] De Deene Y *et al* 2015 *Phys. Med. Biol.* **60** 1543-63
[5] Jirasek A *et al* 2015 *Phys. Med. Biol.* **60** 4399-411
[6] De Gersem W *et al* 2001 *Int. J. Radiation Oncology Biol. Phys.* **51** 1371-88

IOP Conf. Series: Journal of Physics: Conf. Series **1305** (2019) 012060 doi:10.1088/1742-6596/1305/1/012060

# Characteristics of popular photon beam collimators

**Shidong Li**
Department of Radiation Oncology, Temple University, 3401 North Broad Street, Philadelphia, PA 19140

E-mail: Shidong.Li@tuhs.temple.edu

**Abstract**. **Purpose**: To compare the physical and dosimetric aspects of radiation beam collimation systems that may affect the stereotactic radiosurgery and radiotherapy of small lesions. **Methods:** Gamma knife (GK) cones, Cyber-knife Iris/InCise collimators, popular Varian and Elekta multi-leaf collimators (MLCs) were theoretically analyzed. Leaf-edge effects (Inter- and intra- leaf leakages, field penumbras, and dosimetric leaf gap - DLG) and transmission through some collimators were measured with ion chambers and films in phantoms and electronic portal imagers in an extended distance. GK plans and IMRT/VMAT plans were generated for ten patients and conformity index (CI) and dose drop (DD) from target to 5-mm ring were compared. **Results**: New MLCs have improved field shaping conformity, reduced transmission but not field penumbra and DLG. The measured DLG changed with time of usage and varied across the field. GK plans had better CI and DD for small lesions of < 1 cm while MLC-based VMAT/IMRT plans improved CI for large lesions of > 2 cm. **Conclusions:** New thinner and taller MLCs have improved the field shaping conformity and transmission but not leaf edge effects. Cone-based irradiation is still the best to small lesions of < 1 cm while VMAT/IMRT using new MLCs provide more conformal dose to irregular target usually with size of > 2 cm.

## 1. Introduction

I should like to begin by describing MLC leaf-edge effects on delivered dose on which we have some theoretical study twenty-two years ago [1]. As you know that MLC has made IMRT [2] and VAMT [3] possible and new collimation systems have been designed in recent years. To select a machine to treat intracranial and extracranial lesions, there are extensive options of GK, cyber knife, tomo-type and c-arm-type linear accelerators equipped with MLCs for 3DCRT, IMRT, VMAT, SRS and SBRT. Understanding the characteristics of equipped collimators could be the key factor in purchasing the machine. This pilot study is by no means exhaustive to cover all clinical aspects of MLCs but the dosimetric impacts or differences of MLCs.

## 2. Characteristics of Popular Collimation Systems

### *2.1 Summary of popular Collimation Systems*

Table 1 listed the technical specifications of some popular collimation systems with reference [4] for Gamma Knife, [5] [6] for Cyber Knife, [7] for Varian MLC and [8] for Elekta MLC. There are Siemens MLC, Tomotherapy MLC, Brainlab micro MLC, and other MLCs. To address the clinical impacts, an algorithm in round-end leaf control was described in subsection 2.2. Measurements of leaf edge effects

IOP Conf. Series: Journal of Physics: Conf. Series **1305** (2019) 012060 doi:10.1088/1742-6596/1305/1/012060

and their influences on treatment plans were explored in other sections.

**Table 1. Specifications of Popular Collimators (all Length in mm, angle in degree,% dose to that of machine specific reference***

| Machines | Gamma Knife | Cyber Knife | | Varian Machines | | | Elekta Machines | | |
|---|---|---|---|---|---|---|---|---|---|
| Type of Collimation | Cones | Iris MLC | InCise MLC | HD 120 | Millenium 120 | Standard MLC | B.M. MLC | Agility MLC | MLCi/i2 |
| Max Static Field Size (mm) | 18 / 16 | 60 | 100x120 | 400x220 | 400x400 | 400x400 | 160 x 210 | 400x400 | 400x400 |
| Leaf Travel Range (mm) | NA | (0, 30) | (-60,60) | (-200, 201) | (-200,201) | (-200,201) | (-105, 105) | (-150,200) | (-125, 200) |
| Maximal Leaf Speed (mm/s) | NA | | 25 | 25 | 25 | 25 | 30 | 30 | 20 |
| laximal Carriage Speed (mm/s) | NA | NA | NA | 12 | 12 | 12 | NA | 35 | 20 |
| Source to Leaf Tip Level (SCD) | | | | 504 | 504 | 505 | 427.8 | 349.3 | 335.5 |
| ource to Collimeter Base (mm) | 165/140-198 | 400 | 400 | 538 | 538 | 535 | 465.3 | 407.7 | 373 |
| Leaf width (mm) | N/A | Prism-Shaped | 3.85 | Cnt 32 pairs 2.5 Out 28 pairs 5.0 | Cnt 40 pairs 5.0 Out 20 pairs 10 | 40 pairs 10 | 40 pairs 4.0 | 80 pairs 5.0 | 40 pairs 10 |
| Collimator/Leaf height (mm) | 60/120 | 60 | 90 | 68 - 69 | 67 | 60 | 75 | 90 | 75/82 |
| Leaf end radius or shape (mm) | focus | straight | 3-flat | 160 | 80 | 80 | 122 | 170 | 150 |
| Tongue & groove offset | NA | NA | NA | 0 | 0.4 | 0.4 | 0 | 0 | 0.6/0 |
| Leaf material | W | W 95% Cu 5% | W 95% Cu 5% | W 95% | W 92.5% | W 92.5 % | W 95% , Ni 3.75%, Fe 1.25% | | |
| Density ($g/cm^3$) | 19.3 | 18 | 18 | 18.0 -18.7 | 17.15 - 17.85 | 17.15 - 17.85 | | | 18 |
| Average transmission | <2%/<0.2% | <2% | 0.3% | 2.5% | 2.0% | <2.5% | 1.50% | 0.50% | 0.70% |
| Ref. Field Penumbras* | 2 in Z 8 X&Y | <5 | <5 | < 5.5 | < 5.5 | < 6.5 | 6 inp, 8 crp | 7 inp, 8 crp | 7 inp 9 crp |
| Maximal interleaf leakage | | | <0.5% | <3.0% | <2.5% | <3.0% | <6.0% | <2% | <3% |
| nimum Dynamic Leaf Gap (mm) | NA | N/A | <0.5 | >0.5 | >0.5 | >0.5 | >0.5 | 1.0 | >1 |

**GK Ref.: 1.25 MeV 18 mm or 16 mm @ depth 8 cm. MLC Penumbra for 6 MEV 100x100 (or 60 mm Iris circle) field @ depth 100 mm, ~2 mm smaller @ dmax or 20x20 field*

## 2.2 *Theory of Physical Leaf Tip vs Nominal Leaf Lateral Position*

Exact 3D solution for tracing X-rays tangent to a rounded leaf end with the leaf tip position for left and right leaves had been first derived by the author [1] [9] and utilized by AAPM TG 50 report [10].

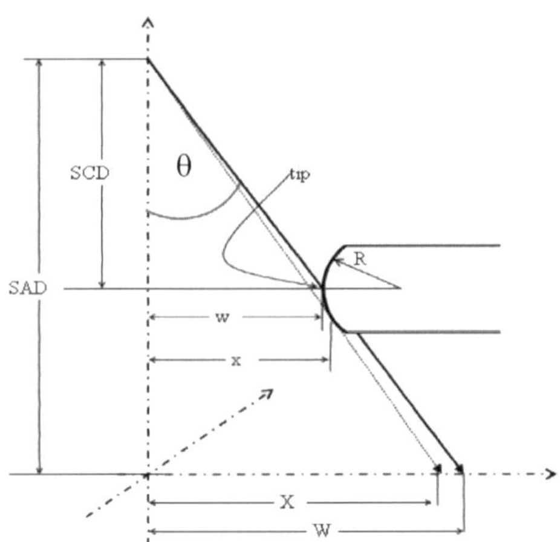


**Figure 1.** A leaf tip position (w) projected onto the SAD plane (W) has the light field edge of X = [W·SCD ± SAD·R(1-cos(θ))]/[SCD ± Rsin(θ)] (-for left leaf).

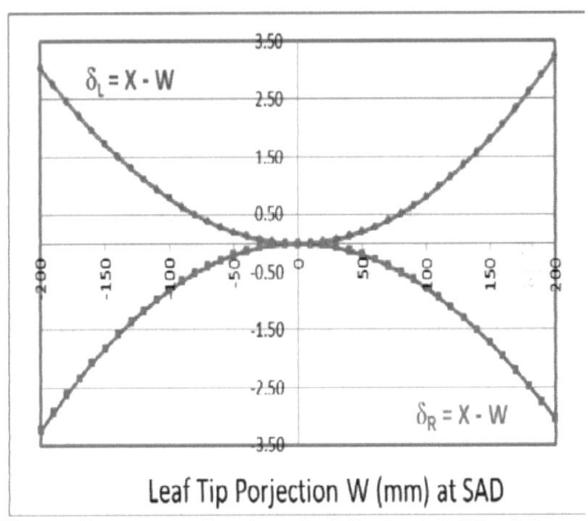


**Figure 2.** An example of standard Varian MLC leaf tip and the field edge difference δ at off-axis distance. Notice that right/left leaf pair collide at W ~ 0.

The field edge position X= W+ δ is the nominal leaf position. Standard Varian MLC leaf tip was pushed away by ~0.2 mm (about 0.4 mm on SAD plane) to avoid collision of closing leaf pair at the central region. Some new MLC chamfers the tip to avoid collision. The δ values for Elekta MLC was experimentally determined with MLC calibration. Theoretically, radiation field edge should be the ray through a half value layer (HVL) chord into the MLC for the 50% transmission. Additional shift for 2.3 HVL chord into the leaf corresponds the increase of penumbra. Non-focus surfaces at the end and sides of a leaf increased scattering and transmission at the edges in comparison with that from the focus jaws. That have been quantified as intra- or inter- leaf leakages under blocking area, enlarged penumbra at the edges and Dosimetry Leaf Gap (DLG) within dynamic IMRT fields [11].

*2.3 Leaf Edge Effects (LEEs) on Dose Delivery*

LEEs are sensitive to the submillimeter leaf position uncertainties. Thus, measurements are required to quantify LEEs particularly for weariness of extensive use or recalibration. Figure 3 and 4 demonstrate that measured LEEs using an on-board electronic portal imager (EPI) after geometric conversion to SAD and dosimetric renormalization to the ion chamber doses at $d_{max}$ = 16 mm. The penumbras at 10 cm depth can be determined with scans in a water tank or films in a solid phantom.

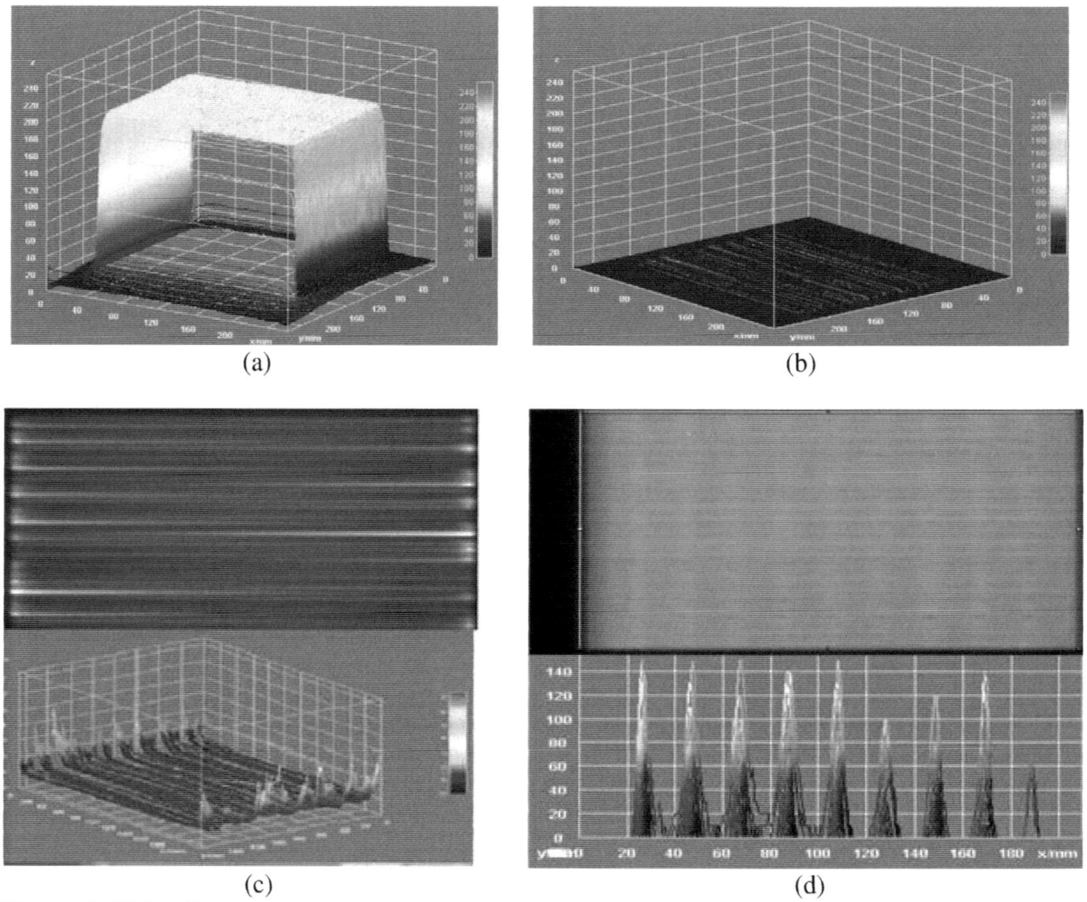


**Figure 3.** EPI uniform response to 6X 21x16 cm open field (a), % transmission in a closing field (b), % interleaf leakage distribution (upper) and 3D plot (c), and normalized intensity distribution in dynamic Picker Fence irradiation (upper) and 10 time % end leakage by subtracting closing field intensity (d).

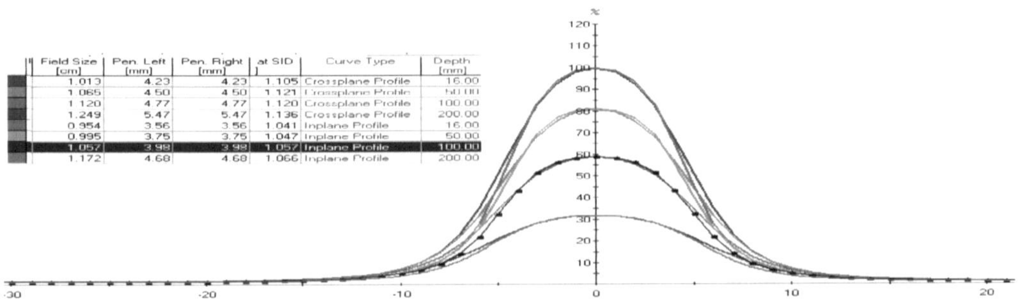


**Figure 4.** Measured cross-plane penumbras 4.8 mm vs in-plane penumbras 4.0 mm for a 1x1 cm field using a vertically mounted pinpoint ion chamber scanned at SSD = 90 cm and depth = 10 cm.

Notice that high interleaf leakage at some points (Fig. 1b) may affect measurement of DLG. A pin-point chamber was placed at the points of interest and doses from 2, 5, 10, and 20 mm width sliding window through a 20-cm square field for Agility (AG) MLC and 21 x 16 cm field for beam modulate (BM) MLC at 90 cm SSD. For 6-MV X-ray beams, the DLG are 0.26, 0.18, and 0.07 mm for BM MLC and 1.7, 0.5, and 1.1 mm for AG MLC. Similar measurement was performed on Varian machine with HD 120 MLC with the DLG of 0.39 and 0.27 mm for 6X and 6X FFF beams, respectively. However, the DLG in Eclipse TPS system had to be 1.1 mm in order to match the plan dose with delivered dose [6]. Both DLG variations across the fields and dose calculation differences support the author's question for the concept of DLG to be briefly described in the next subsection.

### *2.4 What is the "DLG"*

The concept of DLG was introduced by LoSasso *et al* [11] for adding intra-leaf leakage in dynamic IMRT fields. DLG was measured by the distance of the point with a zero dose from linear extension of the plot of dose vs sliding window widths. The dose for zero width of the sliding window should be the convolution of the radiation intensity I(x,y) with the additional scattering and transmission kernels at the leaf ends to the depth of interest. The subtracted profiles for a field defined by those focus jaws from the profiles for the same field defined by the MLC could provide the kernel and that could be approximated by two rectangle triangles (referring to Fig 2(d) for two end edges): one transmission through the leaf tip with a base size of $\tau$ and the height of $k_\tau$ and the other mainly for more scattering from the leaf tip into the open area with a base size of $\sigma$ and height of $k_\sigma$. Integration provided:

$$DLG(x,y) = (k_\tau \tau + k_\sigma \sigma)\, I(x,y) / I_o$$

Where $k_\tau$ and $k_\sigma \in (0,1)$ were ratios of the transmission and scattering peaks to the dose at CAX. If k were a constant, DLG would only change with radiation intensity such as FFF beams. Changes of the leaf end positions and the sides of the leaf as well as dose rate (MU) at the location (X,Y) could all affect the amount of scattering and transmission. Thus, a constant DLG does not truly represent LEEs and measured inter- and intra-leaf leakages should be included in the TPS commissioning.

## 3. MLC effects on Treatment of Small Lesions

### *3.1 Field Shape Conformity Using Various Collimators*

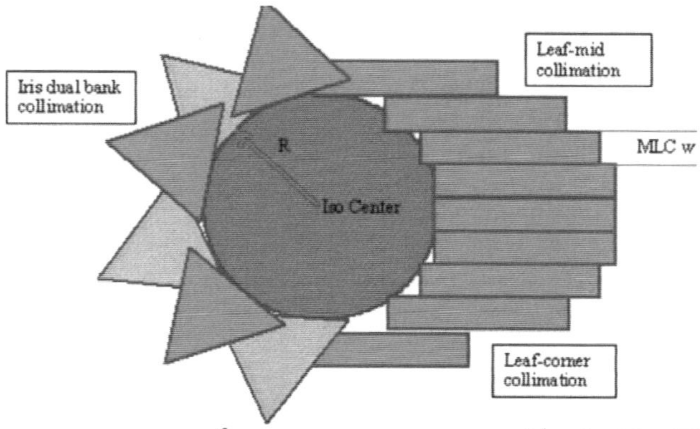


**Figure 5.** Illustration of MLC shaped circular fields by an Iris MLC (left half) or a regular MLC using mid or corner matching to the spherical or circular shape target. The field conformity index (CI) = open area/target area.

*We have* $CI_{iris} \approx \frac{3}{\pi} = 0.95$. $CI_{MLC} = 4w\sum_{i=1}^{R/w}\sqrt{R^2 - i^2w^2}/(\pi R^2)$. For R= 20-mm target, CI are 0.90, 0.79 and 0.55 by using a MLC with width w = 2.5, 5.0 and 10 mm, respectively.

*3.2 Actual Coverage for small spherical targets using Cones or MLCs*

Coverage of small spherical targets by the prescription does using single GK shot, MLC shaped field or arc were validated with film dosimeters. GK cones sizes are circular shapes while accelerators can change the field shape and size through adjusting leaves. The dose distributions from available GK shots and beam-modulate (BM) and agility MLC (AM)-shaped fields and arcs using Gafchromic EBT3 films placed within a spherical phantom as shown in Fig. 6.

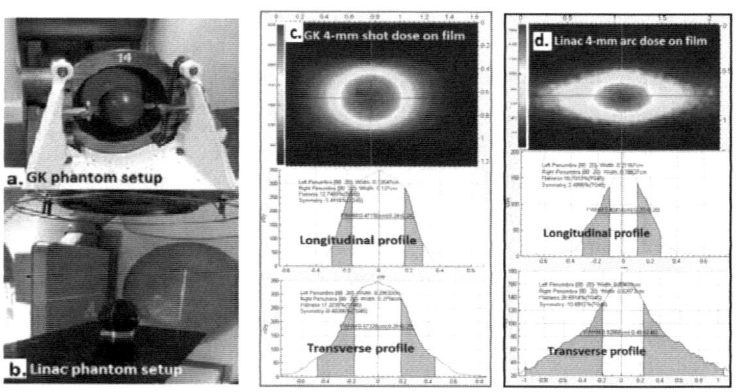


**Figure 6.** A spherical phantom setup on GK 4C unit (a) and on an Linac tabletop with a custom precise support (b), the measured dose distributions and orthogonal profiles from a 4-mm GK shot (c), and the dose and profiles for a MB-shaped 4x4 square full arc (d) are presented here for comparison.

**Table 2.** Results of the measured penumbra, FWHM, symmetry, and peak dose values.

| Mode | Size (mm) | Trasnverse Penumbra[c)] | Transverse FWHM[d)] | Transverse Flatness % | Transverse Symmetry % | Longitudinal Pennubra | Longitudinal FWHM | Longitudinal Flatness % | Longitudinal Symmetry % | Max Dose (cGy)[e)] |
|---|---|---|---|---|---|---|---|---|---|---|
| GK shot[a)] | 4 | 2.7 | 5.6 | 12.3 | 0.1 | 1.2 | 4.7 | 9.4 | 1.1 | 362 |
| | 8 | 4.2 | 10.9 | 9.3 | 0.7 | 1.4 | 9.0 | 7.7 | 2.1 | 389 |
| | 14 | 6.6 | 18.5 | 10.2 | 8.0 | 1.8 | 15.7 | 5.2 | 5.3 | 403 |
| | 18 | 8.2 | 23.5 | 7.1 | 0.1 | 1.9 | 20.0 | 4.3 | 3.8 | 413 |
| BM arc[b)] | 4 | 8.1 | 8.8 | 21.8 | 8.2 | 2.1 | 4.1 | 16.2 | 8.8 | 177 |
| | 8 | 12.5 | 13.2 | 20.0 | 6.2 | 2.4 | 7.5 | 12.1 | 3.3 | 363 |
| | 16 | 15.5 | 24.8 | 14.0 | 0.1 | 2.5 | 15.5 | 8.6 | 3.7 | 424 |
| AM arc[b)] | 5 | 8.3 | 9.2 | 19.2 | 2.5 | 2.6 | 4.6 | 17.8 | 2.6 | 250 |
| | 10 | 12.8 | 15.8 | 16.0 | 4.4 | 2.9 | 9.5 | 10.2 | 0.2 | 383 |
| | 20 | 14.4 | 30.8 | 14.5 | 0.6 | 2.9 | 19.8 | 6.0 | 1.8 | 420 |
| BM Field[b)] | 4 | 2.5 | 4.7 | 16.4 | 3.4 | 1.7 | 3.8 | 15.4 | 0.9 | 192 |
| | 8 | 3.4 | 8.5 | 14.3 | 6.2 | 2.4 | 7.4 | 12.3 | 4.1 | 362 |
| | 16 | 3.6 | 16.5 | 9.1 | 0.3 | 2.5 | 15.5 | 8.4 | 2.2 | 421 |
| AM field[b)] | 5 | 3.0 | 5.4 | 19.4 | 0.8 | 2.4 | 4.8 | 18.2 | 0.5 | 279 |
| | 10 | 3.2 | 10.5 | 10.4 | 3.0 | 2.6 | 9.6 | 8.7 | 1.7 | 372 |
| | 20 | 3.4 | 20.9 | 7.0 | 0.8 | 2.8 | 19.7 | 6.2 | 1.9 | 420 |

a) All GK shots used the same treatment time of 1.25 min for dose of ~410 cGy from 18-mm helemt.
b) All beam-modulate (BM) and agility MLC (AM) arcs/fields were irradiated with 600 MU of 6X.
c) Transverse Penumbras for shots and arcs were larger than corresponding longitudinal penumbras.
d) Full-Width-Half-Maximum (FWHM) were larger than longitudinal values by beam overlapping.
e) The maximal doses on the films were the maximal dose value without using smooth filtering.

*3.3 Comparison of patient SRS plans*

We expanded our study to treatment planning for complete coverage of various targets while sparing nearby function structures. Deformable image registration software (Velocity, Varian Inc.) was used to transfer patients' GK plans with cranial volume MRI, structures, and dose matrix. Patient CT scans were then used for dose calculation for non-coplanar beams or arcs in Pinnacle P3RTP system. If there was no head CT scans, we had overwritten MRI with uniform density of 1 g/cc as Fig. 7.

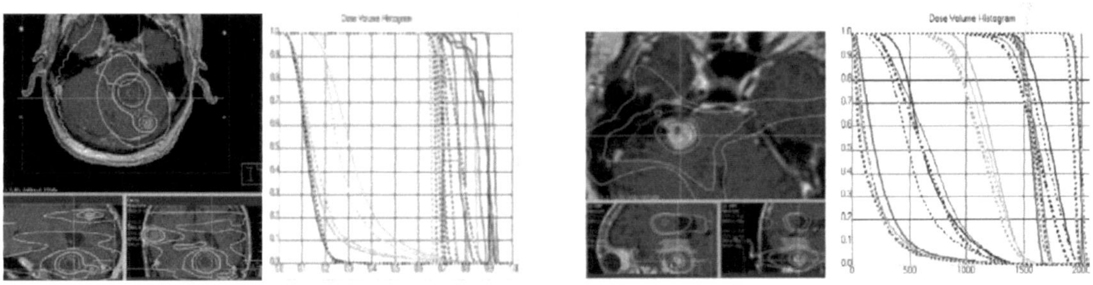

**Figure 7.** A single iso IMRT/VMAT plans and DVHs (left penal) for 6 lesions with sizes from 0.2 to 3.2 cm$^3$ and prescription dose (PD) of 20 Gy and two iso IMRT/VMAT plans and DVHs (right penal) for another patient with a right acoustic neuroma for PD of 15 Gy and a small recurrent GBM in the right parietal lobe for PD of 20 Gy. Dash/Solid DVH lines are for IMRT/VMAT plans, respectively. All plans had coverage of > 95% and optimized conformity index defined as C.I. = coverage$^2$ x Target Volume / Prescription Isodose Volume [12] and the peripheral dose drop-off: DD = % (mean dose of TV – mean dose of 5-mm margin ring) / mean dose of TV.

The results for tested 11 cases (some cases had multiple lesions as a single target) were listed in Table 3 (AM for agility MLC and BM for beam modulator MLC) with red numbers indicating poor values in a standard SRS treatment. For small lesions (TV < 1 cc or size of <1 cm), GK has better CI and DD than that of VMAT or IMRT while IMRT/VMAT provide high CI for large lesions (6 or 8 cc for size of ~2 cm). As we already know if Varian HDMLC or Elekta dynamic MLC with 2.5 mm leaf width are used, CI would be further improved.

| Table 3 | Plan results for | | CI =C^2*TV/PIV (see ref. 3) | | | | | DD = %(MDtv - MD5mm)/MDtv | | | | |
|---|---|---|---|---|---|---|---|---|---|---|---|---|
| Location | PD (Gy) | TV (cc) | GK | AM VMAT | BM VMAT | AM IMRT | BM IMRT | GK | AM VMAT | BM VMAT | AM IMRT | BM IMRT |
| Pons | 15.00 | 0.05 | **0.71** | 0.19 | 0.19 | 0.45 | 0.45 | 84.90 | 26.10 | 36.70 | 46.20 | 46.50 |
| SCLC met | 20.00 | 0.10 | **0.45** | 0.14 | 0.23 | 0.20 | 0.19 | 79.90 | 38.00 | 39.90 | 20.50 | 22.50 |
| Recu. GBM | 18.00 | 0.60 | 0.35 | 0.24 | 0.28 | 0.39 | 0.39 | 51.00 | 18.90 | 15.80 | 15.90 | 12.80 |
| 5 Mets | 20.00 | 1.20 | 0.30 | 0.21 | 0.22 | 0.26 | 0.30 | 51.90 | 56.70 | 55.60 | 54.90 | 54.90 |
| Lt Occip. | 20.00 | 1.30 | 0.43 | 0.48 | 0.56 | 0.58 | 0.47 | 62.20 | 23.60 | 24.60 | 16.80 | 16.20 |
| 5 mets | 20.00 | 1.40 | 0.25 | NA | NA | 0.20 | 0.23 | 47.80 | NA | NA | 21.70 | 22.40 |
| Rt Ac. Neo. | 14.00 | 1.90 | **0.71** | 0.61 | 0.64 | 0.64 | 0.72 | 53.40 | 21.40 | 26.80 | 26.30 | 28.60 |
| Pit Aden. | 18.00 | 2.10 | 0.65 | 0.64 | 0.65 | 0.65 | 0.66 | 57.90 | 27.50 | 27.80 | 25.20 | 24.20 |
| Soli. Met | 15.00 | 5.20 | 0.67 | 0.64 | 0.65 | 0.67 | 0.68 | 33.40 | 29.90 | 29.70 | 33.10 | 33.60 |
| Rt Front Met | 18.00 | 6.40 | 0.28 | 0.33 | 0.35 | 0.51 | 0.52 | 37.70 | 30.40 | 31.80 | 35.00 | 36.50 |
| Sup. Mening. | 18.00 | 8.60 | 0.52 | 0.68 | 0.65 | 0.66 | 0.66 | 42.40 | 33.50 | 33.90 | 33.40 | 36.70 |

## 4. Conclusion

New MLCs had improved field CI but leaf edge effects (LEEs) varied significantly with a long-time usage and/or recalibrations as well as changing across the fields particularly for the new FFF beams. Theoretical comparisons and experimental verification are both useful in quantification of LEEs and the LEEs made GK still the best choice for small brain lesions (<1 cm) while new MLC provided more conformal dose to lesions > 2 cm. DLG concept was created to compensate differences in planning commissioning using the profiles of beams shaped with the focus jaws while real delivered IMRT/VMAT beams with the edges shaped by rounded leaf ends. Thus, conceptually, DLG should not be used if the measured LEEs can be included in the next generation planning systems.

## 5. Acknowledgement

This research was partially supported by a pilot grant from Elekta Inc.

## 6. References

[1] Li S *et al* 1996 *Int. J. Rad. Oncol. Biol. Phys.* **36** 398
[2] Bortfeld T R *et al* 1994 *Int. J. Rad. Oncol. Biol. Phys.* **28** 723-30
[3] Yu C X 1995 *Phys. Med. Biol.* **40** 1435-49
[4] Novotny J *et al* 2008 *J. Neurosurg.* **1009** 8-14
[5] Echner G G *et al* 2009 *Phys. Med. Biol.* **54** 5359-80
[6] Varian Inc. 2014 A comparison of EDGE Radiosurgery and CyberKnife M6 https://www.varian.com/sites/default/files/resource_attachments/EDGEvsCK_RAD10325_Apr2014_secured.pdf
[7] Kim J *et al* 2018 *Phys. and Imaging in Rad. Onc.* **5** 31-6
[8] Elekta Inc. 2017 Specification of Beam Modulator and Agility MLCs
[9] Boyer A L and Li S 1997 *Med. Phys.* **24** 757-62
[10] Boyer A *et al* 2001 AAPM Report No. 72 Medical Physics Publishing, Madison, WI, USA
[11] LoSasso T *et al* 1998 *Med. Phys.* **25** 1919-27
[12] Paddick I 2000 *J. Neurosurg.* **93** 219-22

IOP Conf. Series: Journal of Physics: Conf. Series **1305** (2019) 012061 doi:10.1088/1742-6596/1305/1/012061

# EPIDs and QA of advanced treatments

**B Mijnheer**
Department of Radiation Oncology, The Netherlands Cancer Institute, Plesmanlaan 121, 1066 CX Amsterdam, The Netherlands

E-mail: b.mijnheer@nki.nl

**Abstract**. In this paper I will summarize the various possibilities of EPIDs for their use as tools for QA of advanced treatments. After elucidating the choice of EPIDs for this purpose, I will review the use of EPIDs for pre-treatment and *in vivo* (transit) dosimetry applications. Several solutions became recently commercially available, allowing relative and/or absolute dose verification at points (0D), in 2D, or in 3D. Each of these solutions will be briefly discussed, indicating the differences in possibilities with respect to QA of advanced treatments. I will conclude with some general remarks about the current status of EPID-based QA of advanced treatments and revealing some future developments in 3D pre-treatment and *in vivo* dosimetry using EPIDs.

## 1. Introduction

After the acceptance testing of a treatment unit, specific measurements are required for the commissioning and quality assurance (QA) of that unit, as well as for the verification of treatment techniques performed with that machine. These dose determinations are often performed using point dose (0D), or 1D measurements, e.g., for the determination of output factors, beam profiles and depth dose distributions. The introduction of advanced treatment techniques such as intensity modulated radiation therapy (IMRT) and volumetric modulated arc therapy (VMAT) required in addition 3D dosimetric approaches. For that purpose, various detector systems have been developed: *full-3D*, e.g., using gel or radiochromic detectors, *semi-3D*, using films or detector arrays in a phantom, or *pseudo-3D* using transmission chamber devices attached to the head of the treatment unit, or electronic portal imaging devices (EPIDs). A dose calculation using a linac log file can also be considered as a pseudo-3D technique. In my presentation at the IC3DDose2016 meeting in Galveston I have summarized the possibilities and limitations of these various 3D dosimetry techniques and evaluated those features relative to those of EPID-based techniques [1]. A more extensive comparison of semi-3D and pseudo-3D techniques can be found elsewhere [2, 3]. In this presentation I will discuss only the use of EPIDs as tools for QA of advanced treatments, with the emphasis on EPID-based solutions that became recently commercially available.

## 2. EPID-based dosimetry

### *2.1. The use of EPIDs in modern radiation therapy*

In addition to their original use for patient setup verification during external photon beam therapy, EPIDs are increasingly used for other applications for the following reasons:

1. Most accelerators have an EPID or are sold with an EPID in combination with additional tools for image guidance such as cone-beam CT (CBCT). The role of EPIDs is then to evaluate the congruence between the CBCT image center and the radiation isocenter. Furthermore, EPIDs are used for several other QA tasks such as the verification of MLC leaf positions.
2. EPIDs have many useful dosimetric characteristics such as the (almost) linearity of the response with dose and dose rate, good long-term stability, high spatial resolution, and real-time readout. EPIDs exhibit both high data density and online efficiency as compared to the existing non-EPID based QA tools. However, using EPID dosimetry in clinical practice has also a number of challenges, including its over-response to low energy photons, a possible under-response due to image ghosting, and a contribution to the EPID signal of photons scattered in the EPID and sometimes in the EPID-arm. These characteristics, and methods to correct for deviations from the regular dose-response relationship, have been discussed in detail by McCurdy *et al* [4].
3. Setup time, data analysis and processing time are small and can be automated. These are important prerequisites for large scale clinical implementation of EPID dosimetry. Generally *in vivo* EPID measurements and data analysis can be performed by radiation therapists (RTTs), while alert handling needs input of a medical physicist.
4. One of the most important reasons for performing EPID-based *in vivo* transit measurements, either by comparing EPID images in a relative way or by means of *in vivo* dosimetry, is that this measurement approach has a high effectiveness in detecting errors. In a study by Ford and colleagues it was noted that EPID dosimetry was a very efficient QA tool in detecting high severity incidents in combination with chart checks and reviews [5]. It was also much more effective than the pre-treatment QA procedures performed in the institutes participating in this study. This observation was confirmed in a follow up study concerning the effectiveness of EPID dosimetry in detecting incidents in radiotherapy [6]. In that study it was demonstrated that EPID *in vivo* transit dosimetry during the first fraction was able to detect 74% of the incidents, while pre-treatment EPID dosimetry could only trace 6%.
5. Many semi-3D and EPID-based tools are currently commercially offered for performing pre-treatment dose verification, but EPIDs are the only devices available for 3D *in vivo* dosimetry.

### *2.2. EPID-based QA approaches*

The various EPID-based QA approaches have been discussed in a comprehensive way in a 2008 review by van Elmpt *et al* [7], which has recently been updated by McCurdy *et al* [4]. Briefly, if treatment beams have passed through a patient or phantom before being measured by the EPID, the method is referred to as transit or transmission dosimetry, otherwise, the method can be described as non-transit dosimetry. Note that sometimes the term "exit dosimetry" is used instead of "transit dosimetry", which is not recommended because it has been defined in the past as the determination of the dose at the exit surface of an irradiated phantom or patient. Dose or image comparisons can be made at the EPID level or in the patient/phantom using a forward- or back-projection approach. Forward projection methods compare measured images or 2D dose distributions with predicted images or 2D dose distributions at the EPID level. Back-projection methods can provide point dose, 2D or 3D dose information in a patient or phantom, and the resulting dose values are then compared with planned dose data for that particular patient/phantom. Not many institutions perform EPID-based transit dosimetry with phantoms, and therefore I will discuss in this paper only transit *in vivo* EPID dosimetry.

In this presentation I will focus on the current commercial EPID dosimetry tools for patient-specific QA of advanced treatments. Some of these products became recently available and consequently not many publications on their clinical use are yet available in the literature. Information about the properties of the various EPID-based dosimetry systems discussed in this presentation, is therefore partly based on material provided by the website of the various companies and has therefore to be considered as preliminary info. A detailed report concerning the possibilities and limitations of the various commercial EPID systems for patient-specific IMRT and VMAT QA is in preparation (AAPM TG-307).

### *2.3. Pre-treatment verification*

Before starting clinically with a new treatment technique or class solution, validation of the planning and delivery of that technique should be performed by means of a dedicated measurement program. Such a program generally includes an end-to-end test in which the actual patient treatment procedure is mimicked as much as possible using a full-, semi-, or pseudo-3D system. After finishing such a series of systematic pre-treatment measurements, a patient specific pre-treatment QA measurement is performed before the start of each patient treatment with that technique, frequently in combination with an independent dose or MU check.

The utility of an EPID-based pre-treatment QA measurement depends on the type of software used. The minimum aim of such a procedure is to guarantee that data transfer between systems is correct before patient treatment begins. The Varian Portal Dosimetry system (www.varian.com/course/portal-dosimetry), developed by Van Esch and colleagues [8], is such a system. Basically, EPID images are used to verify if the fluence intensity distribution calculated by the TPS is transferred to and delivered by the linac correctly. It compares measured with predicted EPID images, for both IMRT and VMAT plans calculated and delivered by Varian equipment. Many publications of this product can be found in the literature, including improvements of the original software and the establishment of action levels [4, 7].

The Varian Portal Dosimetry system does, however, not verify the 3D dose calculation for that particular patient irradiation. One step closer to that situation is the Epica software (www.epidos.eu). It converts the EPID image into a dose map at dose maximum in a virtual water phantom, which is then compared with a reference dose distribution in that plane. The software is based on the GLAaS algorithm developed by Nicolini *et al* [9] and can be used for both IMRT and VMAT plans calculated and delivered by Varian equipment.

The perFRACTION Fraction 0 software from Sun Nuclear (www.sunnuclear.com) is also able to perform a 2D dose determination at the EPID level, which can be compared with the dose distribution in that plane calculated with the TPS for independent 2D planar analysis. The EPID image can also be used to determine MLC leaf positions, which can be used as input for a 3D dose calculation on planning CT data using the linac log-file.

The Dosimetry Check (www.lifelinesoftware.com) system is able to perform an independent 3D dose calculation in combination with the non-transit approach. It converts the measured photon fluence at the EPID level into relative number of monitor units per pixel. The 3D dose distribution is then calculated on planning CT data using a forward dose calculation pencil beam algorithm. By exporting the dose distribution from the TPS into the Dosimetry Check system, one is able to make a direct comparison between the calculated and planned 3D dose distribution. It allows DVH analysis and is therefore able to trace errors in the 3D dose calculation for a specific patient, in addition to checking the plan transfer and deliverability. The performance of the system, as well as a comparison with two semi-3D detector systems, has been studied on a Varian accelerator [10], but according to their website the Dosimetry Check software can also be used with EPID images obtained from Siemens and Elekta accelerators.

### *2.4. In vivo verification*

Pre-treatment QA of advanced treatments using EPIDs involves a check of the dose delivery under well-defined conditions, while *in vivo* verification is performed under the actual daily clinical conditions [11]. Pre-treatment non-transit EPID verification approaches will miss errors due to changes in patient anatomy and patient setup, or due to malfunctioning of the treatment machine during the actual patient treatment. Transit *in vivo* EPID-based measurement systems allow day-to-day monitoring and can be used to highlight any major change in the patient anatomy during a treatment series that may have a clinical impact, as often occurs during head and neck or lung cancer treatments.

During transit *in vivo* verification using the forward approach, measured EPID images are compared either with a predicted image, or in a relative way to a reference image. In the first approach not only errors in the data transfer and delivery of the treatment can be determined, but also patient-related errors such as anatomy changes. Comparison with a reference image will only trace deviations during a series of treatments. It should be noted that deviations in transit EPID images observed by systems using the forward approach are not in a simple way related to variations in the dose distribution inside a patient.

PerFRACTION from Sun Nuclear (www.sunnuclear.com) compares EPID images from each field of each fraction n (>1) behind a patient with a reference portal image using 2D gamma analysis. In a sensitivity study of this software it was demonstrated that PerFRACTION, using integrated EPID images, is sensitive enough to identify positional, angular, and dosimetric errors [12]. The EPID image can also be used to determine MLC leaf positions of that fraction, which can be used as input for a 3D dose calculation on planning CT data using the linac log-file. Such an approach should be considered as a pseudo-3D dose verification tool but not as an *in vivo* dosimetry method.

The Adaptivo software (www.standardimaging.com/qa-software/adaptivo/) is also able to compare *in vivo* transit EPID images with a reference EPID image using gamma analysis. According to the website their software can also be used in combination with the MV CBCT system of TomoTherapy.

SOFTDISO from Best Medical Italy has a similar option to compare subsequent daily EPID portal images using gamma analysis. The SOFTDISO software is in addition able to perform a comparison of the planned dose at the isocenter with a back-projection at the isocenter of the transit EPID dose. The software is in use in various Italian centers and has been described in several publications, e.g. [13]. In that paper it was shown that relevant discrepancies were observed using this software in head-and-neck cancer patients treated with VMAT including a setup error and major anatomical variations (weight loss/tumor shrinkage) during the second half of their treatment.

EPIgray (www.dosisoft.com) uses the transmitted signal on the EPID to reconstruct the dose at points of interest within the patient. The dose reconstruction method uses a back-projection of the transit EPID dose onto the patient planning CT. Measured vs calculated dose comparison can be performed in multiple points but not over the whole EPID area. Clinical experience has been described for both Varian [14] and Elekta [15] equipment. Both publications showed that EPIgray is able to pick up a number of anatomical changes and patient setup inaccuracies that could lead to intolerable dose delivery errors.

EPID-based 3D dose verification has been implemented into two commercial systems, each applying a different technique of dose reconstruction. In the Dosimetry Check software (www.lifelinesoftware.com) the photon fluence measured with the EPID is back-projected through the patient to the target of the linac, and then used to calculate the 3D dose distribution in the patient geometry using a forward dose calculation. This approach is similar to the way Dosimetry Check is used for pre-treatment 3D dose verification. The transit dosimetry option has extensively been tested using phantoms [16] but clinical experience has not yet been reported.

iViewDose (www.elekta.com/iViewDose) applies a different approach in which the 2D dose distribution measured with the EPID is back-projected in the 3D patient planning CT data. An example of the possibilities of a prototype of this software is given in Figure 1. The dose calculation algorithm used in this system is a pencil-beam type, as in Dosimetry Check, because it is fast and relatively easy to implement, but the accuracy is inferior to other dose calculation algorithms particularly with respect to tissue heterogeneity corrections. A comparison of iViewDose with a home-made 2D forward-projection *in vivo* EPID dosimetry software package for VMAT prostate treatments has recently been published [17].

IOP Conf. Series: Journal of Physics: Conf. Series **1305** (2019) 012061 doi:10.1088/1742-6596/1305/1/012061

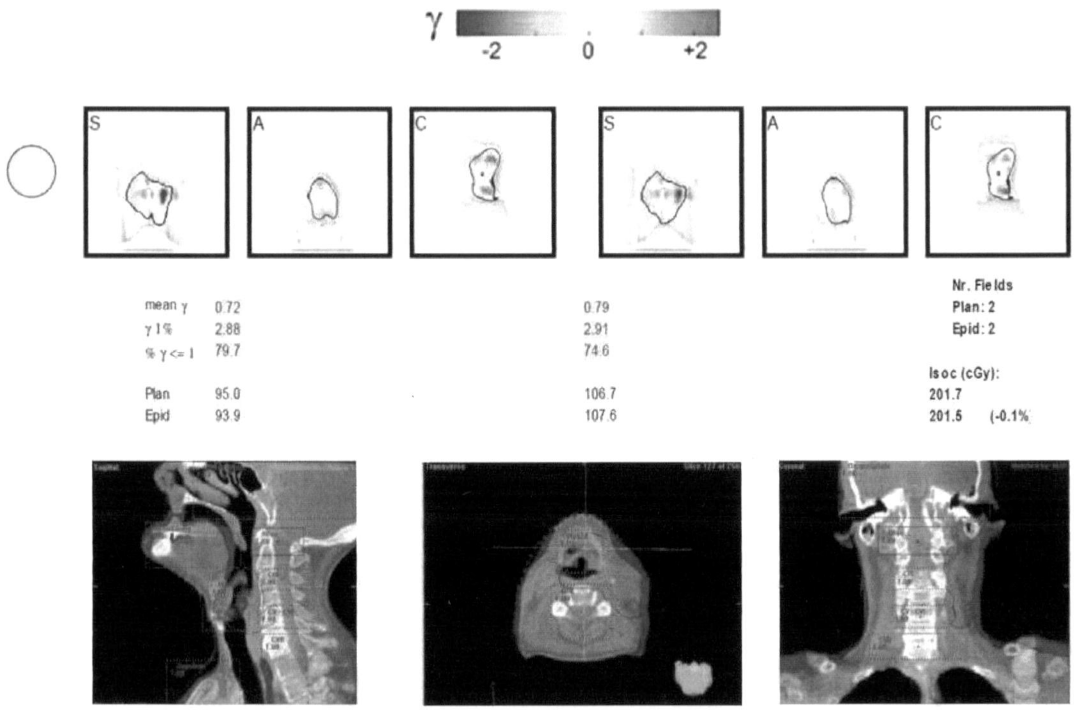


**Figure 1.** The upper part shows the outcome of a 3D EPID-based *in vivo* dose verification of a VMAT treatment of a larynx cancer patient at NKI. Indicated are the results of the 3D gamma evaluation in a sagittal, axial and coronal plane through the isocenter. A signed gamma display is used: the yellow and red color indicate regions where the EPID dose is higher than the planned dose, whereas the green and blue color indicate regions where the EPID dose is lower than the planned dose. The 50% isodose line is shown in black. The yellow dot means that at least one alert criterion is outside the action level. The bottom part shows a CBCT scan (green) made that day compared with the planning CT scan (purple) in the three orthogonal planes, showing a decrease of target volume. Note that the dose at the isocenter did not change.

## 3. Future developments

There are several issues that can be further developed to make EPID software for QA of advanced treatment better designed, more automated and more integrated into the clinical workflow. In addition, new options should be further explored such as the ones discussed in the following sections.

### *3.1. Combination of pre-treatment and in vivo EPID dosimetry*

By combining pre-treatment and *in vivo* EPID-based dose verification data of the same patient treatment, it may be possible to separate patient-related errors from machine, planning and dose reconstruction model errors. A prerequisite for such an approach is that the pre-treatment and *in vivo* dose calculation is performed with the same dose calculation algorithm. Such an approach has recently been developed in our institute [18]. As shown schematically in Figure 2, the primary portal dose distribution at EPID level obtained during a pre-treatment non-transit measurement, can be used in combination with patient CT data to reconstruct 'virtual patient' 3D dose distributions applying the same back-projection algorithm as used during the *in vivo* measurement. In this way pre-treatment and *in vivo* dose distributions can both be analyzed in terms of dose-volume histogram (DVH) parameters, providing a useful indication to radiation therapy staff of the type and impact of detected deviations. Dosimetry Check is in principle also able to combine pre-treatment and *in vivo* 3D dose reconstruction using the same algorithm, but no publication exists about the use of this option. More work is needed to test this approach in clinical practice.

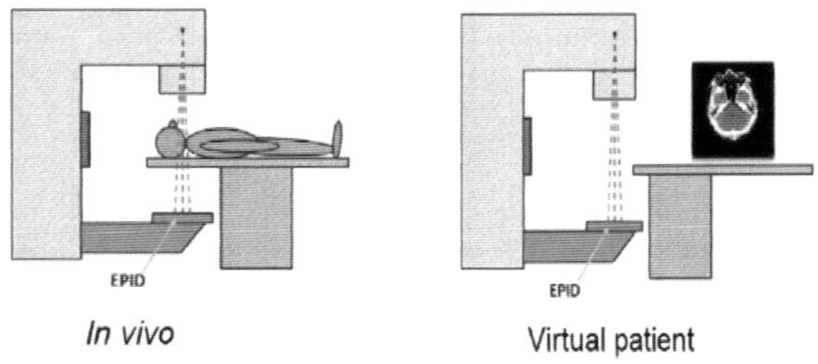


**Figure 2.** *In vivo* and virtual patient EPID measurement configuration allowing *in vivo* and pre-treatment virtual patient 3D dose verification [18].

### *3.2. Combination of CBCT and in vivo EPID dosimetry*

Most, if not all, EPID-based *in vivo* dose verification systems use planning CT data to reconstruct the dose in the patient geometry. An obvious question therefore is why not using the patient anatomy during treatment to determine a more accurate estimate of the dose delivered to a patient? Often a CBCT scan made on the same day is available and a number of errors observed with *in vivo* dosimetry can be understood already qualitatively by visual inspection of the accompanying CBCT scan. Variation in setup, anatomy changes due to changes in atelectasis and pleural effusion, as well as contour variations due to weight loss or other reasons (see Figure 1), can often be observed by CBCT.

There are, however, several concerns that have to be considered when merging the information from *in vivo* 3D dosimetry and CBCT in a quantitative way. First, there are practical issues that make it sometimes difficult to use CBCT scans in the clinic such as non-complete fields of view. Also, there is currently a lack of automation of the integration process requiring manual actions. Furthermore, in case of treatment adaptation, the radiation oncologist often has to define a new CTV in a CBCT scan, which is challenging due to the less good image quality compared to a CT scan, whereas in addition such a procedure is time-consuming. Finally, a more fundamental problem exists: how to incorporate disappearing voxels, for instance due to a shrinking tumor, in deriving the total dose delivered to a changing CTV. Consequently, dose comparisons and alert criteria are less obvious when using CBCT.

Nevertheless, *in vivo* transit EPID dosimetry should not be used in isolation but is most effective in combination with CBCT. For instance, a combination of CBCT and *in vivo* 3D dosimetry will help in selecting those patients that require an adaptive treatment. It will depend on the training of the therapists and the availability of clinical protocols if observed deviations in dose delivery and CBCT are considered to be clinically relevant. A combination of both sets of information will help in choosing the most suitable follow-up action. More work is needed to assess a further incorporation of CBCT when performing *in vivo* transit EPID dosimetry.

### *3.3. Real-time EPID dosimetry*

*In vivo* transit EPID dosimetry should ideally be performed in real-time so that errors can be caught immediately and the treatment interrupted if the delivered dose is not within pre-defined limits of acceptance. This is particularly important for hypofractionated treatments, where detected errors cannot always be compensated for in subsequent fractions. Technically this is possible as demonstrated for a 2D forward dose reconstruction method [19] and a 3D backward dose reconstruction technique [20]. As an example, Figure 3 shows the result of a real-time verification using the latter system during an anthropomorphic phantom irradiation with VMAT rectum plans. The introduced serious delivery errors were detected after 5–10 s irradiation time.

The major challenge when using real-time EPID dosimetry clinically is to assess alert criteria able to detect those deviations in dose delivery that are so large that detecting them with off-line dosimetry, i.e. after completion of the total fraction, would imply unacceptable harm to the patient. The system should therefore be sensitive enough to detect gross errors in real-time, but on the other hand the system needs to be robust, in other words have a low false-positive rate. False positives in real-time dosimetry have serious consequences for the clinical workflow as it interrupts a treatment session leading to time loss and stress. More research is therefore needed to define criteria for halting the linac that should not miss serious errors but have at the same time a small number of false positives. Future work should also include the development of a workflow to integrate real-time dosimetry into routine clinical use.

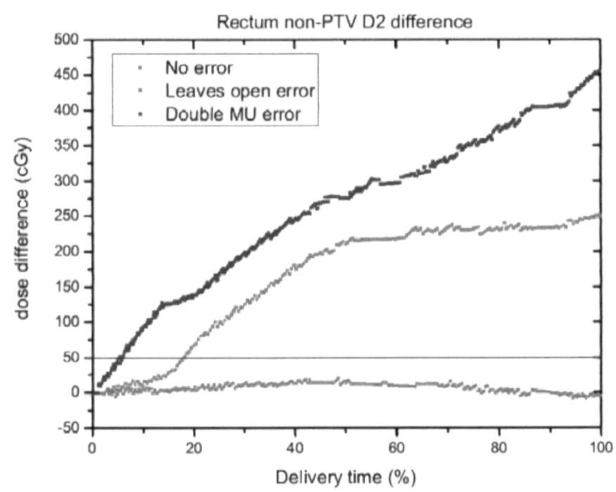


**Figure 3.** Difference between reconstructed and planned near-maximum dose ($D_2$) outside the PTV for three scenarios (no error, all leaves open error, double MU error), as a function of delivery time. The horizontal line indicates a possible detection threshold [20].

## 4. Final remarks

The main purpose of EPID-based pre-treatment verification of advanced treatments is to identify dosimetrically unacceptable errors, i.e. deviations from the "true dose" considered from the physical/technical point of view. The aim of EPID-based *in vivo* dosimetry is often not to measure the "true" dose delivered to a patient, but to detect in a simple and reliable way a deviation between the planned and reconstructed dose at a point, in 2D or 3D, which should be within well-specified criteria. The question then arises when is a difference in dose distribution observed during *in vivo* treatment verification a *clinically unacceptable error*, i.e. when would the treatment outcome in terms of local control or toxicity be affected by this difference? To answer this question, it is necessary to perform for each EPID-based QA system a comprehensive sensitivity analysis, that should result in treatment site specific alert criteria, either gamma- or DVH-based or both, to ensure optimal detectability for various types of errors.

Alert criteria are generally a compromise between the requirement to detect all clinically relevant deviations and the workload to analyze these alerts. Generally, alerts are first reviewed by an experienced medical physicist, and might have many reasons as publicized comprehensively elsewhere [21]. When the error is understood by the physicist, and it was estimated that continuation of the treatment might have negative clinical consequences, the case should be discussed with a radiation oncologist. A decision for corrective action is then generally made by the physicist and radiation oncologist together.

The experience in some centers demonstrates that EPID-based *in vivo* verification of advanced treatments requires a change in attitude to patient specific dose verification. It is no longer purely a physics or technical matter but requires in addition clarifying dose differences observed in the clinic, i.e. incorporating many issues that may happen during the daily treatment of patients. As a result, medical physicists are much more involved in assuring the quality of the actual patient treatment than when only performing a pre-treatment dose verification measurement [3, 21].

## 5. Acknowledgements

The author gratefully acknowledges the numerous discussions with Igor Olaciregui-Ruiz and Anton Mans at the Netherlands Cancer Institute on the many scientific and clinical aspects of EPID-based dosimetry.

## 6. References

[1] Mijnheer B 2017 *J. Phys.: Conf. Ser.* **847** 012024
[2] Moran J *et al* 2017 2D and semi-3D dosimetry systems *Clinical 3D dosimetry in modern radiation therapy* ed B Mijnheer (Boca Raton, FL: CRC Press) chapter 8 pp 203-25
[3] Georg D *et al* 2017 Patient specific QA: pre-treatment 3D dose verification *Clinical 3D dosimetry in modern radiation therapy* ed B Mijnheer (Boca Raton, FL: CRC Press) chapter 17 pp 437-60
[4] McCurdy B *et al* 2017 Electronic Portal Imaging Device (EPID) dosimetry *Clinical 3D dosimetry in modern radiation therapy* ed B Mijnheer (Boca Raton, FL: CRC Press) chapter 7 pp 171-201
[5] Ford EC *et al* 2012 *Int. J. Radiat. Oncol. Biol. Phys.* **84** e263-e269
[6] Bojechko C and Ford EC 2015 *Med. Phys.* **42** 6912-8
[7] van Elmpt W *et al* 2008 *Radiother. Oncol.* **88** 289-309
[8] Van Esch A *et al* 2004 *Radiother. Oncol.* **71** 223–34
[9] Nicolini G *et al* 2006 *Med. Phys.* **33** 2839-51
[10] Narayanasamy G *et al* 2015 *J. Appl. Clin. Med. Phys.* **16** 329–38
[11] Vial P *et al* 2008 *Med. Phys.* **35** 4362-74
[12] Zhuang AH and Olch AJ 2018 *J. Appl. Clin. Med. Phys.* **19** 114-24
[13] Cilla S *et al* 2016 *Phys. Med.* **32** 52-8
[14] Celi S *et al* 2016 *J. Appl. Clin. Med. Phys.* **17** 262-76
[15] Ricketts K *et al* 2016 *Int. J. Radiat. Oncol. Biol. Phys.* **95** 1513-9
[16] Gimeno J *et al* 2014 *Phys. Med.* **30** 954–9
[17] Bedford J *et al* 2018 *Phys. Med. Biol.* **63** 025008
[18] Olaciregui-Ruiz I *et al* 2017 *Phys. Med.* **37** 49-57
[19] Fuangrod T *et al* 2016 *Radiat. Oncol.* **11** 106
[20] Spreeuw H *et al* 2016 *Med. Phys.* **43** 3969-75
[21] Mijnheer B 2017 Patient specific QA: *in vivo* 3D dose verification *Clinical 3D dosimetry in modern radiation therapy* ed B Mijnheer (Boca Raton, FL: CRC Press) chapter 18 pp 461-90

# End to end QA in image guided and adaptive radiation therapy

**L J Schreiner**
Department of Medical Physics, Cancer Centre of Southeastern Ontario at the Kingston Health Sciences Centre, Kingston, Ontario, Canada, K7L 5P9
Departments of Oncology and Physics, Queen's University, Kingston, Ontario, Canada K7L 3N6

E-mail: john.schreiner@kingstonhsc.ca

**Abstract**. It has been suggested that modern radiation therapy could benefit from adopting the *"End to End"* (E2E) type of testing developed originally in computer science to determine whether applications and systems work as required under real-world scenarios. The motivation for adopting E2E techniques for image guided adaptive radiation therapy validation is to extend beyond current common testing using standard physics QA that inherently probes only select points or systems within the IGART schema. E2E methodologies extend the testing to evaluate complete IGART processes, including the complex interchanges that occur during and throughout a patient's treatment as clinical staff interpret and respond to information acquired during the treatment course.

While limited radiotherapy E2E QA may have been adopted periodically by clinics when implementing a new treatment technique, clinical E2E QA has been confined to date mainly to tests mediated by external auditing bodies such as IROC, the Imaging and Radiation Oncology Core in the United States. This testing often includes having the clinic in question irradiate a purpose-built phantom containing dosimeters to specific criteria under protocols set by the auditing body. The auditors then determine off site whether the clinic's treatment process was successful by comparing the dose measurements with the intended dose delivery.

The advance of three dimensional (3D) radiation dosimeters opens the possibility for in-house E2E testing. Approaches for in-house E2E testing have been proposed for over a decade, but such comprehensive internal E2E testing has not been widely adopted. In this presentation the barriers and challenges to the development of clinical in-house E2E QA will be reviewed primarily based on the experience in Kingston.

## 1. Introduction

*"End to End testing"* is a testing method originating in computer science and used to determine whether a software application performs as required in the field: the testing is performed from start to finish under real-world scenarios. End to end (E2E) tests are designed to determine whether an application or system communicates and interacts as required for proper functionality with hardware, networks, databases and other applications. It has been proposed for over a decade [1-4] that modern radiation therapy could benefit from adopting E2E methodology in the clinic, particularly as these modern techniques advance to image guided adaptive radiation therapy (IGART). Many of these proposals have suggested that the potential for clinical E2E is facilitated by the three dimensional (3D) dosimeters and other advanced dosimeters that are being discussed throughout this conference.

The motivation for the development of in-house E2E testing is that these techniques could provide additional quality assurance (QA) that addresses one inherent feature of IGART: IGART techniques incorporate multiple processes (see Figure 1), each of which demands a complex interchange during which a patient's treatment may be modified at various points based on decisions by clinical staff who interpret information acquired during the treatment course. While physics QA would have been performed to validate the hardware used in, or systems in place at, individual steps in the process (e.g., through machine QA) such conventional QA does not by its nature validate the overall treatment process. Establishing regular in-house E2E quality assurance (QA) within treatment clinics would facilitate more regular validation of the processes involved in modern radiation therapy.

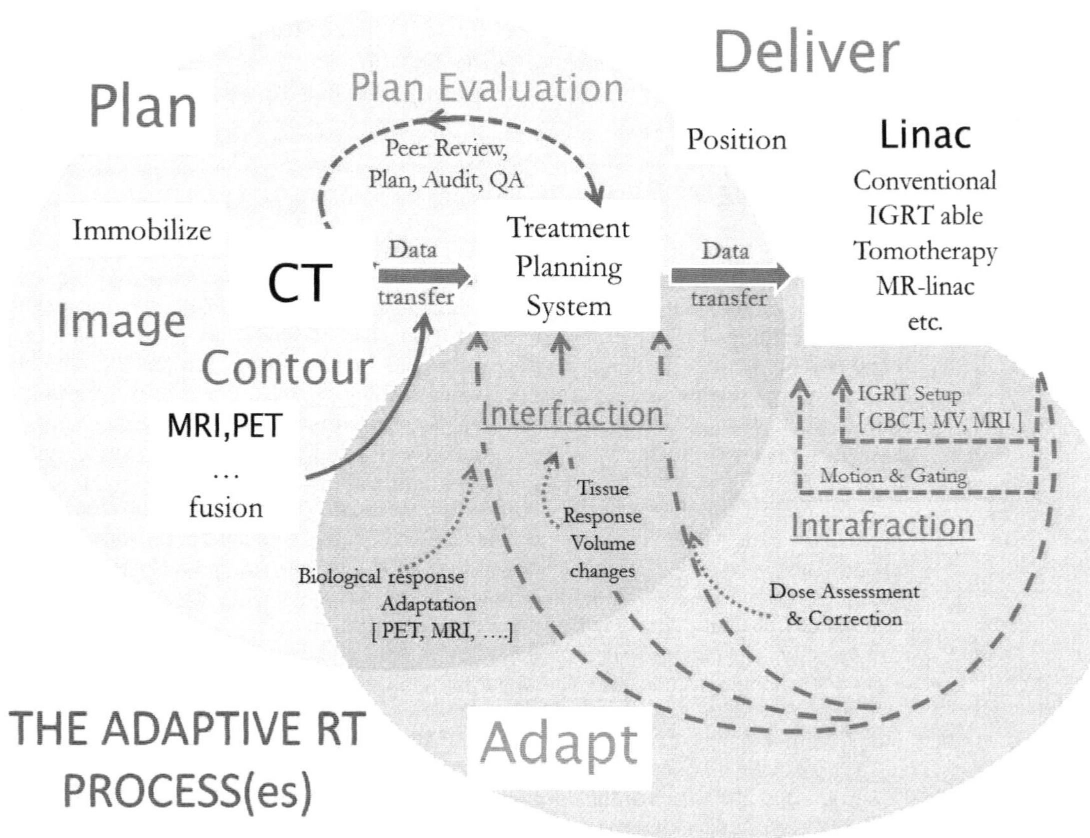


**Figure 1.** An illustration that captures the nature of the processes that can be ascribed to image guided adaptive radiation therapy. Some of the adaptive processes, particularly for intrafraction patient set up correction and motion management, are already in common clinical practice, as are some interfraction corrections, for example, plan adaptation with patient volume changes through the course of the treatment. Other processes, such as adaptive response to deformation of targets and organ at risks and dose correction to account for dose coverage mismatch in previous fractions, are still in development.

## 2. Current E2E Radiotherapy Testing in the Clinic

Currently clinical radiotherapy E2E QA is practiced only selectively in the clinic. One common utilization of E2E testing occurs when a cancer centre joins a cooperative multi-centre clinical trial [5,6]. The validation and acceptance criteria from the clinical trial group will generally include an evaluation and review of the technical components of a clinic's radiation planning and delivery capabilities. This review will generally include an E2E test and audit mediated by an external auditing body such as IROC-H, the Imaging and Radiation Oncology Core-Houston, TX. The IROC-H E2E testing incorporates a component in which specific purpose-built IROC phantoms containing imaging markers and dosimeters are irradiated by the centre to specific criteria under the planning and treatment protocols set by IROC-H. The auditing centre validates that the treatment process was successful by comparing the dosimetry measurements with the intended dose delivery criteria. A detailed analysis of external audits is beyond the scope of this report, but there are excellent reviews in the literature; those by Ibbott [5] and Followill *et al.* [6] are particularly well detailed.

There are some general observations one can make from the experience of these external E2E audits [5,6]. The first is that in spite of the fact that the vast majority of centres coming into the external audits were well established in the IMRT and advanced IGART techniques being audited with well-defined in-house physics QA programs, a significant fraction of centres did not pass the initial IROC-H audits, particularly in the early days of the testing. The E2E testing indicated that some previously unidentified deficiencies still existed somewhere in the complete treatment chain being audited. Secondly, as time went on, the passing rates for the audits with the particular phantoms used for specific treatment sites (e.g., head and neck, lung, prostate, etc.) generally all improved consistently [6]. For example between 2001, when first introduced, and 2011, the pass rate for the IROC-H head and neck phantom improved from 61% to 91% [5]. The evidence from external audits has shown that regular IGRT process QA can help improve the consistency of IGRT delivery within a specific clinic and across institutions that participate in cooperative clinical trials.

A second implementation of clinical E2E testing that occurs more regularly in a clinic is during the development, commissioning and implementation of treatment techniques new to the cancer centre. The implementation of rigorous commissioning and proactive testing has resulted in part because of the treatment communities' experience with the external audits described above. The results from the IROC-H type audits have identified that prior in-house testing using phantom based planning and irradiation to ensure that trial protocols are well understood and deliverable before the external audit help improve audit results. Many in the community have extended such prior testing to the commissioning of new advanced treatment techniques they are adopting in the clinic. An example of such phantom based E2E testing was reported in the very early days of 3D dosimetry [7]. Figure 2 below shows some examples of such E2E approaches in Kingston. We have over many years adopted a strategy of testing new techniques on purpose-built phantoms before a first patient is irradiated with that technique. Depending on the technique the test phantoms may be anthropomorphic, in order to best test required patient set-up procedures or to better mimic internal structures for dosimetry evaluation. In some cases (e.g., for testing fractionated stereotactic radiation therapy (FSRT) delivery), more conventional 'physics type' phantoms have been used because they can better accommodate well positioned dosimeters [10]. In all cases the process has been to CT image, plan and irradiate the phantom with the same procedures as would be used on a patient. Such E2E validation has proved a valuable component of technical commissioning. The testing has helped identify some technical and process challenges and has informed modifications of the planned approaches for the new technique prior to patient use.

The already adopted external and internal commissioning E2E testing described in the last three paragraphs provides a strong indication that clinical programs benefit from E2E testing. However, it has been noted [6] that these current E2E testing approaches only present a validation of the IGART protocol being tested at one specific point in time. Hence it is a snapshot that may not continue to reflect conditions in the clinic as techniques become more familiar and less special, as processes evolve, and as staff change. This suggests that more regular E2E testing, perhaps a few times a year for some of the more involved treatment protocols, may be indicated.

## 3. Advancing E2E Testing in the Radiotherapy Clinic

The discussion above motivates the main points of this brief review: i) that modern radiation therapy would benefit from more regular in-house E2E testing for clinical IGART process quality control, and ii) that such in-house E2E QA is feasible because the systems to enable such testing are now in place.

The availability of improved radiation dosimeters (both 3D and other) that can be inserted into phantoms, the ability of treatment planning systems dose calculation algorithms to accommodate that phantoms used are not perfectly anthropomorphic or tissue equivalent, and the increased availability of phantoms (purchased or manufactured in-house, see Figure 3) all make regular in-house E2E testing achievable. And vendors such as GeVero [Poland, www.polygevero.com], Heuris Pharma [USA, www.presage3d.com], MGS Research Inc. [USA, www.mgsresearch.com], Modus QA [Canada, www.modusqa.com], and RT Safe [Greece, www.rt-safe.com] are now providing tests tools (including dosimeters, phantoms or phantom inserts, and data analysis services) to assist in the regular adoption of E2E within the clinic.

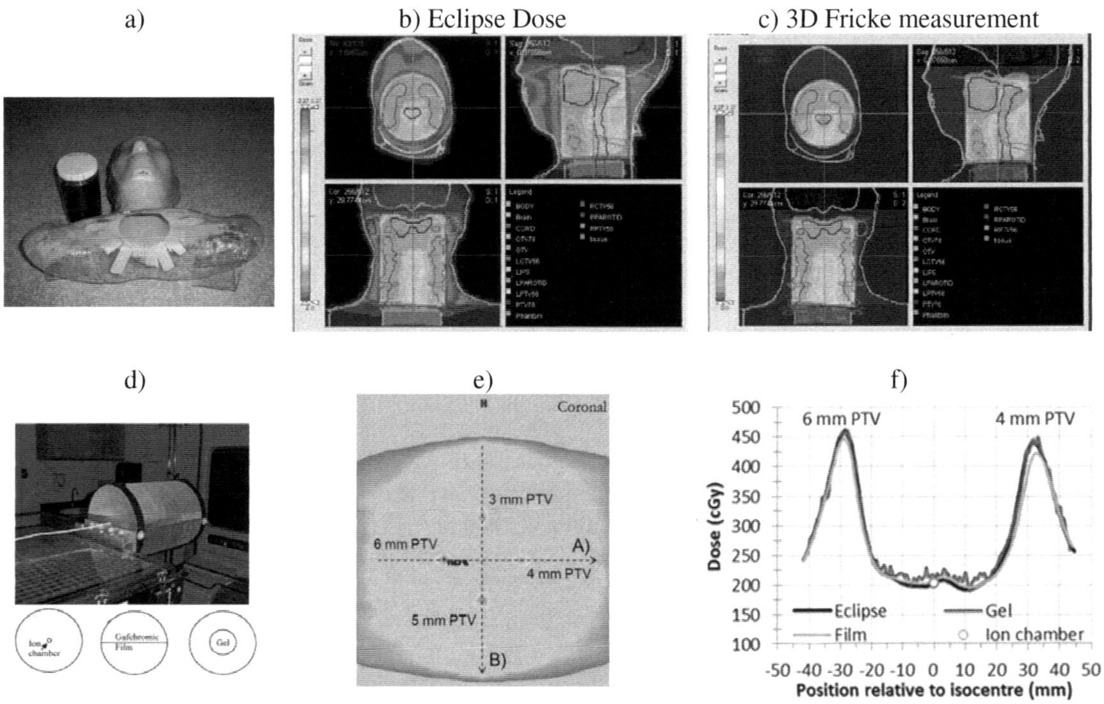


**Figure 2.** Examples of E2E testing performed in Kingston during the implementation and commissioning of two advance radiation therapy techniques (see text). The top row (a-c) shows the in-house built anthropomorphic phantom and sample results of 3D IMRT delivery validation with a Fricke Gel 3D dosimeter [8]. The dose measurements (c) could be compared directly to (b) the planned dose distribution from the Varian Eclipse [Varian Medical, Palo Alto, US] doses. The bottom row shows the commissioning E2E testing for a VMAT enabled fractionated stereotactic radiation therapy (FSRT) delivery [9] using a solid phantom (d) which could hold various detectors including film (e). Panel (f) shows the correlation of the dosimetric measurements used to evaluate the integrity of the dose delivery prior to the first patients being irradiated.

There is one further observation from the E2E testing practices described in the previous section 2 above that should be noted before describing the recommended format for in-house E2E testing that will be proposed below. The E2E testing already adopted as described above is often limited in that the work from imaging, through set-up, to dose delivery is in most cases performed solely by members of the physics department. Thus the testing is, in some regards, artificial as it tests what potentially is achievable in the clinic but does not test the process with the specific personnel who will be treating the patient once the technique is implemented. (Although one might argue that some aspects of the testing, particularly the phantom set-up and on board image verification, might give better results if more experienced treatment staff are involved.) Clinical E2E testing will more appropriately and fully assess that a specific IGART treatment process is working as intended if the very staff involved in the information analysis and treatment adaptation decisions that govern the process during a patient's treatment are the staff involved in the various stages of phantom manipulation in the E2E testing.

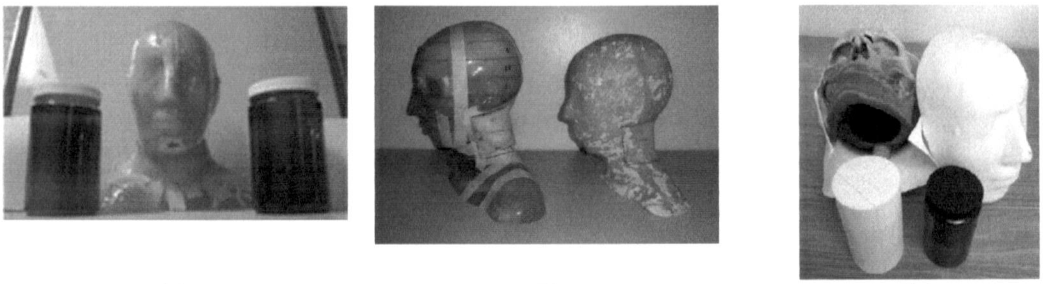

(a) (b) (c)

**Figure 3.** The evolution of head phantoms for various clinical QA and E2E testing development in Kingston. The phantoms are designed to hold various dosimetry systems in geometries relevant to the clinical process being tested. From left to right: a) the initial H&N phantom for IMRT validation [8]; b) a wax phantom to test bolus placement under masks as tumour size changes and gaps appear [11]; and c) our newest phantom 3D printed for ongoing E2E QA. All phantoms mimic a Rando phantom included in the middle panel. The grey skull in the rightmost phantom has been printed with doped material to simulate bone electron density, the surface shell has been designed to obscure the geometric configuration of the interior phantom (which is modifiable) when the phantom is manipulated for imaging or treatment by the radiation therapy staff.

The development of specific in-house E2E tests to provide regular process control for some modern image-guided adaptive radiation therapy treatment commissioning should be guided by the experience from external auditing groups and from past internal treatment commissioning. The E2E test should be designed so that the testing procedure mimics the specific IGART process as it is to be used on a patient. Some phantom with specific testing systems would be prepared in-house, or perhaps purchased from a vendor. (Note, the rest of this description will reflect our experience that the testing system incorporated in the phantom has typically been some dosimeter, either point, 2D, or 3D, depending on the test). In any case the physics staff would prepare or purchase the dosimeter from a commercial vendor (examples of in-house systems would include radiochromic leucodye micelle gel, Fricke-gel or a polymer-gel; commercial systems might include radiochromic film, ClearView gels or PRESAGE radiochromic plastics). The staff would insert the dosimeter into some anthropomorphic or purpose-built phantom which would then be treated by the radiation therapy team as if the phantom were a patient. During this testing, the E2E QA would not be performed solely by the physics staff but rather by the whole clinical treatment team: planning therapists would perform CT simulation; radiation oncologists would contour structures that could be delineated in the target (perhaps using targeting modules that could be inserted in the anthropomorphic phantom); planners or dosimetrists would plan the delivery on the treatment planning system; and treatment therapists would deliver the treatment using the IGART approaches being assessed. This testing could extend to whichever adaptive processes are in regular practice in the clinic: for example, simple setup verification and repositioning under cone beam CT or under MR on an MR-linac, specific high precision treatments such as FSRT, or perhaps an adaptive procedure following tumour deformation over multiple fractions. At the end of the treatment process the physics group would remove and image the 3D dosimeter, register the dose data against the treatment planning dose distribution, and evaluate the results. If the clinical E2E QA passes some previously established scoring metrics (a topic in itself [6]) then the process being tested will have been shown to be running as intended.

Note that the E2E QA tests whether ALL the components of the treatment process are functioning as desired. For this reason the E2E testing cannot fully replace the other conventional or technical physics hardware or systems QA tests, firstly because the treatment hardware and technical systems must be maintained to the high standards required *a priori* for safe patient treatment, and secondly, a full understanding of the technical functioning of the equipment, supported by well-established QA records of the system behaviour, will be needed in the case that the E2E testing fails. The specific E2E QA will have tested the whole process in its complete chain and the E2E testing itself cannot provide sufficient information to specify the exact point of breakdown. Complementary QA information will be required in any root cause analysis [6,12].

While presenting many advantages in the maintenance of high quality IGART, there are significant challenges in initiating the testing. E2E testing is resource intensive; the timing for dosimeter preparation and use presents challenges for incorporation into the regular clinical workflow. And modifying workflow to accommodate the E2E testing has human resource implications that may not be easily accommodated in clinical operation. But these constraints may be better understood and correctable as in-house E2E programs are developed and implemented. Past obstacles to E2E testing were removed when specific attention was directed to their resolution. For example, problems introduced by the timescale required for the readout and analysis of dosimetric data (particularly with 2D and 3D systems) for full analysis and comparison with intended dose delivery as planned in the treatment planning system were minimised in the last few years once specific attention was directed to the development of specific open-source [13,14] and commercial software tools for this function. Thus the resource challenges are not insurmountable barriers.

## 4. Conclusion

In closing, a unique advantage to the E2E testing presented here is the ability to mimic the specific adaptive IGART process as it would be applied to a patient; that is, through the complete chain of required data transfer, analysis and decision making (steps which are often judgment-based and not tested in conventional physics QA procedures). Regular (quarterly, bi-annually, etc.) performance of the E2E QA would ensure sustained adherence to the process protocols at the high levels adopted in the initial commissioning and implementation of the advanced IGART protocols that are advancing patient treatment in our cancer centres. This provides a strong motivation for advancing the clinical implementation of in-house E2E testing at this time. Hopefully this presentation will motivate the community attending this conference to direct some of their efforts and wisdom to establishing practical protocols for clinical end to end testing of specific IGART protocols utilized in their clinics.

## 5. References

[1] Schreiner L J 2009 *J. Phys.: Conf. Ser.* **164** 012001
[2] Schreiner L J and Olding T 2009 In *AAPM Medical Physics Monograph: Clinical Dosimetry Measurements in Radiotherapy* Roger D and Cygler J (ed) Madison, WI: Medical Physics Publishing
[3] Schreiner L J *et al* 2011 A Proposal for Process QA in Modern Radiation Therapy, *Proc. Internat. Symp. Standards, Applications and Quality Assurance in Medical Radiation Dosimetry (IDOS)*; **Vol 2**, 197-205, (IAEA Vienna, Austria)
[4] Knöös T 2015 *J. Phys.: Conf. Ser.* **573** 012001
[5] Ibbott G S and Thwaites D I T 2015 *J. Phys.: Conf. Ser.* **573** 012002
[6] Followill D S *et al* 2018 In *Clinical 3D Dosimetry in Modern Radiation Therapy* Mijnheer B (ed) Boca Raton: CRC Press
[7] Scheib S G *et al* 2001 In *Proc. 2nd Int. Workshop of Radiation Therapy Gel Dosimetry* Baldock C (ed) Brisbane, AU: Queensland University of Technology
[8] Olding T *et al* 2010 *J. Phys.: Conf. Ser.* **250** 012003.
[9] Lalonde M *et al* 2015 Proceedings of the World Congress on Medical Physics and Biomedical Engineering, Toronoto, Canada, D.A. Jaffray (ed.)
[10] Olding T *et al* 2013 *J. Phys.: Conf. Ser.* **444** 012052
[11] Alexander K M *et al* 2017 *J. Phys.: Conf. Ser.* **847** 012038
[12] Huq M S *et al* 2016, *Med. Phys.* **43** 4209-62
[13] Alexander K M *et al* 2015 *J. Phys.: Conf. Ser.* **573** 012042
[14] Alexander K M *et al* 2018 *Biomed. Phys. Eng. Express* **4** 045041

# EPID-based linear accelerator benchmarking using pixel sensitivity map

**Baozhou Sun, Sreekrishna M. Goddu, Sasa Mutic and Bin Cai**
Department of Radiation Oncology, Washington University in St. Louis, St. Louis, MO, USA

E-mail: baozhou.sun@wustl.edu

**Abstract.** Conventional verification of linac performance is conducted by measuring percent depth dose (PDD) as well as beam profiles in a water tank, which is labor intensive and requires expensive beam scanning equipment. We present a novel method to benchmark the linac beam characteristics using pixel sensitivity map (PSM) corrected EPID images and to evaluate the feasibility of EPID-based beam benchmarking. A novel approach was developed to generate a two-dimensional (2D) PSM for EPIDs utilizing an alternating beam and dark-field image acquisition technique. The calculated PSM based on a recursive calculation algorithm was used to correct EPID pixel response. The acceptability of a benchmarked beam is based on its profiles being within predetermined tolerances. The PSM-corrected open field EPID images were compared for three photon energies (6 MV, 10 MV and 6FFF) between three TrueBeam and one Edge linacs with aS1200 imaging panels. The output factors were measured with EPID from 2×2 cm$^2$ to 40×40 cm$^2$. The differences in beam profiles and output factors from EPID were evaluated. The four dosimetrically similar linacs with less than 1% variation in PDD, profiles and output factors measured in water scans showed excellent agreement in the EPID profile measurements. The 1D Gamma of the PSM-corrected profiles between any two linacs showed 100% passing rate for 6 MV and 6 FFF and >97% for 10 MV with 1mm/1% criteria. The maximum difference of output factors was 0.18% among all the measurements except for 2×2 cm$^2$ with 0.6% difference. By preserving the beam dosimetry information, the PSM-corrected EPID images enable the possibility of validating the beam data of linac dosimetry and benchmarking linac performance with the common EPID detector.

## 1. Introduction

The dosimetric characteristics for linacs with the same vendor and nominal energy are known to be very similar [1]. The IROC data measured on over 50 Varian linacs demonstrated that all dosimetric data were within the clinical acceptance range of less than 1% [2]. Benchmarking that two machines are dosimetrically equivalent is a desired methodology that is performed between two or more linear accelerators to ensure patient portability. One of the advantages of beam benchmarking or ensuring that linear accelerators are dosimetrically equivalent is the improved efficiency and flexibility in patient treatments for institutions with multiple linacs where patient treatment can be delivered on any unit whose performance is within clinically acceptable limits without any change in the treatment parameters. Validation of beam performance is usually performed during commissioning and requires acquisition and processing of a significantly large amount of machine beam data [3]. The beam data measurements including percentage depth dose (PDD) and beam profiles are labor intensive, and the entire process is one of the most complex and error-prone in radiation oncology [4]. Incorrect output factors and percentage depth doses were identified as some of the causes of failures [4]. In addition, the beam data measurements require expensive beam scanning equipment.

In this study, we present a novel method to benchmark the linac beam characteristics using pixel sensitivity map (PSM) corrected images measured with electronic portal imaging devices (EPIDs) and to evaluate the feasibility of EPID-based validation of beam properties. With excellent dosimetric characteristics, EPID has been widely used for patient dosimetry measurements and machine QA during the past decades [5]. There have been numerous publications on the use of EPID for QA of virtually every aspect of linac performance, and it has been recently used for all EPID daily QA [6] and rapid acceptance tests of linacs [7]. To utilize EPID for linac beam measurements and benchmarking linac performance, two major questions need to be answered: 1) Can EPID be used to measure the beam energy or quality? 2) Does the beam profile depend on the difference in response of individual pixels of EPID panels, as the EPID equipped on each linac may have a different response?

It is known that profiles or off-axis ratios (OAR) are sensitive metrics for verification of beam energy for both flattened and unflattened photon beams. The OAR or flatness measured either on the orthogonal axes or the diagonals was proposed as a sensitive and efficient way to monitor photon beam energy constancy [8]. Recently we have used EPID profiles for verification of beam energy and correlated the profile changes with the changes in PDD [7]. All these studies have shown that EPID can be used for the constancy check of beam quality. Unlike the conventional flood field calibration of EPID, the PSM approach is independent of beam profiles and separate detector sensitivity and beam profiles. In this study we improved the previously developed PSM approach [9, 10] based on an alternating beam and dark field technique (ABDF) and further removed the uncertainty of pixel response. The PSM-corrected EPID images were used to compare the dosimetric characteristics and for beam benchmarking.

## 2. Materials and Methods

Three Varian TrueBeam linacs and one Edge linac were recently commissioned. All these four linacs are equipped with aS1200 EPID panels. The water scan measurements of PDD, profiles, and output factors were performed to verify that the four linacs were dosimetrically equivalent.

Traditional flood field (FF) calibration is not an ideal method for dosimetric purposes, since it removes the beam profile characteristics from the image. The PSM provides the relative gain correction factor for each pixel and is applied to correct raw images. The PSM method provides several advantages: 1) Separation of detector properties (sensitivity) and beam profile; 2) Calibration is independent of the beam profile (no water tank measurements needed); 3) Calibration is independent of imager position. Figure 1 shows the principle process in the derivation of the PSM by acquiring several sets of EPID images with the panel shifted to different positions between each irradiation. Three open fields with the largest field size were first delivered with EPID shifts along the lateral and longitudinal directions by 50 pixels to calculate an initial estimate of the sensitivity map using the previously developed PSM algorithm [9, 10]. Two EPID images were then acquired at different vertical shifts. The initial PSM was then applied to the two images by considering the inverse square correction. The initial PSM was corrected to minimize the difference of the two PSM-corrected images and to generate the intermediate PSM. When finished, the two PSM-corrected scaled images should be equal (except for the scaling). During Step 3 four small images were acquired with the same open field at four different quadrants of EPID. The four images were corrected using the intermediate PSM. The difference between any "small" images was fitted using a polynomial function for final tuning of the intermediate PSM. When done, the four small images should be equal after PSM correction.

After the PSM was calculated for each linac, an open field of 20×20 $cm^2$ was acquired to validate the profiles, and the raw images were corrected using the final PSM. To quantify the dosimetric equivalence, the 1D gamma analysis with 1mm/1% was compared for these machines. The output factors were also measured with EPID from 2×2 to 40×40 $cm^2$ open fields to compare for all three energies among these four linacs.

It is well known that the PSM procedure relies on the repetition of the same beam for all steps of this process [11]. The PSM procedure assumes no change in beam fluence for subsequent deliveries. Any variation of beam fluence can lead to systematic error propagation. To achieve the highest possible repeatability and avoid errors due to image lag and variation of fluence, the calibration images were acquired with ABDF. XML-scripts (Varian TrueBeam Developer Mode 2.5) were developed to define the imaging acquisition mode named "ABDF technique" that automated the entire acquisition process. During the beam-on time, 2 MUs were delivered for each beam-on image with modulated dose rate and synchronized acquisition to ensure that the maximum signal was derived without saturating the imager. The dark fields taken during the beam-hold period were later subtracted from the raw images to eliminate the background noise and residual signal when radiation had ceased. This process was repeated 100 times. The first 40 pairs of images were discarded to avoid fluence from a previous irradiation field. From the remaining 60 pairs of images, the average image with beam and average image without beam was calculated. The advantage of the ABDF technique was to eliminate the ghosting effects for each frame and therefore reproduce the true pixel signal per frame. A recursive algorithm coded was used to derive the PSM [9, 10].

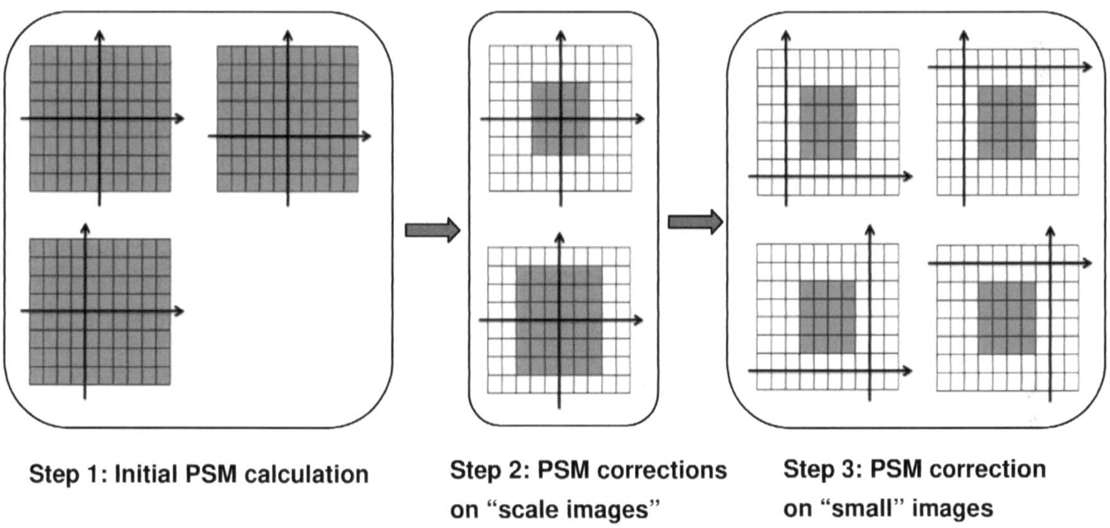


**Figure 1.** Sketch of three steps of calibration to derive PSM. Step 1: The panel along lateral and longitudinal directions for initial PSM calculation; Step 2: The vertical panel shift to correct gradient errors in the sensitivity; Step 3: the panel shifts to acquire images at the four different quadrants to correct any remaining errors.

## 3. Results

Figure 2 shows the comparison of raw image and PSM corrected image acquired with the 20×20 cm$^2$ open field for 6 MV from one of the linacs. The raw image shows stripe patterns from the line readout mechanism along the longitudinal direction. The PSM-corrected image does not show this pattern and is more uniform. Figure 3 shows the beam profiles of the raw image for 6 MV from the four linacs. Though the water scans are equivalent, without correction, the raw beam profiles in both crossline and inline directions do not agree across the four machines. "Wiggling" and "spikes" in the beam profiles due to stripe patterns and noise are present in the profiles.

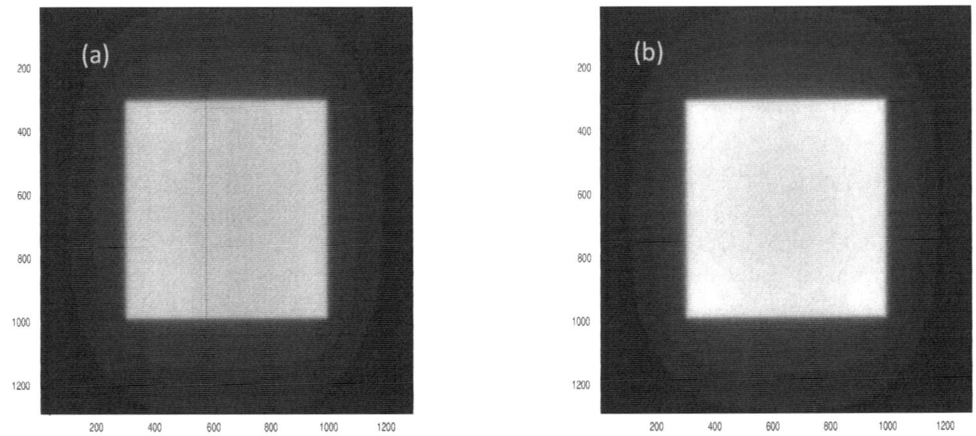


**Figure 2.** Raw image (a) and PSM corrected image (b) for one of the linacs.

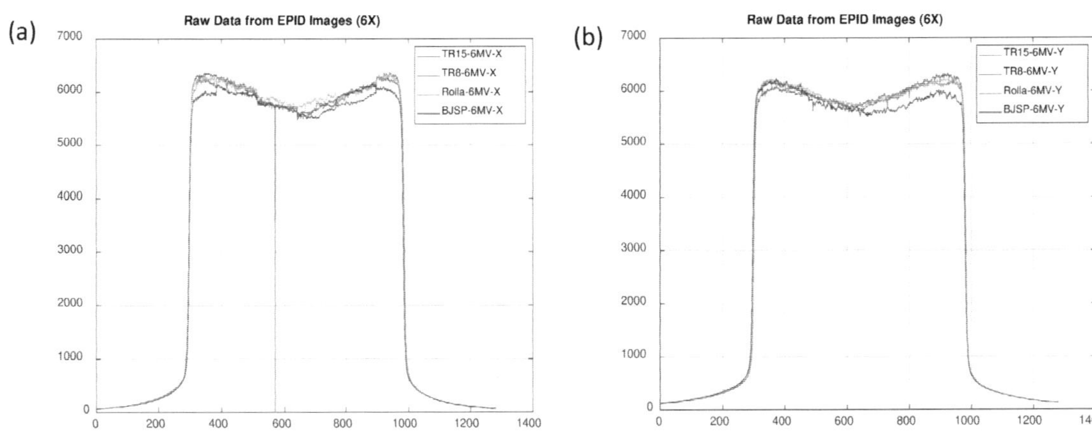


**Figure 3.** Beam profiles of uncorrected EPID images. (a) crossline; (b) inline.

Figure 4 shows the profiles of PSM corrected images with 20×20 $cm^2$ of four linacs for three photon energies: 6 MV(a), 10 MV (b) and 6 FFF (c). The PSM corrected all the artifacts and noise and smoothed out the beam profiles. The 1D Gamma of the PSM-corrected profiles between any two linacs showed 100% passing rate for 6 MV and 6 FFF and >97% for 10 MV with 1mm/1% criteria. The four clinically equivalent linacs with less than 1% variation in PDD, profiles and output factors measured in water scans show excellent agreement in the EPID profile measurements.

The output factors measured from 2×2 $cm^2$ to 40×40 $cm^2$ with EPID are shown in Figure 5. The maximum difference of output factors was 0.18% among all the measurements except 2×2 $cm^2$ with a maximum difference of 0.6%.

The novel PSM calibration approach and EPID-based processes for beam benchmarking take only a few minutes for each beam energy and significantly improve the efficiency compared with water scanning.

IOP Conf. Series: Journal of Physics: Conf. Series **1305** (2019) 012063 doi:10.1088/1742-6596/1305/1/012063

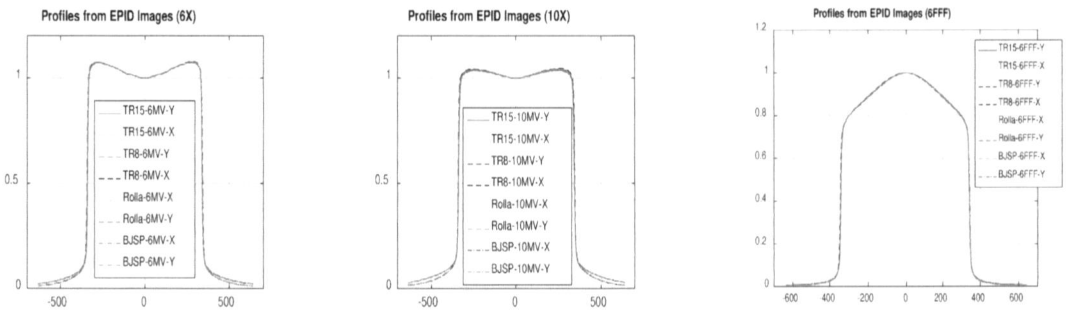


**Figure 4.** Beam profiles of PSM-corrected EPID images. (a) 6 MV; (b) 10 MV; (c) 6 FFF.

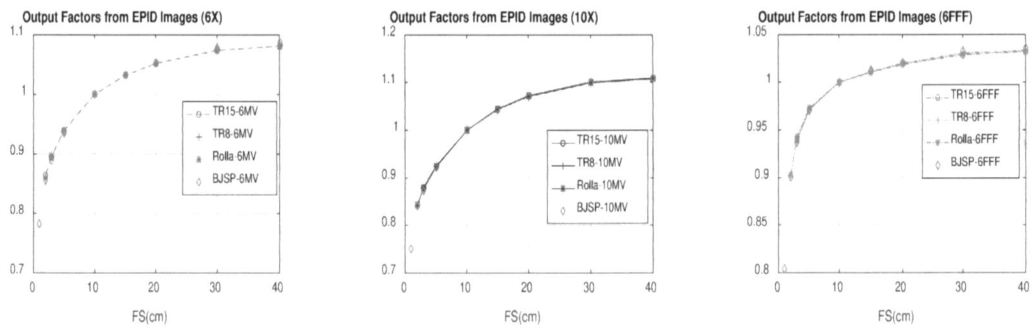


**Figure 5.** Output factor comparison measured with EPID. (a) 6 MV; (b) 10 MV; (c) 6 FFF

## 4. Conclusion
We show that the beam profiles derived from the EPID measurement on four linacs are benchmarked after the PSM normalization. PSM-corrected EPID images enable the possibility of benchmarking linac performance with the common EPID detector.

## 5. References
[1] Jan H *et al* 2007 *Med. Phys.* **34** 2917-27
[2] Hyun C S *et al* 2005 *Med. Phys.* **32** 37-48
[3] Das I J *et al* 2008 *Med. Phys.* **35** 4186-215
[4] Ibbott G S *et al* 2008 *Int. J. Radiat. Oncol. Biol. Phys.* **71** S71-S5
[5] Vial P *et al* 2008 *Med. Phys.* **35** 4362-74
[6] Baozhou S *et al* 2015 *Med. Phys.* **42** 5584-94
[7] Sridhar Y *et al* 2017 *Med. Phys.* **44** 3393-406
[8] Song G *et al* 2016 *J. Appl. Clin. Med. Phys.* **17** 242-53
[9] Sun B *et al* 2015 *J. Phys.: Conf. Ser.* **573** 012041
[10] Cai B *et al* 2018 *J. Appl. Clin. Med. Phys.* **19** 73-85
[11] Boriano A *et al* 2013 *J. Appl. Clin. Med. Phys.* **14** 234-50

# Quantitative evaluation of transmission EPID daily imaging on a Halcyon Linac

**P Jin[1], Y H Xie[1,2], M Huang[1,3] and T C Zhu[1]**
[1]Department of Radiation Oncology, University of Pennsylvania, Philadelphia, PA 19104, USA
[2]Department of Radiation Oncology, Massachusetts General Hospital, Boston, MA 02114, USA
[3]Department of Radiation Oncology, Duke University, Durham, NC 27710, USA

E-mail: Ping.Jin@uphs.upenn.edu

**Abstract.** The recently published AAPM TG100 has advocated streamline IMRT/VMAT QA procedure based on Failure Mode and Effects Analysis (FMEA) criteria, i.e., the product of incidence severity, occurrence probability, and detection probability. Transmission EPID has proven to be able to detect dose errors at a detection probability of 60% vs. those based on pre-treatment patient specific QA, at a detection probability of less than 5%. The objective of this study is to retrospectively evaluate the gamma index (3%, 3mm) using composite transmission EPID images for various treatment sites on a Halcyon Linac. Transmission EPID images are recorded during treatment and compared to first day EPID as reference. Over 1000 treatment fractions, the average gamma index and standard deviation of gamma for various treatment sites are 99.04±1.64, 99.95±0.093, 100±0 and 99.7±0.567 and 95.65±3.57 for head and neck, brain, prostate, pelvis and breast respectively. In conclusion: Our data show very high inter-treatment consistency for most treatment sites except breast. For breast cases, using second day treatment images as reference, the gamma index is much higher and standard deviation is lower than using day1 images as reference. Overall a threshold of gamma passing rate of 95% is a reasonable value.

## 1. Introduction

With the increased complexity introduced by ever increasing popularity of IMRT and VMAT treatments, in-vivo patient QA using on-board EPID during each RT treatment becomes popular [1-7]. The recently published AAPM TG100 has advocated streamline IMRT/VMAT QA procedure-based FEMA criteria, ie, the product of incidence severity, occurrence probability, and detection probability [8]. Transmission EPID [9-11] has proven to be able to detect dose errors at a detection probability of 60% vs. those based on pre-treatment patient specific QA, at a detection probability of less than 5% [12]. The variation of the daily treatment transmission images reflects possible changing in patient set up [13], patient anatomy changes and unexpected errors in beam delivery [14], which could potentially cause significant dose error in the patients. Halcyon Linac records the transmission portal images and exit dosimetry of each field during actual patient RT treatment using on board EPID. A retro-prospective gamma analysis of the composite daily treatment transmission images with day 1 treatment images as reference was conducted in this study for different treatment sites (H&N, Brain, Breast, Prostate and Pelvis).

## 2. Materials and Methods

Varian Halcyon Linear accelerator is equipped with EPID imager which records the transmission images during treatment without any additional effort or time for therapists. Before this study, such existing information were not utilized for patient QA purpose. In this retrospective study, the existing EPID images from daily treatment were used to create the composite images, then they are compared to a reference images, such as day 1 or day 2 EPID images of the plan using gamma index. The Gamma criteria is 3mm and 3%, and it is calculated using Varian Eclipse software. The threshold is 5%, the ROI is MLC CIAO + 1cm, where MLC CIAO +1cm is a complete irradiated area outlined by MLC plus 1cm margin.

## 3. Results and discussion

### *3.1. Head and neck cases*

19 patients with total of 264 fields were examined for head and neck cases. The average gamma of the all head neck plans = 99.04, SD = 1.67. For cases 1 and 2 (see Fig. 1), gamma passing rate variation is mostly due to patients' anatomy change. Excluding case 1 and 2, for the rest of 17 patients, the gamma index passing rate based on 209 treatments is 99.6% with SD 0.3 (see Table1), indicating that head and neck patient treatment reproducibility is very high.

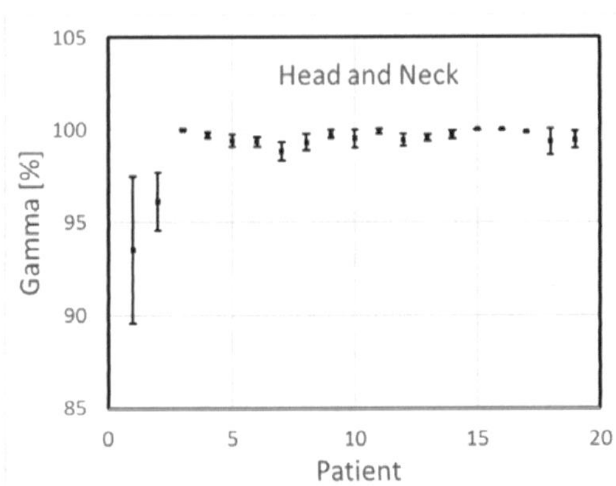


**Figure 1.** Mean gamma passing rate and standard deviation (SD) for patients treated to Head and Neck.

**Table 1**. Summary of Gamma Index for Various Treatment Sites.

| Site | # of Patients | # of Treatments reviewed | Avg. Gamma | SD | Minimum Gamma | Possible Cause of Gamma deviation |
|---|---|---|---|---|---|---|
| Head & Neck | 19 | 264 | 99.04 | 1.67 | 75.5 | Anatomy change |
| Head & Neck 2* | 17 | 209 | 99.58 | 0.32 | 95.9 | *Negligible variation |
| W. Brian + P.B. | 6 | 88 | 99.94 | 0.09 | 98.2 | Negligible variation |
| Breast | 10 | 280 | 95.65 | 3.57 | 75.5 | Setup +breathing |
| Prostate | 1 | 44 | 100.00 | 0.00 | 100.0 | Negligible variation |
| Pelvis | 4 | 108 | 99.70 | 0.57 | 95.5 | Setup + Gas |

*Head and Neck 2 excluded first two patients which have visible anatomy change, so total case number in this group is 17 patients.

### *3.2. A Case study*

We noticed gamma index for case #1 decreased after 17 fractions (see Fig. 2). Patient's left side anatomy had visible changes based on MVCBCT difference images (see Fig.3 a, b). According to the evaluation CT scan and evaluation plan, patient PTV 95% coverage reduced from 7000cGy to 6622cGy (Fig. 4), which is about 6% decreasing. Figure 5 (b) is the setup error trend vs time. The treatment # 2 and treatment # 27 had similar setup errors (less than 3mm in longitudinal, lateral and vertical). However their gamma index have significant difference (Fig. 5a), 100% (Fig. 5c) vs 82.7 %. (Fig. 5d) This indicates that the gamma variation is not from setup error but anatomy change, which is consistency with MVCBCT vs Plan CT images. Re-plan was performed later for that patient.

Suggestion for clinical application: for head and neck case, if the gamma index is below 90%, an evaluation scan and planning would be need.

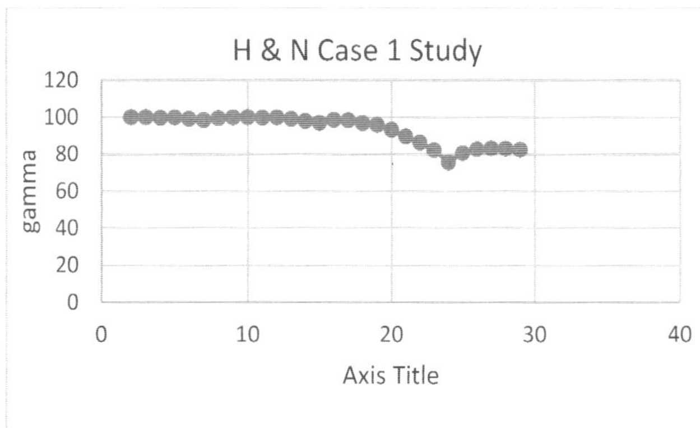


**Figure 2.** Composite gamma vs. time for Patient #1.

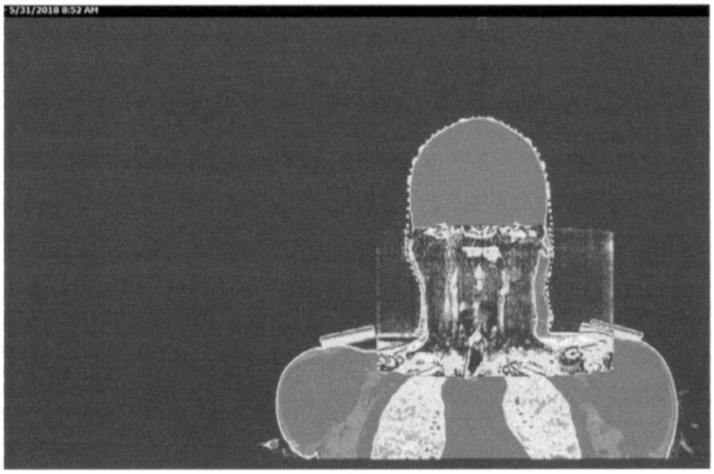


(a)

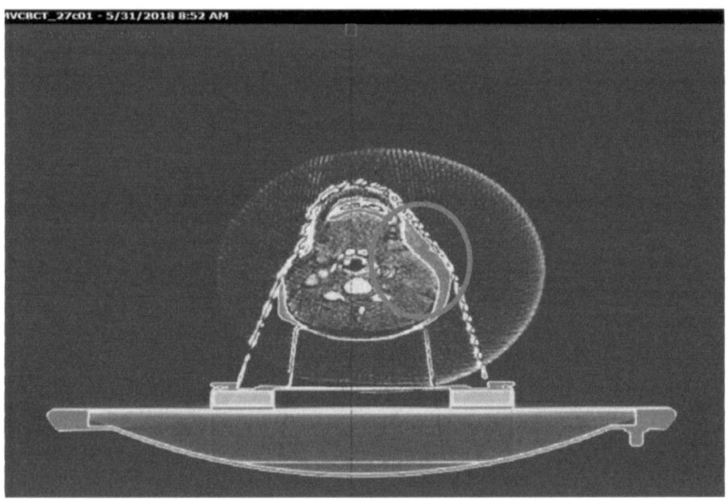


(b)

**Figure 3.** (a) sagittal view of difference images between MVCBCT of treatment 27 and planning CT for patient #1; (b) axial view of difference image between day 27th MVCBCT and planning CT.

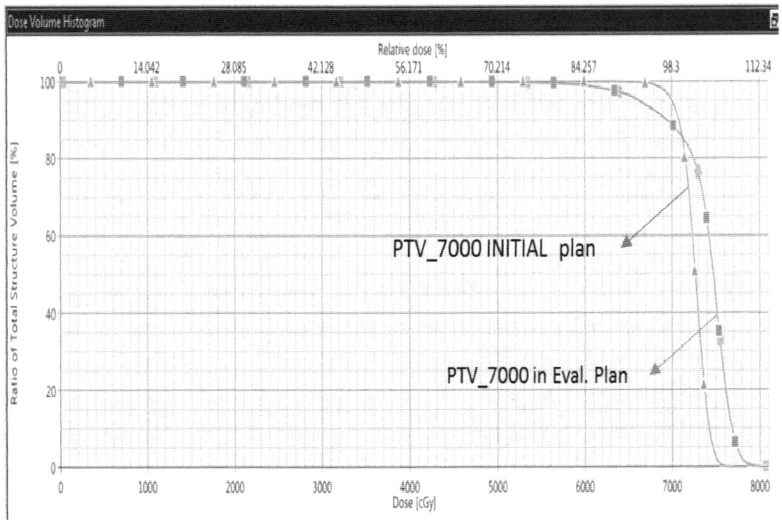


**Figure 4.** DVH of PTV for patient #1 from original CT scan (red color) and PTV from evaluation CT scan (blue color), PTV dose decreased by over 6%.

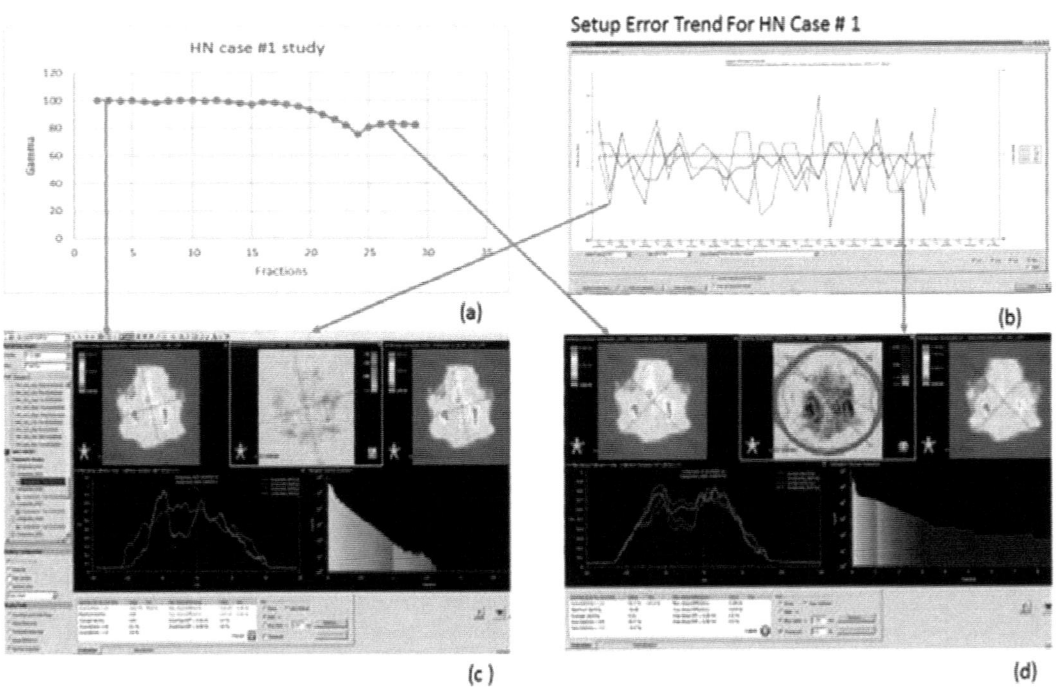


**Figure 5.** (a) Gamma index for each day's composite Gamma compared to day 1 composite Gamma vs. fraction number for patient #1; (b) setup error trend vs. time; (c) Gamma index (100%) of fraction #2; (d) Gamma index (82.7%) of fraction #27.

### *3.3. Brain*

Six patients with whole or partial Brian with 88 fractions were studied (see Table 1). The average gamma passing rate of all the plans is 99.94% with SD= 0.09%. The results indicate high treatment setup reproducibility for the brain cases (see Fig. 6a).

### *3.4. Pelvis and Prostate*

Four pelvis cases with 108 treatments were reviewed (see Table 1). The average gamma for the 108 treatment is 99.69% with SD=0.56%. The overall reproducibility of pelvis treatment is excellent (see Fig. 6b). One prostate patient with 4 plans and 44 treatments were reviewed (see Fig. 6c). The overall average gamma is 100% associated with a 0% SD. This indicated prostate treatment reproducibility between treatments is as excellent as brain cases.

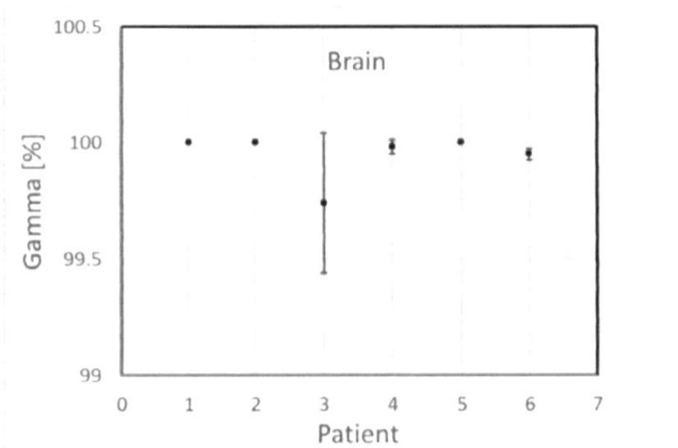


(a)

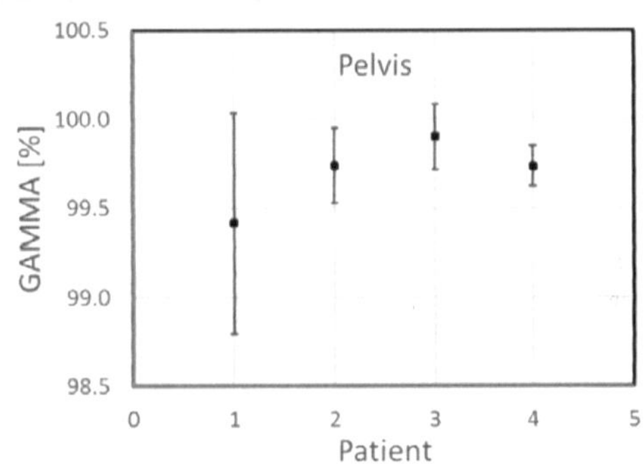


(b)

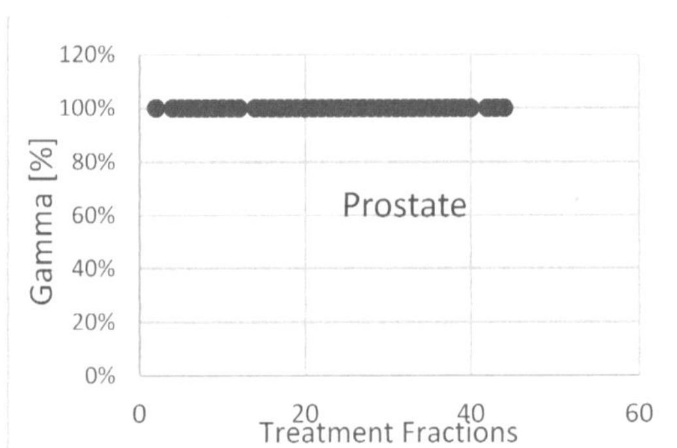


(c)

**Figure 6.** Mean gamma passing rate and standard deviation (SD) for (a) Brain; (b) Pelvis; and (c) gamma index for a prostate case in 44 fractions in 4 plans.

### *3.5. Breast cases*

Ten patients with total of 27 plans and 280 fractions of treatment were evaluated for breast treatment (see Table 1). Nine cases were treated with supine setup and 1 with prone setup. Four patients are with DIBH and six patients are using free breathing. Among all ten patients, three are treated to right breast, three to left breast patients, two to bilateral breasts and two to left chest-wall. The average gamma index for breast treatment is 95.65% with SD= 3.7%. The results show much high variation in the gamma index than in head and neck or brain cases.

Notice that by selecting day 2 images instead of day1 images as reference for breast cases, higher gamma index and lower standard deviation are obtained (see Fig. 7). This indicates that the day 2 treatment is more consistent with the remaining treatment than the day 1 treatment (see Fig. 7).

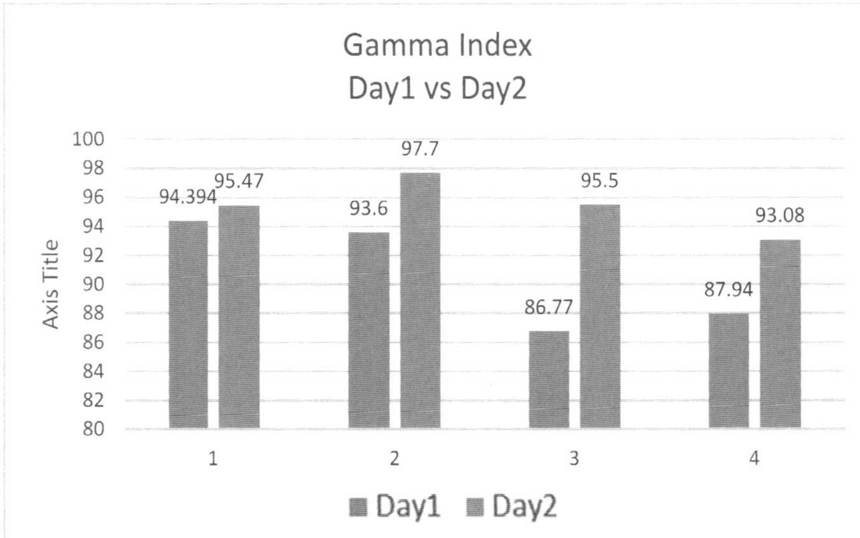


(a)

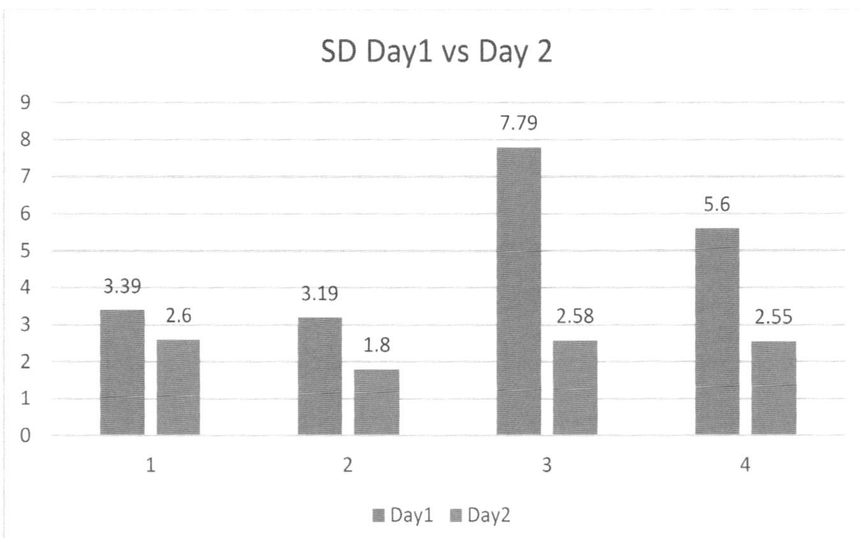


(b)

**Figure 7.** (a) Average Gamma passing rate and (b) Standard Deviation for Breast patients analyzed with Day1 as baseline vs Day2.

There is a slight gamma improvement with free breathing than DIBH with similar amount of standard deviation in breast treatment (see Fig. 8). For all breast cases, the average gamma for right breast is slightly higher than left breast. But there is no significant difference in standard deviation. The same phenomenon was observed in the bilateral breast cases.

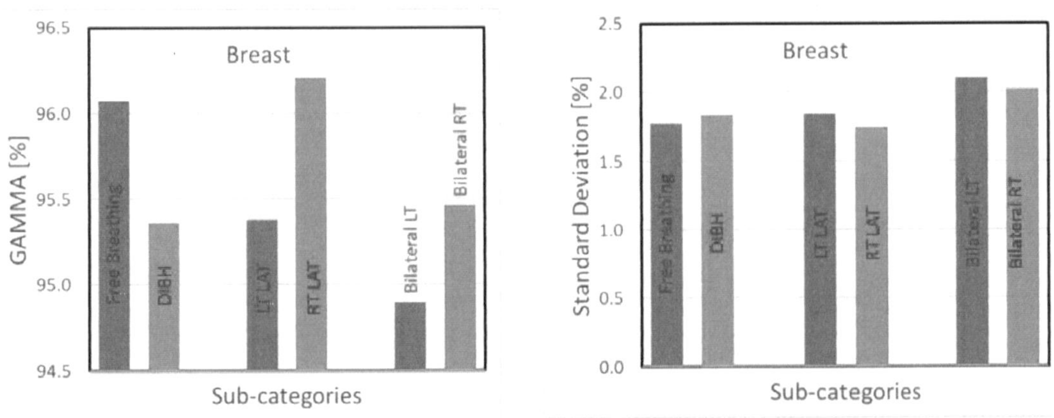


**(a)** **(b)**

**Figure 8.** (a) Average Gamma passing rate and (b) Standard Deviation for breast patients with different immobilization and different laterality.

## 4. Conclusion

We have analyzed gamma index passing rate over 1000 treatment fractions for different sites and found that different sites have very different mean gamma passing rate (See Table 1). We found a good threshold value of gamma passing rate of 95% to be a reasonable value for all sites except breast. For breast, 90% threshold is a reasonable value. Furthermore, second day images are a better reference image than first day of treatment in gamma evaluation for breast. For the two head and neck patients, who had noticeable anatomy variation, the gamma passing rate appears to be lower than the rest of the 17 patients.

## 5. References

[1] Heijmen B J M *et al* 1999 *Phys. Med. Biol.* **40** 1943-55
[2] Bedford J L *et al* 2018 *Phys. Med. Biol.* **63** 025008
[3] Celi S *et al* 2016 *J. Appl. Clin. Med. Phys.* **17** 262-76
[4] Partridge M *et al* 2002 *Med. Phys.* **29** 1847-58
[5] Wendling M *et al* 2006 *Med. Phys.* **33** 259-73
[6] Peca S and Brown D W 2014 *J. Appl. Clin. Med. Phys.* **15** 4752
[7] Vial P *et al* 2008 *Med. Phys.* **35** 4362-74
[8] Huq M S *et al* 2016 *Med. Phys.* **43** 4209-62
[9] Woodruff H *et al* 2015 *Int. J. Rad. Oncol. Biol. Phys.* **93** 516-22
[10] Piermattei A *et al* 2007 *Phys. Med. Biol.* **52** 5101-17
[11] Kavuma A *et al* 2010 *Phys. Med. Biol.* **55** 435-52
[12] Ford E C *et al* 2012 2012 *Int. J. Rad. Onc. Biol. Phys.* **84** 263-9
[13] Li J *et al* 2018 *PLoS* **13** e0192686
[14] Peca S *et al* 2017 *Technol. Cancer Res. Treat.* **16** 956-63

IOP Conf. Series: Journal of Physics: Conf. Series **1305** (2019) 012065 doi:10.1088/1742-6596/1305/1/012065

# A sliding-window approach for improved VMAT dose calculation accuracy

**Haitham N Alahmad, Ji-Yeon Park, Nicholas J Potter, Bo Lu, Guanghua Yan, Chihray Liu and Jonathan G Li**
Department of Radiation Oncology, University of Florida College of Medicine, Gainesville, FL 32610-0385, USA

E-mail: lijong@shands.ufl.edu

**Abstract**. A sliding-window (SW) methodology for VMAT dose calculation was developed. For any two adjacent VMAT control points (CPs) n and n+1, the dose distribution was approximated by a 2-CP SW IMRT beam with the starting MLC positions at CP n and ending MLC positions at CP n+1, with the gantry angle fixed in the middle of the two VMAT CPs. Therefore, for any VMAT beam with N CPs, the dose is calculated with N-1 SW beams. VMAT plans were generated for ten patients in Pinnacle using 4° gantry spacing. For each patient, the VMAT plan was converted to a SW IMRT plan and dose was re-calculated. Another VMAT plan, with 1° gantry spacing, was created by interpolating the original VMAT beam. The original plans were delivered on an Elekta Versa HD and measured with Mapcheck2 using an in-house developed subarc method. For both the isodose distribution and DVH, there were significant differences between the original VMAT plan and either the SW or the interpolated plan. However, they were indistinguishable between the SW and interpolated plans. The average passing rate between the original VMAT plan and measurements was 84%. For both the interpolated and SW plans, the average passing rate was 96%. We conclude that the proposed SW approach improves VMAT dose calculation accuracy without increase in dose calculation time.

## 1. Introduction

Volumetric modulated arc therapy (VMAT) has gained widespread clinical use in recent years due to its comparable and sometimes superior plan quality compared to conventional IMRT with shorter delivery time and the potential to reduce intra-fractional organ motion [1-3]. In VMAT delivery, radiation is always on while the gantry continuously rotates around the patient. In addition, the linac control constantly alters the gantry rotation speed, dose rate, and MLC leaf positions to provide the desired intensity (fluence) modulation.

A VMAT plan consists of an array of control points (CPs). Similar to conventional sliding-window (SW) IMRT, a VMAT beam has a minimum of two CPs. Each CP specifies a machine state, which includes gantry and collimator angles, MLC leaf positions, and the total monitor units (MU) delivered. For any two adjacent VMAT CPs n and n+1, the linac will move from n's machine state to the n+1's machine state with constant speeds and dose rate so that all machine parameters will start and finish at the same time. The MLC movement during a VMAT delivery mimics that of a SW IMRT. The continuous VMAT delivery is usually approximated by discrete static beams for dose calculation in treatment planning systems (TPS). In Pinnacle$^3$ TPS (Philips, Fitchburg, WI), dose from CPs n to n+1 is approximated by two static beams with beam parameters at machine states n and n+1, respectively, each with MUs half of the total MUs to be delivered between the two CPs. Intermediate machine states

IOP Conf. Series: Journal of Physics: Conf. Series **1305** (2019) 012065 doi:10.1088/1742-6596/1305/1/012065

during delivery is therefore not accounted for in dose calculation. Large dose calculation errors can be introduced when coarse gantry spacing or large MLC aperture difference between two adjacent CPs are involved.

To reduce the magnitude of error introduced in VMAT dose calculation, finer gantry spacing is needed. Pinnacle[3] allows for evenly spaced gantry angles of 2°, 3°, or 4° for VMAT planning based on initial user selection. Since dose calculation time is directly proportional to the number of static beams (or CPs) used to approximate a VMAT beam, gantry spacing selection is a trade-off between planning speed and dose calculation accuracy. Feygelman *et al* [4], in their initial dosimetric evaluation of SmartArc, which is the VMAT planning module in Pinnacle[3], concluded that using a gantry spacing of 4° in the optimization provided a good compromise between calculation speed and accuracy. However, there is still considerable variation in the literature concerning the choice of gantry spacing [4]. It may be necessary to interpolate the VMAT plan onto a finer gantry spacing after optimization for final dose calculation, which can result in sub-optimal plans.

We propose a SW approach for improved VMAT dose calculation accuracy. Since the delivery of a VMAT subarc (the arc of a VMAT beam between two adjacent CPs) is intrinsically a SW delivery, albeit with the addition of gantry rotation, we hypothesize that approximating a VMAT subarc with a fixed-gantry SW beam would introduce smaller errors than with two static beams at the start and end of the subarc. Therefore, a VMAT beam with N CPs can be represented by N-1 SW IMRT beams. Since the dose calculation time of a SW beam is similar to that of a static beam, the SW method does not increase the dose calculation time and can be used *during* optimization for improved plan quality and dosimetric accuracy.

## 2. Methods and Materials

### *2.1. Treatment planning*

Five head and neck (HN) and five SBRT spine cases that were previously treated in our institution were chosen for this study. For each case, a VMAT plan with 4° gantry spacing (VMAT4) was generated using the SmartArc module in Pinnacle[3]. Each plan consisted of a single arc with gantry rotating from 170° to 190° (IEC convention) in counter-clockwise direction, resulting in 86 CPs. All the cases were planned with a 6 MV beam. The MLC leaf motion constraint was unchecked to allow more freedom for optimization, which can lead to large leaf travel distances between adjacent CPs. A 20° collimator angle was used to reduce the impact of the tongue and groove effect. For the HN cases, prescription doses were 70, 63, and 56 Gy in 35 fractions to the high-risk PTV (PTVHR), intermediate-risk PTV (PTVIR), and standard-risk PTV (PTVSR), respectively. For the SBRT spine cases, 35 Gy in 5 fractions were prescribed to the PTV with a 22 Gy dose constraint to the spinal cord.

### *2.2. Converting VMAT plan to SW IMRT plan*

Each VMAT4 plan was converted to a SW IMRT plan (SW plan) using an in-house program developed in MATLAB (Mathworks, Natick, MA). Every two adjacent CPs ($CP_n$ and $CP_{n+1}$) of the VMAT arc were replaced by a fixed-gantry SW beam with two CPs ($CP_{sn1}$ and $CP_{sn2}$). The gantry angle of the SW beam was fixed in the middle of $CP_n$ and $CP_{n+1}$. $CP_{sn1}$ and $CP_{sn2}$ had the same apertures as those of $CP_n$ and $CP_{n+1}$, respectively. $CP_{sn1}$ had a zero dose weight while the dose weight of $CP_{sn2}$ corresponded to the dose to be delivered between $CP_n$ and $CP_{n+1}$ in the VMAT4 plan. This resulted in a SW IMRT plan with 85 beams equally spaced every 4° from 168° to 192° for each VMAT4 plan.

### *2.3. Interpolated plan*

To mimic the dynamic delivery of VMAT beams, a VMAT plan with 1° gantry spacing (VMAT1 plan) was also created by interpolating each of the VMAT4 plans. Three additional CPs were equally inserted between every two adjacent CPs ($CP_n$ and $CP_{n+1}$) of the VMAT4 plan, leading to 341 CPs with 1° gantry spacing in the VMAT1 plan. The MLC leaf positions and dose weights of these new CPs were obtained by linearly interpolating those of $CP_n$ and $CP_{n+1}$ using an in-house Matlab program. These VMAT1 plans were used as reference for evaluation.

### 2.4. *Measurement*

The VMAT4 plans were delivered on a VersaHD Linac (Elekta Limited, Crawley, UK) equipped with agility MLC with 0.5 cm leaf width at isocenter. The measurement verification method followed the VMAT patient-specific QA protocol that is currently implemented in our institution [5]. Using an in-house developed program, the VMAT arc was divided into multiple 60-degree subarcs and repositioned with gantry angles extending from 30° to 330°. This method allowed us to verify VMAT plans with a two-dimensional (2D) detector array (MapCheck2, Sun Nuclear Corp., Melbourne, FL) since the MapCheck2 showed minimal angular dependence in this angular range [5]. The measurement was compared with subarc calculations using VMAT4, VMAT1, and SW plans. Gamma criteria of 3%/3 mm were used for the analysis. This is similar to conventional IMRT per-beam verification with 2D detector arrays where each beam is repositioned to gantry angle 0° for verification.

## 3. Results

### 3.1. *Plan comparison*

Figures 1a and 1b show the dose-volume histograms of a representative HN and spine cases, respectively. In both cases, there is evident discrepancy in PTV coverage and spinal cord sparing between the original VMAT4 and the interpolated VMAT1 plans. On the other hand, the proposed SW plans closely match the VMAT1 plans. Dose within the targets becomes more non-uniform for the SW and VMAT1 plans while dose to cord increases significantly. This is observed in all the cases studied.

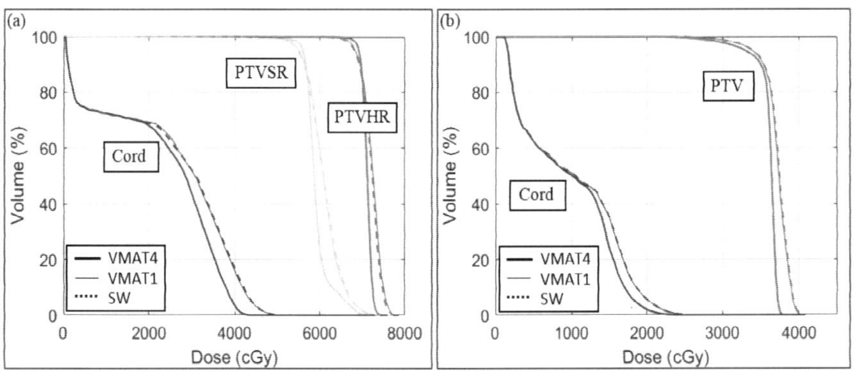


**Figure 1.** Comparison of the dose-volume histograms of a HN (a) and spine (b) cases for the VMAT4, VMAT1 and SW plans.

Tables 1 and 2 compare the dosimetric indices of the VMAT4, VMAT1 and SW plans for the 10 cases. While the differences between VMAT4 and VMAT1 in D95 for the targets are generally on the order of 1-2%, the differences in D20 are higher, reaching 4-5% in some cases. The VMAT1 plans are always hotter and more non-uniform within the targets when compared to the VMAT4 plans. Dose to spinal cord also increased for the VMAT1 plans, with an average increase in the maximum dose to 0.1 cc (D0.1cc) of 12.1% and 5.6% for the HN and spine cases, respectively. On the other hand, the differences between SW and VMAT1 plans are less than 1% in most cases.

### 3.2. *Plan verification*

Table 3 shows the average passing rates for VMAT4, VMAT1, and SW plans when compared with the measurement of VMAT4 delivery. For head and neck cases, the passing rates of the VMAT4 plans are less than 80% while the passing rates of all VMAT1 and SW plans are above 90% and within 2% of each other. For the spine cases, the passing rates for VMAT4 plans increased to above 90% in almost all cases but are still lower than the passing rates of VMAT1 and SW plans. The VMAT1 and SW passing rates are above 98% for all the spine cases.

**Table 1**. Percentage difference in dose indices for the HN cases. D95: dose to 95% of the PTV. D20: dose to 20% of the PTV. D0.1cc: maximum dose to 0.1 cc of the spinal cord volume.

| | | Case # | | | | |
|---|---|---|---|---|---|---|
| Dose Index | | 1 | 2 | 3 | 4 | 5 |
| D95 (PTVHR) | VMAT1 vs VMAT4 | -1.87% | 1.45% | -1.18% | -1.10% | -2.00% |
| | SW vs VMAT1 | 0.28% | -0.34% | -0.32% | -0.60% | -1.34% |
| D95 (PTVSR) | VMAT1 vs VMAT4 | -2.68% | -1.57% | -1.64% | -2.64% | -2.14% |
| | SW vs VMAT1 | -0.62% | -0.25% | -0.36% | 0.18% | -0.66% |
| D20 (PTVHR) | VMAT1 vs VMAT4 | 2.81% | 4.12% | 2.97% | 4.45% | 5.83% |
| | SW vs VMAT1 | 0.18% | -0.03% | -0.40% | -0.47% | -0.38% |
| D0.1cc (Cord) | VMAT1 vs VMAT4 | 11.49% | 10.28% | 15.63% | 11.55% | 11.43% |
| | SW vs VMAT1 | 0.59% | 0.09% | -0.39% | 0.94% | 0.82% |

**Table 2**. Percentage difference in dose indices for the spine SBRT cases. D95: dose to 95% of the PTV. D20: dose to 20% of the PTV. D0.1cc: maximum dose to 0.1 cc of the cord volume.

| | | Case # | | | | |
|---|---|---|---|---|---|---|
| Dose Index | | 6 | 7 | 8 | 9 | 10 |
| D95 | VMAT1 vs VMAT4 | 1.43% | 1.37% | 0.29% | 0.86% | 1.60% |
| | SW vs VMAT1 | -0.28% | 0.51% | 0.11% | -0.28% | 0.06% |
| D20 | VMAT1 vs VMAT4 | 4.29% | 3.05% | 2.71% | 3.82% | 4.75% |
| | SW vs VMAT1 | -0.10% | 0.90% | 0.58% | 0.11% | 0.15% |
| D0.1cc (Cord) | VMAT1 vs VMAT4 | 7.14% | 7.53% | 3.20% | 4.70% | 5.25% |
| | SW vs VMAT1 | 0.00% | 1.25% | 1.14% | -0.37% | -0.20% |

**Table 3.** Average passing rates (%) of VMAT4, VMAT1, and SW plans when compared with the measurement of VMAT4 delivery. Cases 1-5 are HN cases and cases 6-10 are spine SBRT cases.

| Case# | 1 | 2 | 3 | 4 | 5 | 6 | 7 | 8 | 9 | 10 |
|---|---|---|---|---|---|---|---|---|---|---|
| VMAT4 | 74.9 | 78.6 | 78.3 | 74.8 | 70.8 | 87.3 | 95.0 | 97.4 | 91.8 | 94.0 |
| VMAT1 | 94.8 | 94.4 | 96.1 | 94.6 | 94.5 | 98.2 | 99.7 | 99.4 | 99.7 | 98.5 |
| SW | 95.4 | 94.5 | 97.1 | 93.1 | 93.0 | 98.4 | 99.7 | 99.4 | 99.6 | 98.0 |

## 4. Discussion

The low passing rates of most of the VMAT4 plans indicate that the TPS dose calculation does not reflect the actual continuous delivery. This is especially true when MLC aperture shapes are large and change drastically from one CP to the next, as in the HN cases. Approximating continuous delivery of VMAT beams with discrete beams with coarse gantry spacing does not account for the gradual intensity modulation between CPs. Park *et al* [6] concluded that this dose discrepancy due to under-sampling is strongly correlated with the aperture width, the distances that MLC leaves travel between apertures, and the width of the dose kernel. Although the passing rates of VMAT4 plans for the spine cases are higher due to the smaller target sizes, which resulted in smaller apertures and smaller MLC leaf travel distances between apertures, the differences between VMAT4 and VMAT1 plans are still clinically unacceptable (see figure 1(b) and table 2). Using finer gantry spacing for planning would lead to improved passing rates, but would also increase the optimization time significantly, as the dose calculation time is directly proportional to the number of CPs.

Intrinsically, each VMAT subarc is delivered in a SW fashion. Therefore, a SW beam, which accurately accounts for the MLC movement between CPs, would better approximate the continuous delivery of a VMAT subarc than two discrete static beams. This is reflected in the excellent agreement of dosimetric endpoints between the SW and VMAT1 plans and their high passing rates. The approximation made in the SW plans, i.e., by collapsing the 4° gantry rotation within a VMAT subarc to a fixed-gantry beam in the middle of the subarc, seems to introduce negligible error in dose distribution. This is not surprising if we compare the proposed method with that of helical Tomotherapy dose calculation. In Tomotherapy, each full arc is divided into 51 fixed-gantry projections with an even larger gantry spacing (7°) and good dosimetric accuracy is achieved [7-9]. Interestingly, each Tomotherapy projection uses an intensity map resembling that of a step-and-shoot IMRT beam, rather than a SW beam. This can be explained by the fast movement of the binary collimators of the Tomotherapy machine (~23 ms open/close), leading to almost instantaneous MLC aperture change. Therefore, it is sufficient to stack up the intensities of individual apertures within each projection without the need to consider leaf movement. The fluence modulation during a projection is therefore faithfully reproduced in TPS dose calculation and it is afforded a coarser gantry spacing. Whereas in VMAT, because the linac MLC moves in a linear, SW fashion with a much slower speed (6 cm/sec maximum), our method employing SW beams captures the fluence modulation between any two adjacent CPs more accurately, resulting in better dosimetric accuracy.

A further improvement in dosimetric accuracy may be achieved by using the "small-arc approximation' as described by Webb & McQuaid [10]. Under "small-arc approximation", gantry rotation during a VMAT subarc delivery is equivalent to a moving MLC carriage with a fixed source position from the viewpoint of voxels close to the isocenter. They proved that dynamic MLC delivery technique remains valid when the MLC carriage is also moving. The MLC positions of the SW IMRT beams can be modified to incorporate the MLC carriage movement. This approximation breaks down for regions far removed from the isocenter.

Since the dose calculation time of a SW beam is similar to that of a static beam in Pinnacle[3], our method can potentially be used *during* optimization, resulting in improved plan quality and dosimetric accuracy without sacrifice in planning time.

## 5. References

[1] Otto K 2008 *Med. Phys.* **35** 310-7
[2] Rao M *et al* 2010 *Med. Phys.* **37** 1350-59
[3] Holt A *et al* 2011 *Int. J. Radiat. Oncol. Biol. Phys.* **81** 1560-67
[4] Feygelman V *et al* 2010 *J. Appl. Clin. Med. Phys.* **11** 99-116
[5] Park J-Y *et al* 2016 *Med. Phys.* **43** 3525
[6] Park J-Y *et al* 2017 *Med. Phys.* **44** 2096-114
[7] Balog J *et al* 2003 *Med. Phys.* **30** 3097-106
[8] Fenwick J D *et al* 2004 *Phys. Med. Biol.* **49** 2933-53
[9] Gibbons J P *et al* 2009 *J. Appl. Clin. Med. Phys.* **10** 103-19
[10] Webb S and McQuaid D 2009 *Phys. Med. Biol.* **54** 4345-60

IOP Conf. Series: Journal of Physics: Conf. Series **1305** (2019) 012066 doi:10.1088/1742-6596/1305/1/012066

# Multi-scale dosimetry with multi-scale Chinese reference phantoms

**Rui Qiu[1,2], Zhen Wu[3], Chunyan Li[3], Li Ren[1,2], Wenjing Wang[1,2], Ruiyao Ma[1,2], An kang Hu[1,2], Hongyu Zhu[1,2], Junli Li[1,2]**
[1]Department of Engineering Physics, Tsinghua University, Beijing, China
[2]Key Laboratory of Particle & Radiation Imaging (Tsinghua University), Ministry of Education, Beijing, China
[3]Nuctech Company Limited, Beijing, China

E-mail: qiurui@mail.tsinghua.edu.cn

**Abstract**. Recent studies on the development and applications of Chinese reference phantoms are reported in this paper. The Chinese adult reference male (CRAM) and female (CRAF) phantoms had been developed in previous work. A Chinese male phantom library was constructed with 7 different heights ranging from 155 cm to 185 cm and 12 phantoms with different total body masses in each height. Chinese paediatric reference phantoms were constructed for 3 months, 1 year, 5 years, 10 years, and 15 years male and female respectively. Detailed organ models were developed, including a set of microscopic skeletal models, a set of detailed breast models, a detailed human respiratory tract model and a detailed eye model. The phantoms were applied in radiation protection, including dose estimation from external exposure and internal exposure, in normal situations and in accidental situations, and detailed dose distributions in radiation-sensitive organs. The phantoms were also applied in medical imaging field such as dose estimation in mammography, CT and X-ray radiology, and could be potentially used in the secondary cancer estimation induced by radiation therapy.

## 1. Introduction

Stylized phantoms are the first generation of computational phantom, which are comprised of geometric surfaces described by mathematical equations. Though stylized phantoms have advantage in calculation speed, they are unable to provide the fine structure of human organs. Thus, the voxel phantoms have been developed. As the second-generation computational phantom, voxel phantoms can represent detailed body structures, and play important roles in dosimetry calculation. However, the voxel phantom is defined as a 3D-matrix, and it is inflexible to deform, so the boundary representation (BREP) phantoms (also called hybrid or surface phantoms) have been developed as the third-generation computational phantoms. The surface phantom not only has accurate anatomical structures but can also be flexibly adjusted to different body sizes.

The Chinese adult reference male (CRAM) and female (CRAF) phantoms developed by Tsinghua University [1, 2] represent the Chinese reference anatomical and physiological characteristics. The voxel- based CRAM and CRAF models are segmented into 80 and 79 organs and tissues respectively, including all the radiation-sensitive organs with specified weighting factors according to ICRP publication 103 [3]. With the development and applications of Chinese reference phantoms, recent studies on radiation protection, medical imaging, and so on have been reported. Latest development and applications of the Chinese reference phantoms are reviewed in this paper.

## 2. Phantom development

### *2.1. Chinese adult male phantom library*

Based on the CRAM voxel phantom, a Chinese reference adult male polygon surface (CRAM_S) phantom was constructed with the organ models adjusted to match the Chinese reference data. CRAM_S phantom was then transformed to sitting posture (CRAM_Ssit). Referring to the mass and height distribution of the Chinese adult male, a phantom library containing 84 phantoms were constructed by deforming the reference surface phantom as shown in Figure 1. Phantoms in the library have 7 different heights ranging from 155 cm to 185 cm, and there are 12 phantoms with different total body masses in each height.

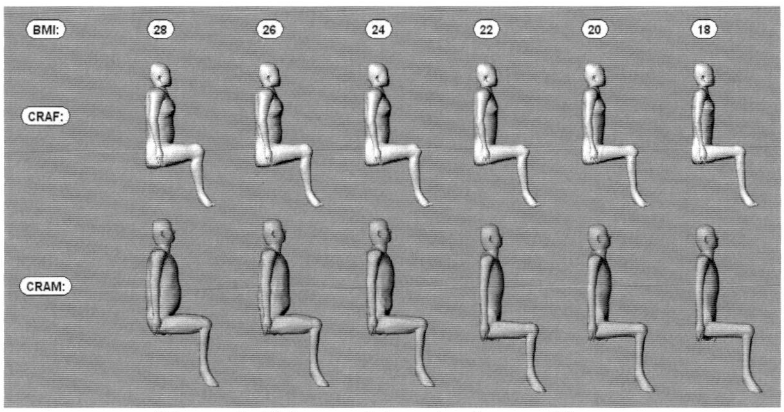


**Figure 1.** Chinese adult male phantom library.

### *2.2. Chinese paediatric reference phantoms*

A series of mesh-type Chinese paediatric reference phantoms for 3 month, 1 year, 5 year, 10 year, and 15 year male and female respectively were constructed based on the CT medical images of different ages as shown in Figure 2. The heights of the six established phantoms were 62 cm, 77 cm, 110 cm, 139 cm, 168 cm and 158 cm, respectively, and the weights were 7 kg, 10 kg, 19 kg, 32 kg, 55 kg and 50 kg, complied with the reference value. Each mesh-type phantom consists of 108 different tissues and organs, which includes all the radiation-sensitive organs.

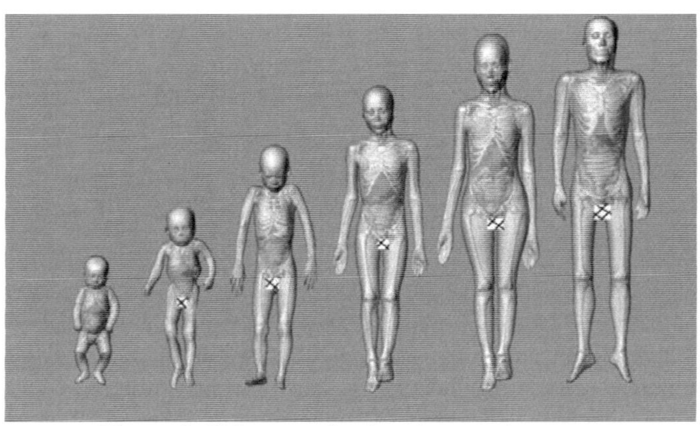

**Figure 2.** Chinese pediatric reference phantoms.

IOP Conf. Series: Journal of Physics: Conf. Series **1305** (2019) 012066 doi:10.1088/1742-6596/1305/1/012066

### 2.3. *Detailed organ models*

Detailed organ models have been developed, including a set of microscopic skeletal models, a set of detailed breast models, a detailed human respiratory tract model and a detailed eye model as shown in Figure 3.

A set of microscopic skeletal models are established based on the micro-CT images. The microscopic spongiosa bone voxel model of Chinese adult male at 32 bone sites was established. Then, CRAM macroscopic spongiosa bone model with microscopic spongiosa bone voxel model was filled and a voxel phantom consisting of micro-voxel in spongiosa bone and macro-voxel in other organs and tissues. Therefore, a multi-scale skeletal model was created.

A set of three-dimensional detailed breast models based on the realistic structures in the breast and the Chinese female breast parameters were built. A mathematical model was established and then converted to a voxel model. First the breast shape was built, and then it was divided into skin, adipose tissue region, and fibroglandular tissue region. Detailed structures such as Cooper's ligaments, intraglandular fats, and ductal lobes were built in different regions.

A mathematical model for human respiratory tract was first established based on the anatomic bronchial parameters of adult Chinese male. Then it was voxelized to build a numerical voxel model and integrated into the CRAM. This model featured by consecutive 16-generation bronchial structures, could represent the structure of entire lung trachea and bronchus structure.

A detailed eye model was built based on the characteristic anatomic parameters of the Chinese adult male. This eye model includes seven main structures, which are scleral, choroid, lens, iris, cornea, vitreous body and aqueous humor. The lens was divided into sensitive volume and insensitive volume based on different cell populations. The detailed eye model was incorporated into the converted polygon-mesh version of the CRAM_S.

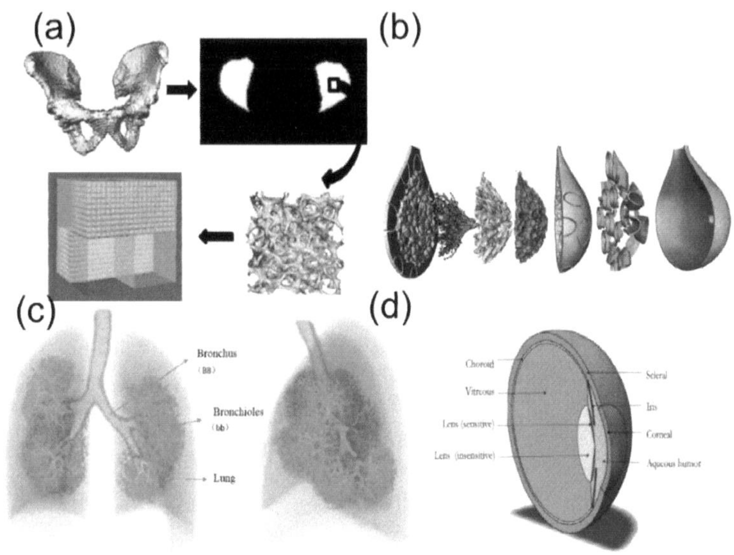


**Figure 3.** Detailed organ models (a)microscopic skeletal models (b) detailed breast models (c) detailed respiratory tract models (d) detailed eye model.

## 3. Applications

### 3.1. *Application in radiation protection*

The CRAM and CRAF phantoms were applied in radiation protection field, such as external exposure, internal exposure to environmental radiation from nuclear facilities under normal or accidental

situations. Conversion coefficients using Chinese adult reference phantoms for air submersion and ground contamination were calculated by Lu *et al* [4]. Absorbed fractions and conversion coefficients for inhalation and ingestion were also calculated. The CRAM_S phantom and the CRAM_Ssit were applied to physical dosimetric reconstruction of radiological accident at Nanjing (China) for clinical treatment. The detailed organ models were also applied in radiation protection field to get knowledge of detailed dose distributions in radiation-sensitive organs. The detailed microscopic skeletal model [5] and the detailed human respiratory tract were used to calculate absorbed fraction in internal exposure. The detailed breast model [6] and the detailed eye model were used to calculate dose conversion coefficients
in external exposure.

*3.2. Application in dose estimation for medical imaging*

The phantoms were applied in medical imaging field as well. A series of organ dose conversion coefficients for dose estimation in CT scanning and X-Ray radiology were calculated with the Chinese adult and pediatric reference phantoms. Figure 4 shows the geometry used in the simulation and experiment set-up in the verification process. A series of glandular tissue dose conversion coefficients for dose estimation in mammography were calculated with the 3D detailed breast models [7]. A voxelized internal radionuclide dosimetry estimation based on a serial quantitative SPECT/CT images were developed using Monte Carol method.

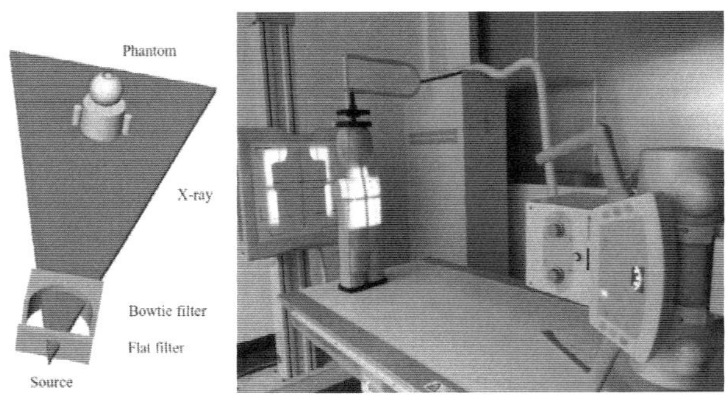


**Figure 4.** Geometry used in the simulation and experiment set-up in the verification process for dose estimation of medical imaging.

## 4. Conclusion

Latest development and applications of the multi-scale Chinese reference phantoms were reviewed in this paper. Base on the CRAM and CRAF phantoms developed in previous work, a Chinese male phantom library was constructed with different heights and different total body masses. Chinese paediatric reference phantoms were constructed for different ages. Detailed organ models including a set of microscopic skeletal models, a set of detailed breast models, a detailed human respiratory tract model and a detailed eye model were also developed. The phantoms were applied in radiation protection, medical imaging, and so on.

## 5. References

[1] Liu L *et al* 2009 *Phys. Med. Biol.* **54** 6645-73
[2] Pan Y *et al* 2014 *Radiat. Prot. Dosim.* **34** 199-205
[3] International Commission on Radiological Protection, The 2007 recommendations of the International Commission on Radiological Protection (ICRP Publication 103) Oxford, Elsevier)
[4] Lu W *et al* 2017 *Phys. Med. Biol.* **62** 2276-314
[5] Gao S *et al* 2017 *Radiat. Prot. Dosim.* **175** 450-59
[6] Qiu R *et al* 2017 *Radiat. Prot. Dosim.* **174** 113-120
[7] Wang W *et al* 2007 *Med. Phys.* **44** 2503-14

IOP Conf. Series: Journal of Physics: Conf. Series **1305** (2019) 012067 doi:10.1088/1742-6596/1305/1/012067

# A chemical evolution of NVP-containing VIPAR-family 3D polymer gel dosimeters – a brief overview

**M Kozicki[1,2], M Jaszczak[3], P Maras[4] and M Dudek[3]**
[1]Department of Man-Made Fibres, Lodz University of Technology, Lodz, Poland
[2]GeVero Co., Lodz, Poland (http://polygevero.com)
[3]Institute of Materials Science and Engineering, Lodz University of Technology, Lodz, Poland
[4]Medical Physics Department, Copernicus Hospital, Lodz, Poland

E-mail: marek.kozicki@p.lodz.pl; gevero@polygevero.com (MK)

**Abstract**. NVP-containing three-dimensional polymer gel dosimeters for radiotherapy dosimetry have been developed for more than 20 years. There have been 11 main modifications of the originally proposed VIPAR polymer gel dosimeter that altogether have amounted to 12 gel dosimeter formulae. This communication is to summarise the main chemical changes made to the VIPAR dosimeter over these years of research. The newest NVP-polymer gel dosimeters are also introduced.

## 1. Introduction

Over the last 25 years, three-dimensional (3D) radiotherapy dosimetry has flourished and offers a range of 3D high-resolution dosimeters [1], which can be divided into two classes: (1) polymer gels and (2) radiochromic dosimeters. Polymer gel dosimeters act upon conversion of vinyl monomers into a polymer and crosslinked polymer, whereas in radiochromic dosimeters, a colour is induced by conversion of a dye or other compound upon irradiation. Examples of polymer gel dosimeters include: BANG, PAG, MAGIC, MAGAT, PAGAS, PAGAT, NIPAM, VIPAR, VIP, VIC, PABIG, PABIG$^{nx}$, and deformable Aquajoint®-based gels [2-19]; however, in the case of radiochromic dosimeters, the following can be mentioned: Fricke-based gels [20], plastic dosimeters [21], deformable FlexyDos3D [22], and radiochromic gels including leuco dye micelle gels [23-25] and leuco dye and tetrazolium salt Pluronic gels [26, 27].

Among these dosimeters are the VIPAR-family gels, which have undergone a range of chemical alterations resulting in 12 *N*-vinylpyrrolidone-containing chemical formulae. This communication is to briefly report on the development of VIPAR gels.

## 2. An overview of NVP–containing dosimeters

In Figure 1, a scheme of VIPAR-family gel dosimeter evolution is shown. The main events of this evolution are presented below. The first polymer gel dosimeter with *N*-vinylpyrrolidone (NVP) was proposed in 1999 by Pappas *et al* [16] and was called VIPAR. In addition to NVP, it also contained 4% *N,N'*-methylenebisacrylamide (MBA or Bis) and 5% gelatine. The composition had to be saturated with argon for elimination of oxygen, which was common to all polymer gel dosimeters at that time (mostly saturated with nitrogen or argon). This allowed for the radical polymerization and crosslinking of NVP and MBA under exposition to ionizing radiation. The main advantage of VIPAR was a high dynamic dose range and saturation at doses greater than 200 Gy. In turn, its main disadvantage was a relatively high dose threshold (assessed in different publications from approximately 2 to 13 Gy). VIPAR exhibited a low dose sensitivity (0.04–0.095 1/(Gy × s)); however, this feature and the high dynamic dose range allowed for its use in measurements of radiation dose distributions for high dose gradients; for instance, in HDR ($^{192}$Ir) brachytherapy or in CyberKnife dosimetry studies [14].

The problem of the VIPAR's high dose threshold was solved by doubling the concentration of NVP; the gelatine concentration had to be increased to 7.5% to maintain proper stiffness of the dosimeter. This gel dosimeter was named VIPAR$^{d}$, where 'd' denotes the double concentration of NVP [28]. The introduction of normoxic gel dosimeters and thus elimination of the necessity of oxygen removal from polymer gel dosimeters [6] led to the proposition of normoxic VIPAR$^{nd}$, where 'nd' denoted normoxic-double [9]. The oxygen scavengers of ascorbic acid (0.007%) and copper sulfate pentahydrate (0.0008%) were used, which facilitated manufacturing of the dosimeters. In turn, VIPAR$^{nd}$ (called also VIP) was commonly used in radiotherapy dosimetry research [14, 29]. The highest dose sensitivity obtained for VIP was equal to 0.0888 1/(Gy x s), whereas the linear dose range was equal to ~0.5–35 Gy and the dynamic dose range was equal to ~0.5–120 Gy (1.5 T magnetic resonance imaging [MRI]) [9].

| | | First published | Dose sensitivity (1/(Gy × s)) |
|---|---|---|---|
| 2018 | VIP3-Pluronic | Jaszczak et al. [38] | 0.101 |
| 2018 | VIC-T | Jaszczak et al. [37] | 0.176 |
| 2018 | VIPET$^{aqua}$ | Maeyama et al. [12] | 0.16 |
| 2017 | AQUAJOINT gel | Takanashi et al. [36] | n.d. |
| 2017 | VIPAR$^{CT}$ (VIC) | Kozicki et al. [11] | 0.171 |
| 2013 | VIPET$^{1v}$, VIPET$^{2v}$ | Papoutsaki et al. [34-35] | 1v: 0.059, 2v: 0.063 |
| 2007 | VIPET | Papadakis et al. [15] | 0.035–0.0761 |
| 2007 | modified VIPAR | Papadakis et al. [15] | ~0.0436 |
| 2007 | VIPAR$^{nd}$ (VIP) | Kozicki et al. [9] | 0.0544–0.0888 |
| 2005 | VIPAR$^{d}$ | Kozicki et al. [28] | ~0.044 |
| 1999 | VIPAR | Pappas et al. [16] | 0.04–0.095 |

**Figure 1**. Scheme illustrating the development of VIPAR-family polymer gel dosimeters. The normoxic dosimeters of the highest dose sensitivity are marked in blue. The dose sensitivities were taken from the following publications: VIPAR [9, 16, 28, 30-32]; VIPAR$^{d}$ [28]; VIP [9, 11, 14, 33]; modified VIPAR [15]; VIPET [12, 15, 34]; VIPET 1v and 2v [35]; VIC [11]; AQUAJOINT gel [36]; VIPET$^{aqua}$ [12]; VIC-T [37]; and VIP3-Pluronic [38].

Papadakis *et al* and Papoutsaki *et al* [15, 35] examined the impact of increasing gelatine concentration in VIPAR and showed that it can be modified to a normoxic gel dosimeter using tetrakis(hydroxymethyl)phosphonium chloride (THPC) following the proposition of De Deene *et al* [5]. Thus, the VIPET gel dosimeter was prepared with 4% MBA, 4% NVP, 7% gelatine, and 5 mM THPC, as proposed previously [39]. VIPET was examined by 1.5 T MRI with some scanning protocols, which lead to dose sensitivities of 0.035, 0.048, and 0.0761 1/(Gy × s). The dose threshold was equal to 2 Gy and the linear dose range was 2-32 Gy (dynamic dose range: ~100 Gy). A reason for this high dose threshold may have been the lack of optimisation of the THPC concentration in VIPET. The next step in VIPET optimisation was to increase the NVP concentration to 8%, as was done previously [28]. This impacted the VIPET dose sensitivity in comparison with VIPET with 4% NVP: 0.063 and 0.059 1/(Gy × s), respectively (1.5 T MRI, PHAPS 32-echo train) [35]. Additionally, VIPET gels were reported to be unstable following preparation and irradiation; however, the dose sensitivity remained unchanged over time after irradiation.

The next significant step in the evolution of VIPAR-family dosimeters was a VIC ($VIPAR^{CT}$) polymer gel [11], comprising 17% NVP, 8% MBA, 12% t-BuOH, 7.5% gelatine, 0.007% ascorbic acid, 0.0008% $CuSO_4 \times 5H_2O$, and 0.02% hydroquinone. The dosimeter exhibited the highest dose sensitivity (0.171 1/(Gy × s), 0.47 T NMR) among all NVP-gel dosimeters (on a par with VIC-T, see below). The VIC performed better than VIP using X-ray computed tomography as a readout method with a dose sensitivity (0.397 HU/Gy) 1.5-fold higher than that of VIP.

Takanashi *et al* [36] and Maeyama *et al* [12] proposed a flexible matrix for NVP-containing gel dosimeters of relatively high dose sensitivity (0.16 1/(Gy × s)). However, this composition requires more detailed analysis.

A substitution of copper sulfate pentahydrate and ascorbic acid with THPC was also examined for the VIC dosimeter [37]. The study resulted in a VIC-T dosimetric composition of analogous to VIC dose sensitivity, a dose threshold of 0.5 Gy, a dynamic dose range of 0.5–50 Gy, and a linear dose range up to 20 Gy (30 Gy for VIC) (derived from 0.47 T NMR measurements). The dose response of VIC-T was independent of the radiation dose rate, type, and energy of radiation for 6 and 15 MV photons and 12 MeV electrons, which is an improvement with respect to VIC, which showed a different dose response for 6 MV photons compared with 12 MeV electrons and 15 MV photons. A feature of VIC-T is its improved/altered thermal properties with respect to other NVP-containing gel dosimeters, as it does not melt in the temperature range of about 0-80°C [37]. The thermal properties of VIC-T are discussed elsewhere [37].

**Table 1**. Comparison of the basic calibration properties of NVP-containing normoxic gel dosimeters of the highest dose sensitivities among all VIPAR-family gels.

| Dosimeter | Dose sensitivity* [1/(Gy × s)] | Intercept [1/s] | $R^2$ | Threshold dose [Gy] | Linear dose range [Gy] | Dynamic dose range [Gy] |
|---|---|---|---|---|---|---|
| **VIP** | 0.089 | 3.447 | n.d. | 0.5–1.5 | <1.5–35 | <1.5–120 |
| **VIC** | 0.171 | 1.783 | 0.999 | 0.5 | 0.9–30 | 0.5–50 |
| **VIC-T** | 0.176 | 1.049 | 0.998 | 0.5 | 0.5–20 | 0.5–50 |
| **VIP3-Pluronic** | 0.101 | 1.249 | 0.999 | 0.5 | 0.5–30 | 0.5–>50 |
| **$VIPET^{aqua}$** | 0.16 | 1.85 | n.d. | n.d. | n.d. | n.d. |

*The highest dose sensitivity is presented. Note that the dose sensitivity may be affected by the type of magnetic resonance imaging and scanning protocol. The calibration parameters are given for different measurements set-up (see references in Figure 1).

A substitution of the natural polymer-based physical gel matrix of gelatine with a synthetic co-polymer physical gel matrix of Pluronic F-127 for NVP-containing polymer gel dosimeter was also proposed [38], which led to the VIP3-Pluronic dosimeter with a dose sensitivity of 0.101 1/(Gy × s), a linear dose range of 0.5–30 Gy, and a dynamic dose range of 0.5–50 Gy (0.47 T NMR). The thermal stability of this gel was improved at temperatures above room temperature due the inclusion of the Pluronic matrix with respect to the NVP-containing polymer gels with gelatine and ascorbic acid and copper sulfate pentahydrate as oxygen scavengers (thermal stability of VIP3-Pluronic based on $T_{gel}$ derived from differential scanning calorimetry, 4.4–56.7°C and VIP, 0–26°C [38]). Table 1 illustrates the main properties of VIPAR-family normoxic gel dosimeters with the highest dose sensitivities, which have been considered promising for further studies.

## 3. Conclusions

This work summarises a 20-year evolution of NVP-containing 3D polymer gel dosimeters. The NVP monomer differentiates this family of gel dosimeters from other 3D dosimeters.

The first proposed VIPAR polymer gel dosimeter underwent several alterations of its composition including changes in monomer concentrations while maintaining the main monomers (NVP and MBA) in each new composition, substituting the gelatine matrix with other matrixes, and the addition of other substances such as oxygen scavengers or co-solvents that allowed for alterations of its mean features. Thus, improvements in dose sensitivity were obtained and the new normoxic gels are easier to prepare. Some of the alterations have also led to improvements and changes of the thermal properties of VIC-T and VIP-Pluronic dosimeters. The newest promising normoxic NVP-containing polymer gel dosimeters of the highest dose sensitivity are VIP, VIC, VIC-T, VIP3-Pluronic, and $VIPET^{aqua}$.

## 4. References


[1] Baldock C *et al* 2010 *Phys. Med. Biol.* **55** R1
[2] Baldock C *et al* 1998 *Phys. Med. Biol.* **43** 695
[3] Brindha S *et al* 2004 *Phys. Med. Biol.* **49** N353
[4] Chain J N M *et al* 2011 *Phys. Med. Biol.* **56** 2091
[5] De Deene Y *et al* 2002 *Phys. Med. Biol.* **47** 3441
[6] Fong P M *et al Phys. Med. Biol.* **46** 3105
[7] Hurley C *et al* 2005 *Appl. Radiat Isot.* **63** 443
[8] Kozicki M 2011 *Radiat. Phys. Chem.* **80** 1419
[9] Kozicki M *et al* 2007 *Macromol. Symp.* **254** 345
[10] Kozicki M *et al* 2014 *Phys. Med. Biol.* **59** 4111
[11] Kozicki M *et al* 2017 *Phys. Med. Biol.* **62** 986
[12] Maeyama T *et al* 2018 *Radiat. Phys. Chem.* **146** 121
[13] Maryanski M J *et al* 1994 *Phys. Med. Biol.* **39** 1437
[14] Pantelis E *et al* 2008 *Med. Phys.* **35** 2312
[15] Papadakis A E *et al* 2007 *Phys. Med. Biol.* **52** 5069
[16] Pappas E *et al* 1999 *Phys. Med. Biol.* **44** 2677
[17] Pappas E et al 2005 *Med. Phys.* **32** 1513
[18] Venning A *et al* 2005 *Nucl. Instrum. Methods Phys. Res.* A **555** 396
[19] Venning A J *et al* 2005 *Phys. Med. Biol.* **50** 3875
[20] Olding T *et al* 2010 *J. Phys.: Conf. Ser.* **250** 012028
[21] Adamovics J and Maryanski M 2006 *Radiat. Prot. Dosim.* **120** 107
[22] De Deene Y *et al* 2015 *Phys. Med. Biol.* **60** 1543
[23] Babic S *et al* 2009 *Phys. Med. Biol.* **54** 6791
[24] Jordan K and Avvakumov N 2009 *Phys. Med. Biol.* **54** 6773
[25] Nasr A T *et al* 2015 *J. Phys.: Conf. Ser.* **573** 012037
[26] Kozicki M *et al* 2017 *Phys. Med. Biol.* **62** 5668
[27] Kwiatos K *et al* 2018 *Phys. Med. Biol.* **63** 095012 (10pp)
[28] Kozicki M *et al* 2005 *Biomed. Tech.* **50** (suppl. vol. 1) 1368
[29] Moutsatsos A *et al* 2010 *J. Phys.: Conf. Ser.* **250** 012060
[30] Karaiskos P *et al* 2005 *Phys. Med. Biol.* **50** 1235
[31] Kipouros P *et al* 2003 *Med. Phys.* **30** 2031
[32] Pappas E *et al* 2003 *Phys. Med. Biol.* **48** N65

[33] Moutsatsos A *et al* 2009 *Med. Phys.* **36** 4277-87

# Dose verification of dynamic MLC-tracked radiotherapy using small PRESAGE® 3D dosimeters and a motion phantom

**Filipa Costa[1], Martin J Menten[1], Simon Doran[2], John Adamovics[3], Ian M Hanson[1], Simeon Nill[1] and Uwe Oelfke[1]**

[1]Joint Department of Physics, The Institute of Cancer Research and The Royal Marsden NHS Foundation Trust, London SM2 5NG, UK

[2]CRUK and EPSRC Cancer Imaging Centre, Institute of Cancer Research, London SM2 5NG, UK

[3]Department of Chemistry and Biology, Rider University, Lawrenceville, NJ 08648, USA

E-mail: filipa.costa@icr.ac.uk

**Abstract.** With the increasing complexity of radiotherapy treatments typical 1D and 2D quality assurance (QA) detectors may fail to detect out-of-plane dose discrepancies, in particular in the presence of motion. In this work, small samples of the PRESAGE® 3D radiochromic dosimeter were used in combination with a motion phantom to measure real-time multileaf collimator (MLC)-tracked radiotherapy treatments. A different sample of PRESAGE® was irradiated for each of three different irradiation scenarios: (1) static: static sample, without tracking (2) motion: moving sample, without tracking and (3) tracking: moving sample, with tracking. Our in-house software DynaTrack dynamically moves the linac's MLC leafs based on the target position. The doses delivered to the samples were reconstructed based on the recorded positions of the MLC and phantom during the beam delivery. PRESAGE® samples were imaged with an in-house optical-CT scanner. Comparison between simulated and measured 3D dose showed good agreement for all three irradiation scenarios (static: 99.2%; motion: 99.7%; tracking: 99.3% with a 3%, 2 mm and a 10% threshold local gamma criterion), failing only at the edges of the PRESAGE® samples (~ 6 mm). Given that the dose distributions deposited using the DynaTrack system have been independently verified, this experiment demonstrates the ability of PRESAGE to measure 3D doses correctly in a tracking context. We conclude that this methodology could be used in the future to validate the delivery of dynamic MLC-tracked radiotherapy.

## 1. Introduction

Dynamic multileaf collimator (MLC) tracking of moving tumors is able to compensate for intra-fractional tumor motion during radiotherapy and preserve target coverage while reducing the dose received by nearby healthy tissues. Lung stereotactic radiotherapy (SBRT) is especially susceptible to intrafracational motion as tumors are usually very small (less than 5 cm in diameter) and have large and variable motion [1]. When compared with conventional radiotherapy, these treatments require a higher level of accuracy due to the high dose delivered in only a few fractions (1 to 8), and steep dose fall-off outside the planning target volume (PTV). Before clinical implementation of these treatments, validation of the delivered dose is needed using anthropomorphic phantoms. 1D and 2D detectors typically used for dose validation might fail to detect disagreements outside the plane in which the detectors are placed. Additional uncertainties can occur as treatment planning system (TPS) simulations are being performed in inhomogeneous media, and previous studies have demonstrated discrepancies of the order of 2-3% between TPS calculations, films and ionization chambers [2, 3]. Ideally, high-resolution 3D dosimeters should be used for a better assessment of the delivered dose.

Polymerizing 3D dosimeters [4] have been successfully used before to verify dynamic MLC-tracked dose delivery using both magnetic resonance imaging (MRI) [5] and optical-CT [6] readout, in homogeneous media. PRESAGE® dosimeters (Heuris Pharma, Skillman, NJ, USA) have been used in inhomogeneous media before to validate gated lung radiotherapy treatments [7]. However, to the best of our knowledge, PRESAGE® samples have not been used before to validate tracked treatments.

In this work, we investigate whether small PRESAGE® 3D dosimeters can be used in combination with a lung motion phantom for dosimetric verification, by demonstrating that PRESAGE® correctly reproduces the results of a previously validated real-time in-house MLC-tracked radiotherapy software [8-10].

## 2. Material and Methods

*2.1 Motion phantom/PRESAGE® dosimeter experimental set-up*

PRESAGE® is a radio-sensitive 3D dosimeter that consists of a radiochromic plastic that shows an optical density (OD) change when irradiated. This change is linear with the delivered dose and can be measured using an optical-CT scanner [11]. To mimic a simplistic human torso with a tumour in the lung, the QUASAR™ MRI4D motion phantom (Modus Medical, London, ON, Canada) was used together with a sample of PRESAGE®. The phantom, which was filled with deionized water, has two cylindrical holes, one central and one offset. A holder was created in-house to place a sample of 3.5 cm diameter and 5 cm length in one of the cylinders. The offset cylinder was filled with deionized water while PRESAGE® was placed in the central insert (Figure 1) to facilitate phantom positioning. Motion can be applied in superior-inferior direction with and without rotation, by attaching one of the inserts to the piezoelectric motor box.

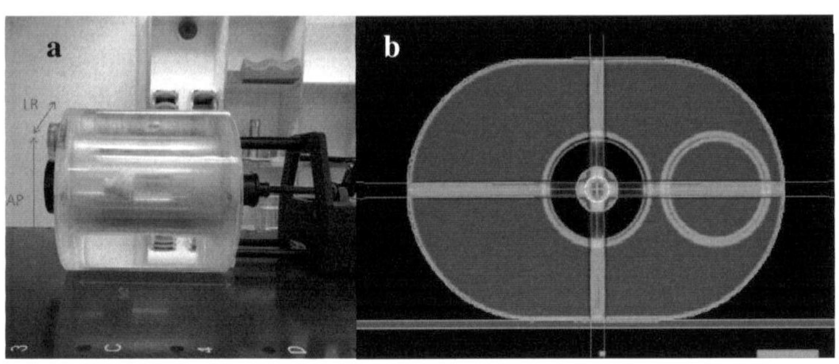


**Figure 1.** (a) Experimental arrangement with an irradiated sample of PRESAGE® placed, with an in-house holder, inside a cylindrical insert of the QUASAR™ MRI4D motion phantom. SI (superior-inferior); AP (anterior-posterior); LR (left-right) are shown. Motion was applied in the SI direction based on a sinusoidal wave with 7.5 mm amplitude and 4 s period. (b) Central axial slice view of the CT scan of the set-up with PRESAGE® sample in place. The treatment plan with four beams is also visible.

*2.2 Experimental set-up workflow*

The phantom/PRESAGE® dosimeter experimental set-up was used to perform an end-to-end test. A computed tomography (CT) scan of this set-up, using a dummy sample of PRESAGE®, was acquired and transferred to the Monaco TPS, research version 5.19.03 (Elekta AB, Stockholm, Sweden). A treatment plan with four equidistant beams and field sizes of 1 × 2.5 $cm^2$ was created and the dose distribution was calculated with a GPU-based Monte Carlo dose calculation algorithm with a maximum dose of 9.7 Gy, 1x1x1 $mm^3$ resolution and 1% statistical uncertainty. The collimator was rotated to 90°, so that the MLC leafs would move in the same direction as the target motion.

Three different plan delivery scenarios were tested on a research Elekta Synergy linac: (1) static: static phantom, without tracking; (2) motion: moving phantom, without tracking; (3) tracking: moving phantom, with tracking. For each scenario a different sample of PRESAGE® was irradiated.

*2.3 MLC-tracking dose delivery and simulations*

Our in-house MLC-tracking software DynaTrack [8] was used to control the research linac and move the MLC leafs based on the position of the moving target, provided by the phantom. A sinusoidal wave of 4 s period and 7.5 mm amplitude was used to move the sample in the superior-inferior direction for scenarios (2) motion and (3) tracking. The target position was provided by the motion phantom piezoelectric motor box. For each irradiation scenario, log-files containing the recorded MLC and phantom positions at 40 ms intervals were obtained at a rate of 25Hz. These allowed reconstruction of the dose in the delivered conditions, allowing a direct comparison of the simulated dose with the PRESAGE® measured dose distributions [12] (see figure 2).

3D dose distributions from both PRESAGE® samples and Monaco simulations were normalized to the average value at the irradiation isocentre and compared using the 3D gamma criterion, for 3% dose difference and 2 mm distance to agreement (3%, 2 mm) [13].

*2.4 Optical-CT/PRESAGE® samples readout*

PRESAGE® samples were imaged 60 min after irradiation with an in-house telecentric optical-CT scanner, a modified version of the one described by McErlean *et al* [14]. Each scan takes approximately 1 min to obtain 1000 projections, each of 320 × 320 pixels, over a 180° rotation to obtain a reconstructed image with a voxel size of 0.2 $mm^3$. Additionally, each sample was scanned before irradiation in order to account for changes due only to dose.

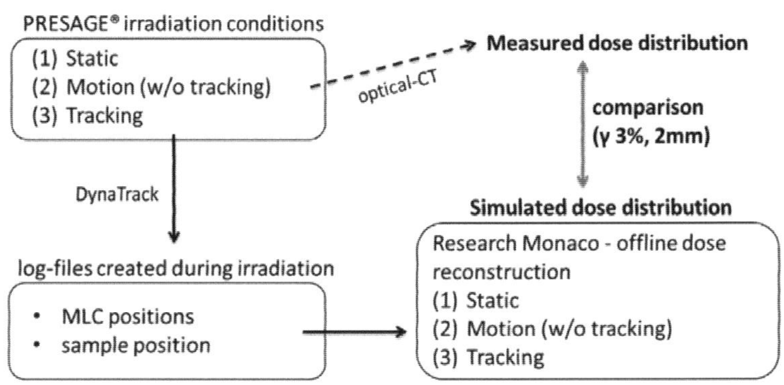


**Figure 2.** Diagram showing the study workflow.

## 3. Results and discussion

The agreement between measured and simulated reconstructed dose distributions is visible by taking a profile at the central slice (Figure 3b, e and h) and by applying the 3D gamma analysis for each irradiation scenario (figures 3c, f and i). For a clinically relevant 3D gamma criterion of 3%, 2 mm and a 10% threshold, a passing rate of 99.2%, 99.7% and 99.3% was obtained for the (1) static, (2) motion and (3) tracking scenarios, respectively. A slightly narrower penumbra is visible for the simulated (3) tracking scenario profile in comparison with the respective measured profile. For purposes of reconstructing the dose delivered, the recorded MLC leaf positions were rounded to their nearest integer millimetre value. This could explain the difference between the measurement and the simulated profiles, as particularly small fields are used.

In figures 3c, f and i, the regions where the gamma criterion fails ($\gamma \geq 1$) are shown in red and occur mainly at the axial edges of the sample. The small disagreement at the edges, which is present in all three irradiated samples, can be visualized in more detail in figure 4b but it is not present in figure 4c. We speculate that this is due to dosimetric inhomogeneities that occur within ~6 mm of the dosimeter surface, for doses ≥30% of maximum dose. Related work (data not shown) suggests that this is caused by the manufacturing process of the PRESAGE® itself. Partial correction is possible and was applied here using data from a previously homogeneously irradiated sample, but further work is needed to develop a robust correction strategy.

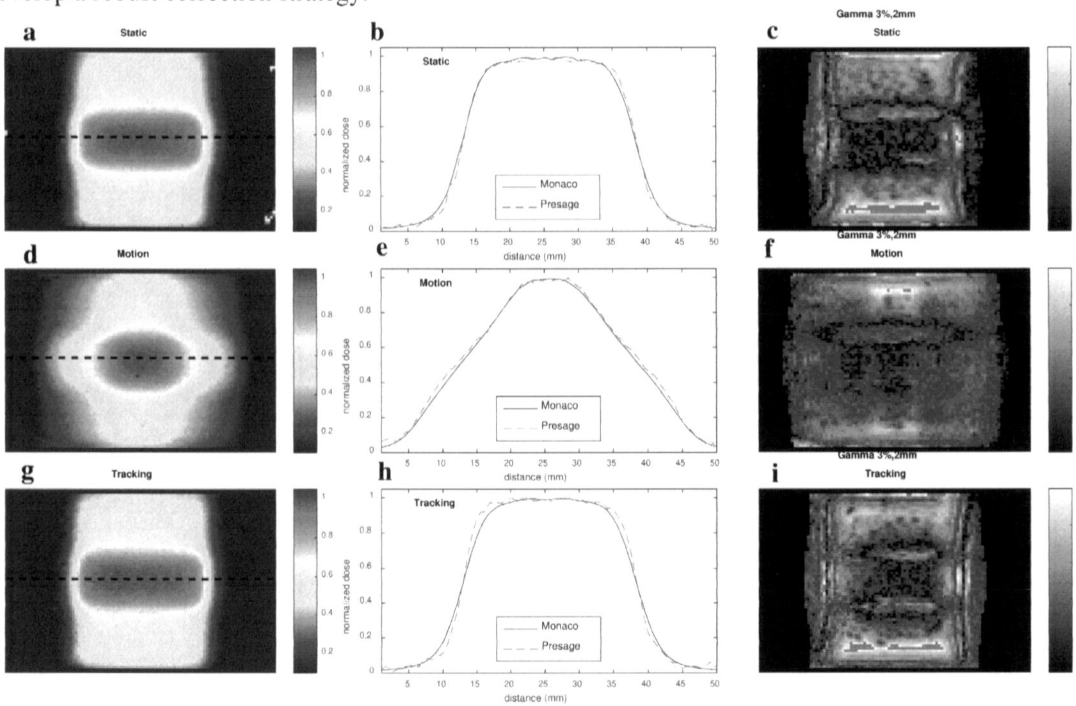


**Figure 3.** Normalized central sagittal slice showing PRESAGE® color-coded dose distributions and the respective profile along the superior-inferior direction for all three studied scenarios. Central sagittal slice of 3D gamma analysis is shown comparing measurements with simulations. Areas where the gamma criterion fails ($\gamma \geq 1$) are shown in red.

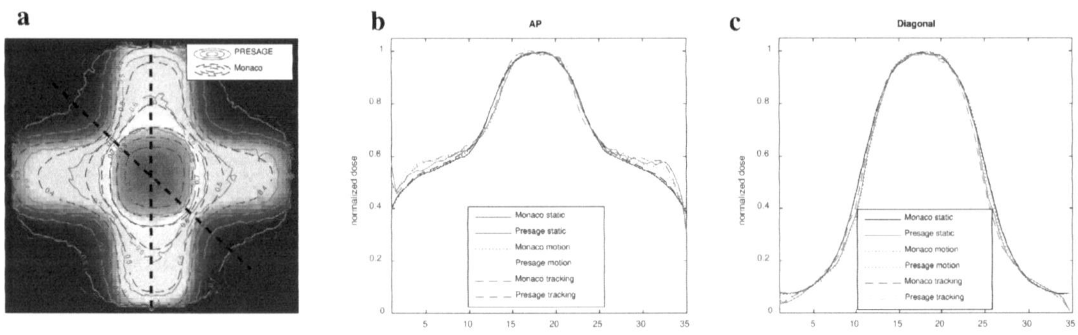


**Figure 4.** (a) Normalized central axial slice color-coded dose distributions and isodose lines showing PRESAGE® under (1) static conditions. Simulated and measured profiles taken at the central axial slice at (b) anterior-posterior (AP) and (c) diagonal directions are shown for (1) static, (2) motion and (3) tracking scenarios.

Both measured and simulated data show the benefit of using tracking to improve dose coverage of moving targets. The difference in the dose distributions obtained for a sample in (2) motion and (3) tracking scenarios are visible in figures 3d and g, respectively. As expected, in the presence of motion without tracking the 3D dose distribution is blurred as opposed to when MLC-tracking is applied. The same conclusions are drawn from figure 5, where the gamma criterion is applied to compare the PRESAGE® sample irradiated in the (1) static scenario with the two other scenarios where PRESAGE® samples are irradiated while moving. The good agreement between the sample in (1) static conditions with the sample in (3) tracking conditions (figure 5b) shows the reproducibility of the samples and the methodology used.

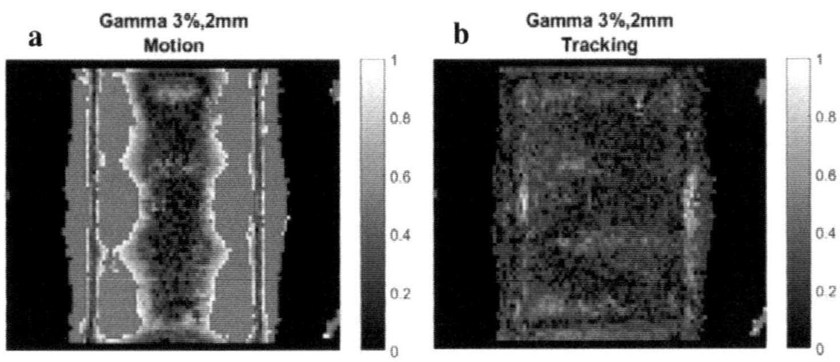


**Figure 5.** Central sagittal slice of the 3D gamma analysis is shown comparing the PRESAGE® sample in static scenario with: (a) motion no tracking PRESAGE® and (b) tracking PRESAGE®. Areas were the gamma criterion fails ($\gamma \geq 1$) are in red.

## 4. Conclusion

To our knowledge, this is the first study using small PRESAGE® dosimeters in combination with a lung motion phantom using real-time tracked MLC radiotherapy treatments. Similar 3D dose distributions obtained for (1) static and (3) tracking conditions highlight the ability of MLC-tracking to improve target coverage. Despite dose discrepancies at the edges of the samples, which need further investigation, the methodology used was reproducible and provided valuable 3D dosimetric information that agreed with the simulated data. This suggests that PRESAGE® samples can be used in the future to validate the delivery of MLC-tracked radiotherapy treatments.

In the future, more complex, clinically relevant dose distributions will be investigated, as well as more physically challenging radiation delivery treatments including rotation on an offset target. We intend to use these dosimeters to verify MLC-tracked treatment on the new Elektra MR-linac, and believe the methodology used here will be easily transferred to our MR-linac.

## 5. Acknowledgements

SJD acknowledges receipt of CRUK and EPSRC support to the Cancer Imaging Centre at ICR and RMH in association with MRC and Department of Health C1060/A10334, C1060/A16464 and the NHS funding to the NIHR Biomedical Research Centre and the Clinical Research Facility in Imaging. The authors acknowledge support from Elekta AB under a research agreement. The ICR is supported by Cancer Research UK under Programme C33589/A19727. We acknowledge Dr Peter Ziegenhein for the gamma analysis and Craig Cummings and Jim Sullivan for the development of the holder.

## 6. References

[1] Sppenwoolde Y *et al* 2002 *Int. J. Radiat. Oncol. Biol. Phys.* **53** 822-34
[2] Clark C H *et al* 2017 *Physica Medica* **44** 171-6
[3] Paudel M R *et al* 2016 *J. Appl. Clin. Med. Phys.* **17** 230-241
[4] Baldock C *et al* 2010 *Phys. Med. Biol.* **55** R1-63
[5] Ceberg S *et al* 2010 *J. Phys.: Conf. Ser.* **250** 235-9
[6] Skyt P S *et al* 2013 *Acta Oncol.* **52** 1445-50

IOP Conf. Series: Journal of Physics: Conf. Series **1305** (2019) 012068 doi:10.1088/1742-6596/1305/1/012068

[7] Brady S L *et al* 2010 *Phys. Med. Biol.* **55** 2187-201
[8] Fast M F *et al* 2014 *Med. Phys.* **41** 111719
[9] Bedford J L *et al* 2015 *Radiother. Oncol.* **117** 491-5
[10] Tacke M B *et al* 2010 *Med. Phys.* **37** 753-61
[11] Guo P Y *et al* 2006 *Med. Phys.* **33** 1338-45
[12] Kamerling C P *et al* 2017 *Med. Phys.* **44** 5997-6007
[13] Low D A *et al* 1998 *Med. Phys.* **25** 656
[14] McErlean C M *et al* 2016 *Phys. Med. Biol.* **61** 320-37

IOP Conf. Series: Journal of Physics: Conf. Series **1305** (2019) 012069 doi:10.1088/1742-6596/1305/1/012069

# A benchtop UV irradiator for 3D dosimetry laboratories with dose considerations in a spinning NMR test tube

**Yves De Deene**
School of Engineering, Macquarie University, North Ryde, Sydney, Australia

E-mail: Yves.DeDeene@mq.edu.au

**Abstract**. Many different chemical radiation dosimeters have been fabricated over the last 20 years. In the search for new dosimeters, next to being sensitive to clinical radiation doses, several other physicochemical characteristics need to be satisfied, such as stability of the dose response, spatial integrity, temperature independence, dose rate independence and tissue equivalence. The development of new dosimeters is often hindered by a limited access to radiation facilities to irradiate hundreds of test tubes or cuvettes to study these physicochemical properties. To facilitate this basic experimental research, we propose the use of an inexpensive UVC irradiator. While care is required in extrapolating the results obtained with UV radiation to high energetic X-rays, for several studies, a UV irradiator is a handy tool for first line investigation of new dosimeters. In this study, we calculated the dose distribution in a cylindrical test tube when being rotated during UV exposure. A quantitative analysis allows the optimization of the set-up to obtain dose rates in the sample in similar order of magnitude that are delivered at a clinical Linac. Regardless the usefulness of a UVC irradiator in the laboratory for preliminary testing, it should not be a complete replacement for measurements with high energetic X-rays.

## 1. Introduction

Several basic experimental studies on new potential 3D dosimeters require the radiation of hundreds of samples [1-26]. Basic experimental studies aim at developing new radiation dosimeters [2-4, 19, 20, 22, 25, 26], obtaining a better understanding of the radiation chemistry of 3D dosimeters [3, 5-8, 17, 18, 21], assessing and optimising the dosimetric performance of the dosimeter in terms of dose sensitivity [2, 9-16, 20, 22, 23], stability [2, 16, 22, 25], temperature dependence [16, 20, 22], dose rate dependence [16, 20, 22-24] and spatial integrity [14, 15, 16, 20, 22, 23, 25]. A fast assessment of the radiation response of new dosimeters can be assessed by exposing the potential dosimeter to UVC light. In order to relate the absorbed dose of a UVC source to the absorbed radiation dose delivered with high energetic X-rays, the light propagation through the vial needs to be considered. In this study, we relate the UVC exposure of a 5 mm NMR tube while spinning to effective radiation dose.

## 2. Methods and Materials

### *2.1. UV irradiator*

A UV irradiator has been fabricated using a 9W germicide UV lamp (Philips TUV PL-S 9W/2P). The UVC power produced by the UV lamp amounts to 2.3 W. To protect the user from accidental UVC exposure, a cover was constructed from white polyurethane plates. The interior of the cover box is coated with aluminium foil tape. A UVC light intensity meter (UVC-254, ISO-9001) is also mounted inside the enclosure to monitor any fluctuations or drift in light intensity output. A cylindrical hole is made in the enclosure for an NMR test tube. To attenuate the light intensity to levels that correspond with the dose rate obtained with a clinical Linac (6 Gy/min), the UVC light bulb is partially shielded with thin microscopy glass plates and only light reflected on the backside wall of the enclosure is used to expose the samples. Shielding glass plates were added until the UVC irradiance measured with the UVC light intensity meter matched the target irradiance and a flat irradiance profile was obtained at the sample location. When switching on the cold UVC light bulb a slow transient in the order of 2 minutes was found (figure 1b). While irradiating the samples, the light bulb was allowed to warm up for at least 5 minutes and the UVC irradiance was recorded at all times and the actual irradiance time course was taken into account while calculating the UVC dose.

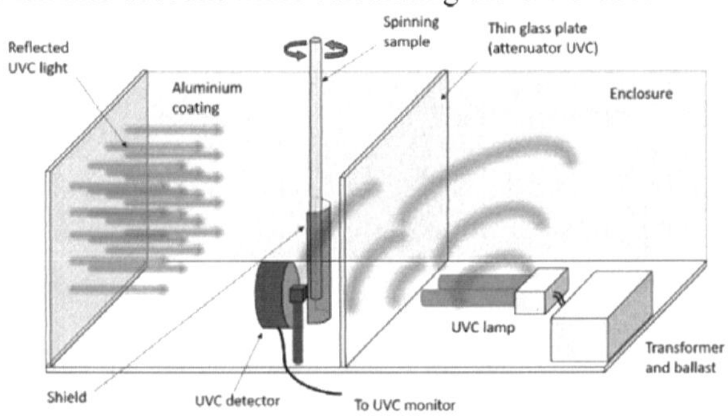


**(a)**

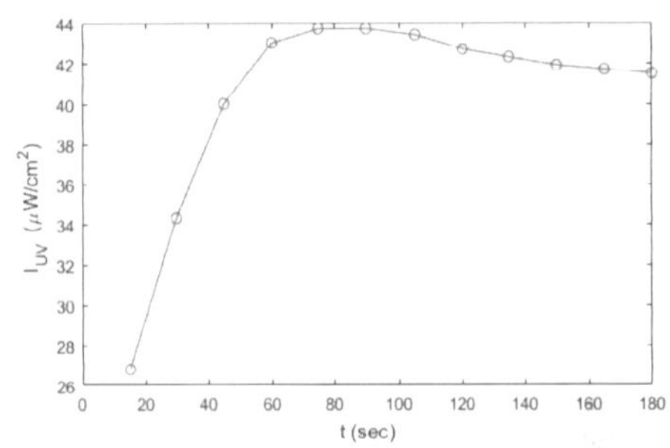


**(b)**

**Figure 1.** UV irradiator set-up (a) and transient UV irradiance of the UVC germicide lamp (b).

*2.2. Relating UV light irradiance to radiation dose*

Irradiance is a radiometric quantity which is the radiant flux $\phi_e$ (in Watts) per unit surface area $A$ (m$^2$) and is thus expressed in W/m$^2$.

$$E_e = \frac{\partial \phi_e}{\partial A} \quad \left(\frac{W}{m^2}\right) \tag{1}$$

In an absorbing medium, the irradiance is a function of depth according to an exponential decay:

$$E_e(x) = E_e(0)\, e^{-\mu x} \tag{2}$$

where $\mu$ is the light attenuation coefficient for UVC light in the medium.

The light energy absorbed in an elementary volume and per unit of time ($\dot{D}_v$) is the difference between incoming and outgoing light (figure 2) and can thus be written as:

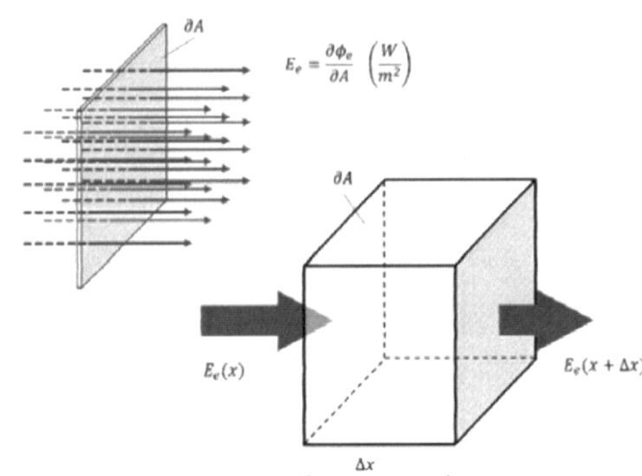


**Figure 2.** Some radiometric quantities and absorbance in an elementary volume element of absorbing medium.

$$\dot{D}_v(x) = \lim_{\Delta x \to 0} \frac{E_e(x) - E_e(x + \Delta x)}{\Delta x} = -\frac{\partial E_e}{\partial x} = \mu \cdot E_e(x) \tag{3}$$

The absorbed dose of UVC light defined as energy absorbed in a medium per unit of mass within an exposure time span $t_{exp}$ is then given by

$$D_{UV}(x) = \int_0^{t_{exp}} \frac{\dot{D}_v(x)}{\rho} d\tau = \int_0^{t_{exp}} \left(\frac{\mu \cdot E_e(x)}{\rho}\right) d\tau \tag{4}$$

where $\rho$ is the mass density of the medium. The term between brackets in equation 4 is the dose rate.

### *2.3. Light tracing simulation*

To calculate the absorbed dose in a 5 mm NMR test tube, the propagation of the light through the test sample needs to be determined for which a ray tracing algorithm was developed in Matlab. A total of 2000 incident rays were considered (note that only 40 rays are drawn in figure 2a and 3a for visibility). Fresnel laws were used to calculate the fraction of transmitted light at every interface between two different media. Snell's law was used to determine the angle of refraction.

The attenuation of the irradiance along each ray is calculated according to an exponential attenuation using the absorption coefficient for the corresponding material (glass, PDMS or gelatine gel). Convergence of light rays will increase the intensity. Once all rays are calculated, the density of rays in every pixel is calculated which results in an irradiance map. The irradiance map can be converted into a dose map, using equation 4.

Two scenarios were considered: 1.) In a first scenario, the sample is static with respect to the light source. The UV light is considered to have a planar propagation outside of the phantom. Only one side of the phantom is exposed to the UV light (figure 1, 2.) In a second scenario, the sample is spinning during exposure at a rate of 500 rotations per minute. In the latter scenario, the irradiance map is first converted to cylindrical coordinates and a radial integration is performed to derive the UV dose.

## 3. Results and Discussion

### *3.1. Dose calculations for the UVC lamp*

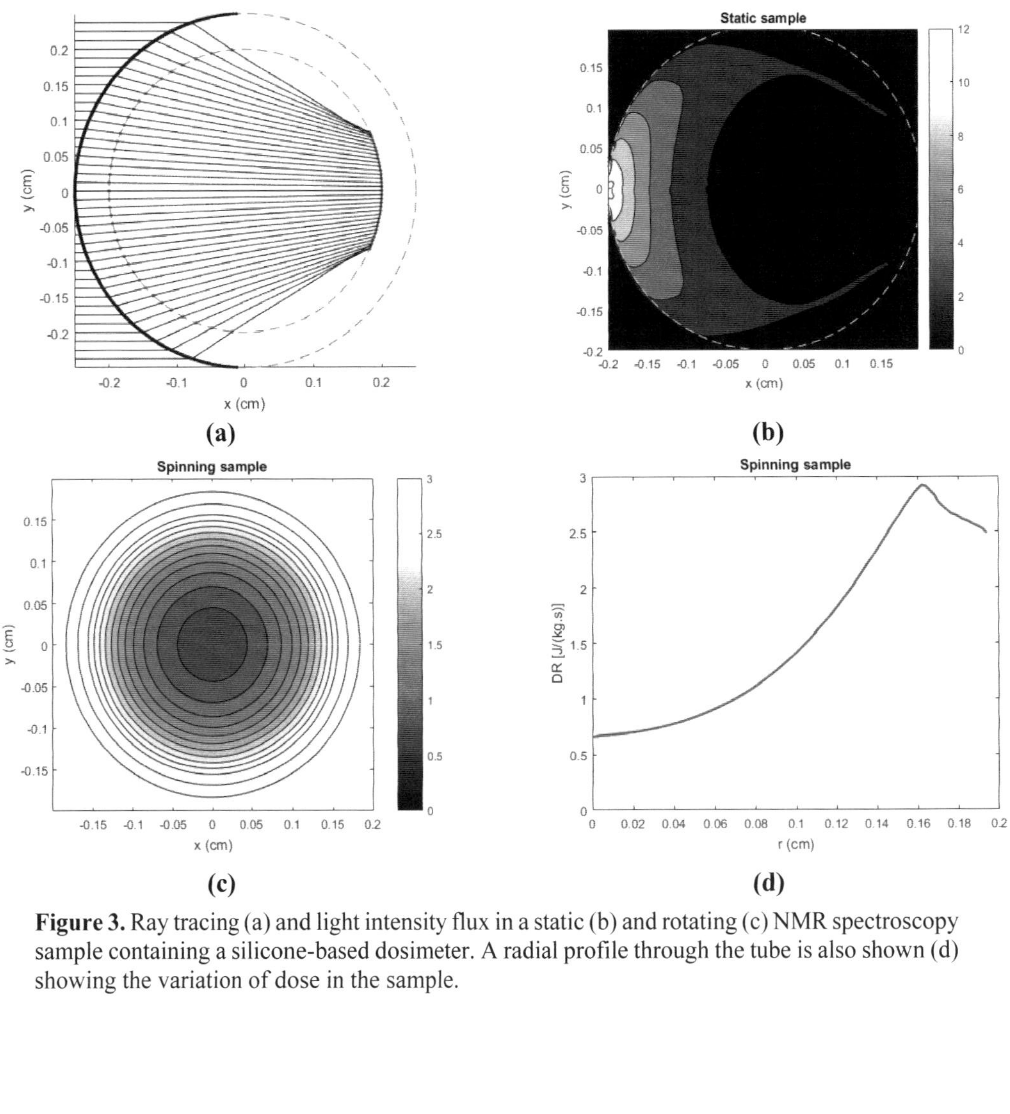


**Figure 3.** Ray tracing (a) and light intensity flux in a static (b) and rotating (c) NMR spectroscopy sample containing a silicone-based dosimeter. A radial profile through the tube is also shown (d) showing the variation of dose in the sample.

The propagation of light rays and UVC dose rate maps for a Flexydos3D sample are shown in figure 2. Refractive indices and attenuation coefficients were based on literature values ($n_{glass} = 1.5$; $n_{PDMS} = 1.418$ [27] and $\mu_{glass} = 3.25$; $\mu_{PDMS} = 16.4$ [28] ).

Because of the high attenuation coefficient, for a non-spinning sample, the dose distribution is very inhomogeneous (figure 2b). Most UVC light is absorbed on the exposed side of the sample but the converging light rays, also create some dose that extends more deeply in the sample. When the sample is spun during exposure, the dose is smeared out more homogeneously inside the sample. The dose rate in the center of the sample is about 20% of the dose rate on the edges.

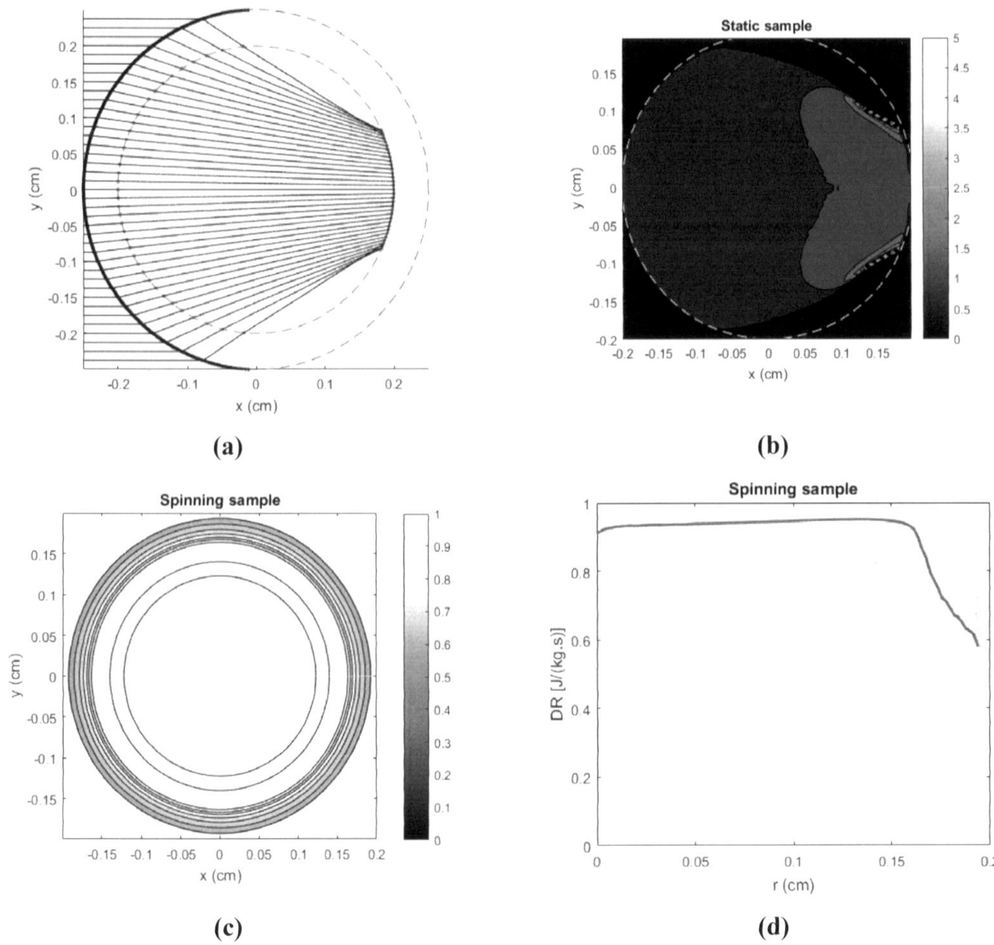


**Figure 4.** Ray tracing (a) and light intensity flux in a static (b) and rotating (c) NMR spectroscopy sample containing a polymer gel dosimeter. A radial profile through the tube is also shown (d) showing the variation of dose in the sample.

For a gelatine gel sample, the dose distribution is more homogeneous (figure 3). The convergence of UVC light rays in the sample create a focussing effect with a slightly higher dose on the side opposite to the exposed surface. For a spinning gelatine gel sample, a more homogeneous dose distribution is obtained (figure 3c – d) because of the smaller attenuation coefficient.

### *3.2. Experimental validation*

Some NMR tube samples were exposed to UVC light with the UVC irradiator for different exposure times and some were irradiated with 6 MV photon beams in the linear accelerator. The glass tubes were then broken and the removed irradiated Flexydos3D silicone samples were scanned in the dual wavelength conebeam optical CT scanner [29]. One sample was exposed to UVC light without spinning.

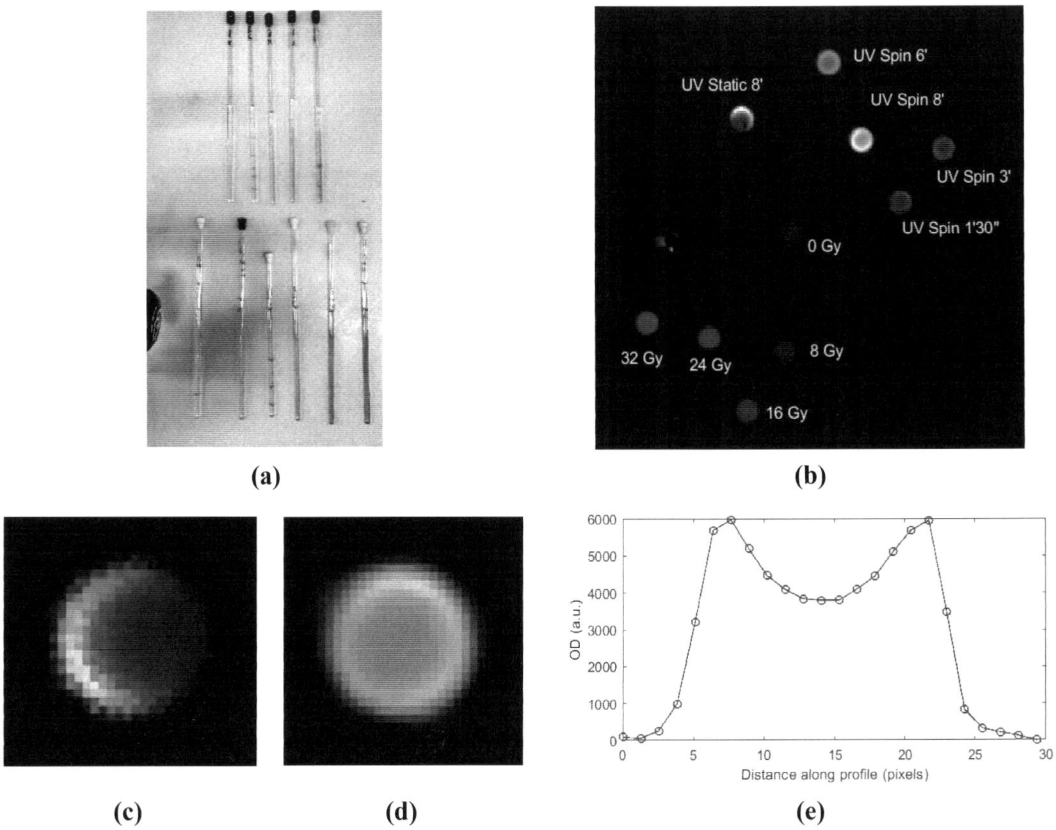


**Figure 5.** Photograph of the exposed samples and optical CT scan (b) through the different samples. Zoomed in image through both a non-spinning UVC irradiated sample (c) and a spinning UVC sample (d). The pixel resolution is $225\,\mu m \times 225\,\mu m$. The diameter of the phantom is 4 mm.

## 4. Conclusions

An inexpensive benchtop UVC irradiator for 3D radiation laboratory's is proposed. The dose rate in the UVC irradiator is adjusted to clinical X-ray dose rates by adding additional shielding. Some improvements in the design of the UVC irradiator will be made to fully optimize the dose delivery. The UVC irradiance meter will be coupled to a PC and the integrated dose will be monitored in real time so that the UVC lamp is switched off when the required dose is reached.

It is shown through computational simulations that by spinning cylindrical samples, a higher homogeneity can be obtained. In polymer gel dosimeter samples, a relatively uniform dose distribution is achievable in a 5 mm test sample while in silicone based samples where the attenuation coefficient for UVC wavelengths is high, the heterogeneity in the sample amounts to 1:5.

It is found that the dose response of dosimeters to UVC light is comparable but not exactly the same as the dose response of dosimeters exposed to high energetic X-rays. While the UVC irradiator is found as a useful tool in the development and characterization of 3D radiation dosimeters laboratory, it should be noted that the radiation chemistry of dosimeters exposed to UVC light may be different from the radiation chemistry in high energetic X-ray beams. It is advisable to perform ultimate tests on the Linac.

## 5. References

[1] Baldock C *et al* 2010 *Phys. Med. Biol.* **55** R1-R63
[2] Maryanski M *et al* 1997 *Phys. Med. Biol.* **42**303-11
[3] De Deene Y *et al* 2000 *Phys. Med. Biol.* **45** 859-78
[4] Chu K C *et al* 2000 *Phys. Med. Biol.* **45** 955-69
[5] Lepage M *et al* 2001 *Phys. Med. Biol.* **46** 1061-74
[6] Lepage M *et al* 2001 *J. Appl. Polym. Science* **79** 1572-81
[7] Jirasek A I and Duzenli C 2001 *Phys. Med. Biol.* **46** 1949-61
[8] De Deene Y *et al* 2002 *Phys. Med. Biol.* **47** 3441-63
[9] Lepage M *et al* 2002 *Phys. Med. Biol.* **47** 1881-90
[10] Hill B *et al* 2002 *Phys. Med. Biol.* **47** 4233-46
[11] Healy B J *et al* 2003 *Med Phys* **30** 2282-91
[12] Gustavsson H *et al* 2004 *Phys. Med. Biol.* **49** 227-41
[13] Hilts M *et al* 2004 *Phys. Med. Biol.* **49**, 2477-90
[14] Hurley C *et al* 2005 *Appl. Radiat. Isotopes* **63** 443-56
[15] Venning A J *et al* 2005 *Phys. Med. Biol.* **50** 3875-888.
[16] De Deene Y *et al* 2006 *Phys. Med. Biol.* **51** 653-73
[17] Jirasek *et al* 2006 *Phys. Med. Biol.* **51** 1891-906
[18] Babic S and Schreiner J 2006 *Phys. Med. Biol.* **51** 4171-87
[19] Chain J N M *et al* 2011 *Phys. Med. Biol.* **56** 2091-102
[20] Vandecasteele J *et al* 2011 *Phys. Med. Biol.* **56** 627-51
[21] Kozicki M 2011 *Radiat. Phys. Chem.* **80** 1419-36
[22] De Deene Y *et al* 2015 *Phys. Med. Biol.* **60** 1543-63
[23] Nasr A T *et al* 2015 *Phys. Med. Biol.* **60** 4685-704
[24] Jirasek A *et al* 2015 *Phys. Med. Biol.* **60** 4399-411
[25] Kozicki M *et al* 2017 *Phys. Med. Biol.* **62** 5668-90
[26] Rabaeh K A *et al* 2017 *Physica Medica* **33** 121-26
[27] Valouch S *et al* 2012 *Optics Express* **20** 28855-61
[28] Stankova N E *et al* 2016 *Applied Surface Science* **374** 96-103
[29] De Deene Y 2015 *J. Phys.: Conf. Ser.* **573** 012058

# AUTHOR INDEX

**Institute of Physics**
Dirac House, Temple Back
Bristol BS1 6BE UK

ISSN: 1742-6588
ISBN 978-1-5108-9970-4
9 781510 899704

# XXV International Conference and Meeting of Departments (CONSTRUMAT 2019)

IOP Conference Series: Materials Science and Engineering
Volume 549

Zuberec, Slovakia
29 – 31 May 2019

IOP
Institute of Physics